Introduction to
Reinforced
Concrete Design

Mohamad Ziad Bayasi
San Diego State University

Linus
Publications, Inc.

Published by Linus Publications, Inc.
Deer Park, NY 11729

ISBN 10 : 1-60797-271-9

ISBN 13 : 978-1-60797-271-6

Printed in the United States of America.

This book is printed on acid-free paper.

Print Numbers   5   4   3   2   1

This book is dedicated to the everlasting memory of my father:
Mohamad Fayez Bayasi.

Table of Contents

Chapter 1

Introduction to Reinforced Concrete

Chapter 2

Reinforced Concrete Materials Plain Concrete and Reinforcing Steel Bars

Chapter 3

Design Procedures for Reinforced Concrete

Chapter 4

Flexural Behavior of Reinforced Concrete Beams

Chapter 5

Analysis and Design of Reinforced Concrete Beams I

Chapter 6

Analysis and Design of Reinforced Concrete Beams II

Chapter 7

Analysis and Design of Reinforced Concrete Beams III

Chapter 8

Analysis and Design of One – Way Slabs

Chapter 9

Analysis and Design of T – Shaped Beams

Chapter 10

Analysis and Design of Doubly Reinforced Concrete Beams

Chapter 11

Shear Behavior of Reinforced Concrete Beams

Chapter 12

Shear Design of Reinforced Concrete Beams

Chapter 13

Bond and Development Length of Steel and Concrete

Chapter 14

Serviceability of Reinforced Concrete Beams

Chapter 15

Introduction to Reinforced Concrete Columns

Chapter 16

Reinforced Concrete Columns Under Axial Force and Bending Moment

Chapter 17

Slender Reinforced Concrete Columns

Chapter 18

Reinforced Concrete Foundation

Chapter 19

Reinforced Concrete Cantilever Retaining Walls

Appendix A

Appendix B

Preface

This concrete textbook is for use in a fundamental reinforced concrete design class in a civil engineering curriculum. It assumes student basic knowledge of strength of materials or solid mechanics and structural analysis. This textbook contains the basic chapters for concrete deign which are flexural, shear, bond, serviceability, columns and footings as well as more advanced topics including retaining walls, torsion, two–way slabs, continuity in reinforced concrete structure, and seismic design and detailing. Additionally, introduction to further topics as prestressed concrete, bridge design and masonry is included. The method of explanation of the concepts of reinforced concrete design in this book depends on sketches, figures as well as mathematical equations. This textbook is simplified yet it contains details of design information that the practicing engineer can use. Such information includes structural details of designed members with reference to applicable code sections. There are appendices in this textbook that simplify reinforced concrete design for the reader. ACI Code is referenced in this book. U.S. units are used in this textbook but a unit conversion table is included for reference. In selected chapters, equations as well as examples and problems are included utilizing the metric units for readers' appreciation.

An important aspect of this book is that it introduces the reader to the concepts of structural design. This is accomplished by review of applicable building codes and methods of determining design load and member strength. Thus, this textbook can also be used for foundation class in structural design principles.

The author of this textbook has a long experience teaching reinforced concrete design in college. He is also a licensed structural engineer in the State of California with a long design experience. The author utilized his experience for this textbook to achieve a well prepared textbook for reinforced concrete.

Mohamad Ziad Bayasi.

Chapter 1

INTRODUCTION TO REINFORCED CONCRETE

1.1 The Principle of Reinforced Concrete

An ideal example for understanding the behavior traits of reinforced concrete is to consider a simply supported homogeneous beam under a uniform gravity load leading to a positive bending moment. Figure 1.1 illustrates the loading condition and deformed shape of this hypothesized beam. As noted, the top portion of the beam is subjected to compression while the bottom portion is subjected to tension as a result of the acting bending moment.

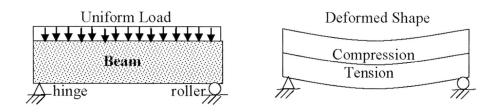

Figure 1.1: Tension and compression in a uniformly loaded homogenous beam.

While concrete is particularly sound in compression, it is vulnerable to failure in tension. As such, concrete can resist the compressive stresses of the beam shown in Figure 1.1. Steel is very tough in tension as well as compression but it is considerably more expensive and vulnerable to environmental factors. Therefore, steel in small amounts may be used in the beam of Figure 1.1 to resist the tensile stresses. As concrete is resilient to environmental factors, the primary advantage of this proposed cooperation between steel and concrete is that concrete can provide protection for steel against corrosion and fire. Based on these principles, a reinforced concrete beam section may look similar to Figure 1.2. Concrete is used in massive amounts to absorb compressive stresses and to provide protection for steel, while steel is used in small ratios adequate for resisting tensile stresses.

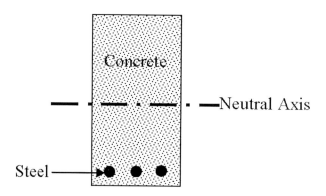

Figure 1.2: A typical reinforced concrete beam section.

1.2 The Advantages of Reinforced Concrete

As reinforced concrete is a widely used construction material, it possesses certain essential advantages that promote its widespread utilization. A summarized list of these advantages is as follows:

- Considerable compressive strength compared with other construction materials.

- Excellent durability against environmental effects.

- Low maintenance.

- Long service life with insignificant reduction in load bearing capacity.

- Inexpensive, particularly for use in footings, slabs, piers, basements, walls, dams and other massive structures or structural elements.

- Can be cast into various shapes with enhanced aesthetic appeal.

- Concrete manufacturing can take advantage of readily available local materials such as sand and rock.

- Typically concrete requires relatively unskilled laborers without a great deal of special training compared with other construction materials.

  Reinforced concrete is currently the main construction material for practically all regions of the world except for places with an abundance of wood.

1.3 The Disadvantages of Reinforced Concrete

In spite of the advantages of reinforced concrete listed in Section 1.2, there are certain disadvantages that need to be addressed including the following items:

- Concrete is relatively weak in tension.

- Concrete is subjected to the effect of shrinkage.

- Due to its tensile weakness and subjection to shrinkage, concrete is exposed to cracking that can result in weakening, vulnerability to environmental effects, and increased maintenance costs. Proper design and construction precautions are essential for this behavioral aspect of reinforced concrete.

- Reinforced concrete typically requires formwork for construction.

- Reinforced concrete is a rather heavy material with a unit weight of 145 lb/ft³ (2.3 kg/m³).

- Concrete strength per unit volume is relatively low, typically resulting in large members.

- Concrete properties can vary widely as raw material properties vary, and can deteriorate in the lack of adequate quality control for mixing, placing and curing.

Appropriate design and construction precautions can limit or eliminate the major disadvantages of reinforced concrete. Such precautions have lessened the effects of concrete disadvantages aiding in its widespread use as a construction material.

1.4 Compatibility of Steel and Concrete

As steel is used as reinforcement for concrete, both elements need to be compatible to achieve the combined mechanical behavior of reinforced concrete. The elements of compatibility of steel and concrete are outlined below.

1.4.1 Bond between steel and concrete

Steel reinforcement for concrete is generally in the form of bars strategically located within the structural element for resistance of tensile stresses. Concrete makes up the mass of structural elements. Upon load application to a structural member, tensile stresses are streamed to the location of steel bars where resistance is provided. These tensile stresses are transferred to the steel bars via the bond of steel and concrete; therefore such bond is essential. Generally, in the analysis of reinforced concrete structural elements (as shown in Chapter 4), perfect bond between steel and concrete is assumed.

The synthesis of bond of steel and concrete is comprised of chemical, friction, and mechanical bonds. Both chemical and friction bonds are considered unreliable. Therefore, prior to the inception of deformed steel bars, hooks were typically implemented to enhance bonding between plain (smooth) steel bars and concrete via an enhanced mechanical bond component as shown in Figure 1.3.

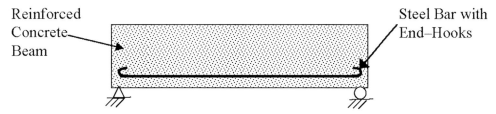

Figure 1.3: Reinforced concrete beam detail prior to the inception of deformed steel bars.

Following the inception of deformed steel bars, the use of hooks in bar detailing has decreased significantly. Deformed steel bars have surface deformations or indentations (as explained in Chapter 2) to improve their mechanical anchorage to concrete. This procedure has not only proven very effective in improving bond strength between steel and concrete, but has also enhanced the overall behavior of reinforced concrete structural members. These improvements stem from a uniform bond across the length of the bar instead of a localized anchorage at the end hooks. Additionally, deformed bars reduce localized stresses and deformations within the concrete due to end anchorage that may cause premature failure. A common type of deformed bars has deformations as shown in Figure 1.4.

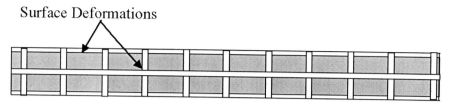

Figure 1.4: Schematic diagram of a deformed steel bar.

1.4.2 Thermal expansion and lateral strains

With similar thermal expansion coefficients and similar poisson's ratios, reinforced concrete will practically behave as one material under the effect of external loading as well as thermal effects assuming that prudent design is provided. Steel and concrete have similar coefficients of thermal expansion about 5.5×10^{-6} in/in/°F (10^{-5} m/m/°C). Such similarity reduces or eliminates thermal stresses within reinforced concrete that could result from temperature variations and aids in enhancing compatibility of steel and concrete.

Steel and concrete also have similar poissons's ratios of about 0.20. This similarity reduces the potential of lateral strains and stresses in reinforced concrete which further enhances the compatibility of steel and concrete.

Test your knowledge

1.   What functions does steel serve in reinforced concrete?

2.   What functions does concrete serve in reinforced concrete?

3.   What is the reason for the decrease of hooked steel bar detailing in reinforced concrete recently?

4.   In your opinion, what are the main two advantages of reinforced concrete?

5.   In your opinion, what are the main two disadvantages of reinforced concrete?

6.   In addition to the factors listed in this chapter, what other factors can be beneficial for compatibility of steel and concrete?

7.   Based on your personal observations, list in order the most common construction materials in your area.

Chapter 2

REINFORCED CONCRETE MATERIALS PLAIN CONCRETE AND REINFORCING STEEL BARS

Prior to analyzing the overall behavior of reinforced concrete, it is essential that one understands the behavior of each individual component. Reinforced concrete is a composite material consisting of two elements, plain concrete and reinforcing steel bars. The behavior of plain concrete under stresses and other factors needs to be realized. The mechanical behavior of steel also needs to be studied. Plain concrete is a relatively complicated material since it is a composite material consisting of several constituents each of which has an important effect on their collective group behavior. The behavior of steel bars is considered rather simple compared with the behavior of plain concrete. In the following parts of this chapter, an overall review of plain concrete and steel bars for reinforced concrete from a material standpoint is presented.

2.1 Plain (Unreinforced) Concrete

2.1.1 Components of Concrete

Plain concrete is, in essence, a composite material since it consists of several constituents combined in a monolithic mass. The primary concrete components are:

1.   Portland cement (7 to 15% of concrete volume).

2.   Water (14 to 21% of concrete volume).

3.   Aggregate (65 to 79% of concrete volume) consisting of:
      Fine aggregate or sand (24 to 30% of concrete volume), and
      Coarse aggregate consisting of crushed rock (stone) or gravel (31 to 51% of concrete volume).

      Concrete basically consists of a binder or matrix, termed cement paste, which constitutes 10 to 30% of total volume, granular inclusions or aggregate that constitute(s) 65 to 85%, and air voids that constitute the balance to 100%. The binder or cement paste is formed by the chemical reaction of portland cement and water or hydration. Aggregate is comprised of a mixture of variable size particles classified into two categories: fine aggregate (sand) and coarse aggregate (crushed rock or gravel). Each component plays an important role in achieving the desired design concrete properties; including workability, strength, durability and economy. Figure 2.1 illustrates the various components of concrete.

Coarse Aggregate

Cement Paste

Fine Aggregate
(Sand)

Air Voids

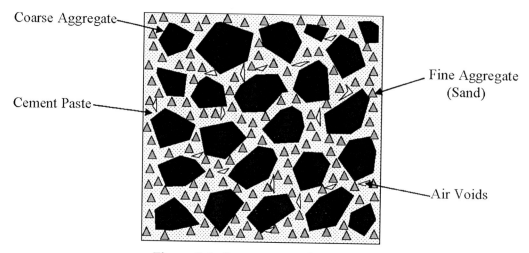

Figure 2.1: Components of concrete.

As one can see in Figure 2.1, the voids among coarse aggregate particles are filled with a mixture of sand and cement paste (mortar). The voids among sand particles are filled with cement paste. Therefore, concrete is viewed as a mixture of multi-size materials with complementary gradation leading to the cooperative formation of a solid material with minimal air voids. Air voids are minimized by proper placement of concrete.

Cement paste is the binding element of concrete. It is formed by the reaction of portland cement and water. Portland cement is a powder made by baking limestone and clay. Its particle size ranges between 0.4 and 2.4 $\times 10^{-3}$ in (10 and 60 μm). Calcium silicate $[(CaO)_n.(SiO_2)_m]$ is the main compound of cement where calcium and silicone are contributed by the limestone and clay, respectively. Portland cement combines with water in a chemical reaction termed hydration to form cement paste:

$$K[(CaO)_n.(SiO_2)_m] + LH2O \rightarrow P(CaO).Q(SiO_2).R(H_2O) + S[Ca(OH)_2] \qquad (2.1)$$

Portland Cement + Water → Hydrated Cement Paste (Calcium Silicate Hydrate + Hydrated Lime)

Calcium Silicate Hydrate (CSH) makes up about 60% and hydrated lime makes up about 25% of the total volume of cement paste while other compounds make up the balance of the volume. CSH is the main contributor to strength and durability of cement paste and, consequently, concrete. Hydrated lime is a relatively weak compound compared to CSH; and the abundance of hydrated lime can be detrimental to strength and durability of cement paste. The formation of hydrated lime is encouraged by higher water amounts within cement paste.

2.1.2 Phases of Concrete

Concrete goes through three stages/phases of formation:

1. The <u>fresh state</u> occurs after the addition of water to the dry mix ingredients. Hydration commences upon contact between portland cement and water. Concrete is considered fresh after the addition of water to portland cement prior to setting. While hydration is in progress, fresh concrete possesses properties that facilitate material handling, transporting, placing, forming, consolidating and finishing. Such properties are collectively termed workability or fluidity of fresh concrete. Fresh concrete has to be adequately flowable for molding during construction. It is a plastic material that is typically cast or poured into formwork; consequently after hardening, concrete takes the shape of formwork. Within concrete composition, water is the lubricant and increasing the water amount results in enhanced flowability or workability. Using excessive amounts of water in fresh concrete can result in reduction of plasticity needed

to achieve a homogeneous mixture of constituents. Plasticity or stickiness is needed so that concrete components would adhere together and form a homogeneous mix. As a result, control of concrete water content is pivotal.

2.   As hydration and water evaporation continue to take place, concrete gradually loses its workability. During the <u>setting stage</u>, working with or reforming concrete is impractical and generally disallowed due to its low workability and potential negative effect upon hardened material properties. In this stage, concrete strength or resistance is low and any disturbance could cause permanent damage or deformation of the impending hardened concrete.

3.   <u>The hardened state</u> starts directly after setting state ends. In the hardened state, concrete can withstand increasing levels of external and handling stresses. Hardened concrete, however, is not considered mature until the age of 28 days where it can resist the design stresses and can fully perform its function.

2.1.3 Water/Cement Ratio of Concrete

Concrete is generally designated by its water/cement ratio. Water/cement ratio needed to complete the reaction of Equation (2.1) is 20 to 22%. In practice, water/cement ratio ranges from 50 to 80% to facilitate sufficient workability by increasing its flowability. Water in excess to portland cement hydration demand, however, evaporates, creating voids and cracks within the structure of concrete. The presence of cracks and voids in concrete is detrimental to concrete properties as these cracks and voids cause stress concentrations that result in brittleness and strength decrease. While compressive strength is somewhat unaffected by such cracks and voids, tensile strength decreases dramatically. Thus, the main property that is usually measured for brittle materials such as concrete is compressive strength. Cracks behave quite differently under the effect of tension and compression stresses (Figure 2.2). While cracks simply close under compression, they open at a fast rate under tension due to high stresses at the crack tips.

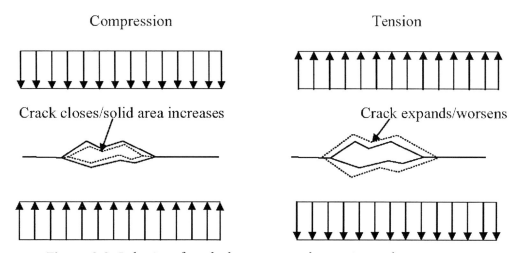

Figure 2.2: Behavior of cracked concrete under tension and compression.

2.1.4 Void System of Concrete

In addition to the solid components, voids are present within fresh and hardened concrete. Two different types of void system can be defined, a system made up of interconnected voids and another system of isolated voids. Interconnected voids consist of three types:

1.   Interlayer voids of CSH are typically too small to affect concrete properties.

2. Capillary voids are created due to the partial evaporation of fresh concrete mixing water in excess to portland cement hydration demand. Capillary voids are irregular in shape and orientation, and range between 4×10^{-5} to 4×10^{-2} in (10^{-6} to 10^{-3} m) in size depending on concrete quality. They are typically harmful to concrete strength and impermeability. Capillary voids are partially filled with water under normal conditions. Evaporation of water in capillary voids causes shrinkage and/or shrinkage cracking of cement paste and, consequently, concrete. Use of steel reinforcing bars is an effective procedure against shrinkage effects on concrete. The presences of pre-existing cracks as well as drying shrinkage are inherent characteristics of concrete that have a significant effect upon its structural behavior.

3. Entrapped air voids represent another type of interconnected void system within concrete. During mixing of fresh concrete, air or water pockets are entrapped due to incomplete consolidation or compaction. Vibration of fresh concrete that facilitates consolidation is employed to reduce entrapped air voids. Entrapped air voids are harmful to concrete strength and impermeability and should be reduced by following good concreting practice.

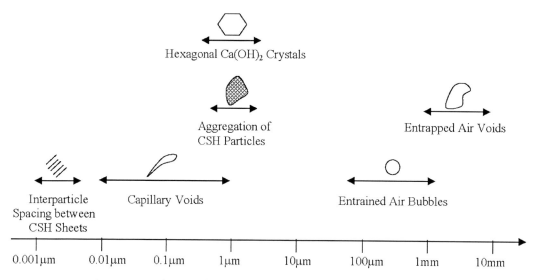

Figure 2.3: depicts the sizes of solids of and voids in concrete.

The second type of void system is isolated air voids in concrete, which are purposely induced by the application of an air-entraining agent to fresh concrete (typically added as a powder or dissolved in water). As a result, isolated air bubbles are created within the cement paste with sizes ranging between 2 and 8×10^{-3} in (50 and 200 μm). Air entrainment of concrete is very beneficial for protection against water expansion due to freezing. While interconnected voids are readily filled with water during concrete saturation (e.g. by rain water), entrained air bubbles remain to a considerable extent unfilled. As such, upon water expansion due to freezing, expanded water or ice finds the air bubbles as relief or extra space that may be occupied rather than causing fractures and cracks in concrete to forcefully create extra space. With the absence of entrained air, further water saturation of concrete fills the newly created voids and cracks by the previous freezing expansion. A consequent freezing cycle exacerbates the problem until complete disintegration of concrete. As such, air entrainment is the preferred method for concrete protection against repeated cycles of freezing and thawing in cold weather. Figure 2.4 schematically illustrates the role of entrained air in protection against expansion due to water freezing.

Air entrainment can also be used for protection against expansive chemical attack, particularly sulfate attack, upon concrete. While capillary voids are irregular in shape, entrained air voids are spherical. As they help concrete durability, however, entrained air voids damage concrete strength and permeability.

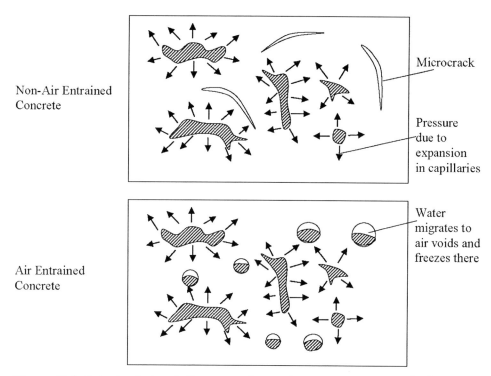

Figure 2.4: Entrained air role for protection against water expansion by freezing.

2.1.5 The Transition Zone of Concrete

During mixing of fresh concrete, aggregate acquires a moisture content termed absorption capacity. The water absorbed by aggregate is considered idle during mixing as it does not contribute to the workability of fresh concrete. As such, it is not considered a part of the required fresh concrete mixing water and is adjusted for during concrete mix design. Extra water is added if aggregate moisture content is below absorption capacity, while the mixing water is reduced if aggregate moisture exceeds absorption capacity. The balance of water added to fresh concrete plus aggregate moisture should equal to the required fresh concrete mixing water plus aggregate absorption capacity. With such precaution, aggregate moisture and water absorption will not interfere with fresh concrete mixing water, consequently, eliminating or reducing potential harmful effects on fresh concrete hardened concrete properties.

Absorption capacity of aggregate is defined as the moisture condition when, for each aggregate particle, all internal voids are filled with water while all external surfaces are dry. Absorption capacity of concrete aggregate is typically referred to as the saturated-surface-dry condition. Figure 2.5 illustrates the different moisture conditions of concrete aggregate.

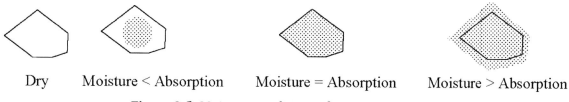

Dry   Moisture < Absorption   Moisture = Absorption   Moisture > Absorption

Figure 2.5: Moisture conditions of concrete aggregate.

As the water quantity within concrete decreases due to evaporation and hydration of portland cement, the increasingly–dry cement paste tends to absorb the water contained within the aggregate or the water previously absorbed by aggregate during fresh concrete mixing. This reversal of water absorption occurs while concrete is still fresh resulting in a localized increase in water content or water/cement ratio in the area directly surrounding the aggregate particles especially coarse aggregate particles due to their larger size. As such, the area surrounding aggregate particles is the weakest link in the cement paste structure due to its higher water/cement ratio compared with the bulk of cement paste. This area is called the transition zone as it is the transitional link between aggregate particles and cement paste. It is typically about 2×10^{-6} in (50 µm) in thickness. Another phenomenon that causes the transition zone to be weaker than the rest of cement paste (bulk cement paste) is the relatively looser particle packing of portland cement next to aggregate particle surfaces or the wall effect. This localized increase of water to cement ratio in addition to the relatively looser particle packing result in higher void ratios and higher concentrations of calcium hydroxide in the transition zone as schematically illustrated in Figure 2.6.

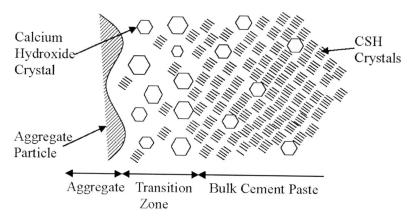

Figure 2.6: The transition zone of concrete.

2.1.6 Concrete Ingredients

In the design of a concrete mixture, one needs to select the appropriate ingredients followed by selection of the proper proportions. Concrete mix ingredients need to follow the standards of the American Society for Testing and Materials (ASTM). Significant deviation from these standards could cause concrete properties (strength and durability) to deteriorate.

For concrete aggregate, ASTM standards are utilized for aggregate particle size distribution (gradation), compatibility within the alkaline environment of cement paste (cement paste pH = 12 – 14 due to the presence of sodium oxide in portland cement), and durability against exterior effects such as freezing and thawing, vehicle tire abrasion (particularly for pavements), as well as resistance to potential chemical attacks including sulfate attack and others. Furthermore, ASTM standards illustrates the procedures used for determining aggregate unit weigh, specific gravity, moisture and absorption capacity necessary for mix design.

ASTM contains standards for portland cement (ASTM C150). There are several different types of portland cement:

Type I	–	Ordinary Use
Type II	–	Moderate Chemical Resistance (to sulfate attack and alkali – aggregate reaction)
Type III	–	Fast Strength Gain
Type IV	–	Low Heat of Hydration
Type V	–	High Sulfate Resistance
White	–	For decorative purposes

Furthermore, using the letter "A" in the designation of a portland cement type indicates that it contains an air entraining agent that it is suitable for use in cold environments with repeated freezing and thawing. For example, Type IA is an ordinary type with added air entraining agent. A common type for Southern California is Type II low alkali portland cement. It has a moderate resistance for chemical attack (especially sulfate attack) and it contains a low quantity of sodium oxide which typically increases cement paste alkalinity. Increased quantities of sodium oxide can potentially cause a chemical reaction between portland cement and alkali – susceptible aggregate.

It has to be noted that utilizing a low water/cement ratio is generally the most effective procedure for enhancing concrete strength and durability. However, certain aspects of durability may require further precautions in addition to reducing water/cement ratio, such as using a chemical admixture or special cement type.

2.1.7 Concrete Admixtures

In addition to the aforementioned standard concrete mix ingredients, there are certain admixtures that are typically utilized with concrete to enhance its properties. These admixtures are also covered in ASTM standards. The most common admixtures are briefly reviewed in the following parts of this section.

Air Entraining Admixtures induce isolated air bubbles within the cement paste of concrete resulting in enhanced resistance to repeated freezing and thawing cycles. Such isolated air bubbles work as relief chambers against water expansion caused by freezing, thereby significantly lessening the damage sustained by concrete. With these air bubbles being isolated and extraordinarily abundant, they are in no way completely filled with water and an adequate number of them are always present to provide protection for concrete.

Plasticizers and water reducers are utilized to enhance concrete workability necessary for construction. This enhancement is done in the absence of increasing water/cement ratio which is harmful to concrete properties. Plasticizers and water reducers decrease concrete water demand for a sought workability up to about 10%. High range water reducers and superplasticizers can reduce concrete workability water demand up to 30%.

Set retarders and accelerators for concrete are also available. Set retarders and accelerators are utilized to decelerate and accelerate setting and hardening of concrete. Other chemical admixtures such as corrosion inhibitors are also common for concrete.

Another family of concrete admixtures is mineral admixtures which include fly ash, silica fume, slag, and other pozzolans. All these pozzolans trigger the pozzolanic reaction within concrete; which is the reaction of the amorphous silica of the pozzolans and calcium hydroxide of cement paste, resulting in calcium silicate hydrate (CSH) as shown below:

$$K[(SiO_2)_m] + L[Ca(OH)_2] + M(H2O) \rightarrow K(CaO).L(SiO_2).M(H_2O) \tag{2.2}$$

Amorphous Silicate + Calcium Hydroxide + Water $\rightarrow$ Calcium Silicate Hydrate (CSH)

Thus, pozzolans improve the micro-structure of concrete via replacing the weak calcium hydroxide with the strong CSH. Other benefits or side effects of a pozzolans vary among types. Fly ash typically enhances workability of concrete, while silica fume damages workability. Certain types of pozzolans with contain sulfuric acid (SiO_2) may adversely affect the chemical resistance of concrete.

2.1.8 Standard Concrete Workability Test Procedure

Favorable workability of fresh concrete is essential for successful construction. Concrete workability is primarily affected by water content. However, increasing concrete water content can adversely affect hardened material properties. Such effect stems from the increased volume of voids and cracks left consequent to excess water evaporation. Concrete mix design targets reaching stability in water content or water/cement ratio that would result in favorable workability without significantly harming mature concrete strength. The slump test, specified in ASTM C 143 standards, is commonly used to assess workability. The slump cone and a tamping rod (Figure 2.7) are utilized. The slump cone is made up of a 12 in (300 mm) high hollow frustum with a top diameter of 4 in (100 mm) and a base diameter of 8 in (200 mm). It has two foot pieces and two handles. The slump testing procedure is performed as illustrated in Figure 2.8. The slump test operator stands on the foot pieces to hold the cone firmly in place. The tamping rod is about 24 in (600 mm) – long and 5/8 in (16 mm) in diameter. The slump cone is dampened and placed on a flat moist nonabsorbent rigid surface. The operator stands on the foot pieces and fills the slump cone with fresh concrete in three layers. Each layer's thickness is approximately equal to 1/3 of total cone's height. The operator rods each layer 25 times with the tamping rod. Rod strokes should be uniformly distributed over the surface of each layer without penetrating the underlying layer significantly. For the top layer, fresh concrete should heap above the cone prior to and during rodding. Consequent to rodding, excess fresh concrete heaping above the mold is stricken off utilizing the tamping rod. The operator removes the mold in a steady vertical motion using the two handles within 5 ± 2 seconds. The mold is placed upside down on the rigid surface next to the slumping fresh concrete frustum. The tamping rod is placed horizontally across the cone and slump is measured. Collapsed (sheared) concrete cone results in incorrect slump measurement and should be disregarded. Typically a slump of 1 – 2 inch (25 – 50 mm) is considered acceptable workability, 3 – 4 inch (75 -100 mm) is considered moderate workability, 5 – 6 inch (125 – 150 mm) is considered good workability, and 7 – 8 inch (175 – 200 mm) is considered excessive workability. More than 8 inch slump may be considered excessively fluid for homogeneous mixing and special design may be required.

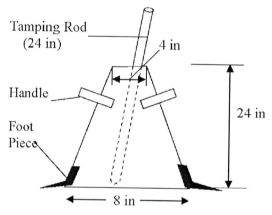

Figure 2.7: The slump cone.

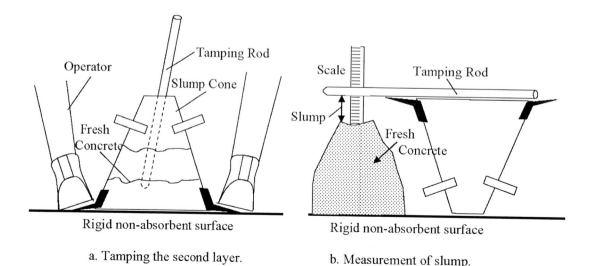

a. Tamping the second layer.   b. Measurement of slump.

Figure 2.8: The slump cone test procedure.

2.1.9 Standard Concrete Compression Samples

The aforementioned discussion illustrates that, due to cracks and voids, concrete tensile strength is relatively small compared to its compressive strength. As such, compressive strength is used as the standard quality control procedure for hardened concrete as described by ASTM Standard C 39. ASTM specifies standard concrete compression test sample as 6-inch (150-mm) diameter x 12-inch (300-mm) high cylindrical specimens or 4-inch (100-mm) diameter x 8-inch (200-mm) high cylindrical specimen. They are cast using plastic cylinders with internal dimensions identical to sample size. Fresh concrete is mixed then poured/placed in the cylinder in 4-inch (100 mm) thick layers. Each layer is rodded, using near vertical strokes of a 24-inch (600-mm) long x 5/8-inch (16-mm) diameter rod, until adequate packing or consolidation is achieved. Such consolidation is important to reduce entrapped air voids within fresh concrete. High quantities of entrapped air results in decreased strength as explained earlier.

Fresh concrete is kept in a covered plastic cylinder for about 24 hours to prevent evaporation of surface moisture and early shrinkage cracking (plastic shrinkage cracking). The plastic cylinder is removed upon concrete gaining adequate strength (typically 24 hours). Standard curing of concrete is performed for 28 days under 100% RH (relative humidity) and 77°F; which are favorable conditions for cement hydration. Curing conditions and/or period may vary depending on the project engineer's recommendations and project conditions. To obtain a statistically significant average of concrete compressive strength, ASTM specifies two or preferably three 6x12 in (150x300 mm) or three or preferably five 4x8 in (100×200 mm) standard compression samples for each mixing batch of 150 yd^3 (140 m^3).

Due to statistical variations in concrete mix ingredients, proportioning, mixing and curing, two concrete compressive strengths are realized, the required compressive strength (f'_{cr}) and the design compressive strength (f'_c). f'_{cr} is used by the materials engineer as a sought compressive strength for mix proportions, ingredient selection, and choice of mixing and curing procedures. f'_{cr} is the anticipated test result average of the standard concrete compression test cylindrical samples. f'_c is the design compressive strength of concrete and is utilized for structural design for members and whole structures. f'_c is typically lower than f'_{cr} taking into account statistical variation of mixing ingredients, mixing proportions as well as construction and curing procedures. Generally, f'_c is about 10% lower than f'_{cr}. In the following portions of this textbook, the term concrete compressive strength will be used to refer to the design compressive strength of concrete, f'_c.

2.1.10 Compression Stress – Strain Behavior Concrete

Assuming a successful mix design of concrete, concrete ultimate compressive strength would match (or exceed) the design (or specified) compressive strength. As such, in the following parts of this textbook, f'_c is used to designate the ultimate compressive strength of concrete. It is essential to understand the compressive stress strain behavior of concrete for its structural applications. Figure 2.9 illustrates the traits of concrete behavior under compressive stress. An initial linear elastic stage reaching a stress of about $f'_c/3$ is first noted. The following stage is nonlinear to a peak compressive stress (strength), f'_c, at a strain of about 0.002. A failure stage follows that concludes with a minimum crushing failure strain between 0.003 and 0.0035.

Figure 2.10 shows the variations of concrete compressive stress – strain behavior due to the variations of f'_c. As noted, ultimate stress or strength occurs for all concrete types at a strain of about 0.002. Furthermore, higher strength concretes are more brittle than conventional concretes. The failure stage is abrupt rather than gradual in high strength concrete as compared with conventional concrete. As such, higher safety factors are typically assigned to high strength concrete as will be explained in later chapters.

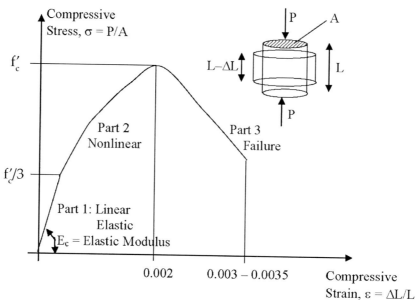

Figure 2.9: The stress–strain behavior of concrete.

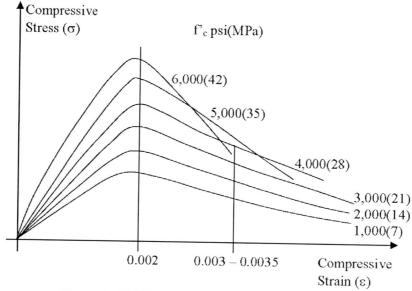

Figure 2.10: The stress–strain behavior of concrete.

2.1.11 Concrete Static Modulus of Elasticity

Although concrete is not elastic, it is generally considered elastic under compression for stresses below $f_c'/3$. In this semi–elastic region of concrete compressive behavior, the static modulus of elasticity can be determined or measured experimentally. Typically, one of two empirical approximate equations can be used to determine the elastic modulus of concrete based on its compressive strength, f_c' (psi), as shown in equations 2.3 and 2.4:

$$E_c = 57,000\sqrt{f_c'} \qquad (2.3)$$

$$E_c = w_c^{1.5} \times 33 \times \sqrt{f_c'} \qquad (2.4)$$

It should be noted that equations 2.3 and 2.4 are correct only when using U.S. units. Equation 2.3 is generally more popular. Concrete elastic modulus can also be determined experimentally according to ASTM C 469 test method.

Where:

E_c (psi) = the static modulus of elasticity of concrete in compression, and

w_c = the unit weight of concrete in lb/ft^3.

Concrete elastic modulus is determined experimentally according to ASTM C 469.

2.1.12 Tensile Strength of Concrete

It is difficult to determine the tensile strength of concrete since it has a relatively small value compared to its compressive strength. Furthermore, the presence of microcracks in concrete results in sensitivity to local stresses including the stresses caused by the grips themselves. Failure of concrete samples in tension commonly occurs at the grips. Within the ASTM standards, there are three different procedures for determining the tensile strength of concrete depending upon the structural application. They are the direct tensile strength, splitting tensile strength and flexural strength or modulus of rupture. Each testing procedure is briefly explained in the following sections.

2.1.12.1 The Direct Tensile Strength of Concrete (ASTM C 192)

This procedure includes applying a direct tension force to a briquette – shaped concrete sample with a critical cross-section of 1×1 in (25×25 mm) as shown in Figure 2.11.

Due to the small size of samples in Figure 2.11, the direct tensile strength is generally applied to mortar (concrete with no coarse aggregate) or cement paste with no aggregate. To determine the direct tensile strength experimentally, the maximum load sustained by the sample, P_{max}, is divided by the critical sample cross sectional area, A (1×1 in or 25×25 mm). Equation 2.5 is an empirical equation that is typically used to determine the value of direct tensile strength of concrete, mortar, or cement paste based on the compressive strength, f'_c (psi):

$$f_c = 5.6\sqrt{f'_c} \qquad (2.5)$$

Where f_c (psi) is the direct tensile strength of concrete, mortar or cement paste.

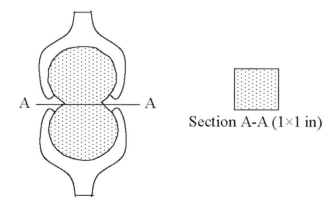

A ——— A

Section A-A (1×1 in)

Figure 2.11: Direct tensile strength testing procedure.

2.1.12.2 The Splitting Tensile Strength of Concrete (ASTM C 496)

This procedure involves applying a line load to two opposite sides of a concrete cylinder placed in a horizontal position as shown in Figure 2.12. Typically, 4×8 in (100×200 mm) cylindrical samples are used for this test. The stress distribution at the critical section of the sample is shown in Figure 2.12. The compressive stresses at the top and bottom of the section are significantly larger than the tensile stress within the middle portion of the sample. Nevertheless, the sample fails in tension since concrete is weak in tension and strong in compression. Equation 2.6 can be used to calculate the splitting tensile strength of concrete:

$$f_{ct} = 2P_{max}/\pi DL \tag{2.6}$$

Where:

f_{ct} is the splitting tensile strength of concrete,

P_{max} is the maximum load sustained by the sample,

D is sample diameter, and

L is sample length.

Equation 2.7 is an empirical equation (using U.S. units) that is typically used to determine the value of splitting tensile strength of concrete, f_{ct} (psi), based on its compressive strength, f'_c (psi):

$$f_{ct} = 6.7\sqrt{f'_c} \tag{2.7}$$

Generally, the splitting tensile strength is considered a good approximation of concrete tensile strength for structural applications.

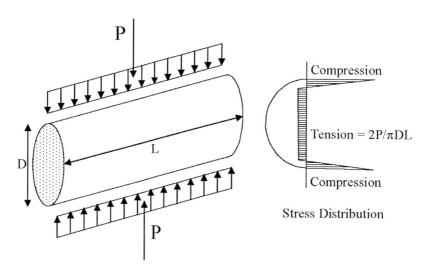

Figure 2:12: The splitting tensile strength testing procedure.

2.1.12.3 The Bending Tensile (Flexural) Strength of Concrete (ASTM C 78)

This concrete strength is determined by subjecting a concrete beam to bending with four–point (1/3–point) loading. The beam specimen has a span of 12 in (300 mm) and nominal dimensions of 4×4×14 in (100×100×350 mm) as shown in Figure 2.32. The maximum value of load P is attained when the concrete sample cracks at the bottom fiber of the section where tension stress is highest. Based on the principles of solid mechanics, the maximum bending stress (tension or compression) occurring in the sample can be determined using Equation 2.8:

$$f_r = M_{max}/S = M_{max}/(bh^2/6) = P_{max}L/(bh^2) \tag{2.8}$$

Where:

f_r is the flexural strength or the modulus of rupture of concrete.

M_{max} is the maximum bending moment sustained by the sample.

P_{max} is the maximum load sustained by the sample.

b is sample width.

h is sample depth.

Equation 2.8 assumes a linear elastic distribution of stresses throughout the section without redistribution potential due to the nonlinear behavior of concrete. Both of these assumptions are inaccurate, which results in an approximate value for the modulus of rupture of concrete (Figure 2.12).

Equation 2.9 is an empirical equation (using U.S. units) that is typically used to determine the value of modulus of rupture of concrete, mortar, or cement paste f_r (psi) based on the compressive strength, f'_c (psi):

$$f_r = 7.5\sqrt{f'_c} \tag{2.9}$$

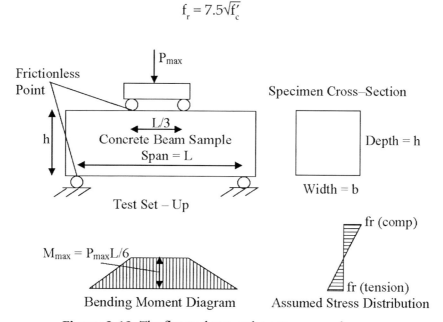

Figure 2:13: The flexural strength testing procedure.

2.1.13 Dimensional Stability of Concrete

2.1.13.1 Drying Shrinkage

As water retained by hardened concrete evaporates gradually, capillary voids empty and contract in size resulting in drying shrinkage of concrete. Concrete shrinkage occurring when concrete is fresh is termed plastic shrinkage, while shrinkage of hardened concrete is termed drying shrinkage. To protect against plastic shrinkage, proper placement, consolidation, finishing, and curing procedures must be followed for fresh concrete, particularly for protection against rapid water evaporation due to high temperature, wind, and direct exposure to sunlight. Such protection is typically achieved by misting or fogging. For hardened concrete, drying shrinkage is generally in the range of 600 to 800 µstrain (0.06 to 0.08%). A timeline history of concrete drying shrinkage is shown in Figure 2.14. To reduce the effect of drying shrinkage, certain precautions may be taken including:

1.   Decrease of concrete mixing water,

2.   Application of proper curing (at least 7 days of moist curing),

3.   Use of good quality aggregate that can work as shrinkage arrestors within concrete. Furthermore, aggregate with low absorption capacity can help reduce shrinkage of concrete,

4.   Utilization of proper concrete mix design without increased amounts of cement paste (the component causing shrinkage),

5.   Use of construction joints (with reduced section depth) so that shrinkage cracking is confined to the joints and, thus, is controlled, and

6.   Use of shrinkage reinforcement to reduce shrinkage cracks width and frequency.

In the absence of design and construction precautions against plastic and drying shrinkage, parallel cracks spaced at about 10 ft would occur in concrete slabs on grade and other types of concrete flat work.

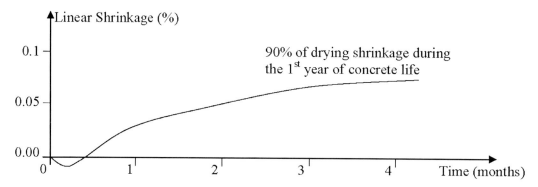

Figure 2:14: Timeline history of concrete drying shrinkage.

2.1.13.2 Creep

After setting, in addition to shrinkage, water discharges from concrete pores and cracks by external pressure or compression resulting in stress–induced dimension shrinkage, phenomenon termed creep or plastic flow. Creep damages the microstructure of concrete, causing a reduction of f'_c that can reach as high as 25%. With the effect of creep, the final deformation (compressive strain) of concrete can be 2 – 3 times initial deformation. Creep deformations occur over a period of a few years with about 80% of total creep taking place within the first year after load application. With load removal, creep strains (deformations) are typically irrecoverable while elastic deformations are recoverable. Figure 2.15 illustrates the phenomenon of creep under the effect of compressive stresses on concrete.

To reduce the effect of creep, certain precautions may be taken including:

1.   Completion of adequate curing (at least 7 day moist curing) prior to loading. With longer curing periods and increased age, concrete microstructure matures and becomes more resilient to the effect of creep,

2.   Using concretes with higher strengths which have less creep than conventional concrete due to their enhanced microstructure,

3.   Decrease of concrete mixing water so that fewer voids and microcracks are created,

4.   Use of good quality aggregate that can work as creep arrestors within concrete.

5.   Utilization of proper concrete mix design without increased amounts of cement paste (the component causing creep), and

6.   Use of steel reinforcement to reduce creep effect upon concrete members.

With increased temperature and reduced humidity, creep and shrinkage are worsened since such conditions encourage water evaporation.

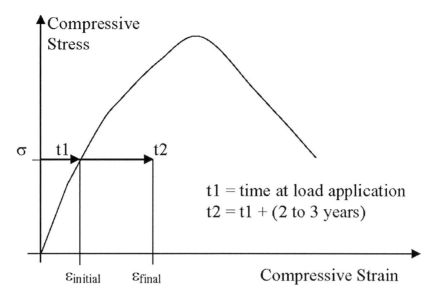

Figure 2.15: Creep effect on the compressive behavior of concrete.

2.2 Reinforcing Steel Bars

As mentioned in Chapter 1, the use of smooth reinforcing bars has been deserted in practice. With the abandonment of plain (smooth) steel bars, deformed steel bars currently represent the most common procedure for concrete reinforcing. The bond between steel bars and concrete consists of three components:

1. Chemical bond caused by the reaction of cement paste alkaline substances and steel,

2. Physical bond caused by friction and encouraged by concrete shrinkage, and

3. Mechanical bond or anchorage resulting from the interlocking of steel bar surface corrugations or deformation and concrete. The purpose of using deformed bars with surface deformations is to enhance the effectiveness of mechanical bond of steel bars and concrete.

Mechanical bond is typically considered reliable for stress and load transfer between steel and concrete and is relied on for structural design. Chemical and physical bond, however, are not considered reliable for structural design. The utilization of hooks with smooth bars can be considered an archaic method of activating mechanical anchorage of steel bars. A sketch of a deformed reinforcing steel bar is shown in Figure 1.4. Another example of a deformed bar is outlined in Figure 2.16. Deformed steel bars for use as concrete reinforcement must comply with ASTM standard A 615 so that adequate bond strength can be developed with concrete and steel bars.

Surface Deformations

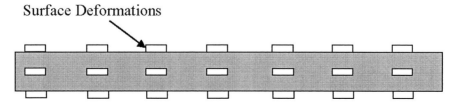

Figure 2.16: Schematic diagram of deformed steel bars.

Steel bar designations, diameters and required deformations are provided in Table 2.1 as per ASTM A615. As noted in Table 2.1, steel bars are designated by their diameter in multiples of 1/8 inch (3 mm). As such bar no. 3 has a diameter of 3/8 inch. Note that bar sizes jump from No. 11 to No. 14 to No. 18 where sizes 12, 13, 15, 16 and 17 are not used. In addition to the diameter and area of each bar size, Table 2.1 lists information about their surface deformations. Such information includes spacing, height, and gap of the deformation pattern.

Table 2.1: Summary of ASTM A615 Requirements.

Bar No.	Diameter in (mm)	Area (in²)	Deformation Requirements		
			Max Ave Spacing (in)	Max Ave Height (in)	Max Gap (in)
3	$^3/_8$ (9.5)	0.11	0.26	0.015	0.14
4	$^1/_2$ (13)	0.20	0.35	0.020	0.19
5	$^5/_8$ (16)	0.31	0.44	0.028	0.24
6	$^3/_4$ (19)	0.44	0.52	0.038	0.29
7	$^7/_8$ (22)	0.61	0.61	0.044	0.33
8	1 (25)	0.79	0.70	0.050	0.38
9	$1^1/_8$ (29)	1.00	0.79	0.056	0.43
10	$1^1/_4$ (32)	1.27	0.89	0.064	0.49
11	$1^3/_8$ (35)	1.56	0.99	0.071	0.54
14	$1^3/_4$ (44)	2.25	1.19	0.085	0.65
18	$2^1/_4$ (57)	4.00	1.60	0.102	0.86

1 in² = 6.45 cm²

A typical stress–strain behavior curve of steel is shown in Figure 2.17. Additionally, Figure 2.17 contains a simplified stress–strain diagram of steel. The simplified stress – strain diagram of steel represents an elastic–perfectly plastic material that reduces design complexity to a large extent. Typically, post–yield stress (f_y) steel experiences strain hardening leading up to its ultimate stress (f_u). In the simplified diagram, after yielding, steel maintains its yield stress indefinitely. Such simplification may introduce a small error in the design; however, such error is to the safe side. The simplified stress – strain diagram of steel is generally used in reinforced concrete design and is adopted by the American Concrete Institute (ACI) Code. In Figure 2.17:

1. f_y = yield strength of steel,

2. f_u = ultimate stress of steel,

3. E_s = modulus of elasticity of steel = 29×10^6 psi (200 MPa),

4. $\varepsilon_y = f_y/E_s$ = yield strain of steel, and

5. ε_r = rupture strain of steel where tension failure or collapse occurs.

As noted in Figure 2.17, yielding of steel typically continues to a strain of 3 to 5%. Rupture of steel in tension generally takes places at a strain of 15 to 20%.

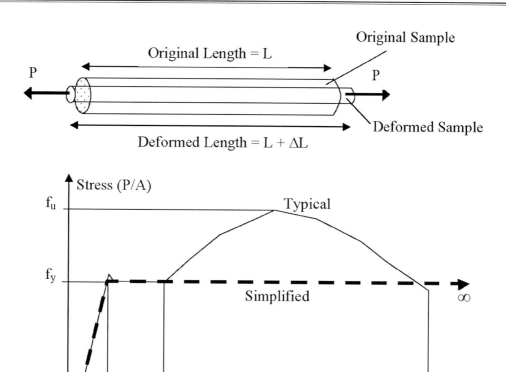

Figure 2.17: Stress – Strain diagrams of steel.

In addition to deformed reinforcing steel bars, Welded Wire Fabric (WWF) is also used for concrete reinforcement, especially for slabs and shells. WWF is made of two welded perpendicular layers of cold-drawn steel wires with yield strength of 75 to 80 ksi (520 to 550 MPa). Welding is provided at the points of intersections of wires. WWF can be manufactured of smooth or deformed wires. Wire size and spacing as well as specification vary as dictated by the design requirements. Welded wire fabric is supplied in sheets or rolls that are generally used in slabs or flat works.

WWF type designation is as follows:

$$a \times b - Sc \times S'd \tag{2.10}$$

Where:

a,b = spacing (in) of wires in the longitudinal and transverse directions, respectively. Commonly, WWF spacing is 4 or 6 in (100 or 150 mm),

S,S' = designation of wires in the longitudinal and transverse directions, respectively. There are only two designations, W and D, referring to smooth or deformed wire, respectively, and

c,d = cross sectional area (in²) of wires in the longitudinal and transverse directions, respectively.

Table 2.2 presents the common sizes or types of WWF.

Table 2.2: Common Sizes/Types of Welded Wire Fabric.

WWF Designation	Steel Area (in²/ft)		Weight (lb/100 ft²)
	Longitudinal	Transverse	
Rolls			
6 × 6 – W1.4 × W1.4	0.028	0.028	21
6 × 6 – W2 × W2	0.040	0.040	29
6 × 6 – W2.9 × W2.9	0.058	0.058	42
6 × 6 – W4 × W4	0.080	0.080	58
4 × 4 – W1.4 × W1.4	0.042	0.042	31
4 × 4 – W2 × W2	0.060	0.060	43
4 × 4 – W2.9 × W2.9	0.087	0.087	62
4 × 4 – W4 × W4	0.120	0.120	86
Sheets			
6 × 6 – W2.9 × W2.9	0.058	0.058	42
6 × 6 – W4 × W4	0.080	0.080	58
6 × 6 – W5.5 × W5.5	0.110	0.110	80
4 × 4 – W4 × W4	0.120	0.120	86

$1 \text{ in}^2 = 6.45 \text{ cm}^2$

Test your knowledge

1. Based on the stress – strain diagram of concrete, which type of concrete appears more brittle? Estimate the toughness index for low – strength and high – strength concrete (toughness index = area under stress – strain curve / peak stress).

2. What is the main parameter that determines the mechanical properties of concrete?

3. Why is the modulus of rupture of concrete is higher than its splitting tensile strength?

4. What are the main two quality control procedures for concrete?

5. Why is tensile strength of concrete low?

6. List the factors that contribute to compatibility of steel and concrete.

7. What is the value of water – to – cement ratio needed for hydration? Why is higher water – cement ratio used in practice?

8. What are the disadvantages of using high or excessively high water – to – cement ratio on fresh and hardened concrete?

9. What is concrete segregation and honeycombing and what causes these phenomena?

10. What are the methods of testing the tensile strength of concrete? In your opinion which procedure is most representing of field conditions?

11.   What are the differences between the simplified and the actual stress – strain diagram of steel?

12.   A reinforced concrete member has a reinforcement ratio, ρ_g, where ρ_g = area of steel/total area of reinforced concrete member. If plain concrete shrinkage strain is ε_{sh}, and compressive strength = f'_c, what is the resultant shrinkage strain of the member and the stresses in concrete and steel for the following cases:

a. $\rho_g = 1\%$, $\varepsilon_{sh} = 0.06\%$, $f'_c = 4{,}000$ psi,   b. $\rho_g = 1\%$, $\varepsilon_{sh} = 0.1\%$, $f'_c = 4{,}000$ psi,

c. $\rho_g = 0.5\%$, $\varepsilon_{sh} = 0.06\%$, $f'_c = 4{,}000$ psi,   d. $\rho_g = 1\%$, $\varepsilon_{sh} = 0.06\%$, $f'_c = 5{,}000$ psi.

Chapter 3

DESIGN PROCEDURES FOR REINFORCED CONCRETE

3.1 Design Standards and Codes

Reinforced concrete design is governed by a set of standards and codes listed as follows:

3.1.1 American Society for Testing and Materials (ASTM) Standards

The American Society for Testing and Materials is headquartered in Philadelphia, Pennsylvania. ASTM standards cover material specifications and test procedures and do not extend to cover structural analysis and/or design or concrete mix design. As such, ASTM standards provide the specifications for materials including portland cement, coarse aggregate selection and gradation, fine aggregate selection and gradation, concrete additives, as well as steel reinforcing bars. In addition to the specifications for materials, ASTM standards include the standard test methods and standard practice for testing a material to determine compliance with the material standards or suitability for the sought application. Typical material specifications for reinforced concrete may read:

Concrete materials shall comply with the following ASTM standards:

Portland Cement Type II according to ASTM C150

Coarse Aggregate Type 57 according to ASTM C127

Fine Aggregate (Sand) with Fineness Modulus 2.8 according to ASTM C128

Chemical Admixture according to ASTM C494

Reinforcing Steel Bars with yield strength = 60 ksi according to ASTM C615

Optional: All aggregate must achieve less than 0.01% alkali – silica reactivity expansion according to ASTM C289

Batching of concrete shall be according to ASTM C685

Quality control testing of concrete shall include:

Slump according to ASTM C143

Compressive Strength Test according to ASTM C38

Flexural Strength Test according to ASTM C78

Materials specifications are typically contained in the first page of the project structural design documents or plans. The design engineer for a project may include other ASTM specifications as the situation requires.

3.1.2 The American Concrete Institute Building Code Requirements for Structural Concrete (ACI 318)

The American Concrete Institute is located in Detroit Michigan. ACI 318 Code discusses target concrete strength for mix design f'_{cr} as a function of the design or specified compressive strength of concrete f'_c, as discussed in Chapter 2, as well as other material specifications. ASTM standards are used in ACI 318 as references for material specifications. Construction considerations are also discussed.

ACI 318 focuses on the <u>structural design procedures for reinforced concrete structural elements such as beams, columns, footings, joints, etc as well as buildings</u>. The structural design method for concrete utilized in ACI 318 is <u>the strength design method</u>. The assumptions and utilization of the strength design method are stated. As ACI 318 details the design procedure and structural design standards for various structural elements and structures constructed of reinforced concrete, it is considered the primary reference for reinforced concrete design. ACI 318 Code does not cover design loads or material specifications in detail. It refers to other codes regarding these topics. Nevertheless, ACI 318 provides its users with load combinations used for member or building design utilizing the strength design method.

ACI also publishes a large number of standards, books, journal, reports and symposium and conference proceedings about aspects of concrete behavior including materials, construction, design, as well as research findings. In particular, the scientific advances in concrete are periodically published by ACI that are instrumental for code updates and structural design of structures. As such, ACI acts as a clearing house for various issues concerning concrete design, construction, service and durability.

Typical structural detailing specifications for reinforced concrete may read:

Reinforcing steel concrete cover must comply with the minimum standards of ACI 318 Code.

Concrete design compressive strength, f'_c (defined per ACI 318), must not be less than

2,500 psi   Footings

3,000 psi   Columns and Beams

Minimum steel reinforcement and reinforcement detailing standards according to ACI 318 must be used.

Excerpts from ACI 318 may also be contained in the structural detailing specifications for a building.

Structural specifications are typically contained in the first page of the project structural design documents or plans.

3.1.3 The American Society for Civil Engineers/Structural Engineering Institute (ASCE/SEI) Minimum Design Loads for Buildings and Other Structures Standard ASCE/SEI 7

The American Society for Civil Engineers is headquartered in New York. The American Society for Civil Engineers (ASCE) Minimum Design Loads for Buildings and Other Structures Standard ASCE/SEI 7 provides minimum load requirements for the design of buildings and other structures. Furthermore, load combinations used for design are also provided by ASCE 7.

Buildings are generally subjected to various types of loads including self weight, weight of building occupants, soil pressure, hydraulic pressure, wind load, seismic (earthquake) load, snow load, and others. Some of these loads can be determined by the design engineer such as self weight based on unit weight of components. Soil pressure can be determined based on soil properties utilizing appropriate engineering principles. Hydraulic pressure can be also determined utilizing sound engineering methods. However, load of building occupants (typically referred to as live load), wind load, earthquake load as well as snow load are challenging to determine for the design engineer and would require significant research. Furthermore, design engineers could easily disagree upon the required load; as such, ASCE 7 serves two purposes. By defining these loads, this standard greatly assists design engineers in assimilating a good value for live, wind, seismic, and snow loads. Also, it serves as a clearing house among engineers for the general consensus of the values of these loads.

In selecting the load combination a building should be designed for, it is important to be prudent, realistic and scientific. It is unrealistic to assume that a building, in addition to its weight, is subjected to its full live load, severe wind load (like a hurricane), severe seismic load (major earthquake), and full snow load added to soil and water pressure at the same time. Therefore, ASCE/ SEI 7 defines realistic load combinations for which a building should be deigned. Typically, wind or earthquake load is included. Also, some loads are magnified while others are reduced to achieve a critical combination of load. ASCE 7 depends on sound scientific principles in determining these load combinations typically using statistics.

One may ask what happens if all the foreseen loads are applied to a building. While a building may not be designed for such load, it would sustain damage. However, by utilizing other safety measures in the design procedures, it is forecasted that the building would not experience complete collapse that may result in loss of life.

3.1.4 The International Building Code (IBC)

The International Building Code (IBC) is published by the International Code Council located in Whittier California. It contains a collection of building code requirements from various sources including ASCE 7, ASTM Standards, ACI 318, American Institute for Steel Construction (AISC), The Masonry Society (TMS), Occupational Safety and Hazards Administration (OSHA), and others. In many cases portions of these codes and standards are simply reprinted in the IBC. The goal is a document which contains all building requirements including access, materials, and design provisions. Designs with wood, steel, concrete, masonry and other materials are also included. As such, the IBC is a comprehensive document for building design in the U.S. Other countries have also adopted the IBC.

Chapter 19 of the IBC covers concrete structural design. It opens by stating that, while there are differences, ACI 318 is the main reference for this chapter. There are chapters in the IBC for different facets of the building industry. In some instances, the IBC for may not be sufficient and other references may be needed. As an example, design loads are not included. Instead, the IBC refers designer is referred to ASCE 7 for the values of design loads.

The IBC can be considered the parent code for all states in the United States as well as other countries that use it. It is comprehensive and broad and it covers various design conditions for buildings including: severe wind (hurricanes), major quakes, extreme heat and extreme cold temperatures.

3.1.5 The California Building Code (CBC)

While some states and municipalities in the U.S. have elected to adopt the IBC without additions and changes, others have decided to use the IBC with addendums. Other states or municipalities publish their own codes and standards that depend largely on the IBC but contain particulars for that state or municipality. Such particulars may be more or less strict than the IBC. One may refer to the IBC as the parent code, while states' or municipalities' codes are referred to as focus or branch codes. A focus code is particular to one state or one municipality whose conditions require more specific design parameters.

The State of California adopts the California Building Code (CBC) which is a focus or branch code of the IBC. California, with its large size and earthquake susceptibility, has a code devoted to the particular conditions of California. Although there are differences between the CBC and the IBC, for the most part they are equivalent. Some of the differences between these codes can be traced to state vs. national regulations. To maintain similarity, Chapter 19 of the CBC is also devoted to reinforced concrete.

Use of the CBC for building design is mandatory in the State of California. As a result, the CBC is considered a legal document.

3.1.6 Other Organizations

Other reputable organizations publish scientific studies that are referred to in the code due to their importance. A notable mention in the field of concrete is for the Portland Cement Association (PCA) in Chicago, Illinois. Significant scientific advances have been published by PCA and are utilized in various codes.

Each code or standard publishing entity has an organizational structure. Typically, each code publishing entity is a non–profit organization that uses member support as well as government funding for continued existence and function. Codes and standards are updated periodically; typically, every three to five years.

3.2 Loads

Loads affecting reinforced concrete structures can be listed and defined as follows:

a. **Dead Load (D):** The self weight of the structure plus the weight of all added fixtures permanently attached to the structure. Permanent fixtures may include doors, windows, ducts, ceilings, tile, roof cover, wall finish, air conditioning or heating unit, etc. Dead loads are typically determined by the design engineer based on the self weight of each building component.

b. **Live Load (L):** The weight of building occupants including people, furniture and other movable objects. Design values for live load may be obtained from the governing code based on the type of building. For example, for residential buildings living spaces, live load is 40 pounds per square foot (psf).

c. **Roof Live Load (Lr):** The weights of service or repair crew for roofs. It is also specified in the code. A value of 20 psf is widely used.

d. **Snow Load (S):** Caused by the weight of snow. Snow load is typically specified in the code. Some cities and counties specify their own snow load for buildings as well.

e. **Rain Load (R):** Caused by the weight of rain water. Rain load is typically specified by cities and counties.

f. **Soil Pressure (H):** Lateral load caused by soil or underground water on portions of structures below ground level. Soil pressure can be determined by the engineer based on soil data and building configuration.

g. **Hydrostatic Fluid Pressure (F):** Lateral load caused by hydrostatic fluid pressure. This load can also be determined using the scientific principles of fluid mechanics.

h.   **Wind Load (W):** Pressure due to wind forces is generally specified in the code. Nevertheless, some cities and counties specify higher loads for extra safety.

i.   **Seismic or Earthquake Load (E):** Seismic loads on buildings require a sophisticated method of analysis detailed in ASCE 7. Due to such complexity, all municipalities use ASCE 7 for seismic load.

j.   **Temperature or Self Restraining Load (T):** Due to temperature variation, a structure may experience extra forces that require special attention in design. The IBC specifies the types of buildings that require calculations for such load and the method of including such load in design.

3.3 Design Methods

Typically two design methods have been used for reinforced concrete. They are the Working Stress Method (WSD) and the Strength Design Method also referred to as the Load and Resistance Factors Design method (LRFD).

3.3.1 The Working Stress Design Method (WSD)

The working stress method analyzes a building or a structural element under service conditions. It compares the service stress with a designated allowed stress. The structure or building is analyzed under all load combinations representing service conditions (ASCE 7). The maximum stresses are determined and compared with the allowed stress. As such, the main design equation for WSD is:

$$f_{service} < f_{allowed}$$ (3.1)

Where:

$f_{service}$ = maximum stress under service conditions using the load combinations in ASCE 7, and

$f_{allowed}$ = allowed maximum stress under service conditions as specified by ACI 318 for reinforced concrete.

WSD is referred to in ACI 318 as the alternate design method and is included in an appendix. LRFD is and has been the dominant design method for reinforced concrete for about 40 years. The primary disadvantage of WSD is its limited coverage of structural behavior, where it only encompasses service conditions. Under extreme or ultimate conditions of excessive load (such as a large hurricane or a massive earthquake) or unpredicted structural deficiency, WSD does not provide any information about structural failure behavior. Furthermore, loads of different sources are treated equally. For example, dead loads that can be determined with a high degree of certainty are weighed the same as wind or earthquake loads which are difficult to accurately predict.

To study a structural member under service conditions, one needs to superimpose the different loads of section 3.2 on the building or element. Consequently, the element or structure is analyzed to determine the stresses under service conditions; however, it is statistically impossible that all the loads of section 3.2 would be applied simultaneously. Thus, ASCE/SEI 7 presents a number of reasonable load combinations that need to be applied concurrently for analysis under service conditions. These service load combinations are not included in this text.

3.3.2 The Strength Design or Load and Resistance Factors Design Method (LRFD)

The basic equation for LRFD or the strength design method is:

$$\phi R_n \geq U \tag{3.2}$$

Where:

R_n = nominal member resistance. It is termed nominal since it is based on assumptions and structural analysis method selected by the designer; as such, there is a degree of uncertainty in its value. For reinforced concrete, these methods are included in ACI 318,

ϕ = capacity or resistance reduction factor utilized to account for unforeseen circumstances, and

U = Ultimate or maximum applied load as specified by ACI 318 or ASCE 7.

R_n represents member flexural resistance, column axial capacity, or beam shear resistance depending on the aspect of member or structural behavior being studied. The value of ϕ changes depending on loading conditions. For bending ϕ varies between 0.83 and 0.9. For shear ϕ is 0.75. U is the ultimate applied load as defined by code. In the equations for U, code takes into account the accuracy of determining each load, the frequency of every possible load and the importance or effect of every possible load. Based on a statistical analysis, the following equations have been introduced by ACI 318 for the different possibilities of ultimate or maximum load on a structural element of a building:

U1 = 1.4(D + F)

U2 = 1.2(D + F + T) + 1.6(L + H) + 0.5 (Lr or S or R)

U3 = 1.2D + 1.6(Lr or S or R) + (L or ±0.8W)

U4 = 1.2D ± 1.6W + L + 0.5(Lr or S or R)

U5 = 1.2D ± 1.0E + L + 0.2S

U6 = 0.9D ± 1.6W + 1.6H

U7 = 0.9D ± 1.0E + 1.6H

A prudent designer would calculate the value of U1 through U7 and adopt the largest value of U in design and investigating each structural element or building. Typically, one or two values of Ui are applicable. It is a common exercise for students to calculate the values of U for different values of loading. Wind and seismic load are applied twice in each relevant equation with a positive value and negative value since wind and earthquake directions cannot be predicted.

LRFD can overcome the disadvantages of WSD mentioned earlier. Particularly, LRFD studies structural behavior under extreme conditions or circumstances namely failure scenario thereby ensuring ductile failure essential for building design. Ductile/gradual rather than catastrophic failure is highly recommended for structural design so that lives can be saved and buildings repaired or sustained following extreme events. Furthermore, the strength design method presents a more realistic approach to reinforced concrete since concrete is not an elastic material and failure, rather than service analysis lends itself better to reinforced concrete design.

3.3.3 Structural Safety

There are four sources of structural safety in LRFD or the strength design method:

a.   **Material Safety:** While design concrete strength is f'c, the concrete mix is designed for a higher compressive strength value termed f'cr > f'c for safety as explained earlier. Also, steel design yield strength is fy. However, actual steel yield strength is higher than fy by a margin of material safety specified in ASTM,

b.   **Load Factors:** Load factors are specified by code and take into account load variability and uncertainty. Load factors are typically magnification factors to increase the applied load for safety concerns,

c.   **Resistance Factors:** Resistance factors are reduction factors for member and building resistance to external load. These factors take into account loading type (flexure, shear, axial, etc.), accuracy of analysis methods, as well as assumptions, and

d.   **Failure Scenario:** LRFD, as an important safety measure, primarily aims at achieving a gradual or ductile failure of members and buildings accompanied by the absorption of a large quantity of energy. Such gradual failure scenario allows for taking remedial actions if failure was eminent. Remedial action could include evacuation, repair, retrofit, or a number of other preservation measures. Gradual failure allows the element or structure to absorb significant amounts of energy resulting in better resistance to catastrophic loads; the structure may sustain significant damage, but complete collapse is unlikely.

In the strength design method, structural safety is sorted into four categories. Each category is analyzed separately to achieve a comprehensive and thorough structural design methodology encompassing all aspects of safety.

3.4 Inspection of Concrete Structures

Prior to construction of a concrete structure, one must obtain a permit from the jurisdictional municipality such as the city, county or state. Furthermore, the IBC and CBC codes dictate that concrete structures be inspected prior to placement of fresh concrete. There are two types of inspections:

1.   Municipality inspection where a designated structural inspector visits the site prior to fresh concrete pouring. The inspector verifies member dimensions, reinforcement placement as well as other conditions specified in the design documents or plans. This activity is followed by one of two outcomes: clearance for pouring or a requested list of corrections prior to issuing such clearance, and

2.   Special inspection where a certified special inspector confirms compliance of site conditions (member dimensions, steel reinforcement etc) with design documents or plans. Also, the special inspector follows any instructions for special inspections indicated in the design documents or plans such as checking air content in cold climates. The special inspector observes fresh concrete placement to ensure that proper concreting practice is followed. Furthermore, the special inspector measures the slump of fresh concrete and collects compression cylinders for testing to verify achieving code specified compressive strength.

While the municipality inspection is paid for as part of the permit fee, the special inspection is an added expense to the project. Engineers may specify a concrete compressive strength, $f'_c = 2,500$ psi (17 MPa) where the IBC and the CBC codes do not require special inspection. However, these codes dictate special inspection for important members regardless of the specified f'_c. An example is members that are part of the lateral load resisting system of the structure.

Reinforced concrete coverage presented herein utilizes the main reference code for concrete design, ACI 318 Code, and uses the strength design method or LRFD. Furthermore, focus is placed on element or member design.

Test your knowledge

1. What is the main reference code for reinforced concrete design?

2. Which standards do engineers use for material specifications in reinforced concrete design?

3. What is the role of ASCE/SEI 7 in structural design?

4. Is there duplication between the IBC and the CBC? Explain the relationship between these two codes.

5. What are the two design methods for reinforced concrete? Which method is the more dominant?

6. What are the disadvantages of the WSD method?

7. What are the advantages of the strength design method?

8. What are the structural safety sources in the strength design method?

9. What is the difference between the strength design method and the load and resistance factors design method?

10. In the classroom, classify the following as dead or live load: a. concrete slab, b. flooring material, c. suspended ceiling, d. utility pipes, e. desks, chairs and table, and f. instructor and students.

11. Determine the value of U_{max} and U_{min} for a structural member under the following loads:

    D = 80   L = 40   F = 15   Lr = 20

    R = 20   H = 14   S = 0   W = 90

    E = 120

Chapter 4

FLEXURAL BEHAVIOR OF REINFORCED CONCRETE BEAMS

4.1 Flexural Behavior of Reinforced Concrete

The strength design method focuses on extreme or failure conditions occasionally termed ultimate conditions. Thorough understanding of ultimate conditions requires that one study the flexural behavior of beams under increasing moment leading to failure. A typical reinforced concrete beam is shown in Figure 4.1. The beam in Figure 4.1 is simply supported and uniformly loaded with a progressively increasing distributed load labeled "w".

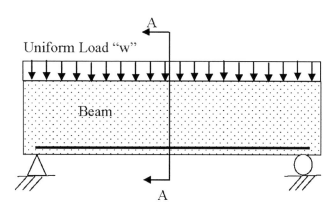

Figure 4.1: Reinforced concrete beam.

Certain assumptions are needed to facilitate the analysis of the reinforced concrete beam in Figure 4.1:

1.   Plane sections before bending remain plane after bending as stated in Bernoulli's beam bending theory. This implies that the strains in a deformed section are linearly proportional to the distance from the Neutral Axis, and

2.   Steel and concrete are perfectly bonded. As such, the strain in steel bars is equal to the strain in neighboring concrete portion provided that they are at an equal distance to the Neutral Axis (NA).

Cross section A-A of the beam in Figure 4.1 is shown in Figure 4.2 representing a typical reinforced concrete beam section. The following notations are shown in Figure 4.2:

b = beam width.

h = beam thickness or height.

d = beam effective depth = distance between the highest point of section (extreme compression fiber) to the centroid of reinforcing bars.

A_s = reinforcing steel bar area.

ρ = beam reinforcement ratio according to the following equation:

$$\rho = A_s/bd \qquad\qquad (4.1)$$

N.A. = the Neutral Axis.

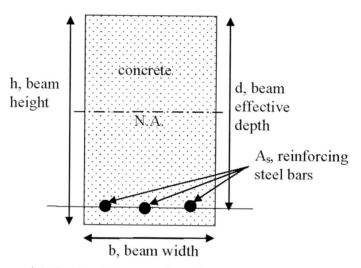

Figure 4.2: Section A-A – typical reinforced concrete beam section.

Under increased loading, the bending moment as well as stresses and strains increase in Section A-A. Based on the value of bending moment or load, Section A-A goes through several phases of behavior as explained below.

Phase 1: Elastic Section Behavior – Uncracked Concrete

Phase 1 is experienced when the loads or moments are small. In phase 1, the flexural tensile strength of concrete (f_r) has not been exceeded and concrete is elastic to a large extent as shown in Figure 4.3. Also, the tensile cracking or rupture strain of concrete in bending ($\varepsilon_r = f_r/E_c$) has not been exceeded. As noted, the strain distribution is linear with respect to the location of the N.A. which has no strain. The strain in concrete adjacent to the steel bars is identical to the strain in steel bars. However, the stress in concrete next to steel bars is significantly less than the stress in steel bars. Under identical strains, ε, the ratio of the stress in steel, σ_s, to the stress in concrete, σ_c, is termed "n" determined according to the following equation:

$$n = E_s/E_c \qquad\qquad (4.2)$$

Where Equation 4.2 may be derived as follows. The stresses in steel and concrete under the effect of an identical strain (ε) can be calculated as:

$$\sigma_s = \varepsilon.E_s \qquad\qquad \sigma_c = \varepsilon.E_c$$

Thus, the ratio of stress in steel to stress in concrete due to ε is:

$$\sigma_s/\sigma_c = E_s/E_c = n \text{ (typically } 7-9)$$

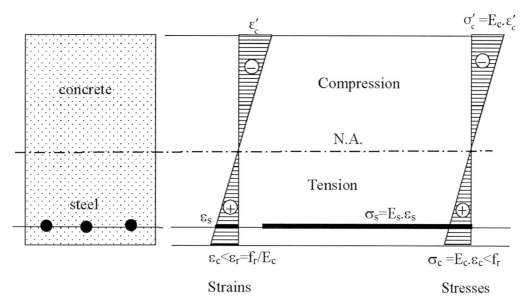

Figure 4.3: Phase 1 of flexural behavior.

To determine stresses in phase 1 of behavior, strains are multiplied by the modulus of elasticity of the perspective material. Both steel and concrete are considered elastic in phase 1 since the strains in the section are small.

Phase 2: Elastic Behavior – Cracked Concrete

Phase 2 follows phase 1 and is characterized by cracking of concrete in tension. Tensile strains in the section exceed ε_r: however, the behavior of concrete in compression remains somewhat elastic. In phase 2, the cracking strain of concrete in bending ($\varepsilon_r = f_r/E_c$) and the flexural tensile strength of concrete (f_r) have been exceeded causing concrete cracking on the tension side as shown in Figure 4.4. The dashed lines of the strain and stress diagrams in Figure 4.4 represent hypothetical distributions within cracked concrete. The strain distribution across the section remains linear with respect to the location of the N.A. that has no strain per Bernoulli's theorem. The main tensile stresses/forces in the section are resisted by steel bars reflecting the cooperation between concrete in compression and steel

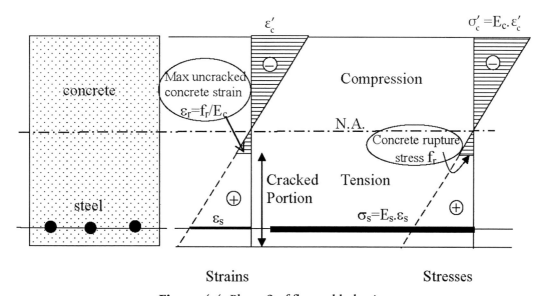

Figure 4.4: Phase 2 of flexural behavior.

in tension in reinforced concrete. Concrete remains elastic in compression since compression stresses and strains in the section remain fairly minor.

Phase 3: Inelastic Behavior

Phase 3 follows phase 2 and is characterized by increased stresses and strains. With such increase, concrete becomes inelastic in compression and steel may become inelastic in tension.

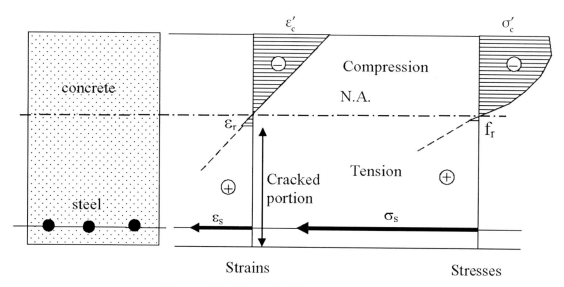

Figure 4.5: Phase 3 of flexural behavior.

Similar to phase 2, in phase 3, the cracking strain of concrete in bending ($\varepsilon_r = f_r/E_c$) and the flexural tensile strength of concrete (f_r) are exceeded. The dashed lines of the strain and stress diagrams in Figure 4.5 represent a hypothetical strain or stress distribution in cracked concrete. Linear strain distribution with respect to the location of the N.A. per Bernoulli's theorem remains. The main tension in the section is due to steel.

Phase 4: Failure in Bending

Phase 4 represents the failure stage of reinforced concrete in bending. Although, disagreement may arise regarding the definition of failure, ACI 318 Code's definition is the one most generally adopted. Failure of reinforced concrete in bending (as well as other loading situations) is defined by concrete reaching its crushing strain in compression = 0.003.

In a properly designed reinforced concrete beam, the strain in steel, ε_s, exceeds the yield strain $\varepsilon_y = f_y/E_s$ in Phase 4. As such, the stress in steel, σ_s, equals the yield stress, f_y (see Chapter 2). This is a necessary precaution to insure a slow or gradual failure of reinforced concrete beams imparting toughness and enhancing damage tolerance. While steel yields, it experiences large deformation causing significant and visible damage as it absorbs significant amounts of energy prior to complete structural collapse. Remedial action may take place prior to failure, and complete collapse is unlikely in the case of a catastrophic event. The compressive stress in concrete at the extreme compression fiber in the section is about $0.85f_c'$ based on the stress strain diagram of concrete in Chapter 2. f_c' is reached at a strain of about 0.002 which occurs within the compression area or zone of the concrete beam, but not at the point of maximum strain. As discussed in Chapter 2, f_c' is the design compressive strength of concrete that is relied on for structural design. In the failure phase in bending, in spite of

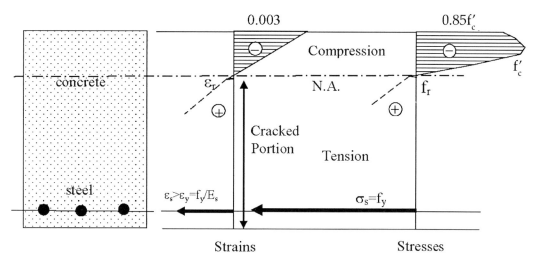

Figure 4.6: Failure in bending.

concrete and steel inelasticity, flexural analysis of concrete beams assume that Bernoulli's theorem of plane sections remaining plane continues to be applicable.

As loading is increased, the location of the NA rises in the section and crack depth continues to increase. This is due to the reduction of tangent modulus of elasticity and the inelastic behavior of concrete with increased stresses (see Chapter 2). In phase 4, the N.A. location is dictated by equilibrium rather than Bernoulli's theorem. The tension force in steel must equal the compression force in concrete to achieve force equilibrium.

4.2 Simplified Analysis of Flexural Failure Behavior of Reinforced Concrete

Figure 4.7 illustrates a simplified procedure for flexural failure analysis of reinforced concrete beams. Two more simplifying assumptions are utilized:

1.   Concrete tensile strength in bending $f_r = 0$, and

2.   Concrete compressive stress distribution is rectangular according to Whitney's simplified rectangular compression block.

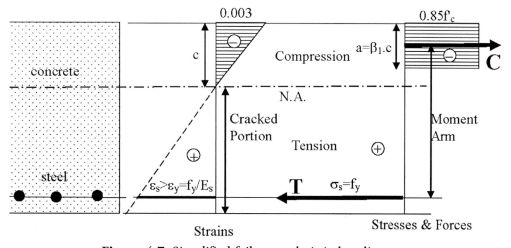

Figure 4.7: Simplified failure analysis in bending.

In Figure 4.7, "c" is the depth of concrete compression zone while "a" is the depth of Whitney's rectangular compression block. The area of Whitney's compression block "$0.85f'_c$ a" must be equal to the area of the irregular compression stress block occurring at failure in reinforced concrete beams shown in Figure 4.6. Since Whitney's block adopts a uniform compression stress of $0.85f'_c$, then changing the value of "a" based on concrete type (f'_c) achieves this sought equality. In Chapter 2, the compressive stress–strain diagrams for various concretes are illustrated. For high strength concrete (high f'_c), the stress–stain diagram includes steep lines indicating large variations of stresses with strain increases (high stress gradient = $\Delta\sigma/\Delta\varepsilon$) and leading to a relatively small average stress compared with f'_c. Such is not the case for conventional and low strength concrete resulting in a higher average stress compared with f'_c. Whitney has adopted the following equation using appropriate values for β_1 to compensate for the variations of stress behavior of concrete with f'_c:

$$a = \beta_1 \times c \tag{4.3}$$

Where:

    a = depth of Whitney's compression block.

    c = depth of compression zone.

    β_1 = 0.85 for $f'_c \leq 4{,}000$ psi (28 MPa)

        0.80 for $f'_c = 5{,}000$ psi (35 MPa)

        0.75 for $f'_c = 6{,}000$ psi (42 MPa)

        0.70 for $f'_c = 7{,}000$ psi (49 MPa)

        0.65 for $f'_c \geq 8{,}000$ psi (56 MPa)

In Figure 4.7, it is assumed that the reinforced concrete beam is properly designed so that the reinforcing steel yields prior to concrete crushing in compression at a crushing strain of 0.003. While yielding of steel is slow, gradual, and accompanied by significant absorption of energy, concrete crushing is catastrophic, sudden and does not involve significant energy absorption. Reinforced concrete beams sustain damage such as cracking and excessive deflection due steel yielding, but complete collapse does not occur until concrete crushes. With this design philosophy, design safety is enhanced under catastrophic loading, deficient materials, or poor workmanship events.

Example 4.1

Determine the nominal bending moment capacity of a concrete beam utilizing the simplified failure analysis by Whitney with the following information:

    b = 14 inch   d = 27 inch $A_s = 4$ in^2

    $f'_c = 4{,}000$ psi   fy = 60,000 psi

Solution

1.   Determine the depth of compression block:

    Based on force equilibrium, since the beam is subjected to bending moment only not subjected to axial forces, then:

    $\Sigma F = 0$ $\Longrightarrow$ Compression in concrete = Tension in Steel

    $C = T$ $\Longrightarrow$ $0.85f'_c \times a \times b = A_s \times f_y$ (4.4)

                            $\Longrightarrow$ $a = A_s . f_y / 0.85f'_c \times b$ (4.5)

$$\Longrightarrow \qquad a = (4 \times 60{,}000)/(0.85 \times 4{,}000 \times 14)$$

$$\Longrightarrow \qquad a = 5.04 \text{ in}$$

2.   Determine the value of the moment arm for C & T (see Figure 4.7):

$$\text{Moment Arm} = d - a/2 \qquad\qquad (4.6)$$

$$\text{Moment Arm} = 27 - 5.04/2 = 24.5 \text{ in}$$

3.   Determine the value of the bending moment capacity of the beam, M_n:

$M_n = $ C or T multiplied by the moment arm

$$M_n = A_s.f_y.(d - a/2) = 0.85f'_c.a.b.(d - a/2) \qquad\qquad (4.7)$$

$$M_n = 4 \times 60{,}000 \times 24.5 = 0.85 \times 4{,}000 \times 5.04 \times 14 \times 24.5 = 5{,}880{,}000 \text{ lb.in}$$

$$M_n = 5{,}880{,}000 \text{ in.lb} = 490 \text{ k.ft}$$

$$M_n = 490 \text{ k.ft}$$

The beam in this example is realistic. Note that the depth of compression zone in this beam, $c = a/\beta_1$, is about 6 inches. Only 6 inches of the beam are not cracked under failure analysis while the remaining 21 inches are cracked. Under service conditions, it is typical that 1/3 of the beam is not cracked while the other 2/3 are cracked.

There are two core concepts in the analysis of reinforced concrete beams: strain compatibility and equilibrium. Strain compatibility indicates that Bernoulli's theorem for strain distribution governs as in Figure 4.7. The solution of Example 4.1 applies force equilibrium utilizing the stresses based Whitney's rectangular compression block. Strain distribution compatibility and equilibrium generally result in matching solutions. The values of a, c and ε_t can be determined using either method.

Example 4.2

Determine the nominal bending moment capacity of a concrete beam utilizing the simplified failure analysis by Whitney with the following information:

b = 400 mm	d = 700 mm	A_s = 2,600 mm²
f'_c = 28 MPa	fy = 420 MPa	

Solution

1.   Determine the depth of compression block:

$a = A_s.f_y/0.85\ f'_c.b = (2{,}600 \times 420)/(0.85 \times 28 \times 400) = 115$ mm

2.   Determine the value of the moment arm for C & T:

Moment Arm = $d - a/2 = 700 - 115/2 = 643$ mm

3.   Determine the value of the bending moment capacity of the beam, M_n:

$M_n = $ C or T multiplied by the moment arm

$M_n = A_s \times f_y \times (d - a/2) = 0.85f'_c \times a \times b \times (d - a/2)$

$M_n = 2{,}600 \times 420 \times 643 = 0.85 \times 28 \times 115 \times 400 \times 643 = 703 \times 10^6$ N.mm

$M_n = 703$ kN.m

Test your knowledge

1. What are the assumptions used in the analysis of reinforced concrete beams in bending?

2. What is the stress beyond which concrete ceases to be elastic?

3. What is the stress beyond which steel ceases to be elastic?

4. How is slow or gradual failure achieved in reinforced concrete beams?

5. What phase of behavior of concrete beams is closest to service conditions?

6. List the four phases of behavior of reinforced concrete beams and the theme for each phase.

7. Explain the reason for the different values of β_1 for different values of f'_c in Whitney's rectangular compression block.

8. Why is it considered prudent in reinforced concrete design to make certain that steel yields prior to concrete crushing?

9. Redo the example included in this chapter for b = 14, 20, 26, and 32 in. Draw a diagram showing the changes in M_n with b.

10. Redo the example included in this chapter for d = 17, 27, 37, and 47 in. Draw a diagram showing the changes in M_n with d.

11. Redo the example included in this chapter for f'_c = 2000, 4000, 6000, and 8000. Draw a diagram showing the changes in M_n with f'_c.

12. Redo the example included in this chapter for A_s = 4, 8, 12, and 16 in². Draw a diagram showing the changes in M_n with A_s.

13. Redo the example included in this chapter for f_y = 40000, 60000, 80000, and 100000 psi. Draw a diagram showing the changes in M_n with f_y.

14. Based on your answer to questions 9 – 13, rank the factors that affect the value of M_n in the order of influence. The most influential factor first and the least influential last.

15. Redo problems 9 – 14 using SI units.

ANALYSIS AND DESIGN OF REINFORCED CONCRETE BEAMS I

5.1 Load Factors and Magnified Bending Moment

In order to analyze and design a reinforced concrete beam for bending, it is essential to determine the value of the factored design bending moment, M_u. Reduced beam bending moment capacity, ϕM_n, is compared with the ultimate or factored design bending moment, M_u, as explained in Chapter 3.

The design bending moment according to the strength design method or LRFD is the factored bending moment according to ASCE 7 typically termed M_u.

5.1.1 Method 1: Factored Moments

The first method for determining M_u is outlined herewith. The method described in this section assumes that:

— The location(s) of beam or structure critical section(s) is(are) known.

The factored bending moment affecting a reinforced concrete beam is determined via structural analysis for determining the values of the following moments:

a. M_D = the bending moment affecting the beam caused by dead load.

b. M_L = the bending moment affecting the beam caused by live load.

c. M_{Lr} = the bending moment affecting the beam caused by roof live load.

d. M_S = the bending moment affecting the beam caused by snow load.

e. M_R = the bending moment affecting the beam caused by rain load.

f. M_H = the bending moment affecting the beam caused by soil pressure.

g. M_F = the bending moment affecting the beam caused by fluid pressure.

h. M_W = the bending moment affecting the beam caused by wind load.

i. M_E = the bending moment affecting the beam caused by seismic load.

j. M_T = the bending moment affecting the beam caused by shrinkage and temperature change.

The beam or structure is loaded with each load discussed in Chapter 3 individually, and then analyzed. The effect of each individual load can then be ascertained including the values of bending moments listed above. Consequently, the following equations are utilized to determine the maximum and minimum factored bending moments acting on the critical section(s) beam in question:

$M_{u1} = 1.4(M_D + M_F)$

$M_{u2} = 1.2(M_D + M_F + M_T) + 1.6(M_L + M_H) + 0.5 (M_{Lr}$ or M_S or $M_R)$

$M_{u3} = 1.2M_D + 1.6(M_{Lr}$ or M_S or $M_R) + (M_L$ or $\pm 0.8M_W)$

$M_{u4} = 1.2M_D \pm 1.6M_W + M_L + 0.5(M_{Lr}$ or M_S or $M_R)$

$M_{u5} = 1.2M_D \pm 1.0M_E + M_L + 0.2M_S$

$M_{u6} = 0.9M_D \pm 1.6M_W + 1.6M_H$

$M_{u7} = 0.9M_D \pm 1.0M_E + 1.6M_H$

The values of M_{u1} through M_{u7} are determined for all beam's critical sections that require design or investigation. For each beam section, the maximum and the minimum values of bending moment among M_{u1} thru M_{u7} are then adopted for the reinforced concrete beam design. It should also be remembered that wind and seismic loads are applied twice once with a positive value once and another time utilizing a negative value due to their alternating direction. Also, the equations of M_{u2}, M_{u3} and M_{u4} require the use of different loads to achieve the maximum and minimum values of M_u. Thus a level of complexity is recognized in determining the design values of M_u. Since this textbook is focused on member design primarily subjected to gravity load (dead and live load), the following equations of M_u will be used frequently:

$$M_{u1} = 1.4M_D \tag{5.1}$$

$$M_{u2} = 1.2M_D + 1.6M_L \tag{5.2}$$

5.1.2 Method 2: Factored Loads

Another procedure termed factored loads to determine M_u may be followed. According to this procedure, the beam or structure is loaded with the factored loads U_1 through U_7 (based on the equations in Chapter 3) and analyzed for each different loading situation. The values of M_u for each analysis case are compiled and the beam is designed for the most critical values (maximum and minimum values among M_{u1} through M_{u7}).

The above two methods may appear tedious or lengthy, however, for each beam or element, only one or two loading combinations may be critical while the others are simply not applicable or not critical.

Example 5.1

Analyze the cantilever beam shown in the following figure to determine M_u according to ASCE 7 if $W_D = 2$ k/ft, $P_L = 30$ kips and span = 20 ft.

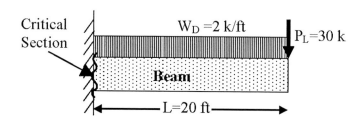

Solution – Method 1: Factored Moment

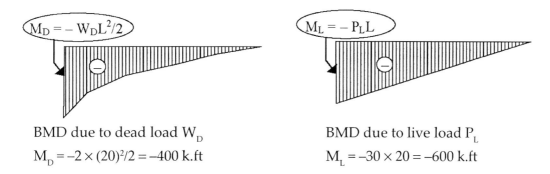

BMD due to dead load W_D BMD due to live load P_L

$M_D = -2 \times (20)^2/2 = -400$ k.ft $M_L = -30 \times 20 = -600$ k.ft

The sought value of M_u for structural design is needed for sizing the beam and determining the area and distribution of reinforcing steel bars. While dead load is always applied to the beam, live load may not be applied. For this example, since both bending moments caused by dead as well as live load are negative, then $M_u < 0$ ft.k. This beam needs to be designed for unidirectional bending moment rather than a bending moment with alternating direction. Design is required for worst case scenario or for the maximum possible value of M_u. For the beam in Example 5.1, $M_u = M_{umax}^-$. By examination of the equations of M_{u1} thru M_{u7} listed in page 51, one may conclude that the relevant equations are M_{u1} and M_{u2} (Equations 5.1 and 5.2) due to absence of all loads except for dead and live load. The value of M_u can be determined by one of the following equations that include dead load or dead load and live load:

$M_{u1} \quad = 1.4(M_D + M_F) = 1.4\,M_D = -560$ k.ft

$M_{u2} \quad = 1.2(M_D + M_F + M_T) + 1.6(M_L + M_H) + 0.5\,(M_{Lr}$ or M_S or $M_R)$

$\quad\quad\quad = 1.2M_D + 1.6M_L = -1{,}440$ k.ft

$\Longrightarrow \quad M_u = -1{,}440$ k.ft

Solution – Method 2: Factored Loads

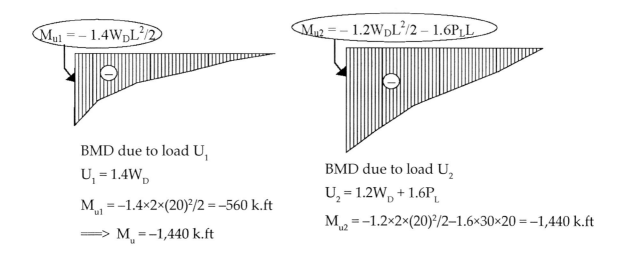

BMD due to load U_1

$U_1 = 1.4W_D$

$M_{u1} = -1.4 \times 2 \times (20)^2/2 = -560$ k.ft

$\Longrightarrow M_u = -1{,}440$ k.ft

BMD due to load U_2

$U_2 = 1.2W_D + 1.6P_L$

$M_{u2} = -1.2 \times 2 \times (20)^2/2 - 1.6 \times 30 \times 20 = -1{,}440$ k.ft

The critical loads for the cantilever beam in Example 5.1 are U_1 and U_2. U_1 is simply the dead load multiplied by 1.4. Utilizing U_2, dictates that the beam is loaded simultaneously with the dead load multiplied by 1.2 and the live load multiplied by 1.6. So the beam is loaded once with U_1 and another time with U_2. The bending moment diagram is drawn for each case and the value of M_u is determined.

The direction of the design bending moment is essential in determining the location of steel bars within the section. Positive moments cause tension below the N.A. of the beam, thus, requiring bars in the bottom portion of the section and vice versa. If a section is subjected to a unidirectional bending moment, then it needs to be designed for the maximum value of the bending moment. If a section is subjected to an alternating direction bending moment, then it needs to be designed for the maximum positive and maximum negative values of M_u since each would require different steel bars in size and, more importantly, location. Then, two layers of steel are needed. The first layer is at below the N.A. near the bottom of the section for resistance of positive moments. The other layer is at above the N.A. near the top of the section for resistance of negative moments. Design bending moments with alternating direction are typical of wind and seismic loading as these loads are also of alternating direction.

Example 5.2

Analyze the simply beam shown in the following figure to determine M_u according to ASCE 7 if $W_D = 20$ kN/m, $P_L = 90$ kN and span = 6 m.

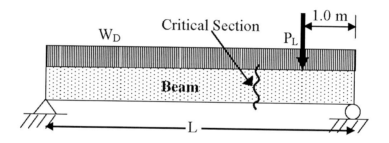

Solution – Method 2: Factored Loads

Structural analysis is needed to determine the critical section location at maximum bending moment or at zero shear. If procedure 1 is followed, then shear diagrams from each load are superimposed to determine the location of zero shear or maximum bending moment which could complicate the solution. Therefore, it is preferred to follow procedure 2 in this case. The ultimate load is placed on the beam and the shear force diagram is drawn.

First U_1 is used. In the absence of live load, dead load is multiplied by 1.4 resulting in a uniformly distributed load of $1.4W_D = 1.4 \times 20 = 28$ kN/m as shown below.

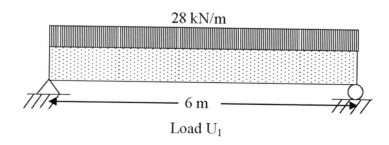

Load U_1

Utilizing U_1, the critical section is located at midspan and $M_{u1max} = M_{u1CL} = 1.4W_D L^2/8$:

Therefore for U_1: $M_u = M_{u1max} = M_{u1CL} = 28 \times (6)^2/8 = 126$ kN.m

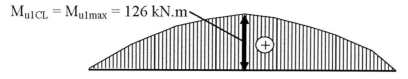

$M_{u1CL} = M_{u1max} = 126$ kN.m

Bending Moment Diagram for U_1

Then U_2 is used. Dead load is multiplied by 1.2 resulting in a uniformly distributed load of $1.2W_D = 1.2 \times 20 = 24$ kN/m as shown below. Live load is multiplied by 1.6 resulting in a concentrated load of $1.6 \times 90 = 144$ kN. Load U_2 is shown below. Using solid mechanics principles, shear force and bending moment diagrams for U_2 are drawn as shown below.

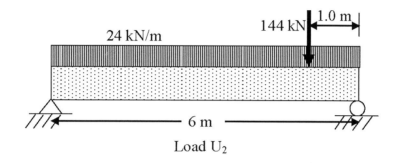

Load U_2

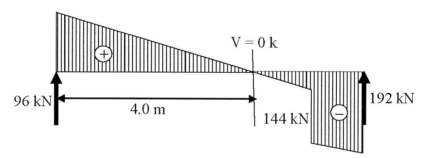

Shear Force Diagram for U_2

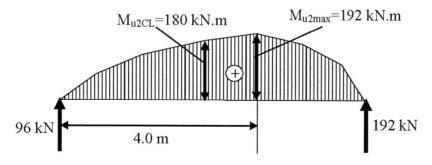

Bending Moment Diagram for U_2

Therefore, for U_2: $M_u = M_{u2max} = 192$ kN.m

The governing moment is the moment with the larger value for U_1 and U_2. Thus:

$M_u = 192$ kN.m

The maximum bending moment occurs at a section at a distance 4.0 m from left support at the beam. This is the section that requires design with M_u = 192 kN.m. The section at midspan is subjected to M_u of 180 kN.m. To achieve a practical design, M_u = 192 kN.m is frequently adopted for both sections.

5.1.3 Method 3: Factored Load Addition

Factored load addition is used to determine M_u for beam design if loading configurations of different load types are identical. U_{max} is determined among U_1 thru U_7 based on the equations in Chapter 3. The beam or structure is loaded with U_{max} and analyzed to determine M_u. A simply supported beam with uniformly distributed dead load W_D and uniformly distributed live load W_L throughout the entire span is an example. Ultimate or factored load is dictated by U_2:

$$U \text{ or } W_U = 1.2W_D + 1.6W_L \tag{5.3}$$

The beam is loaded with a uniformly distributed load equal to W_U, the bending moment diagram is drawn and M_u is determined for all critical sections. Only one critical section for bending moment is noted at midspan. M_u is determined as follows:

$$M_u = W_U.L^2/_8 \tag{5.4}$$

Figure 5.1: M_u for a uniformly loaded beam.

Another example is a cantilever beam subjected to a concentrated dead load P_D and a concentrated live load P_L both applied at the free end. Then:

$$U \text{ or } P_U = 1.2P_D + 1.6P_L \tag{5.5}$$

The beam is then loaded with a concentrated load equal to P_U, the bending moment diagram is drawn and the value of M_u is determined for all critical sections. In this case, only one critical section for bending moment occurs at the fixed end. M_u may be determined as follows:

$$M_u = -P_U.L \tag{5.6}$$

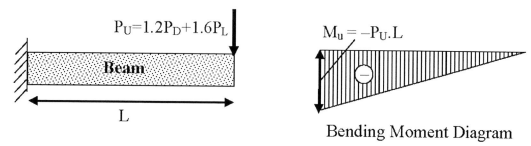

Figure 5.2: M_u for a cantilever beam with concentrated force.

5.2 Capacity Reduction Factor or Resistance Factor, ϕ

ACI 318 code relates structural safety of reinforced concrete beams to the strain in steel reinforcing bars at failure which is defined by concrete crushing ($\varepsilon_c = 0.003$). If the strain in reinforcing steel bars (ε_t) at failure equals or exceeds 0.005, then $\phi = 0.9$. The reinforced concrete beam is then considered adequately safe since steel has stretched significantly to allow for absorption of significant amount of energy resulting in the sought gradual failure. Reinforced concrete beam toughness and energy absorption is considered suitable. With decreasing steel tension strain at failure, ε_t, failure safety lessens as steel stretch and section deformations prior to concrete crushing are reduced. The desirable characteristics for reinforced concrete beams of gradual failure and toughness are also reduced. To compensate for this safety reduction, ACI 318 code dictates a smaller value of resistance factor ϕ according to the following equation.

$$\phi = 0.48 + 83\varepsilon_t \tag{5.7}$$

Beams with no axial loading, the subject matter of this chapter, are considered tension controlled sections where the tension strain is presumed to exceed the compression strain at failure. For this type of sections, ACI 318 code does not allow a value of ε_t less than 0.004 to achieve adequate safety. In compliance with ACI 318 code, Table 5.1 lists the value of ϕ for ε_t ranging between 0.004 to more than 0.005.

Table 5.1: The value of φ in relation with ε_t.

ε_t	ϕ
≥ 0.0050	0.900
0.0049	0.887
0.0048	0.878
0.0047	0.870
0.0046	0.862
0.0045	0.854
0.0044	0.845
0.0043	0.837
0.0042	0.829
0.0041	0.820
0.0040	0.812

Example 5.3

Investigate the adequacy of the reinforced concrete beam shown in the Figure 5.3.

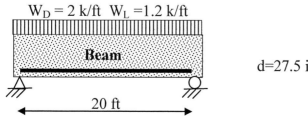

DL includes self weight.

$f'_c = 3,000$ psi

$f_y = 60,000$ psi

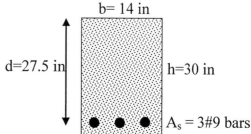

Beam's Cross-Section

Figure 5.3: Simply supported concrete beam with uniformly distributed load.

Solution

1. Determine the value of M_u (use Figure 5.1):

    $M_u = W_u L^2/8 = (1.2{\times}2 + 1.6{\times}1.2){\times}(20)^2/8 = 216$ k.ft

2. Determine the value of a = the depth of compression block (use Figure 4.7)

    Tension = Compression $\Longrightarrow$ T = C

    $A_s.f_y = 0.85f'_c.a.b$

    A_s = area of reinforcing steel bars = 3.00 in² (see Table 2.1)

    $\Longrightarrow$ a = 3.00×60,000/(0.85×3,000×14) = 5.04 in

3. Determine the value of c = the depth of compression zone (use Figure 4.6)

    $c = a/\beta_1 = 5.04/0.85 = 5.93$ in (the value of β_1 from Chapter 4)

4. Determine the value of ε_t = strain in tension steel at failure based on the diagram shown:

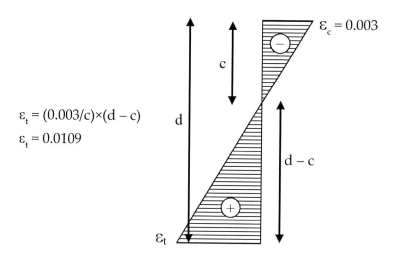

$\varepsilon_t = (0.003/c){\times}(d - c)$

$\varepsilon_t = 0.0109$

5. Determine the value of ϕ based on the Table 5.1:

    $\phi = 0.9$

6. Determine the value of reduced nominal bending moment capacity of the beam, ϕM_n:

    $\phi Mn = \phi.A_s.f_y.$(moment arm) = $\phi.A_s.f_y.$(d – a/2) (see Figure 4.7)

    ϕMn = 0.9×3.00×60,000×(27.5 – 5.04/2) = 4,048x10³ lb.in

    ϕMn = 4,048x10³ lb.in/12,000 = 337 k.ft (12,000 is unit conversion factor from lb.in to ft.k)

7. Compare ϕM_n with M_u

    if $\phi M_n \geq M_u$ Beam is safe

    if $\phi M_n < M_u$ Beam is unsafe

    For Example 5.3, the beam is safe.

Solution of Example 5.3 using SI Units

L = 6.1 m W_D = 29.5 kN/m W_L = 17.7 kN/m b = 356 mm   d = 699 mm A_s = 1,935 mm²

f'_c = 20.7 MPa f_y = 414 MPa

$M_u = W_u L^2/8$ = (1.2×29.5 + 1.6×17.7)×(6.1)²/8 = 296 kN.m

a = 1,935×414/(0.85×20.7×356) = 128 mm

c = a/β_1 = 128/0.85 = 150 mm

ε_t = (0.003/c)×(d − c) = (0.003/150)×(699 − 150) = 0.0109 > 0.005 ==> ϕ = 0.9

$\phi M_n = \phi.A_s.f_y.(d − a/2)$ = 0.9×1,935×414×(699 − 128/2) = 458x10⁶ N.mm = 458 kN.m

$\phi M_n \geq M_u$ ==> 458 kN.m ≥ 296 kN.m ==> Beam is safe

Test your knowledge

Determine the value(s) of M_u for the beams shown:

1.

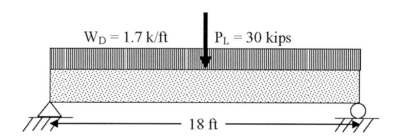

2.

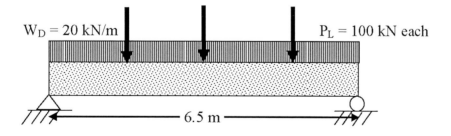

3.

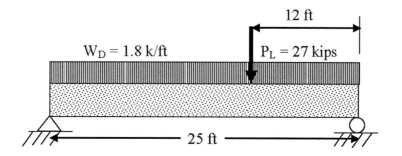

4.

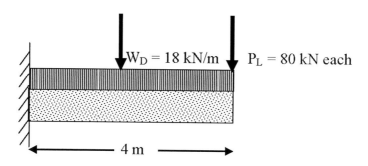

$W_D = 18$ kN/m $P_L = 80$ kN each

4 m

5.

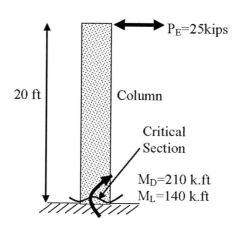

$P_E = 25$ kips

20 ft   Column

Critical
Section

$M_D = 210$ k.ft
$M_L = 140$ k.ft

6.

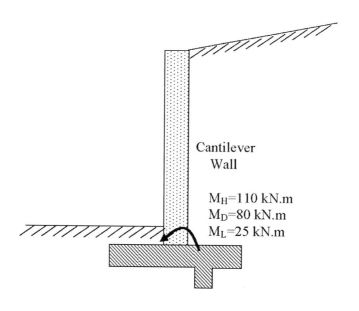

Cantilever
Wall

$M_H = 110$ kN.m
$M_D = 80$ kN.m
$M_L = 25$ kN.m

Determine the value of the reduced bending moment capacity ϕM_n for the beam sections shown using $f_y = 60$ ksi and $f'_c = 4,000$ psi:

7.   8.

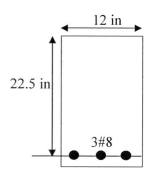   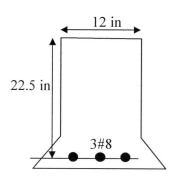

Determine the value of the reduced bending moment capacity ϕM_n for the beam sections shown using f_y = 420 MPa and f'_c = 30 MPa:

9.   10.

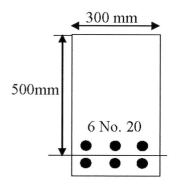   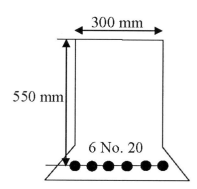

No. 20 bar is 20 mm in diameter.

Determine the value of the reduced bending moment capacity ϕMn for the beam sections Table at the end of chapter for problem 12

Beam No.	b (in)	d (in)	Bars
11	14	23	5#11
12	300mm	450mm	3 No. 12
13	18	25	4#9 + 3#10
14	21	28	2#5

No. 12 bar is 12 mm in diameter.

Chapter 6

ANALYSIS AND DESIGN OF REINFORCED CONCRETE BEAMS II

6.1 Beam Reinforcement Ratio

The beam reinforcement ratio, ρ , was defined in Chapter 4 as the area of steel reinforcing bars, A_s, divided by the effective area of concrete beam. The effective area of a reinforced concrete beam is the width, b, multiplied by the effective depth, d. d is the distance between the extreme compression fiber of the section and the centroid of reinforcing steel bars. As such, the reinforcement ratio can be written as $\rho = A_s/bd$. Failure strain and simplified stress distributions in a reinforced concrete beam section are shown in Figure 6.1.

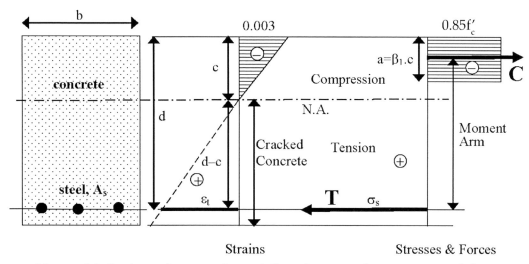

Figure 6.1: Strains and stresses in a reinforced concrete beam section at failure.

The depth of the compression block, a, can be determined utilizing the following equation as described in Chapter 4:

$a = A_s \cdot f_y / 0.85 f_c' \cdot b$

The above equation may be rewritten as follows:

$$a = A_s \cdot f_y / 0.85 f_c' \cdot b = \rho \cdot b \cdot d \cdot f_y / 0.85 f_c' \cdot b = \rho \cdot d \cdot f_y / 0.85 f_c'$$

$$a/d = \rho \cdot (f_y / 0.85 f_c') \tag{6.1}$$

Based on equation 6.1, the ratio of depth of compression block to effective beam depth is, to a large extent, controlled by the beam reinforcement ratio, ρ. Furthermore, the beam reinforcement ratio, ρ, also controls ε_t, the tension strain in steel upon failure (see Equation 6.2 that follows). This added importance to ρ clearly displays the significance of steel reinforcement ratio:

$$\varepsilon_t = (0.003/c) \cdot (d - c) = [0.003/(a/\beta_1)] \cdot [d - (a/\beta_1)]$$

$$\varepsilon_t = \{0.003/[\rho \cdot d \cdot (f_y/0.85\ f_c')/\beta_1]\} \cdot \{d - [\rho \cdot d \cdot (f_y/0.85 f_c')/\beta_1]\}$$

$$\varepsilon_t = 0.003\{1/[\rho \cdot (f_y/0.85\ f_c')/\beta_1] - 1\} = 0.003(0.85 f_c' \cdot \beta_1/\rho \cdot f_y - 1)$$

$$\varepsilon_t = 0.003(0.85 f_c' \cdot \beta_1/\rho \cdot f_y - 1) \tag{6.2}$$

6.2 Balancing Reinforcement Ratio

An important characteristic value of ρ is $\rho_{balanced}$ or ρ_b. In reinforced concrete beams with ρ equal to ρ_b, steel begins to yield when concrete reaches its crushing strain of 0.003. As such, steel strain is equal to $\varepsilon_y = f_y/E_s$ (yield strain) at concrete crushing. ACI 318 Code defines by crushing of concrete at a compression strain of 0.003. Gradual failure is not achieved in beams with $\rho = \rho_b$. The preferred type of failure is when steel yields and experiences significant deformation prior to concrete crushing; thus, beams with $\rho = \rho_b$ are undesirable. One may consider that steel and concrete are exhausted simultaneously if $\rho = \rho_b$. To determine the value of ρ_b the following method may be employed:

$$\varepsilon_t = 0.003(0.85 f_c' \cdot \beta_1/\rho \cdot f_y - 1) \text{ based on Equation 6.2}$$

$$\text{Special case, } \varepsilon_t = \varepsilon_y = f_y/E_s, \text{ and } \rho = \rho_b$$

$$\varepsilon_y = f_y/E_s = 0.003(0.85 f_c' \cdot \beta_1/ \rho_b \cdot f_y - 1)$$

$$\rho_b = [0.003E_s/(0.003E_s + f_y)] \cdot (0.85 f_c' \cdot \beta_1/f_y) \tag{6.3}$$

A reinforced concrete beam with the balancing reinforcement ratio shown above may be considered to have compression concrete and tension steel of equal strength. Increased loading on a beam with the balancing reinforcement ratio would lead to concrete reaching its crushing strain and steel reaching its yielding strain simultaneously. Failure of such beams occur as concrete crushes and steel commences to yield. Under–reinforced beams have reinforcement ratios smaller than ρ_b, resulting in steel yielding prior to concrete crushing strain; the converse is true for over–reinforced beams. Over–reinforced sections experience catastrophic failure. Only significantly under–reinforced sections are allowed in ACI 318 Code to assure gradual failure of reinforced concrete beams.

6.3 Maximum Reinforcement Ratio

For a prudent structural design of concrete beams, ACI 318 Code dictates that the strain in tension steel at failure, ε_t, be equal or greater than 0.004. This precaution is adopted to ensure slow or gradual failure of reinforced concrete beams which enhances toughness and energy absorption capacity. Substituting this maximum value of ε_t in Equation 6.2 results in a code imposed maximum value of reinforcement ratio, ρ_{max}:

$$0.004 = 0.003(0.85 f_c' \cdot \beta_1/ \rho_{max} \cdot f_y - 1)$$

$$\rho_{max} = 0.36 f_c' \cdot \beta_1/f_y \tag{6.4}$$

Although the above value of ρ_{max} is allowed by code, the preferred value is the one leading to $\varepsilon_t = 0.005$ at beam failure. This is an added precaution to achieve a gradual failure in bending. Thus, the preferred maximum value of reinforcement ratio is:

$$\rho_{maxp} = 0.32 f'_c \cdot \beta_1 / f_y \qquad (6.5)$$

6.4 Minimum Reinforcement Ratio

Achieving a slow or gradual failure mode is an essential part of the strength design method for reinforced concrete. This failure mode can be achieved via

Yielding of Steel Prior to Crushing of Concrete

Therefore, failure of a reinforced concrete beam must include yielding of steel and cannot result from only crushing of concrete. A reinforced concrete beam with low reinforcement ratio is susceptible to sudden failure by crushing of concrete without yielding of steel. Recall from Chapter 4 the behavior of a reinforced concrete beam under increasing bending moment and its four phases. Phase 1 is when concrete is uncracked and both steel and concrete are elastic. Phase 2 is when concrete is cracked but steel and concrete still remain elastic. Steel is lightly loaded or ineffective prior to cracking since concrete helps resist tensile stresses. This situation changes after concrete cracks, however, as the tensile stresses or forces are transferred to the reinforcing steel. The transition from phase 1 to phase 2 loads the steel with these forces in the section consequent to cracking of concrete. Steel must be able to resist these forces and consequently yield under increased loading so that the section can achieve gradual failure. In order to reach phases 3 and 4 under gradual failure conditions, a beam must pass through phase 2 under increasing bending moment. As such, the cracked beam bending moment capacity must be greater than its uncracked beam bending moment capacity under phase 1 where steel is ineffective; in other words, a cracked beam that would fail gradually must be stronger than an uncracked beam that would fail catastrophically.

Uncracked beam capacity in bending (steel is ignored)
$$M_{nplain} = f_r s \approx f_r bd^2/6$$
$$M_{nplain} = 7.5\sqrt{f'_c} bd^2/6$$

Cracked beam capacity in bending
$$M_{ncracked} \approx A_s f_y d = \rho b d^2 f_y$$

$$M_{ncracked} > M_{nplain} \Longrightarrow \rho b d^2 f_y > 7.5\sqrt{f'_c} bd^2/6 \Longrightarrow \rho > 1.25\sqrt{f'_c} / f_y$$

ACI 318 Code has two equations for ρ_{min}, taking into account a safety factor, where ρ_{min} may not be less than either value:

$$\rho_{min} = 3\sqrt{f'_c}/f_y \qquad (6.6)$$

The value of ρ_{min} from Equation 6.6, however, may not be less than:

$$\rho_{min} = 200/f_y \qquad (6.7)$$

f'_c and f_y in Equations 6.6 and 6.7 are is psi units. Table 6.1 lists the values of ρ_{min}, ρ_{maxp} ρ_{max} and ρ_b for various values of f'_c and f_y according to ACI 318 Code.

Table 6.1: Minimum and maximum reinforcement ratios (%) for concrete beams according to ACI 318 Code.

f_c' psi (MPa)		3,000 (21)	4,000 (28)	5,000 (35)	6,000 (42)	7,000 (49)	8,000 (56)
f_y psi (MPa)							
40,000 (280)	ρ_{min}	0.50	0.50	0.53	0.58	0.63	0.67
	ρ_{maxp}	2.04	2.72	3.20	3.60	3.92	4.16
	ρ_{max}	2.30	3.06	3.60	4.05	4.41	4.68
	ρ_b	3.71	4.95	5.82	6.55	7.13	7.57
60,000 (420)	ρ_{min}	0.33	0.33	0.35	0.39	0.42	0.45
	ρ_{maxp}	1.36	1.81	2.13	2.40	2.61	2.77
	ρ_{max}	1.53	2.04	2.40	2.70	2.94	3.12
	ρ_b	2.14	2.85	3.35	3.77	4.11	4.36
75,000 (520)	ρ_{min}	0.27	0.27	0.28	0.31	0.33	0.36
	ρ_{maxp}	1.09	1.45	1.71	1.92	2.09	2.22
	ρ_{max}	1.22	1.63	1.92	2.16	2.35	2.50
	ρ_b	1.55	2.07	2.43	2.74	2.98	3.16
90,000 (630)	ρ_{min}	0.22	0.22	0.24	0.26	0.28	0.30
	ρ_{maxp}	0.91	1.21	1.42	1.60	1.74	1.85
	ρ_{max}	1.02	1.36	1.60	1.80	1.96	2.08
	ρ_b	1.18	1.58	1.86	2.09	2.27	2.41

6.5 Reinforced Concrete Beam Detailing

Detailing of reinforced concrete beam includes section configuration and geometry. While d, b and A_s are determined during analysis and design, steel bar selection, reinforced concrete cover over steel bars, bar spacing etc. need to be addressed.

6.5.1 Concrete Cover Over Steel Bars

A protective concrete layer over steel reinforcing bars is necessary for:

1. The combined action of steel and concrete. As stated in Chapter 4, analysis of beams in bending assumes perfect bond between steel and concrete. Such bond requires that steel bars are adequately encased with a layer of concrete for stress and load transfer.

2. Protection of steel reinforcing bars against corrosion and fire.

ACI Code 318 states that reinforcing steel bars in concrete shall have the following minimum concrete clear cover:

1. Bars in concrete not exposed to weather = 1.5 in (38 mm).

2. Bars in concrete exposed to earth/soil or weather = 2 in (50 mm).

3. Bars in concrete cast against and in permanently exposed to earth/soil = 3 in (75 mm).

Figure 6.2 illustrates the details of a reinforced concrete beam section with one layer of reinforcing bars. As noted, two other types of reinforcement are used in beams in addition to flexural tension reinforcement. Stirrups that consist of steel bars bent around the perimeter of the beam section.

Stirrups are used to resist shear and will be covered in later chapters in this book. Also, longitudinal steel reinforcement on the compression side, A_s' is used for tying or anchoring the stirrups within the beam. Typically, A_s' is about 10% of A_s or two steel bars of appropriate size. Based on analysis of Figure 6.2, it is concluded that $h - d \approx 2.5$ to 3 in (63 to 75 mm) for a beam with one layer of steel bars.

Figure 6.3 shows the details of a reinforced concrete section with two layers of reinforcing steel bars. The effective depth of such a beam, as shown in Figure 6.3, is the weighted average of the effective depth of the lower layer and the effective depth of the upper layer based on steel bar area. ACI 318 Code specifies that the bars in the upper layer be placed directly above the bars in the lower layer. Based on analysis of Figure 6.3, it is concluded that $h - d \approx 4$ to 5 in (100 to 125 mm) for a beam with two layers of steel bars.

Table 6.2 lists the minimum required beam width, as a function of the size and number of reinforcing steel bars, to satisfy code requirements. Table 6.2 assumes a single layer of steel bars, ¾ in (19 mm) maximum aggregate size, 3/8 in (9 mm) stirrup diameter and interior exposure of concrete (minimum clear cover = 1.5 in = 38 mm).

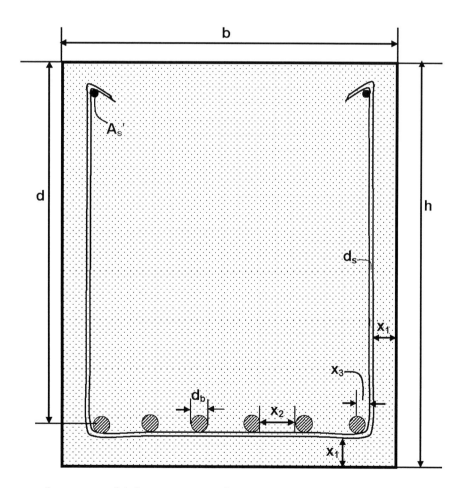

x_1 = concrete clear cover thickness = 1.5 in (38 mm) min
x_2 = clear distance between bars = max of 1 in (25 mm), d_b and 4/3×max aggregate size
x_3 = distance from the inner side of stirrup to centroid of first bar = max of $2d_s$ and $d_b/2$
d_b = diameter of reinforcing bar
d_s = diameter of stirrup = 3/8 or ½ in (10 or 12 mm)
n = number of steel bars
h = total beam depth or thickness
d = effective beam depth $\Longrightarrow$ $d = h - x_1 - d_s - d_b/2$
b = beam width $\geq nd_b + (n - 1)x_2 + 2(x_3 + d_s + x_1)$

Figure 6.2: Details of a reinforced concrete beam section with one layer of steel bars.

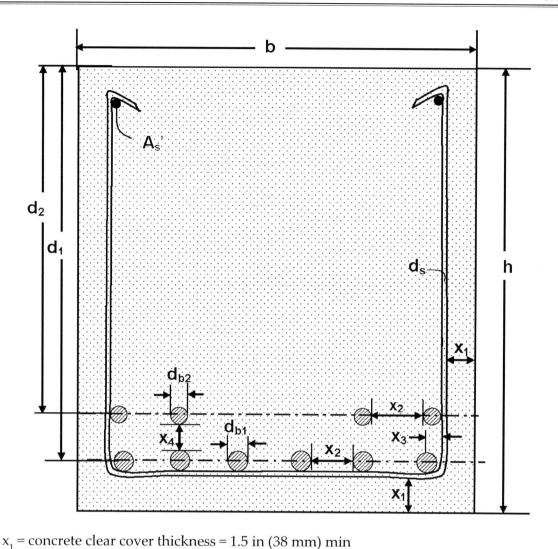

x_1 = concrete clear cover thickness = 1.5 in (38 mm) min

x_2 = clear distance between bars = max of 1 in (25 mm), d_b and 4/3×max aggregate size

x_3 = distance from the inner side of stirrup to centroid of first bar = max of $2d_s$ and $d_b/2$

x_4 = clear distance between lower and upper bars = max of 1 in (25 mm) d_{b1}, and d_{b2}

d_{b1} = diameter of reinforcing bar in the lower layer

d_{b2} = diameter of reinforcing bar in the upper layer

d_s = diameter of stirrup = 3/8 or ½ in (10 or 12 mm)

h = total beam height

d_1 = effective beam depth for the lower layer = $h - x_1 - d_s - d_{b1}/2$

d_2 = effective beam depth for the upper layer = $h - x_1 - d_s - d_{b1} - x_4 - d_{b2}/2$

A_{s1} = the total area of steel bars in the lower layer

A_{s2} = the total area of steel bars in the upper layer

d = effective depth for the beam as a whole = $(A_{s1} \cdot d_1 + A_{s2} \cdot d_2) / (A_{s1} + A_{s2})$

Figure 6.3: Details of a reinforced concrete beam section with two layers of steel bars.

Table 6.2: Minimum required beam (in) width per ACI 318 to fit steel bars.

# of Bars Bar Size	2	3	4	5	6	7	8	Added for One Extra Bar
4	6.8	8.3	9.8	11.3	12.8	14.3	15.6	1.5
5	6.9	8.5	10.2	11.8	13.4	15.0	16.7	1.7
6	7.0	8.8	10.5	12.3	14.0	15.8	17.5	1.8
7	7.2	9.0	10.9	12.8	14.7	16.5	18.4	1.9
8	7.3	9.3	11.3	13.3	15.3	17.3	19.6	2.0
9	7.5	9.8	12.0	14.3	16.5	18.9	21.0	2.3
10	7.8	10.3	12.8	15.3	17.8	20.3	22.8	2.5
11	8.0	10.8	13.5	16.3	19.0	21.8	24.5	2.8
14	9.0	12.5	16.0	19.5	23.0	26.5	30	3.5
18	10.5	15.0	19.5	24.0	28.5	33.0	37.5	4.5

Minimum beam width to fit steel bars can be evaluated using Equation 6.8 shown below:

$$b_{min} = 3 \text{ in or } 150 \text{ mm} + (N - 1) \cdot (d_b + x) + 2d_s + 2y \qquad (6.8)$$

Where:

b_{min} = minimum required beam width.

N = number of steel bars.

d_b = diameter of steel bars.

d_s = diameter of stirrup.

x = 1 in (25 mm), d_b and 4/3×max aggregate size whichever is the largest.

y = $2d_s$ or d_b whichever is larger.

6.5.2 Steel Bar Spacing

Spacing or distance between neighboring steel bars is an important detailing parameter. Adequate spacing of steel bars is essential for bond between steel and concrete so that the effectiveness of each steel bar is not compromised. A sufficient area of concrete is required to encase each steel bar for proper bond. Furthermore, clear spacing between steel bars must be adequate for fresh concrete to infiltrate and fill the formwork completely including the spaces among bars. ACI 318 Code specifies that the clear spacing of reinforcing steel bars in concrete may not be less than:

1.   1 in (25 mm).

2.   The diameter of steel bar d_b.

3.   4/3 times the maximum aggregate size in concrete.

ACI 318 Code allows the use of bundled (grouped) steel bars in concrete beams; however, the maximum number of bars in a bundle not to exceed four according to ACI 318 Code. Bundled bars are, typically, two or three reinforcing steel bars bunched or grouped together to act as a single reinforcing bar. Bundled bars may be stacked horizontally or vertically, in a triangular or rectangular/square arrangement depending on the number of bars in a bundle. Bundled bars are treated as an individual steel bar with a cross–sectional area equal to the sum of the areas of all the bars in the bundle. Furthermore, all bundled bars must be enclosed within stirrups or ties (closed stirrups).

6.5.3 Steel Bar Size Differential

Transfer of stresses between steel and concrete is linked to steel bar surface area. A smaller bar has a large surface area compared with its cross–sectional area contrary to a larger bar. As such, with two sizes of bars in a concrete beam, smaller bars may be loaded at a different rate than larger bars. Hence, smaller bars may experience yielding prior to larger bars. Such mismatch of stresses among steel bars is not beneficial to reinforced concrete. Therefore, ACI 318 Code specified that steel bars used to resist tension caused by bending may not be different from each other by more than two numbers or ¼ in = 6 mm in diameter (e.g. #6 and #8 is ok but #6 and #9 is not). It is also recommended that no more than two different sizes of bars be used for tension reinforcement.

6.5.4 Stirrup Bending

Bending of stirrups must also comply with ACI 318 Code. Typically, bend radius for stirrups is $4d_s$ where d_s is stirrup diameter. Furthermore, extension beyond the bend for stirrups is typically 2½ inch. The value of x_3 in Figures 6.2 and 6.3 is based on such requirements.

6.6 Beam Deflection

The strength design method is concerned with member or structure's behavior at ultimate condition or failure. Therefore, ACI 318 Code has certain requirements for beams so that their service conditions are acceptable. Among such requirements is that beam deflections are within acceptable code limits. ACI 318 Code provides a table for minimum beam depth, h, where beam deflection are within acceptable limits and deflections need not be computed to determine conformance with code limits. Table 6.3 is a partial reproduction of this ACI table, and assumes normal weight concrete (unit weight, γ = 145 lb/ft³ = 23 kN/m³) and yield strength of steel, f_y = 60,000 psi = 420 MPa. For light weight concrete and steel with yield strength different than 60,000 psi, the numbers in Table 6.3 need to be multiplied by $(1.65 - 0.005\gamma)$ and $(0.4 + fy/100,000)$, respectively, using US units or $(1.65 - 0.032\gamma)$ and $(0.4 + fy/700)$ using SI units, respectively. Also, Table 6.3 should not be used for members supporting construction that is likely to be damaged by large deflections. This table is also duplicated in Appendix A (Table A.12).

Table 6.3: Minimum thickness (h) for reinforced concrete beams unless deflections are computed.

Members	Simply Supported	One-End Continuous	Both-Ends Continuous	Cantilever
Beam	Span/16	Span/18.5	Span/21	Span/8

6.7 Beam Self—Weight

It is common that the self weight of a rectangular reinforced concrete beam requires calculation since it adds to the dead load. Figure 6.4 illustrates the procedure of calculating self weight. It is typically calculated as a load per linear foot of beam's length. The product of beam's cross section and the unit weight of reinforced concrete (145 lb/ft^3 = 23 kN/m^3) is equal to the beam's self weight in pound per linear foot as illustrated in Equation 6.9:

$$\text{Beam Self Weight} = [bh(in^2)/144] \cdot 0.145 \text{ lb/ft} \qquad \text{US units} \qquad (6.9)$$

$$\text{Beam Self Weight} = [bh(mm^2)/10^6] \cdot 23 \text{ kN/m} \qquad \text{SI units}$$

Where:

    b = beam width.

    h = beam thickness or height.

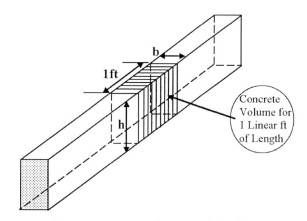

Figure 6.4: Concrete beam's self weight determination.

6.8 Other Code Requirements

Upon being subjected to compression, the concrete compression zone of beams may tend to buckle. ACI 318 Code requires that lateral support for beams be provided at spacing equal to 50b where b is beam's width. Lateral support is generally provided by concrete slabs or intersecting beams (Figure 6.5).

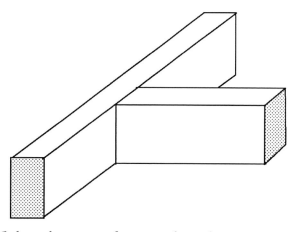

Figure 6.5: lateral support of concrete beam by an intersecting beam.

ACI 318 Code also requires skin against excessive cracking. Skin reinforcement is required for beams with depth greater than 36 in (910 mm) and is located in the bottom half of the beam. Skin reinforcement consists of two vertical layers of steel bars running the whole length of the beam. Spacing of these bars shall be the least of the following three parameters:

1.   Beam effective depth divided by 6 = d/6.

2.   12 in (300 mm).

3.   The area of a single bar used for skin reinforcement × 1000/(d – 30).

Further code requirements include that total area of skin reinforcement need not exceed half of the computed steel area required for flexural reinforcement (A_s). Also, skin reinforcement may be included in steel area for resistance of bending moment.

6.9 Analysis of Reinforced Concrete Beams

Analysis of a reinforced concrete beam in bending assumes that the beam exists and it is required to investigate its adequacy for the applied moment or load. Loading, beam dimensions and configuration as well as material properties are known. As such, it is only required to determine if the beam can support the load imposed on it. The procedure for analysis is outlined as follows:

Analysis of Singly Reinforced Concrete Beams in Bending

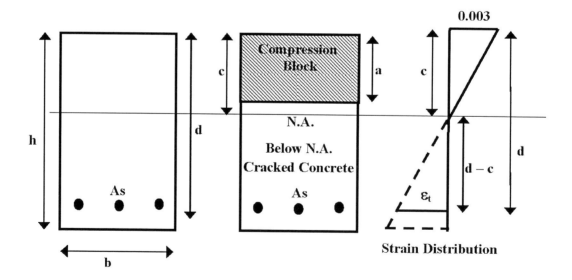

Known

1.   Section design including b, d, h, A_s.

2.   Material properties or f'_c and f_y.

3.   Applied magnified moment M_u is known or determined according to Chapter 5.

Step 1

Calculate the depth of the compression block and compression zone:

Utilizing equilibrium: Tension = Compression ==> $A_s \cdot f_y = 0.85 f'_c \cdot b \cdot a$

Then: $a = \dfrac{A_s \cdot f_y}{0.85 f'_c \cdot b}$ Also, $c = a/\beta_1$

Step 2

Check that the strain in tension steel is greater than 0.004 and the section is within code limits

Strain in tension steel: $\varepsilon_t = \dfrac{0.003 \times (d - c)}{c}$

If $\varepsilon_t > 0.004$ Then $\rho < \rho_{max}$ and section is within code limits. $\varepsilon_t > 0.005$ is preferred

Step 3

Check if the reinforcement ratio in the beam is higher than the minimum required by the code:

$\rho = A_s/bd > 3\sqrt{f'_c}/f_y$ Also, $\rho = A_s/bd > 200/f_y$

Step 4

Check other code requirements for reinforced concrete beams including deflection, bar detailing, skin reinforcement, etc.

Step 5

Calculate the nominal bending moment capacity M_n

$M_n = A_s \cdot f_y \cdot (d - a/2)$

Step 6

Calculate ϕ based on ε_t in step 2 using the equation:

$\phi = 0.483 + \varepsilon_t \cdot (83.3)$ if $\varepsilon_t > 0.005$ then $\phi = 0.9$

Step 7

Compare ϕM_n with M_u

If $\phi M_n > M_u$ Then section is OK

If $\phi M_n < M_u$ Then section is NG

Example 6.1

Investigate the adequacy of the section shown given that: $M_u = 150$ k.ft, $f'_c = 5{,}000$ psi and $f_y = 60{,}000$ psi.

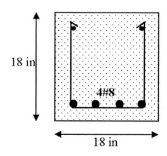

Solution

Determine the value of a:

$a = 4 \times 0.78 \times 60/(0.85 \times 5 \times 18) = 2.46$ in
where 0.78 in² is the area of #8 bar (Chapter 2 or Table A.5)

Determine the value of c:

$c = 2.46/0.80 = 3.08$ in
where $\beta_1 = 0.80$ as in Chapter 4

Determine the value of ε_t:

$\varepsilon_t = 0.003 \times (15 - 3.08)/5.61 = 0.012$ so section is within code limits and $\phi = 0.9$
where $d \approx 18 - 3 = 15$ in as explained in this chapter

Check that $\rho > \rho_{min}$

$\rho = 4 \times 0.78/(18 \times 15) = 0.0116$

$\rho_{min1} = 3 \times \sqrt{5,000}/60,000 = 0.0035$ $\rho_{min2} = 200/60,000 = 0.0033$

ρ_{min} = the larger of ρ_{min1} and ρ_{min2} = 0.0035

$\Longrightarrow \rho > \rho_{min}$ OK

Determine the value of M_n:

$M_n = 4 \times 0.78 \times 60 \times (15 - 2.46/2) = 2,594$ k.in = 216 k.ft

Compare ϕM_n with M_u:

$\phi M_n = 0.9 \times 216 = 195$ k.ft $> M_u = 150$ k.ft $\Longrightarrow$ OK

Thus the beam section of Example 6.1 is within code limits and considered safe.

Example 6.2

Investigate the adequacy of the section shown given that: $M_u = 200$ kN.m, $f'_c = 35$ MPa and $f_y = 420$ MPa.

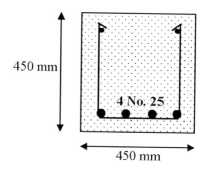

450 mm

4 No. 25

450 mm

Solution

Determine the value of a:

$a = 4 \times 490 \times 420/(0.85 \times 35 \times 450) = 61.5$ mm
where 490 mm² is the area of No. 25 bar (diameter – 25 mm)

Determine the value of c:

$c = 61.5/0.80 = 77.0$ mm
where $\beta_1 = 0.80$ as in Chapter 4

Determine the value of ε_t:

$\varepsilon_t = 0.003 \times (375 - 77.0)/77.0 = 0.012$ so section is within code limits and $\phi = 0.9$
where $d \approx 450 - 75 = 375$ mm as explained in this chapter

Check that $\rho > \rho_{min}$

$\rho = 4 \times 0.78/(18 \times 15) = 0.0116 > \rho_{min} = 0.0035$ (Table 6.1)   OK

Determine the value of M_n:

$M_n = 4 \times 480 \times 420 \times (375 - 61.5/2)/10^6 = 278$ kN.m

Compare ϕM_n with M_u:

$\phi M_n = 0.9 \times 278 = 250$ kN.m $> M_u = 200$ kN.m   ===>   OK

Thus the beam section of Example 6.2 is within code limits and considered safe.

Test your knowledge

Investigate the adequacy of the beams shown below to resist bending moment. All indicated dead load does not include self weight:

1.

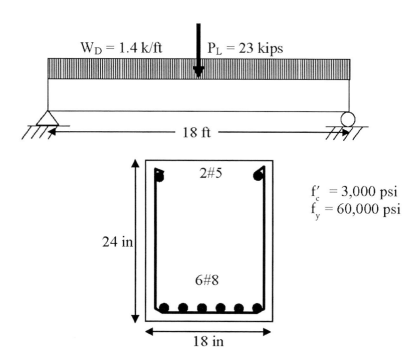

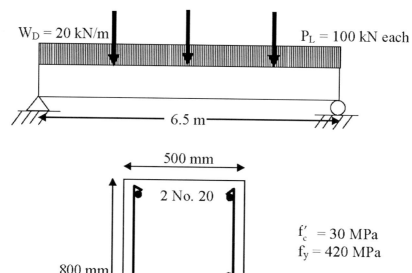

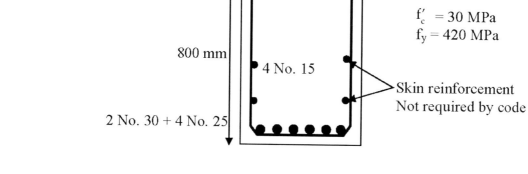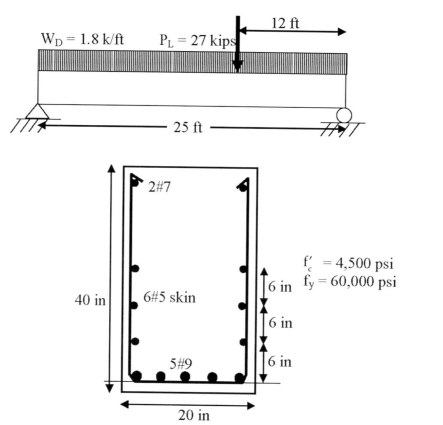

2.

$W_D = 20$ kN/m $P_L = 100$ kN each

6.5 m

500 mm

2 No. 20

$f'_c = 30$ MPa
$f_y = 420$ MPa

800 mm

4 No. 15

Skin reinforcement
Not required by code

2 No. 30 + 4 No. 25

3.

$W_D = 1.8$ k/ft $P_L = 27$ kips   12 ft

25 ft

2#7

$f'_c = 4,500$ psi
$f_y = 60,000$ psi

6 in

40 in   6#5 skin   6 in

6 in

5#9

20 in

4.

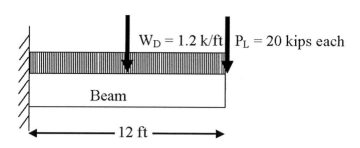

$W_D = 1.2$ k/ft $P_L = 20$ kips each

Beam

12 ft

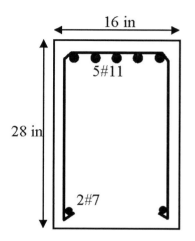

16 in

5#11

28 in

2#7

$f'_c = 4,000$ psi
$f_y = 60,000$ psi

5.

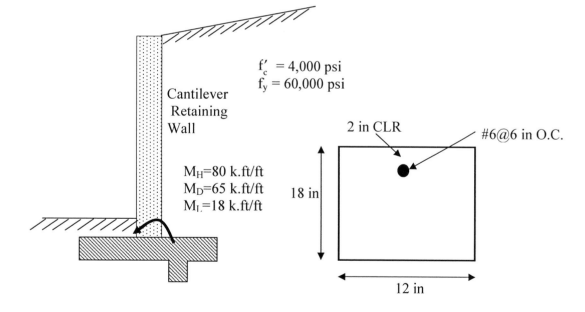

$f'_c = 4,000$ psi
$f_y = 60,000$ psi

Cantilever
Retaining
Wall

$M_H = 80$ k.ft/ft
$M_D = 65$ k.ft/ft
$M_L = 18$ k.ft/ft

2 in CLR

#6@6 in O.C.

18 in

12 in

Retaining Wall Profile   Retaining Wall Section

6.   Determine the value of ϕM_n for the following beam section using f'_c = 4,000 psi (28 MPa) and f_y = 60,000 psi (420 MPa).

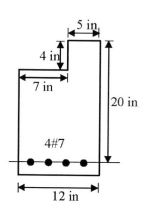

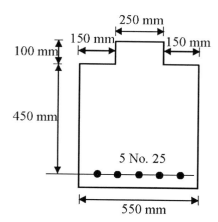

ANALYSIS AND DESIGN OF REINFORCED CONCRETE BEAMS III

7.1 Introduction to Reinforced Concrete Beam Design in Bending

This chapter covers the flexural design of reinforced concrete beam cross–sections. Prior to beginning the design process, the following information are required:

1. Beam configuration including span length, support conditions (simply supported, cantilever, etc), and exposure conditions (interior, exterior, earth, etc).

2. Loads on the beam including dead load, live load, earth pressure load, wind and seismic load.

3. Other design limitations such as the maximum value of beam depth (h) allowed for the building or the maximum deflection allowed for the type of structure.

The main parameter for a reinforced concrete beam cross–section design subject to bending is M_u which can be determined as shown in Chapter 5 based on load factors of ASCE 7. Other code imposed limitations on beam cross–sectional geometry or reinforcement are also design considerations. These are obtained from the beam configuration as well as the type of building or structure of the beam.

The flexural design of a reinforced concrete beam cross–section yields a detailed beam section with the following information:

1. Material properties, primarily f_c' and f_y.

2. Beam depth = h.

3. Beam width = b.

4. Steel bar area, A_s, and distribution (number, size and spacing of bars).

5. Section detailing including concrete cover over steel bars.

6. Other design considerations required by code.

Mathematical design output for a reinforced concrete beam can be summarized as the following parameters:

1. Concrete design compressive strength = f_c'.

2.   Steel yield strength = f_y.

3.   Beam width = b.

4.   Beam effective depth = d.

5.   Area of reinforcing steel bars = A_s.

Generally, the beam's total depth, h, can be determined based on the equations in Figures 6.2 and 6.3 of Chapter 6. Steel bar size, distribution and spacing can be determined using Table 2.1 of Chapter 2, and Section 6.5 and Table 6.2 of Chapter 6 or the tables in Appendix A which are based on ACI 318 Code requirements for cover and spacing.

7.2 Important Parameters for Reinforced Concrete Beam Design

As explained in Section 7.1, there are five (5) fundamental unknowns in reinforced concrete beam design for flexure. However, Equation 4.5 is the only irredundant equation for analysis of reinforced concrete beam in bending. As such, only one unknown of the above five unknowns can be determined using this equation. As will be explained in the following, a different method is used to evaluate the other four unknowns. Beam section design is an iterative process based on the following conceptual equation procedure:

Design = Initial Assumptions for Beam Section + Analysis of Proposed Section under Design Inputs + Modification of Assumptions + Analysis of Proposed Section + Modification of Assumptions + ... Until a satisfactory design of beam section is achieved

The initial assumptions for beam section can be based on the design equations covered in Chapters 4 through 7 or formulas devised by the engineer. Knowledge of reinforced concrete and experience are essential in achieving a fewer number of iterations and a more efficient design. Design is an art as well as being a science. One should strive to accomplish an efficient design from the stand points of structural strength as well as simplicity of construction termed efficient construction design.

As stated above, there are five unknowns and one mathematical equation for flexural design. A number of these unknowns, however, may be assumed. The yield strength of steel, f_y, can be selected from a handful of standard values: 40, 60, 75 or 90 ksi (280, 420, 520, 630 MPa). Market availability also influences the selection of f_y. For example, based on U.S. market availability, f_y = 60 ksi (420 MPa) is the most common type of reinforcing steel bars while 40 ksi (280 MPa) bars are somewhat uncommon. A prudent design engineer can assume yield strength for steel of 60 ksi initially; however, a larger value of yield strength may be selected if needed for design. With the adoption of yield strength of steel as part of the initial design assumptions, one can reduce the number of unknowns from five (5) to four (4).

The process of steel yield strength selection can be used for concrete compressive strength, f'_c. Based on market availability, the design engineer typically assumes f'_c as well; although, based on analysis results f'_c may be modified. Typical values of f'_c in the U.S. are 2,500 to 6,000 psi (17 to 42 MPa) with usual increments of 500 psi with special cases using 7,000 psi or 8,000 psi (49 to 56 MPa). Other values of f'_c can be adopted by the design engineer if properly justified. With the selection of f'_c, the number of unknowns is reduced to three (3). For slab on grade and foundations, f'_c = 2,500 to 3,000 psi (17 to 21 MPa) is common. For small size projects, f'_c = 4,000 psi (28 MPa) is also common. For demanding projects (e.g. tall buildings), the specified concrete compressive strength can equal or exceed 6,000 psi (42 MPa).

Furthermore, for an economic and efficient design, it is typical to utilize the following equation by designers:

$$h/b = 1.5 \text{ to } 2.0 \quad \text{or} \quad b/h = {}^1/_2 \text{ to } {}^2/_3 \tag{7.1}$$

With the employment of Equation 7.1, the number of equations for flexural design of beam sections has been increased to two (2) while the number of unknowns remains at three (3). To resolve this dilemma, the equation of the nominal resistance factor, R_n, can be utilized. The derivation of this equation is shown below:

$$M_n = A_s \cdot f_y \cdot (d - a/2) \tag{4.5}$$

Utilizing Equation 4.1: $\rho = A_s/bd$

Then $\quad M_n = \rho \cdot b \cdot d \cdot f_y \cdot (d - a/2)$

Utilizing Equation 4.3: $a = A_s \cdot f_y/0.85 f_c' \cdot b$

Then $\quad M_n = \rho \cdot b \cdot d \cdot f_y \cdot (d - \rho \cdot d \cdot f_y/1.7 f_c')$

$\quad M_n/bd^2 = \rho \cdot f_y \cdot (1 - \rho \cdot f_y/1.7 f_c')$

$$R_n = \rho \cdot f_y \cdot (1 - \rho \cdot f_y /1.7 f_c') \tag{7.2}$$

or

$$\rho = (0.85 f_c'/f_y) \cdot [1 - \sqrt{1 - 2R_n/0.85 f_c'}] \tag{7.3}$$

Equations 7.2 and 7.3 link the beam reinforcement ratio ρ with R_n which is a function of material properties, f_c' and f_y, and independent of beam sectional dimensions b and d. As will be explained in the following parts of this chapter, R_n plays an important role in beam design. With the application of Equation 7.2 or 7.3, the number of usable equations increases to three (3) to match the number of unknowns, thus making beam design possible. R_n may only be used for rectangular–shaped beams since Equations 7.2 and 7.3 are for rectangular sections.

R_n is considered a design aid. Equation 7.2 for R_n is a function of f_c', f_y as well as ρ. Tables A.14 through A.23 tabulate the value of R_n versus ρ to assist in the design of reinforced concrete beams. In each table f_c' and f_y are constants while ρ is variable. Furthermore, these tables contain the values of ρ_{max}, ρ_{maxp} and ρ_{min}. The use of Tables A.14 through A.23 can greatly facilitate flexural design of reinforced concrete beam sections.

There are two procedures for beam cross–section design for bending:

1. Beam sectional dimensions, b and h, are known due to architectural design. The structural designer is required to determine the value of ρ, select steel bars, and detail the beam section based on code requirements. Selection of bars should be reasonable to facilitate construction. Experience with concrete design is important in this case. One should avoid using a large number of bars in a cross–section (e.g. 20 bars) unless absolutely necessary. The minimum number and minimum size of steel bars for tension reinforcement are 2 and #4 (No. 12), respectively. One should use steel bar sized reasonable for the beam section [e.g. #18 bars (55 mm diameter) for a beam with a 12×15 in (300×400 mm) cross–section is not realistic]. One should avoid mixing and matching many different sizes of bars for the beam (e.g. using #4, #5, #6, #7 in two layers of steel). ACI 318 Code allows mixing two different bar sizes for tension bending reinforcement. Furthermore, these bar sizes may not be more than 2 bar numbers or ¼ in = 6 mm apart (e.g. #4 and #6, #5 and #7, or #8 and #10 are okay but #4 and #7, #5 and #10, or #6 and #11 are not). Steel reinforcement is required to have symmetrical bar placement about the vertical centerline. Section details are also important to facilitate construction where designer's experience is key.

2. Beam sectional dimensions, b and h, are not known. The structural designer can assume a value for the reinforcement ratio, ρ, which leads to an economic and efficient design. Among the commonly utilized values of ρ are $\rho_b/2$ and $(0.2 f_c'/f_y)$. Based on the selection of ρ, beam dimensions b and d can be determined and the beam section detailed as explained above. Equation 7.1 is helpful for achieving efficient beam section design. In the selection of b and h, values reasonable for construction need to be used. Typically, values for b and h are multiples

of 2 and 3 in (50 to 75 mm). Since concrete requires formwork, one needs to investigate the availability of standard dimension forms for concrete for design optimization.

Both procedures listed above are iterative in nature. Further iterations refine and simplify the beam's cross–sectional design. Furthermore, both procedures above utilize Equations 7.2 and 7.3 as explained in detail in the following sections.

Other design procedures are also utilized in addition to the ones discussed above. Regardless, any adopted design method must comply with all code design requirements and must satisfy the following equation:

$$\phi M_n \geq M_u \qquad\qquad (7.4)$$

As previously mentioned, instead of using Equations 7.1 through 7.4 as design aid for beams, one can utilize Tables A.14 through A.23 in Appendix A. Each table is specific to the values of f'c and fy. The value of R_n as a function of ρ is tabulated in these tables for f'_c = 2,500 psi 3,000 psi 4,000 psi 5,000 psi and 6,000 psi and for f_y = 40,000 psi 60,000 psi and 80,000 psi.

Design Procedures for Reinforced Concrete Beam Sections under Bending

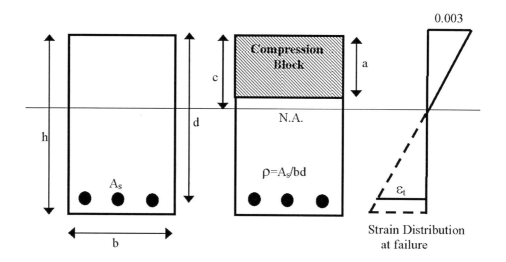

Case 1

Beam dimensions are pre–determined

Known

1.   Material properties or f'_c and f_y

2.   Beam cross–sectional dimensions (b and h)

3.   Beam configurations and loading

Step 1

Determine the value of M_u using Chapter 5 including beam's self weight.

Step 2

Determine the value of M_n assuming that $\phi = 0.9$

Since $\phi M_n \geq M_u$ Use $M_n = (M_n)_{minimum} = M_u/\phi$

Step 3

Approximate the value of d based on the following equations:

d = h – (2.5 to 3 inch)   for one layer of steel

d = h – (4 to 5 inch)   for two layers of steel

One needs to make an educated guess if the steel in the beam would need one or two layers.

Step 4

Calculate R_n as follows:

$R_n = M_n/(bd^2)$

Step 5

Calculate the value of reinforcement ratio, ρ, based on the following equation:

$\rho = (0.85f'_c/f_y)\cdot[1 - \sqrt{1-2R_n/0.85f'_c}]$

or use the tables of Appendix A to determine the value of ρ.

Step 6

Calculate the value of the required steel area A_s using the following equation:

$A_s = \rho bd$

Step 7

Select the required steel bars and detail the beam section using Chapters 4 through 7.

Step 8

Perform beam analysis using the procedure of Chapter 6 to determine the value of ϕM_n that is required to satisfy the following equation:

$\phi M_n \geq M_u$ Where the value of M_u is obtained from Step 1.

Step 9

Repeat steps 1 to 8 until a satisfactory beam section design is achieved.

Case 2

Beam cross–sectional dimensions are not provided

Known

1.   Material properties or f'_c and f_y

2.   Beam configurations and loading

Step 1

Make an initial assumption for h greater than h_{min} (Table 6.3) so that the beam is within the code limitations for deflections.

Step 2

Assume a value for b based on the following equation:

b = h/2 to h/1.5

Step 3

Determine the beam's self weight:

Self weight = $(h/12) \times (b/12) \times (0.15)$ k/ft³

Step 4

Determine the value of M_u using Chapter 5 including beam's self weight.

Step 5

Determine the value of M_n assuming that $\phi = 0.9$

Since $\phi M_n \geq M_u$ Use $M_n = (M_n)_{minimum} = M_u/\phi$

Step 6

Assume a value for ρ for efficient design

e.g. $\rho = \rho_b/2$ or $\rho = 0.2f_c'/f_y$

Step 7

Determine the value of R_n using the following equation:

$R_n = \rho \cdot f_y \cdot (1 - \rho \cdot f_y/1.7f_c')$

Or use the tables in Appendix A to determine the value of R_n.

Step 8

Calculate bd² based on the following equation:

$bd^2 = (M_n)/R_n$

Step 9

Reselect b and d by trial and error. A good assumption is that

h/b = 1.5 to 2 or (d + 3)/b = 1.5 to 2.

Step 10

Estimate the value of h based on the following equations:

h = d + (2.5 to 3 inch)   for one layer of steel

h = d + (4 to 5 inch)   for two layers of steel

One needs to estimate if the steel would need one or two layers.

Step 11

If the values of b and h determined in steps 9 and 10 match or are close to the assumed values in steps 1 and 2, go to step 12, otherwise repeat steps 1 through 10 with the values of b and h calculated in step 10.

Step 12

Determine the value of the required steel area A_s using the following equation:

$A_s = \rho bd$

Step 13

Select the required steel bars and detail the beam section using Chapters 4 through 7.

Step 14

Perform beam analysis using the procedure of Chapter 6 to determine the value of ϕM_n that is required to satisfy the following equation:

$\phi M_n \geq M_u$ Where the value of M_u is obtained from Step 1.

Step 15

Repeat steps 1 through 14 until a satisfactory beam design is achieved.

Example 7.1

Design the reinforced concrete beam shown below if the beam is not exposed to weather (interior exposure with min. clear cover = 1.5 in) and $f'_c = 4,000$ psi and $f_y = 60,000$ psi.

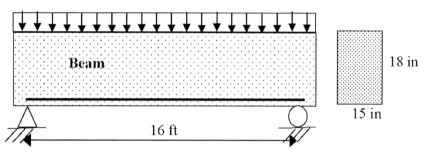

$W_D = 2$ k/ft including self weight $W_L = 1.3$ k/ft

Beam

18 in

15 in

16 ft

Solution

Since the beam dimensions are given then one needs to use Case 1 conditions:

$W_u = 1.2\,W_D + 1.6\,W_L = 1.2 \times 2 + 1.6 \times 1.3 = 4.48$ k/ft

$M_u = W_u L^2/8 = 4.48 \times (16)^2/8 = 143$ k.ft

$M_n = M_{nmin} = M_u/\phi = 143/0.9 = 159$ k.ft

$d = 18 - 2.5 = 15.5$ in

$R_n = M_n/(bd^2) = 159 \times 12,000/[15 \times (15.5)^2] = 530$ psi

where 12,000 is for unit conversion from k.ft to lb.in

$\rho = (0.85 f'_c/f_y)\cdot[1 - \sqrt{1-2R_n/0.85f'_c}]$

$\rho = 0.85 \times (4,000/60,000) \times \{1 - [1 - 2 \times 530/(0.85 \times 4,000)]^{1/2}\}$

$\rho = 0.0097 = 0.97\%$ ===> $A_s = 0.0097\times15\times15.5 = 2.25$ in^2

Appendix A9 ===> $\rho_{min} < \rho < \rho_{max}$ O.K.

Use 4#7 bars, $A_s = 2.41$ in^2 (Table 2.1) & $b_{min} = 10.2$ in (Table 6.2) < 15 in OK

Using #3 stirrups (typical), $d = 18 - 1.5 - 3/8 - (7/8)/2 = 15.7$ in > 15.5 in OK

The beam can be detailed as follows:

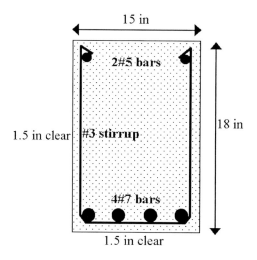

Beam Section Detail

Beam Analysis:

$a = (2.41 \times 60)/(0.85 \times 4 \times 15) = 2.84$ in

$c = 2.84/0.85 = 3.34$ in

$\varepsilon_t = (0.003/3.38) \times (15.7 - 3.34) = 0.0111 >> 0.005 ==> \phi = 0.9$ and $\rho < \rho_{maxp}$

$\phi M_n = 0.9 \times 2.41 \times 60 \times (15.7 - 2.84/2)/12 = 154.9$ k.ft $> M_u = 143$ k.ft OK

From Table A.9 $\rho_{min} = 0.33\% ==> \rho = 0.97\% > \rho_{min}$ O.K.

With 4#7 bars $b_{min} = 10.2$ in (Table 6.2) < 15 in OK

Also, $h = 18$ in is greater than h_{min} from Table 6.3 = span/16 = 1ft = 12 in. As such, according to ACI 318 Code, deflections need not be computed or verified.

Example 7.2

Design the reinforced concrete cantilever beam shown below if the beam is exposed to weather (min. clear cover = 2 in) and $f'_c = 4,500$ psi and $f_y = 60,000$ psi. Beam size has not been determined by the architect.

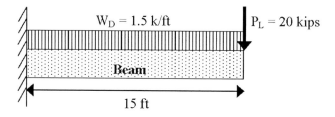

Solution

Since the beam dimensions are not given then one needs to utilize Case 2:

Assume beam dimensions based on Table 6.3 so that deflections need not be computed:

Initial assumption: $h = 30$ in $>$ Span/8 per Table 6.3

$b = 16$ in within the preferred range of b/h = ½ to ²/₃

Beam self weight = (30×16/144) × 0.15 = 0.5 k/ft

where 144 is for unit conversion from square inches to square foot

The critical section for this beam is at the fixed end. For this problem, the moment is negative and tension steel should be placed near the top of the beam section. To facilitate the solution of this problem, absolute values of moments are used. Magnified design bending moment for this critical section:

$$M_u = 1.2\ M_D + 1.6\ M_L = 1.2×(1.5 + 0.5)×(15)^2/2 + 1.6×20×15 = 750\ k.ft$$

Now, the designer needs to adopt a reinforcement ratio for this beam. $\rho = \rho_b/2$ is chosen for this example using Equation 6.3:

$$\rho = \rho_b/2 = [0.003E_s/(0.003E_s + fy)]×(0.85f'_c\beta_1/f_y)/2$$

$$\rho = \{0.003×(29×10^6)/[0.003×(29×10^6) + 60,000)]\}×(0.85×4,500×0.825/60,000)/2$$

$$\rho = 0.0156 = 1.56\% \quad \text{Table A.9 Appendix A could also be used for } \rho_b$$

Note that in the equation for ρ, the value of β_1 used was 0.825 due to $f'_c = 4,500$ psi in this example.

Based on Tables A.18 – A.19, R_n is approximately 820 psi and that $\phi = 0.9$ and $\rho < \rho_{maxp}$ With this, the value of bd² can be determined:

$$bd^2 = M_{nmin}/R_n = (M_u/\phi)/R_n = (750×12,000/0.9)/820 = 12,195\ in^3$$

where 12,000 is for unit conversion from k.ft to lb.in

With b = 16 inch ===> d = (12,195/16)^½ = 27.6 in

One may conclude that the assumed beam of 16×30 is borderline. Then, new beam dimensions can be assumed: h = 32 in and b = 18 in.

Repeating the same procedure as above with b =18 in and h = 32 in:

Beam self weight = 0.6 k/ft $M_u = 764$ k.ft

$A_s = 0.0156 × 16 × 29 = 7.24\ in^2$ Use 7#10 bars $A_s = 8.89\ in^2$

The beam can be detailed as follows:

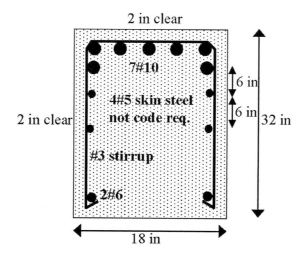

Beam Section Detail

Although the ACI Code 318 does not require skin reinforcement for the above beam, but its depth is close to the 36 in threshold for skin reinforcement. As such, one may elect to use skin reinforcement to be prudent. After detailing the section, one needs to analyze the proposed section to determine the value of M_n and to verify code compliance.

The minimum required width to fit 5#10 bars in Table 6.2 based on 1.5 in clear cover is 15.3 in with the increase of clear cover due to exposure conditions, the minimum width would become 16.3 in < 18 in provided O.K. The effective depth can be determined based on the weighted average as follows where 1.27 in is the diameter of #10 bar:

$$d = \{5 \times [32-2- 3/8-(1.27)/2] + 2 \times [32-2-3/8-(1.27)- (1.27)-(1.27)/2]\}/7$$

$$d = 28.3 \text{ in}$$

$$a = A_s \cdot f_y/(0.85 \cdot f_c' \cdot b) = 8.89 \times 60 / (0.85 \times 4.5 \times 18) = 7.8 \text{ in}$$

$$c = a/\beta_1 = 7.7/0.825 = 9.4 \text{ inch where } \beta_1 = 0.825 \text{ for } f_c' = 4{,}500 \text{ psi (Table A.7)}$$

$$\varepsilon_t = 0.003/9.4 \times (28.3 - 9.4) = 0.006 > 0.005 ===> \phi = 0.9$$

$$\phi M_n = 0.9 \times 8.89 \times 60 \times (28.3 - 7.7/2)/12 = 977 \text{ k.ft} > M_u = 764 \text{ k.ft O.K.}$$

Example 7.3

Design the reinforced concrete beam shown below if the beam is not exposed to weather (interior exposure with min. clear cover = 38 mm) and $f_c' = 30$ MPa and $f_y = 420$ MPa.

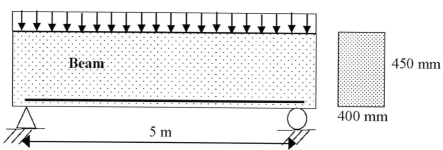

$W_D = 25$ kN/m including self weight $W_L = 18$ kN/m

Beam

450 mm

400 mm

5 m

Solution

Since the beam dimensions are given then one needs to use Case 1 conditions:

$$W_u = 1.2 W_D + 1.6 W_L = 1.2 \times 25 + 1.6 \times 18 = 59 \text{ kN/m}$$

$$M_u = W_u L^2/8 = 59 \times (5)^2/8 = 184 \text{ kN.m}$$

$$M_n = M_{nmin} = M_u/\phi = 184/0.9 = 205 \text{ kN.m}$$

$$d = 450 - 65 \approx 380 \text{ mm}$$

$$R_n = M_n/(bd^2) = 205 \times 10^6/[400 \times (380)^2] = 3.55 \text{ MPa } 10^6 \text{ is for unit conversion}$$

$$\rho = (0.85 f_c'/f_y) \cdot [1 - \sqrt{1-2R_n/0.85 f_c'}]$$

$$\rho = 0.85 \times (30/420) \times \{1 - [1 - 2 \times 3.55/(0.85 \times 30)]^{1/2}\}$$

$$\rho = 0.0091 = 0.91\% ===> A_s = 0.0091 \times 400 \times 380 = 1380 \text{ mm}^2$$

Use 5 No. 20 bars, $A_s = 1570$ mm² (Appendix A) and No. 9 stirrups (9 mm)

$$b_{min} = 150 + 4 \times (20 + 25) + 2 \times 10 + 2 \times 20 = 396 \text{ mm (Equation 6.8)}$$

$$b = 400 \text{ mm} > b_{min} = 384 \text{ mm OK}$$

The beam can be practically detailed as shown:

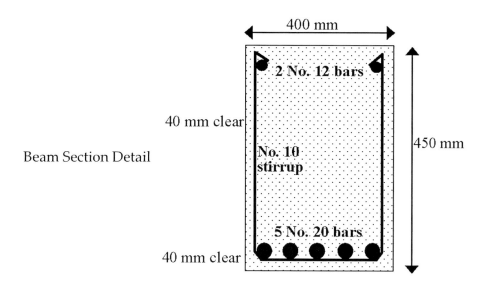

Beam Section Detail

Beam Analysis:

$a = (1570 \times 420)/(0.85 \times 400 \times 30) = 65$ mm

$c = 65/0.85 = 76$ mm

$\varepsilon_t = (0.003/76) \times (380 - 76) = 0.0111 \gg 0.005 \Longrightarrow \phi = 0.9$ and $\rho < \rho_{maxp}$

$\phi M_n = 0.9 \times 1570 \times 420 \times (380 - 65/2)/10^6 = 206$ kN.m $> M_u = 184$ kN.m OK

$\rho_{min} = 3\sqrt{f'_c} f_y > 200/f_y$ Equations 6.6 and 6.7 where f'_c and f_y are in psi

$\rho_{min} = 3\sqrt{30/6.95}/(420/6.95) = 0.0021 > 200/(420/6.95) = 0.0033$

$\rho = 0.97\% > \rho_{min} = 0.33\%$ O.K. where 6.95 above is a conversion factor

Also, h = 450 mm is greater than h_{min} from Table 6.3 = span/16 = 313 mm. As such, according to ACI 318 Code, deflections need not be computed or verified.

Test your knowledge

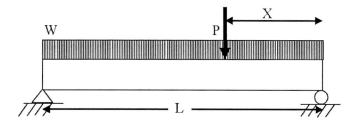

Design and detail a reinforced concrete beam to support the bending moment caused by the loads according to the diagram shown above utilizing the strength design method.

Beam	L (ft)	W_D (k/ft)	W_L (k/ft)	P_D kips	P_L kips	X (ft)	f'_c (psi)	f_y (ksi)	h (in)	b (in)	ρ (%)
1	12	1.2	0.9	0	0	0	3,000	60	15	12	N/A
2	15	1.3	1.4	0	0	0	3,000	60	20	15	N/A
3	18	1.6	1.7	0	0	0	4,000	60	24	15	N/A
4	25	2.2	1.8	0	0	0	4,000	60	32	18	N/A
5	30	3.5	2.5	0	0	0	5,000	60	36	20	N/A
6	12	1.2	0.9	10	15	6	3,000	60	20	15	N/A
7	15	1.3	1.4	15	14	7.5	3,000	60	24	15	N/A
8	18	1.6	1.7	18	20	9	4,000	60	24	15	N/A
9	25	2.2	1.8	15	0	12.5	4,000	60	32	18	N/A
10	30	3.5	2.5	0	25	15	5,000	60	36	20	N/A
11	14	1.5	1.2	0	0	0	2,500	60	N/A	N/A	1.0%
12	18	1.6	1.4	15	12	9	4,000	60	N/A	N/A	$\rho_b/2$
13	24	1.5	1.3	12	15	10	4,500	60	N/A	N/A	1.2–1.8
14	28	1.5	0	0	30	14	4,000	60	N/A	N/A	1 – 2

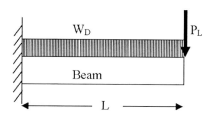

Beam	L (ft)	W_D (k/ft)	W_L (k/ft)	P_D kips	P_L kips	f'_c (psi)	f_y (ksi)	h (in)	b (in)	ρ (%)
15	8	1.2	0	10	8	3,000	60	18	12	N/A
16	10	1.3	0	8	6	3,000	60	20	15	N/A
17	12	1.6	0	12	0	4,000	60	24	15	N/A
18	15	2.2	0	10	0	4,000	60	32	18	N/A
19	18	3.5	0	12	0	5,000	60	36	20	N/A
20	10	1.2	0.9	10	15	4,000	60	24	18	N/A
21	12	1.3	1.2	12	10	4,000	60	28	18	N/A
22	10	1.6	1.7	18	20	4,000	60	N/A	N/A	N/A
23	14	1.5	1.0	0	10	4,000	60	N/A	N/A	N/A
24	20	1.7	1.4	0	25	5,000	60	N/A	N/A	N/A

The reader is encouraged to practice some of the problems in Chapter 7 in SI by converting the units using the following conversion factors:

1 ft = 0.305 m

1 in = 25.4 mm

1 k/ft = 14.59 kN/m

1 k = 4.45 kN

1 ksi = 6.95 MPa

Chapter 8

ANALYSIS AND DESIGN OF ONE – WAY SLABS

8.1 Introduction to Slabs

Slabs are flat horizontal surfaces with two large dimensions and a small thickness. They are intended to be foundation for living or activity areas, and are supported on two or four sides by beams or walls as shown in Figures 8.1 and 8.2. The slabs discussed in this chapter are elevated slabs, which are different from slabs on grade.

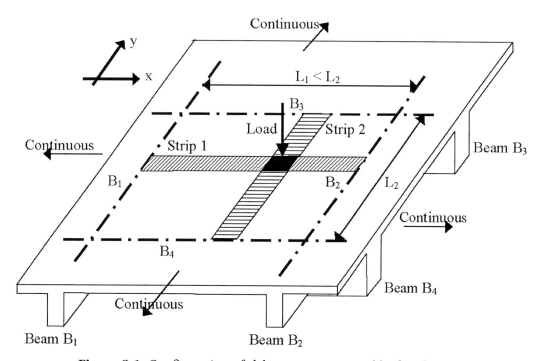

Figure 8.1: Configuration of slab system supported by four beams.

Figure 8.1 above illustrates a slab supported by four beams on four sides. Slabs may be supported by other means such as walls or columns. One bay of the slab is shown in the figure with planner dimensions of L_1 and L_2. The darkened square in Figure 8.1 shows a gravity load applied upon a representative area of the concrete slab with planner dimensions of 1 ft × 1ft. The load is transferred to the supporting beams by 1 ft–wide strips 1 and 2 parallel to x and y axes, respectively. The portions of load transferred to strips 1 and 2 depend upon strip lengths. Deflections of strips 1 and 2 are equal at load application point. Based on the relative rigidity or stiffness of each strip, the load is proportionally divided between them. Assuming that the slab is of a constant thickness within bay $L_1 - L_2$ and that both strips are of equal width (1 ft), strip relative rigidity becomes a direct function of its span length. A special case occurs if:

$$L_2/L_1 \geq 2.0 \tag{8.1}$$

Slab that meet this condition are termed one–way slab. This terminology indicates that the rigidity of strip 1 (the shorter strip) is significantly larger than the rigidity of strip 2 (the longer strip). As such, the load supported by strip 2 is negligible in comparison with strip 1, and strip 1 may be assumed to support the load in its entirety. Given this assumption, beams B_1 and B_2 become primarily responsible for supporting the slab. Transfer of load in one–way slabs occurs almost exclusively in the short direction. Bending is also applied primarily in the short direction. In Figure 8.1 this direction is the x direction.

Figure 8.2 represents a concrete slab supported by only two beams B_1 and B_2. With the absence of support in y direction, the slab is a one–way slab supported in the x direction by strips similar to strip 1, each with length of L_1 and thickness equal to slab thickness.

Therefore, a concrete slab can be a one–way if the ratio of its dimensions (L_2/L_1) is greater than 2.0 or if it is supported in one direction only using beams or walls. The previous discussion is intended for a bay of the slab (bay $L_1 - L_2$) but may be generalized to an entire slab consisting of several bays. This analysis should be extended to all the bays of the slab to determine if one–way slab analysis is applicable.

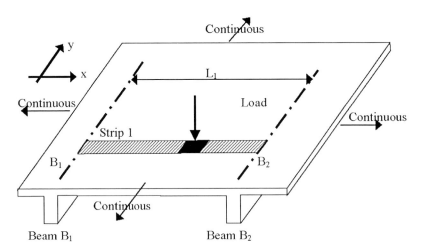

Figure 8.2: Configuration of slab system supported by two beams.

This chapter covers the analysis of simply supported one–way slabs. The theory for such analysis is shown below in Figure 8.3. It contains a one–way slab supported by two beams with a span of L and thickness of h. The dead load, W_D, and live load, W_L, are as shown as liquidated loads over the entire slab surface area represented in units of pound per square ft. A 12–inch representative strip (termed loading strip) spanning between the supports is shown. For one–way slabs subjected to bending moment, the analysis and discussion for beams in Chapters 4 thru 7 are largely applicable.

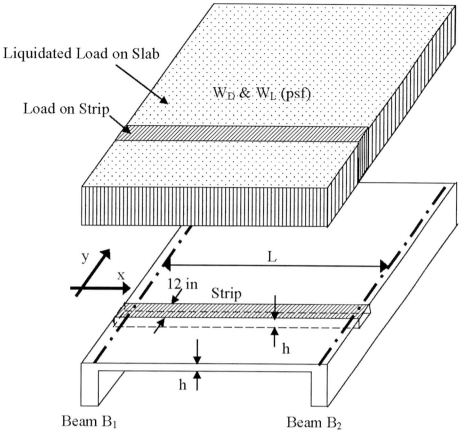

Liquidated Load on Slab

Load on Strip

W_D & W_L (psf)

y

x

L

12 in   Strip

h

h

Beam B_1 Beam B_2

Figure 8.3: One – way slab.

Figure 8.4 shows a two–dimensional and a three–dimensional representation of the loading strip of Figure 8.3. The liquidated load on the slab in psf is converted to a linear load in lb/ft of the same magnitude applied to a 1 foot–wide loading strip. It is important for one to realize that the one foot width of the representative loading strip is only selected as a mathematical tool to simplify calculations. As such, the slab strip is a beam with a uniformly distributed line load equal to the liquidated load on the slab in psf. This beam's thickness is h = slab thickness and width = 12 inch. For SI units, the loading strip width is 1 m and the liquidated load is in kN/m^2.

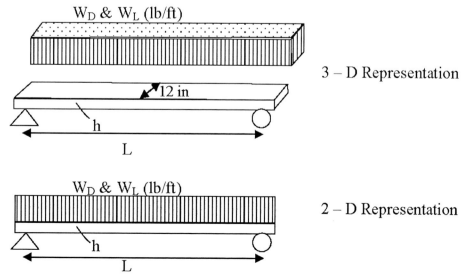

W_D & W_L (lb/ft)

3 – D Representation

12 in

h

L

W_D & W_L (lb/ft)

2 – D Representation

h

L

Figure 8.4: Representative loading strip for design of one – way slabs.

Analysis and design of the loading strip portrayed as a beam in Figure 8.4 is similar to the design and analysis of conventional reinforced concrete beams with the following differences that are dictated by ACI 318 Code:

1.   Slab thickness requirements so that deflections need not be calculated are according to the following table (duplicated in Appendix A as Table A.12):

Table 8.1: Minimum thickness (h) for reinforced concrete slabs unless deflections are computed.[1]

Members	Simply Supported	One-End Continuous	Both-Ends Continuous	Cantilever
Solid One - Way Slab	Span/20	Span/24	Span/28	Span/10

[1]   Normal weight concrete (unit weight, γ = 145 lb/ft³ = 23 kN/m³) and f_y = 6,000 psi = 420 MPa yield strength of steel are assumed. Multiply by (1.65 – 0.005γ) and (0.4 + fy/100,000), for light weight concrete and for fy ≠ 60,000 psi, respectively. Use (1.65 – 0.032γ) and (0.4 + fy/700) for SI units of kN/m³ and MPa. Does not apply to slabs supporting construction that is likely to be damaged by large deflections.

2.   Other steel reinforcement considerations in concrete slabs are as follows:

   a.   There are two curtains of reinforcement in the slab: one for resistance of the bending moment termed main reinforcement (typically in the short direction), and another termed secondary steel that is placed is in the transverse direction (typically the long direction). It is often referred to as shrinkage and temperature steel and is for reducing cracks due to shrinkage and temperature changes,

   b.   The two curtains of steel are placed one on top of the other with the main steel placed to result in a larger value of effective depth, d. As such, main steel is on the bottom and shrinkage steel is on top if the bending moment is positive, and vice versa if the slab is subjected to negative bending moment.

   c.   Maximum spacing of bars of main steel shall not exceed three times the thickness of slab (3h) or 18 in (450 mm) whichever is smaller. Maximum spacing of bars of shrinkage steel shall not exceed five times the thickness of slab (5h) or 18 in (450 mm) whichever is smaller.

   d.   Clear cover over steel bars in concrete slabs may not be less than ¾ in (20 mm) according to ACI 318 Code.

3.   Minimum reinforcement ratio in slabs is different than beams. Minimum reinforcement must be provided for concrete slabs in both directions against shrinkage and temperature variation. The value of minimum reinforcement is summarized as follows based on ACI 318 Code:

   a.   For f_y < 60,000 psi (420 MPa), ρ_{min} = 0.0020 based on the full cross sectional area of the slab. As such, A_{smin} = 0.0020×12×h (in²/ft),

   b.   For f_y = 60,000 psi (420 MPa), ρ_{min} = 0.0018 based on the full cross sectional area of the slab ⟹ A_{smin} = 0.0018×12×h (in²/ft), and

   c.   For f_y > 60,000 psi, ρ_{min} = 0.0018×(60,000/f_y) based on the full cross sectional area. For f_y units in MPa, ρ_{min} = 0.0018×(420/f_y).

4. Design of concrete slabs entails determining slab thickness and the area of main steel (in²/ft) and transverse steel (in²/ft). Although the analysis depends on a 12 in wide representative strip, steel placement need not follow such strips. Strips are a mathematical tool for simplifying analysis and design. Steel distribution is normally designated by steel bar number and spacing such as #4@10 in, #6@7 in or #5@4 in. One may need to convert the area of the selected steel reinforcement in slabs to an equivalent area per foot wide strip or per linear foot. This conversion may be achieved using the following equation:

$$A_s = \text{area of selected steel bar x 12 in / selected bar spacing} \qquad (8.2)$$

Where A_s = representative area of steel bars in in²/linear ft.

Table 8.2 can be utilized for steel bar area conversion to in²/ft in lieu of using Equation 8.2.

Table 8.2: Equivalent area of steel bars in concrete slabs (in²/ft).

Bar # Spacing	3	4	5	6	7	8	9	10	11
3	0.44	0.79	1.23	1.77	2.41	3.14	4.00	5.07	6.25
4	0.33	0.59	0.92	1.33	1.80	2.36	3.00	3.80	4.68
5	0.27	0.47	0.74	1.06	1.44	1.88	2.40	3.04	3.75
6	0.22	0.39	0.61	0.88	1.20	1.57	2.00	2.53	3.12
7	0.19	0.34	0.53	0.76	1.03	1.35	1.71	2.17	2.68
8	0.17	0.29	0.46	0.66	0.90	1.18	1.50	1.90	2.34
9	0.15	0.26	0.41	0.59	0.80	1.05	1.33	1.69	2.08
10	0.13	0.24	0.37	0.53	0.72	0.94	1.20	1.52	1.87
12	0.11	0.20	0.31	0.44	0.60	0.79	1.00	1.27	1.56
14	0.095	0.17	0.26	0.38	0.52	0.67	0.86	1.09	1.34
16	0.083	0.15	0.23	0.33	0.45	0.59	0.75	0.95	1.17
18	0.074	0.13	0.20	0.29	0.40	0.52	0.67	0.84	1.04

1 in²/ft = 2,115 mm²/m

5. The designed thickness of concrete slabs and spacing of steel bars need to be construction friendly. As such, thickness of slabs is selected in ½ in (or 10 mm) increments up to a thickness of 6 in (150 mm) followed by 1 in (25 mm) increments up to a thickness of 12 in (300 mm) after which 2 in (50 mm) increments are used. Spacing of steel also needs to be construction friendly. One may use 1 in (20 mm) increments up to 10 in (120 mm) spacing followed by 2 in (50 mm) increments up to 18 in (450 mm).

To determine the loading on a concrete slab, one needs to include primarily dead and live load. While dead load may be computed, live load on slabs is specified by ASCE/SEI 7 standards. One needs to compute the dead load on a concrete slab using self weight and other permanent attachments, and then include the standardized live load in analysis and design. Other loads may also be included as the situation dictates. There are two types of live loads that slabs need to be designed for: uniform load and concentrated load. The distributed load is applied over the whole area of the slab. The concentrated load is applied on a 2.5 ft × 2.5 ft square area located so that it generates maximum effect (moment or shear) in the structure. Therefore, in design of the slab 1 ft strip, it is loaded with the dead load, and then (1) it is loaded with the distributed live load and maximum moment and shear are determined, and (2) the distributed live load is removed and 40% (1 ft strip width divided by 2.5 ft side length of concentrated load square) of the code dictated concentrated live load is placed at midspan (or critical location) and maximum moment and shear are determined. The maximum moment and shear of cases 1 and 2 are used for design. Table 8.3 is a summary of live load for slab design by ASCE/SEI 7.

Table 8.3: Summary of live load on slabs.

Use of Structure	Uniform Live Load (psf)	Concentrated Live Load (lb) on 2.5 ft Square Area
Balconies	100	N/A
Decks	100	N/A
Residential Dwellings	40	N/A
Offices	50	2,000
Office Corridors & Computers	100	2,000
Schools	40	1,000
Corridors	100	1,000
Light Storage	125	N/A
Heavy Storage	250	N/A
Retail Store	100	1,000
Whole Sale Store	125	1,000
Light Manufacturing	125	2,000
Heavy Manufacturing	250	3,000
Restaurant and Ballroom	100	N/A
Hospitals	80	1,000
Hotel – Rooms	40	N/A
Corridors and Public Rooms	100	N/A
Passenger Garage	40	N/A
Sidewalks and Driveways	250	8,000
Roofs	20	N/A

Example 8.1

Investigate the adequacy of the simply supported concrete slab shown in the figure if its use for passenger garage and driveway. $f'_c = 2,500$ psi and $f_y = 60,000$ psi.

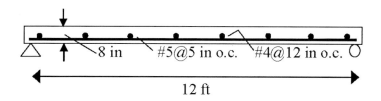

12 ft

Solution

1.   Min. slab thickness (Table 8.1) = span/20 = 144/20 = 7.2 in < 8 in available O.K.

2.   Dead load due to self weight = (8/12)×150 pcf (self weight of concrete) = 100 psf
      Distributed live load = 250 psf (Table 8.3 for driveway)
      Concentrated live load = 8,000 lb (Table 8.3 for driveway)

3.   Check slab under distributed live load
      W_u = 1.2×100 + 1.6×250 = 520 psf
      1 ft strip – Distributed load, W_u = 0.52 k/ft
      $(M_u)_{max}$ = $W_u L^2/8$ = 0.52×(12)²/8 = 9.36 k.ft

Slab under uniformly distributed live load

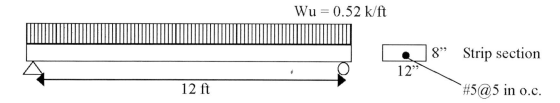

Beam similitude of 1 ft slab strip

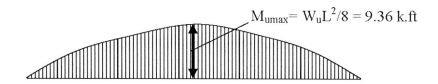

Bending moment diagram for 1 ft strip

4.   Check slab under concentrated live load

      W_u = 1.2×100 = 120 psf (dead load only)
      P_u = 1.6×8,000 = 12,800 lb (concentrated live load only)
      1 ft strip – Distributed load, W_u = 0.12 k/ft
              Concentrated load at midspan = 0.4×12,800 = 5,120 lb = 5.12 kips
      $(M_u)_{max}$ = $W_u L^2/8 + P_u L/4$ = 0.12×(12)²/8 + 5.12×12/4 = 17.52 k.ft

Slab under uniformly concentrated live load

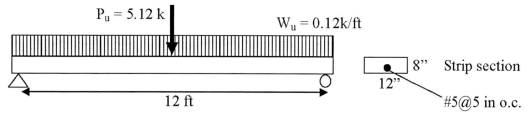

Beam similitude of 1 ft slab strip

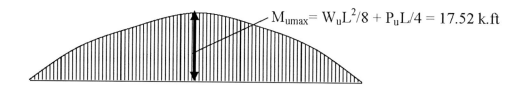

$$M_{umax} = W_u L^2/8 + P_u L/4 = 17.52 \text{ k.ft}$$

Bending moment diagram for 1 ft strip

Therefore, for bending moment, the effect of the concentrated live load on slab is more critical than the effect of distributed live load. The magnified bending moment to be used for slab design is 17.52 k.ft

5.   Adopt M_n = the lower bound of M_n or $(M_n)_{min} = M_u/\phi$ (remember that $M_n > M_u/\phi$) $M_n = 17.52/0.9 = 19.47$ k.ft

6.   Determine the value of d as follows
     d = slab thickness – clear cover – diameter of main steel bars/2
     d = 8 – ¾ – (5/8)/2 = 6.94 in

7. $R_n = M_n/(b.d^2) = 19.47 \times 12,000/[12 \times (6.94)^2] = 404$ psi
     Based on Table A.17, $\rho = 0.0075 \Longrightarrow A_{sreq} = 0.0075 \times 12 \times 6.94 = 0.62$ in²/ft
     $A_{sprov} = \#5@5$ in o.c. Based on Table A.24 $\Longrightarrow A_{sprov} = 0.74$ in²/ft
     $A_{sprov} = 0.74$ in²/ft $> A_{sreq} = 0.62$ in²/ft   O.K.

8.   Verify that slab has minimum reinforcement or better in both directions
     $A_{smin} = 0.0018 \times 12 \times 8 = 0.173$ in²/ft
     Main steel $A_s = 0.74$ in²/ft > 0.173 in²/ft   O.K.
                     Spacing = 5 in < Max spacing = 3h or 18 in whichever is smaller

     Shrinkage steel $A_s = \#4@12$ in o.c = 0.20 in²/ft (Table A.24) > 0.173 in²/ft   O.K.
                     Spacing = 12 in < Max spacing = 5h or 18 in whichever is smaller

Example 8.2

Design the cantilever concrete slab shown in the diagram if it is used as a balcony with railing applies a line load of 200 lb/ft. f′c = 4,000 psi and fy = 60,000 psi.

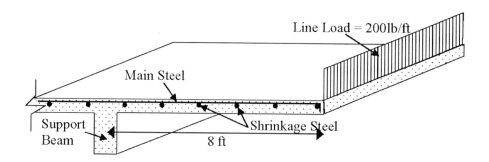

Solution

1.   Min. slab thickness (Table 8.1) = span/10 = 8x12/10 = 9.6 in – Use h = 10 in

2.   Self weight = (10/12)×150 pcf (self weight of concrete) = 125 psf
     Concentrated dead load (railing line load) = 200 lb/ft
     Distributed live load = 100 psf (Table 8.3 for balcony)
     Concentrated live load = 0 lb (Table 8.3 for balcony)

3. Since ASCE/SEA 7 (Table 8.3) does not require a concentrated load for balconies. Thus, one only needs to check slab under dead load + distributed live load
$W_u = 1.2 \times 125 + 1.6 \times 100 = 310$ psf Line Load $P_u = 1.2 \times 200 = 240$ lb/ft
For 1 ft strip Distributed dead load, $W_u = 0.30$ k/ft
        Concentrated dead load (railing), $P_u = 0.24$ kips
$(M_u)_{max} = -W_u L^2/2 - P_u L = -0.30 \times (8)^2/2 - 0.24 \times 8 = -11.84$ k.ft

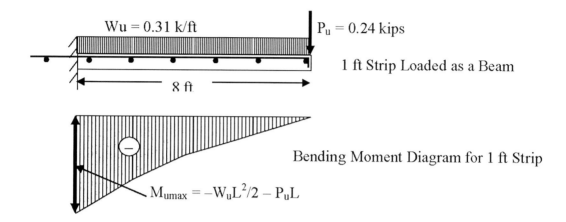

Bending Moment Diagram for 1 ft Strip

4. Adopt M_n = the lower bound of M_n or $(M_n)_{min} = M_u/\phi$ (remember that $M_n > M_u/\phi$) $M_n = -11.84/0.9 = -13.2$ k.ft

5. Determine the value of d as follows
d = slab thickness – clear cover – diameter of main steel bars/2
Assume main steel bars are #6
d = 10 – ¾ – (6/8)/2 = 8.88 in

6. $R_n = -M_n/(bd^2) = -13.2 \times 12,000/[12 \times (8.88)^2] = -167$ psi. Use steel in top portion of the slab
Based on Table A.19, $\rho = 0.0029 \Longrightarrow A_{sreq} = 0.0029 \times 12 \times 8.88 = 0.31$ in²/ft
A_{sprov} = #6@12 ino.c. Based on Table A.24 $\Longrightarrow A_{sprov} = 0.44$ in²/ft
$A_{sprov} = 0.44$ in²/ft > $A_{sreq} = 0.30$ in²/ft   O.K.

7. Verify that slab has minimum reinforcement or better in both directions
$A_{smin} = 0.0018 \times 12 \times 10 = 0.22$ in²/ft
Main steel $A_s = 0.44$ in²/ft > 0.22 in²/ft O.K.
        Spacing = 12 in < Max spacing = 3h or 18 in whichever is smaller

Shrinkage steel A_s = #5@12 in o.c. = 0.31 in²/ft (Table A.24) > 0.22 in²/ft   O.K.
        Spacing = 12 in < Max spacing = 5h or 18 in whichever is smaller

Example 8.3

Investigate the adequacy of the simply supported concrete slab shown in the figure if its use for passenger garage and driveway. $f'_c = 20$ MPa and $f_y = 420$ MPa.

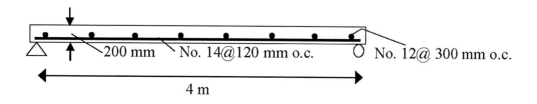

Solution

1. Min. thickness (Table 8.1) = span/20 = 4,000/20 = 200 mm ≤ 200 mm O.K.

2. Self weight = (200/1,000)×23 kN/m³ (self weight of concrete) = 4.6 kN/m
    Distributed live load = 250 psf = 12.0 kPa (Table 8.3 for driveway)
    Concentrated live load = 8,000 lb = 36 kN (Table 8.3 for driveway)

3. Check slab under distributed live load
    W_u = 1.2×4.6 + 1.6×12.0 = 24.7 kN/m
    1 m strip – Distributed load, W_u = 24.7 kN/m
    $(M_u)_{max}$ = $W_u L^2/8$ = 24.7×(4)²/8 = 49.4 kN.m

    Slab under uniformly distributed live load

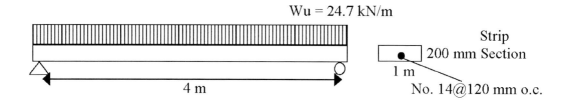

    Beam similitude of 1 m slab strip

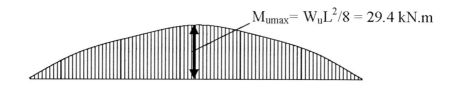

    Bending moment diagram for 1 m strip

4. Check slab under concentrated live load
    W_u = 1.2×4.6 = 5.5 kN/m (dead load only)
    P_u = 1.6×36 = 58 kN (concentrated live load only)
    1 m strip –   Distributed load, W_u = 5.5 kN/m
                    Concentrated load at midspan = 58 kN since 1 m > 2.5 ft
    $(M_u)_{max}$ = $W_u L^2/8 + P_u L/4$= 5.5×(4)²/8 + 58×4/4 = 69 kN.m

    Slab under uniformly concentrated live load

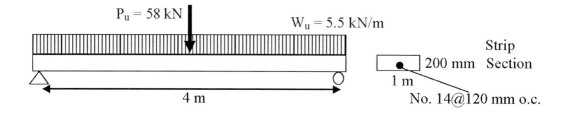

    Beam similitude of 1 m slab strip

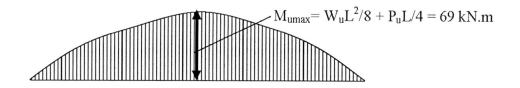

$M_{umax} = W_u L^2/8 + P_u L/4 = 69 \text{ kN.m}$

Bending moment diagram for 1 m strip

Therefore, for bending moment, the effect of the concentrated live load on slab is more critical than the effect of distributed live load. The magnified bending moment to be used for slab design is 69 kN.m

5.   Adopt M_n = the lower bound of M_n or $(M_n)_{min} = M_u/\phi$ (remember that $M_n > M_u/\phi$) $M_n = 69/0.9 = 77 \text{ kN.m}$

6.   Determine the value of d as follows
     d = slab thickness – clear cover – diameter of main steel bars/2
     d = 200 – 20 – (14)/2 = 170 mm

7. $R_n = M_n/(b.d^2) = 77 \times 10^3/[1,000 \times (170)^2] = 2.66 \text{ MPa}$
     Equation 7.3, $\rho = 0.0069 ===> A_{sreq} = 0.0069 \times 1,000 \times 170 = 1,173 \text{ mm}^2/\text{m}$
     $A_{sprov} = \text{No. 14 @120 mm o.c.} ===> A_{sprov} = 1,283 \text{ mm}^2/\text{m}$
     $A_{sprov} = 1,283 \text{ mm}^2/\text{m} > A_{sreq} = 1,173 \text{ mm}^2/\text{m}$ O.K.

8.   Verify that slab has minimum reinforcement or better in both directions
     $A_{smin} = 0.0018 \times 1,000 \times 200 = 360 \text{ mm}^2/\text{m}$
     Main steel $A_s = 1,283 \text{ mm}^2/\text{m} > 360 \text{ mm}^2/\text{m}$ O.K.
                 Spacing = 120 mm < Max spacing = the smaller of 3h or 450 mm

     Shrinkage steel $A_s = \text{No. 12@300 mm o.c.} = 377 \text{ mm}^2/\text{m} > 360 \text{ mm}^2/\text{m}$ O.K.
                 Spacing = 300 mm < Max spacing = the smaller of 5h or 450 mm

Test your knowledge

1.   What is the definition of a one–way slab?

2.   What is the difference between main and secondary steel in concrete slabs?

3.   How is main steel placed in concrete slabs?

4.   Why is minimum steel reinforcement in concrete slabs different than concrete beams?

5.   What phase of behavior of concrete slabs is closest to service conditions?

6.   Why is a 12 in–wide strip selected for the design of concrete slabs, and does steel reinforcement need to be spaced at 12 in apart? Explain.

     Design the following concrete slabs

	Type	Span	Use	f'_c & fy	Line Load
7.	Simple Span	10 ft	Residential	4 and 60 ksi	200 lb/ft (P_D @ CL)
8.	Simple Span	12 ft	Light Industrial	3 and 60 ksi	200 lb/ft (P_L @ CL)
9.	Simple Span	15 ft	Education	4 and 60 ksi	0
10.	Cantilever	6 ft	Residential	4 and 60 ksi	200 lb/ft (P_D @ Edge)
11.	Cantilever	5 ft	Balcony	4 and 60 ksi	250 lb/ft (P_D @ Edge)
12.	Cantilever	8 ft	Office	3 and 60 ksi	150 lb/ft (P_D @ Edge)
13.	Simple Span	2 m	Residential	28–420 MPa	4 kN/m (P_D @ Edge)
14.	Simple Span	4 m	Light Industrial	20–420 MPa	3 kN/m (P_L @ CL)
15.	Simple Span	5 m	Education	28–420 MPa	0
16.	Cantilever	2 m	Residential	28–420 MPa	2.5 kN/m (P_D @ Edge)
17.	Cantilever	3 m	Residential	35–420 MPa	3 kN/m (P_D @ Edge)

Chapter 9

ANALYSIS AND DESIGN OF
T – SHAPED BEAMS

9.1 Reinforced Concrete Floor Systems

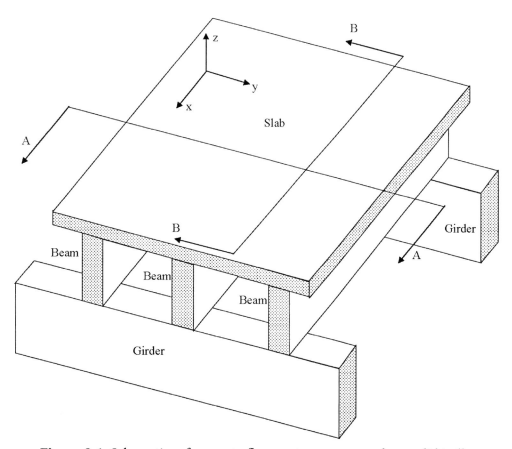

Figure 9.1: Schematics of concrete floor system not poured monolithically.

Figure 9.1 illustrates a reinforced concrete floor system consisting of a concrete slab supported by beams which are in turn supported by girders. The coordinate axes are also depicted. The different elements in Figure 9.1 (i.e. slab, beams and girders) are not poured (fresh concrete placement) monolithically. The girders are poured or placed first, followed by the beams and lastly the slab. For further illustrations, floor system sections A-A and B-B are presented in Figure 9.2. Based on Figures 9.1 and 9.2, the concrete slab spans along the y–axis and is subjected to a moment about x (M_x). Likewise, the beams span along the x–axis and are subjected to a moment about y (M_y). Note that sections A–A and B–B are perpendicular to one another.

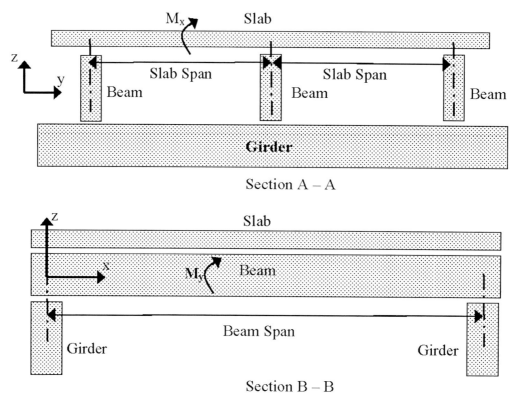

Figure 9.2: Cross – section of concrete floor system.

Using three separate pours for a floor system is not efficient; however, the differentiation in Figures 9.1 and 9.2 is solely for illustration purposes. Figures 9.1 and 9.2 are intended primarily to show the structural members involved in the construction of a floor system. The slab is supported by the beams. The beams are perpendicular to the girders. The girders provide support for the beams. The girders are supported by other girders or columns. The slab span is the distance between beam centerlines and the beam span is the distance between the girder centerlines. Slab span and beam span are clearly shown in Figures 9.1 and 9.2.

Efficient construction of a slab–beam–girder system involves one simultaneous pour of fresh concrete, referred to as the Monolithic Pour Construction Method. The monolithic concrete pour includes portions common among the slab, beams and girders as shown in Figure 9.3.

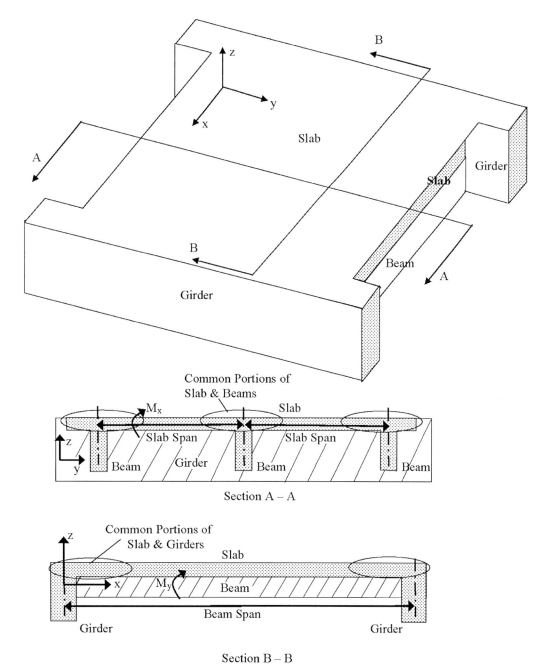

Figure 9.3: Monolithic concrete floor system.

Note that Figures 9.1, 9.2, and 9.3 are not to scale but intended to clearly show all elements of the floor system; furthermore, all cross–sections are dotted for ease of distinction. The subject of this chapter is the analysis and design of T–shaped (T–beams for short) reinforced concrete beams with cross–sections as in section A–A of Figure 9.3. With a monolithic pour, the slab, beams, and girders share portions. The common portions of slab and beams are subjected to M_x and M_y simultaneously. Such double function of concrete is acceptable by ACI 318 Code. Research has shown that concrete behaves satisfactorily under such conditions.

For better visualization of the layout of a concrete floor system, its plan view can be utilized. It is a two dimensional view of the floor system on a horizontal plane with the beams and girders depicted along with the slab shown as a horizontal flat surface. To comprehend the plan view, one needs to envision looking at a floor system from the underside to see the slab along with the beams and girders. An example of such a floor system is shown in Figure 9.4. In this figure, beams are referred to as T-beams considering the shape of their cross–section (Figure 9.3 Section A-A). Concrete floor systems that include T–beams are commonly referred to as ribbed floor systems where the webs of T–beams are considered ribs in otherwise flat plate.

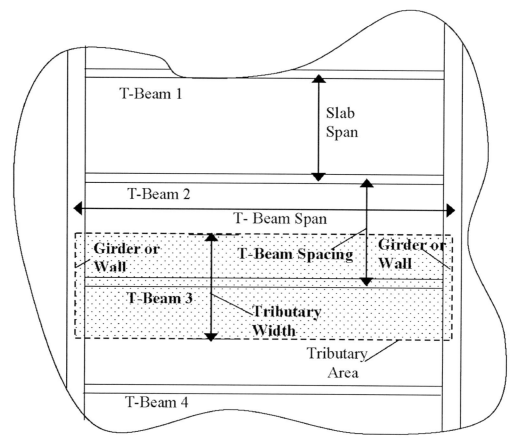

Figure 9.4: Floor system plan view.

Figure 9.4 introduces two new terms: spacing and tributary width of T-beams. They can be defined as follows:

T–Beam Spacing: is the center–to–center distance between adjacent T-beams. Typically the spacing of T–beams in concrete floor systems is constant.

T–Beam Tributary Width: is the length between middle of the two adjacent spacings of the T–beam in question. One may think of the tributary width as the width in which all loads are supported by (tributary to) the T–beam in question.

T–Beam Tributary Area: is the area equal to tributary width × beam span. An alternative definition is the area within which all loads are supported by (tributary to) the T-beam in question.

Each T–beam of a floor system has two spacings and one tributary width. If such spacings are of equal magnitude, then the tributary width and spacing distance are also of equal magnitude. T–beam spacing in floor systems is typically constant, and thus spacing is typically equal to tributary width.

9.2 Effective T-Beam Cross Section

T-beams' effective cross section requires definition and determination since they have common areas with the concrete slab in floor systems. The effective cross section of a T–beam may be defined as:

The beam section that contributes to (or effective in) resisting external force effects including the common areas with the concrete slab allowed by ACI 318 Code.

Figure 9.5 illustrates the effective cross section of T–beams with the following parts:

Web or stem = the vertical part of T–beam cross section.

Flange = the horizontal part of T–beam cross section. The flange constitutes the common area between the T–beam and concrete slab.

b_w = web width.

h_f = flange thickness which is equal to concrete slab thickness.

b = flange width as allowed by ACI 318 Code which specifies that:

b = the smallest value among b_1, b_2 and b_3,
    b_1 = T–beam span/4   (Figure 9.3 Section B-B)
    b_2 = tributary width   (as explained above)
    $b_3 = b_w + 16h_f$

d = distance between the top of flange and the centroid of steel reinforcing bars.

A_s = area of steel reinforcing bars.

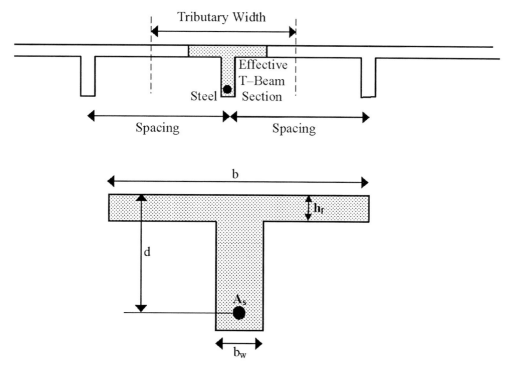

Figure 9.5: Effective T-beam cross –section.

Special Cases

1.   For a beam with a concrete slab on one side only (termed L–beam), the equations for b are:

b = the smallest value among b_1, b_2 and b_3,
b_1 = (span of L–beam)/12 + b_w
b_2 = tributary width
b_3 = b_w + 6h_f

2.   For isolated T–beams (not a part of a floor system) where the beam T–shaped section is a special design feature, the following code provisions need to be satisfied for the beam's cross section:

$h_f \geq b_w/2$ and $b \leq 4b_w$

9.3 Simplified Analysis of T–Beams under Positive Bending Moments

As explained earlier, analysis is investigating the ability of a known section to resist an identified bending moment. Positive moments cause tension at the bottom portion of the beam and compression at the top portion of the beam.

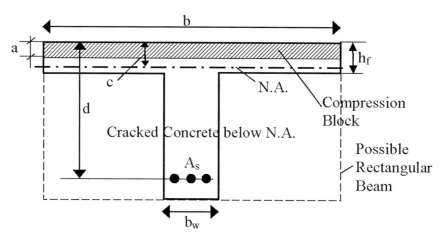

Figure 9.6: T–beam acting as a rectangular beam under the effect of a positive bending moment.

It is generally a good assumption that the area of Whitney's compression block does not exceed the area of the flange. As such, the compression block is completely contained in the flange and the T–beam may be treated as a rectangular beam as shown in Figure 9.6. In effect, due to cracking, the concrete section shape below the neutral axis is immaterial to beam analysis and does not contribute to member's flexural resistance. Due to the large concrete area contained within the flange, it is infrequent that the compression block in T–beams extends into the web. As such, it is generally a good assumption that most T–beams may be analyzed as rectangular beams. A simplified analysis for similar T–beams derived from rectangular beam analysis may be used as follows:

1.   Determine the value of magnified design bending moment acting on the T–beam (M_u). The T–beam's self weight is needed to compute M_u. Self weight of the concrete slab is a part of the floor system dead load. The web of T–beam's weight may be calculated as shown for rectangular beams (Section 6.7) assuming a width of b_w and a depth of $(h - h_f)$.

2.   Determine the value of effective flange width (b) as specified by ACI Code and explained in Section 9.2.

3.   Determine the value of "a" assuming that a < h_f based on the following equation:

$a = A_s f_y/(0.85f'_c b)$

4.   If a < h_f then assumption is valid and the section is acting as a rectangular beam in bending. If not, go to Section 9.4 where further explanation is provided.

5.   Determine the value of tensile strain in steel (ε_t) similar to the procedure followed for rectangular beams utilizing the following equation:

$\varepsilon_t = (0.003/c) \times (d - c)$ Where $c = a/\beta_1$

6.   If $\varepsilon_t \geq 0.005$, then $\phi = 0.9$. Alternatively, based on the value of ε_t, determine the value of capacity reduction factor (ϕ) using the following equation or Table A.8:

$\phi = 0.48 + 83\varepsilon_t \leq 0.90$

7.   Determine the value of the nominal bending moment capacity (M_n) as follows:

$M_n = A_s \cdot f_y \cdot (d - a/2)$

8.   Compare the value of A_s with minimum steel area dictated by ACI Code:

$$A_s \geq \rho_{min} \cdot b_w \cdot d \qquad (9.1)$$

ρ_{min} is the same minimum reinforcement ratio required by ACI Code for rectangular sections (Equations 6.6 and 6.7). One should note that ρ_{min} is multiplied by the effective area of the beam web ($b_w \cdot d$) and not the total beam area which includes the flange. The flange of T–beam is also a part of the concrete slab. Slab design includes minimum reinforcement for the flange. Therefore, for T–beams it is typical to compare ρ_w with ρ_{min} to check code compliance, where ρ_w is based on b_w rather than b and can be determined as follows:

$$\rho_w = A_s/b_w \cdot d \qquad (9.2)$$

9.   Compare the value of ϕM_n with M_u to determine if the beam section is adequate:

$\phi M_n \geq M_u$ O.K.   T–beam is adequate
$\phi M_n < M_u$ N.G.   T–beam is in adequate

Example 9.1

Investigate the adequacy of the T-beam section shown in the following diagram provided that $f'_c = 4,000$ psi, $f_y = 60,000$ psi, and the simple span of T–beam is 25 ft.

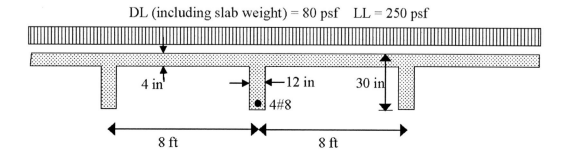

DL (including slab weight) = 80 psf   LL = 250 psf

Solution

1. The load applied to one T–beam in the floor system is the total load within the tributary width. Thus:

    W_D (per linear ft) = dead load on slab per ft² × tributary width = 80×8 = 640 lb/ft
    T–beam web self weight = [12×(30 – 4)]×150/144 = 325 lb/ft
    W_L (per linear ft) = live load on slab per ft² × tributary width = 250×8 = 2,000 lb/ft

    ===> $W_D = 0.64 + 0.33 \approx 1.0$ k/ft $W_L = 2.0$ k/ft

    ===> $W_u = 1.2W_D + 1.6W_L = 1.2×1.0 + 1.6×2.0 = 4.4$ k/ft

    ===> $M_u = W_u L^2/8 = 4.4×(25)^2/8 = 344$ k.ft

2. Effective flange width, b = the smallest value among b_1, b_2 and b_3:

    b_1 = T-beam span/4 = 25/4 = 6.25 ft = 75 in <=== Governs
    b_2 = tributary width = 8 ft = 96 in
    $b_3 = b_w + 16h_f = 12 + 16×4 = 76$ in

3. Assuming that $a < h_f$ then:

    $a = A_s \cdot f_y/(0.85f'_c \cdot b) = 3.16×60/(0.85×4×75) = 0.74$ in $< h_f$ O.K. – Valid assumption

4. Tensile strain in steel (ε_t):

    $c = a/\beta_1 = 0.74/0.85 = 0.87$ in
    $d \approx 30 – 3 = 27$ in
    $\varepsilon_t = 0.003(d – c)/c = 0.003×(27 – 0.87)/0.87 = 0.090 > 0.005$ ===> $\phi = 0.9$

5. Nominal bending moment capacity (M_n):

    $M_n = A_s \cdot f_y \cdot (d – a/2) = 3.16×60×(27 – 0.74/2)/12 = 421$ k.ft

6. Minimum steel area dictated by ACI Code:

    $A_s = 3.16$ in² $> A_{smin} = \rho_{min} \cdot b_w \cdot d = 0.0033×12×27 = 1.07$ in² OK

7. Compare ϕM_n with M_u to determine if the beam section is adequate:

    $\phi M_n = 0.9×421 = 379$ k.ft $> M_u = 344$ k.ft   OK

Detail of the T- beam of Example 9.1 is shown in Figure 9.7.

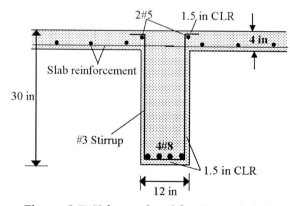

Figure 9.7: T–beam detail for Example 9.1.

A review of the tables and standards in previous chapters reveal that b_w is adequate for 4#8 bars and skin reinforcement is not needed.

Example 9.2

Investigate the adequacy of the T-beam section shown in the following diagram provided that $f'_c = 30$ MPa, $f_y = 420$ MPa, and the simple span of T–beam is 8 m.

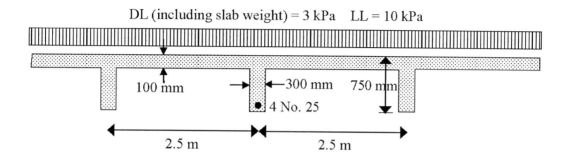

Solution

8. The load applied to one T–beam in the floor system is the total load within the tributary width. Thus:

    W_D (per linear m) = dead load on slab per m² × tributary width = 3×2.5 = 7.5 kN/m
    T–beam web self weight = [300×(750 – 100)]×23/10⁶ = 4.5 kN/m
    W_L (per linear ft) = live load on slab per ft² × tributary width = 10×2.5 = 25 kN/m

    ⟹ $W_D = 7.5 + 4.5 = 12$ kN/m $W_L = 25$ kN/m

    ⟹ $W_u = 1.2W_D + 1.6W_L = 1.2×12 + 1.6×25 = 54.4$ kN/m

    ⟹ $M_u = W_u L^2/8 = 54.4×(8)^2/8 ≈ 435$ kN.m

9. Effective flange width, b = the smallest value among b_1, b_2 and b_3:

    b_1 = T-beam span/4 = 8/4 = 2 m = 2,000 mm
    b_2 = tributary width = 2.5 m = 2,500 mm
    $b_3 = b_w + 16h_f = 300 + 16×100 = 1,900$ mm ⟸══ Governs

10. Assuming that $a < h_f$ then:

    $a = A_s·f_y/(0.85f'_c·b) = 1,960×420/(0.85×30×1,900) = 17$ mm $< h_f$ O.K.

11. Tensile strain in steel (ε_t):

    $c = a/\beta_1 = 17/0.85 = 20$ mm
    $d ≈ 750 – 80 = 670$ mm
    $\varepsilon_t = 0.003(d – c)/c = 0.003×(670 – 20)/20 = 0.098 > 0.005$ ⟹ $\phi = 0.9$

12. Nominal bending moment capacity (M_n):

    $M_n = A_s·f_y·(d – a/2) = 1,960×420×(670 – 17/2)/10⁶ = 545$ kN.m

13. Minimum steel area dictated by ACI Code:

    $A_s = 1,960$ mm² $> A_{smin} = \rho_{min}·b_w·d = 0.0033×300×670 = 663$ mm² OK

14.   Compare ϕM_n with M_u to determine if the beam section is adequate:

$\phi M_n = 0.9 \times 545 = 491$ kN.m $> M_u = 423$ kN.m   OK

Practical detail of the T- beam of Example 9.2 is shown below.

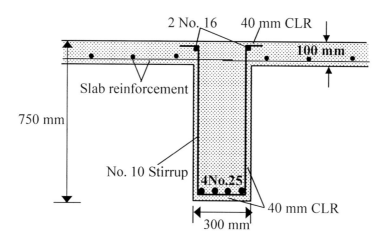

9.4 Detailed Analysis of T-Beams under Positive Bending Moments

Occasionally, the area of flange is not adequate for Whitney's compression block and an extra area of the web is needed to generate a compression force that counteracts the tension force in steel bars as shown in Figure 9.8. In spite of shape change of the compression zone and compression block, Equation 4.3 ($a = \beta_1 \times c$) remains applicable.

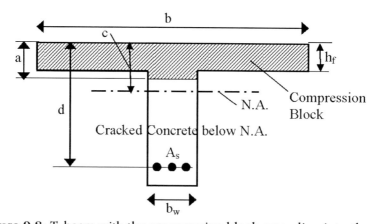

Figure 9.8: T–beam with the compression block extending into the web.

Due to this geometric change, T–beam analysis requires further detail as compared with the previous example. The analysis procedure for this case is as follows:

1.   Determine the value of magnified design bending moment acting on the T–beam (M_u) including web self weight as explained earlier.

2.   Determine the value of effective flange width (b) per ACI Code (Section 9.2).

3.   Determine the value of "a" assuming that $a < h_f$ based on the following equation:

$a = A_s.f_y/(0.85f'_c.b)$

4. If $a < h_f$ then assumption is valid and the section is acting as a rectangular beam in bending. Thus, the analysis presented in Section 9.3 can be used. However, if $a > h_f$ then go to Step 5.

5. Hypothetically split the T–beam section into two beam sections (Figure 9.9). The first beam consists of the concrete in the flange overhang plus a portion of tension steel bars. The second beam consists of the concrete in the web plus the remainder of tension steel bars.

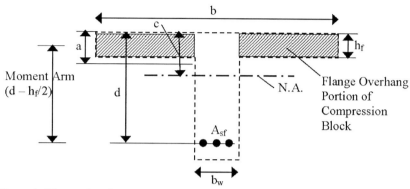

Beam 1: Flange Overhang Portion of Compression Block and Corresponding Steel Area

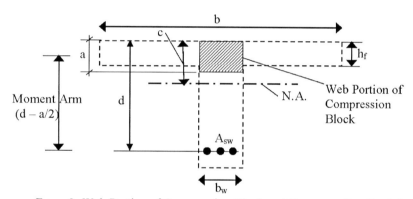

Beam 2: Web Portion of Compression Block and Corresponding Steel Area

Figure 9.9: Conjectural division of T–beam with compression block extending to the web.

6. Determine the value of A_{sf}, the corresponding steel in Beam 1 (Figure 9.9). A_{sf} is required to generate a tension force to counteract the compression force generated by the concrete of the flange overhang. As such:

$$A_{sf}{\cdot}f_y = 0.85f_c'{\cdot}(b - b_w){\cdot}h_f$$

or

$$A_{sf} = [0.85f_c'{\cdot}(b - b_w) {\cdot}h_f]/f_y \tag{9.3}$$

7. Determine the value of A_{sw}, the corresponding steel in Beam 2 (Figure 9.9). A_{sw} is required to generate a tension force to counteract the compression force generated by the concrete of the web. As such:

$$A_{sw}{\cdot}f_y = 0.85f_c'{\cdot}a{\cdot}b_w$$

or

$$A_{sw} = [0.85f_c'{\cdot}a{\cdot}b_w]/f_y \tag{9.4}$$

However, since "a" is not known at this stage, the following procedure may be used for calculation of A_{sw}:

$$A_s = A_{sf} + A_{sw} \implies A_{sw} = A_s - A_{sf}$$

8.   Determine an updated value of "a" utilizing Beam 2 (Figure 9.9) and Equation 9.8:

$$a = A_{sw} \cdot f_y / (0.85 f_c' \cdot b_w)$$

9.   Determine the value of tensile strain in steel (ε_t) similar to the procedure followed for rectangular beams utilizing the following equation:

$$\varepsilon_t = (0.003/c) \times (d - c) \qquad \text{Where } c = a/\beta_1$$

10.   If $\varepsilon_t \geq 0.005$, then $\phi = 0.9$. Alternatively, based on the value of ε_t, determine the value of capacity reduction factor (ϕ) using the following equation or Table A.8:

$$\phi = 0.48 + 83\varepsilon_t$$

11.   Determine the values of the nominal bending moment capacities (M_{nf} and M_{nw}) for Beams 1 and 2 as follows:

$M_{nf} = A_{sf} \cdot f_y \cdot (d - h_f/2)$
Where $d - h_f/2$ is the moment arm for Beam 1 (Figure 9.9)

$M_{nw} = A_{sw} \cdot f_y \cdot (d - a/2)$
Where $d - a/2$ is the moment arm for Beam 2 (Figure 9.9)

$M_n = M_{nf} + M_{nw}$

12.   Compare the value of A_s with minimum steel area dictated by ACI Code:

$$A_s \geq \rho_{min} \cdot b_w \cdot d$$

13.   Compare the value of ϕM_n with M_u to determine if the beam section is adequate:

$\phi M_n \geq M_u$ \qquad O.K. \quad T–beam is adequate

$\phi M_n < M_u$ \qquad N.G. \quad T–beam is in adequate

Example 9.3

Investigate the adequacy of the precast T–beam section shown in the following diagram provided that $f_c' = 5{,}000$ psi, $f_y = 60{,}000$ psi, and $M_u = 680$ ft.k.

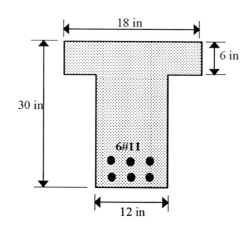

Solution

Check that the beam section dimensions are within code limits for isolated T–beams according to the following equations introduced in Section 9.2:

$h_f \geq b_w/2$ and $b \leq 4b_w$
$h_f = 6$ in $\geq b_w/2 = 12/2 = 6$ in ====> O.K.
$b = 18$ in $\leq 4b_w = 48$ in ====> O.K.
====> Section dimensions are within code limits

Determine the value of "a" assuming that the compression block is within the flange:

$a = A_s \cdot f_y/(0.85f'_c \cdot b) = 9.36 \times 60/(0.85 \times 5 \times 18) = 7.35$ in $> h_f = 6$ in   N.G.
Where: $A_s = 6\#11$ bars $= 9.36$ in^2 (Table A.5)

Since "a" is greater than h_f, then the procedure outlined in Section 9.4 needs to be followed:

$A_{sf} = [0.85f'_c \cdot (b - b_w) \cdot h_f]/f_y = [0.85 \times 5,000 \times (18 - 12) \times 6]/60,000 = 2.55$ in^2

====> $A_{sw} = A_s - A_{sf} = 9.36 - 2.55 = 6.81$ in^2

$a = A_{sw} \cdot f_y/(0.85f'_c \cdot b_w) = 6.81 \times 60,000/(0.85 \times 5,000 \times 12) = 8.01$ in

Determine the value of ε_t to verify that the beam is within code limits:

$\varepsilon_t = 0.003(d - c)/c = 0.003 \times (26 - 10.01)/10.01 = 0.0048$

Where $c = a/\beta_1 = 8.01/0.80 = 10.01$ in   and $d = 30 - 4 = 26$ in

Since $\varepsilon_t > 0.004$ ====> Beam is within code limits O.K.

$\phi = 0.48 + 83\varepsilon_t = 0.48 + 83 \times 0.0048 = 0.878$

$\rho_w = A_s/(b_w \cdot d) = 9.37/(12 \times 26) = 0.03 = 3\% >> 0.35\% = \rho_{min}$ from Table A.9

====> Section is within code limits

Determine the value of M_n:

$M_n = M_{nf} + M_{nw} = A_{sf} \cdot f_y \cdot (d - h_f/2) + A_{sw} \cdot f_y \cdot (d - a/2)$

$M_n = 2.55 \times 60 \times (26 - 6/2) + 6.81 \times 60 \times (26 - 8.01/2) = 12,500$ k.in $= 1,042$ k.ft

Determine the value of ϕM_n and compare it with M_u to verify that the beam can support the ultimate design bending moment acting on it:

$\phi M_n = 0.878 \times 1,042 = 915$ k.ft $> M_u = 680$ k.ft ====> O.K. Beam is satisfactory

9.5 Another Analysis Procedure for T–Beams under Positive Bending Moments

Instead of hypothetically dividing the T–beam with the compression block extending into the web, one may handle the compression block as a singular area. For this, the moment arm for the beam needs to be mathematically determined. Figure 9.10 graphically illustrates this alternative method.

In Figure 9.10, "z" is the moment arm between the centroid of steel bars and the centroid of concrete compression block. It can be expressed as follows:

$$z = d - [(b - b_w) \cdot h_f^2/2 + b_w \cdot a^2/2]/A_c \qquad (9.5)$$

Where A_c is the area of compression block determined as follows:

$$A_c = A_s \cdot f_y / 0.85 f_c' \qquad (9.6)$$

And where "a" is the depth of compression block determined as:

$$a = [A_c - h_f \cdot (b - b_w)]/b_w \qquad (9.7)$$

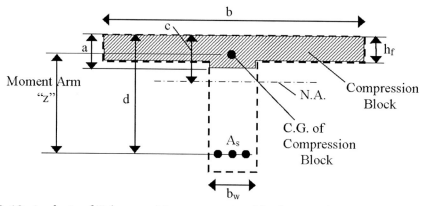

Figure 9.10: Analysis of T–beam with compression block extending to the web using the integral method.

To follow this alternative analysis, steps 1 through 4 are identical to the steps in Section 9.4. If it is determined that $a > h_f$ in step 4, the following steps may be followed to complete the solution and determine ϕM_n for the T–beam:

1. Determine the area of compression block using Equation 9.6.

2. Determine the value of "a" using Equation 9.7.

3. Determine the value of "z" utilizing Equation 9.5.

4. Determine the value of tensile strain in steel (ε_t) similar to the procedure followed for rectangular beams utilizing the following equation:

    $\varepsilon_t = (0.003/c) \times (d - c)$ \qquad Where $c = a/\beta_1$

5. Based on the value of ε_t, determine the value of capacity reduction factor (ϕ) using the following equation or Table A.8:

    $\phi = 0.48 + 83\varepsilon_t \leq 0.9$

6. Compare the value of A_s with minimum steel area dictated by ACI Code:

    $A_s \geq \rho_{min} \cdot b_w \cdot d$

7. Determine the value of ϕM_n using the following equation:

    $$M_n = A_s \cdot f_y \cdot \{d - [(b - b_w) \cdot h_f^2/2 + b_w \cdot a^2/2]/A_c\} = A_s \cdot f_y \cdot z \qquad (9.8)$$

8. Compare the value of ϕM_n with M_u to determine if the beam section is adequate:

    $\phi M_n \geq M_u$ \qquad O.K. \quad T–beam is adequate

    $\phi M_n < M_u$ \qquad N.G. \quad T–beam is inadequate

Example 9.4

Redo example 9.3 utilizing the alternative method outlined in Section 9.6, and compare the results.

Solution

In this solution, the pertinent information included in Example 9.3 will not be duplicated. Only the steps different from Example 9.3 will be listed.

Determine the value of A_c = the area of compression block using Equation 9.7:

$A_c = 9.37 \times 60/(0.85 \times 5) = 132.1$ in²

Determine the value of "a" = the depth of compression block using Equation 9.6:

$a = [132.1 - 6 \times (18 - 12)]/12 = 8.01$ in

Determine the value of "z" = moment arm for bending in T–beam using Equation 9.5:

$z = 26 - [(18 - 12) \times (6)^2/2 + 12 \times (8.01)^2/2]/132.1 = 22.3$ in

Determine the value of M_n using Equation 9.8:

$M_n = 9.36 \times 60 \times 22.3 = 12,524$ k.in $= 1,044$ k.ft

Determine the value of ϕM_n:

$\phi M_n = 0.879 \times 1,044 = 916$ k.ft $> M_u = 680$ k.ft ===> O.K.

9.6 T–Beams under Negative Bending Moments

Negative bending moments cause tension in the upper portion of the beam and compression in the lower portion of the beam. Therefore, T–beams under negative moment act as rectangular beams since the compression zone is rectangular. An equivalent rectangular beam may be utilized using b_w as the beam's width. The difference in the analysis of T–beams subjected to negative moment compared with rectangular beam subjected to negative moment is that the tension steel mat be distributed within the flange width (b), rather than the web width (b_w). All other equations, code standards, and limitations applied to rectangular beams apply equally to T–beams under negative bending moment. Such equations and standards include the limits on ε_t and ρ using b_w as the beam's width. Figure 9.11 illustrates the concepts of T–beam subjected to negative moment.

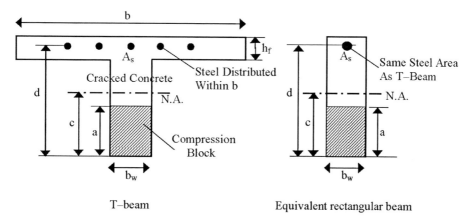

T–beam   Equivalent rectangular beam

Figure 9.11: Analysis of T–beam under negative bending moment.

9.7 Design of T-Beams

The following information is required for beam design:

1. Beam or floor system configuration including span length, support conditions (simply supported, cantilever, etc), and exposure conditions (interior, exterior, earth, etc). Furthermore, prior to T–beam design, the design for concrete slab is assumed completed. Thus, flange thickness (h_f) and possible spacing of T–beams are known.

2. Loads on the beam or floor including dead load, live load, seismic load, etc.

3. Design limitations including maximum allowed beam or floor depth (h) and recommended spacing of T-beams.

The design or magnified bending moment, M_u, which can be determined as shown in Chapter 5 based on load factors specified within ASCE 7, is the most significant design input. An estimate of beam's self weight is needed to compute M_u. Self weight of the concrete slab is a part of the floor system dead load. The stem of T–beam's weight may be calculated as shown for rectangular beams assuming a width of b_w and a total depth of ($h - h_f$). Thus one needs an initial estimate for b_w and h for T–beam design. Other code imposed limitations on beam cross–sectional geometry or reinforcement may also apply.

The flexural design of a reinforced concrete T–beam cross–section yields a detailed beam section with the following information:

a. Material properties, mainly, f'_c and f_y.

b. Beam or ribbed floor depth = h.

c. T–beam spacing = s (if not given as part of slab design or it requires modification); and other floor design details.

d. T–beam web width = b_w.

e. Concrete cover over steel bars.

f. Steel bar area, A_s, and distribution (#, size and spacing of bars).

g. Other design considerations required by code.

The design output for a reinforced concrete T–beam includes all the above parameters. Nevertheless, one needs to realize that design is an iterative process based on assumptions and verifications followed by re-assumptions and re-verifications until a satisfactory design is reached. The design equations are included in the verification of assumptions. Attention in regard to making reasonable initial assumptions can greatly facilitate the design process. Experience is key in yielding an efficient design that facilitates construction.

As explained in rectangular beam design, there is only one significant irredundant design equation based on equilibrium. Thus, six of the above unknowns need to be determined via other methods. With this consideration, the design procedure for T–beams can be outlined as follows:

1. Specified material strength f'_c and f_y can be initially reasonably assumed based on availability.

2. T–beam spacing (s) is generally predetermined based on the concrete slab design. The concrete slab spans between T–beams where the span is equal to the spacing.

3. Beam or ribbed floor depth (h) may be initially assumed using Table 6.3 or Table A.12.

4. T–beam web width (b_w) in essence serves one function only: to contain the reinforcing steel bars. Thus, b_w may be assumed initially utilizing one's experience and modified later after computing the value of A_s and detailing the steel bars.

5. Concrete cover over steel bars is selected based on exposure conditions as discussed in Section 6.5.1 of Chapter 6.

As such, the only remaining unknown is the area of tension steel bars A_s that can be computed determined as follows:

6. Steel bar area (A_s) may then be estimated using the following equation:

$$A_s = (M_u/\phi)/(0.9d \times f_y) \qquad (9.9)$$

Where the following approximations are used in Equation 9.9:

ϕ = capacity reduction factor = 0.9 as an initial assumption.

d = h – 2.5 to 3 inch for one layer of bars.

d = h – 4 to 5 inch for two layers of bars.

0.9d = an initial estimate of the T–beam moment arm (distance between the centroid of steel bars and the centroid of compression block).

With A_s evaluated, one can determine if ϕ = 0.9 is appropriate and estimate if one or two layers of bars are needed.

7. The reduced bending moment capacity ϕM_n is determined for the initial design of T–beam achieved in Step 6.

8. If ϕM_n is greater than M_u with a comfortable margin of safety acceptable to the designer, then the T–beam design achieved in steps 6 and 7 is detailed based on the ACI 318 Code. Potentially, the initial design may require minor adjustments to achieve construction suitable parameters.

9. If ϕM_n is not greater than M_u with a comfortable margin of safety acceptable to the designer, then changes or modifications to the initial T–beam design are made and iteration is performed.

10. The final distribution is then detailed (#, size and spacing of bars), stirrups, b_w, A_s well as other design considerations required by code.

11. Analysis of the final design is conducted to confirm all the design input and output.

The design of precast or isolated T–beams follows the same procedure outlined above, but takes into account the special design inputs and code limitations specific for these types of beams.

Example 9.5

Design the T–beams or ribs for the floor system shown for use in a parking structure with slab thickness = 6 in, proposed slab span = 8 ft, DL = self weight + 50 psf, LL = 250 psf, floor system span = 60 ft, and f'_c = 4,000 psi and fy = 60,000 psi.

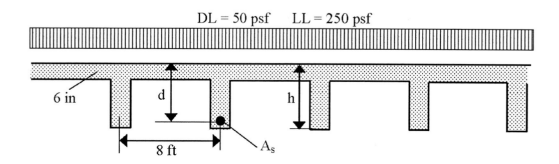

Solution

Determine the value of M_u:

Self weight of slab and T–beam web:

Self Weight $=$ weight of slab within Tributary width + web weight

$=$ [Tributary width × slab thickness + (T–beam height – slab thickness) × web width] x concrete unit weight

$=$ [Tributary width × h_f + $(h - h_f) × b_w$] × γ_c

$= (6/12)×8×150 + [(48 - 6)/12]×(18/12) ×150 = 1,388$ lb/ft

Where:

γ_c = concrete unit weight = 150 pcf
h = 48 in as an initial assumption using Tables 6.3 or A.12, and
b_w = 18 in is utilized as an initial assumption.

Superimposed load $=$ (Superimposed DL + LL on the slab) x Tributary width
$= 50 × 8 + 250 × 8 = 400$ lb/ft DL + 2,000 lb/ft LL

$W_u = 1.2DL + 1.6LL = 1.2×(1,388 + 400) + 1.6×(2,000) ≈ 5,350$ lb/ft = 5.35 k/ft

$M_u = W_u ×$ (T–beam span)^2/8 assuming a simply supported beam set–up.

$M_u = 5.35 × (60)^2/8 = 2,408$ k.ft

Estimate the value of steel bar area A_s as shown in Equation 9.19:

$A_s = (M_u/\phi)/(0.9d×f_y) = (2,408×12/0.9)/(0.9×43×60) = 13.8$ in^2

Where:

$d = h - 5$ in = 43 in assuming two layers of steel bars

In the opinion of the author, the area of steel bars is excessive. Revisiting the initial assumptions is warranted.

The design of the floor system can be modified as follows:

Tributary width = 8 ft, beam height, h = 56 in = 4.5 ft and beam effective depth = 51 in Redoing steps 1 through 3 above, the following results can be obtained:

$M_u ≈ 1,990$ k.ft   As ≈ 9.6 in^2

Analysis of the T–beam obtained after the second iteration using Section 9.3 reveals the following information:

$b = 72$ in, $a = 2.35$in, $\phi = 0.9$ and $M_n = 2,392$ k.ft $\Longrightarrow \phi M_n = 2,152$ k.ft $> M_u$ O.K.

With the above satisfactory results, the proposed T – beam can be detailed as shown below:

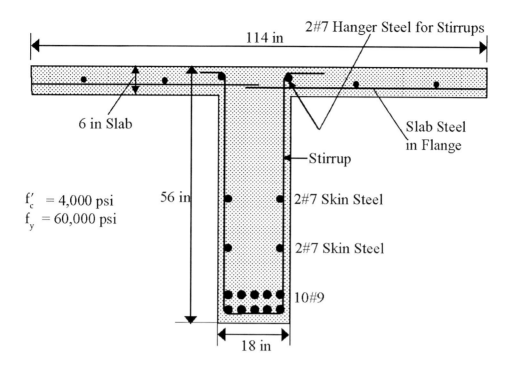

Analysis of the final design detail should follow to confirm that all design aspects were within the ACI 318 Code limitations and requirements.

Example 9.6

Design the T–beams or ribs for the floor system shown for use in a parking structure with slab thickness = 150 mm, proposed slab span = 2.5 m, DL = self weight + 2.5 kPa, LL = 12 kPa, floor system span = 18 m, and f'_c = 30 MPa and fy = 420 MPa.

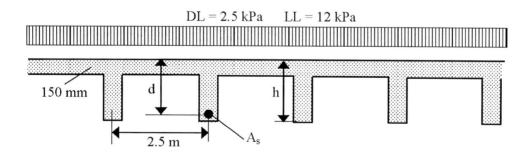

Solution

Determine the value of M_u:

Self weight of slab and T–beam web:

Self Weight = weight of slab within Tributary width + web weight

= [Tributary width × slab thickness + (T–beam height – slab thickness) × web width] x concrete unit weight

= [Tributary width × h_f + (h – h_f) × b_w] × γ_c

= (0.15)×2.5×23 + [(1.5 – 0.15)]×(0.5)×23 = 24 kN/m

Where:

γ_c = concrete unit weight = 23 kN/m³
h = 1.5 m as an initial assumption using Tables 6.3 or A.12, and
b_w = 0.5 m is utilized as an initial assumption.

Superimposed load = (Superimposed DL + LL on the slab) x Tributary width
= 2.5 × 2.5 + 12 × 2.5 = 6.25 kN/m DL + 30 kN/m LL

W_u = 1.2DL + 1.6LL = 1.2×(24 + 6.25) + 1.6×(30) ≈ 84 kN/m
M_u = W_u × (T–beam span)²/8 assuming a simply supported beam set–up.
M_u = 84 × (18)²/8 = 3,400 kN.m

Estimate the value of steel bar area A_s as shown in Equation 9.19:

A_s = (M_u/ϕ)/(0.9d×f_y) = (3,400/0.9)×10⁶/(0.9×1,375×420) = 7,300 mm²
Where:

d = h – 125 mm = 1,375 mm assuming two layers of steel bars

With the above satisfactory results, the proposed T – beam can be detailed as shown below:

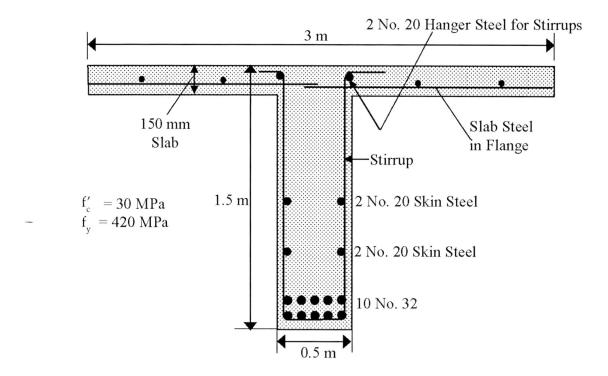

9.8 Concluding Remarks T–Beams

As shown in this chapter, T–beams are classified into two major categories:

1.   T–beams that can be treated as rectangular beams where the compression block is within the flange. The analysis and design in this case can be simplified, and

2.   T–beams that require treatment as T–shaped beams where the compression block extends into the web. Thus, the analysis and design of such T–beams is detailed.

For the above cases, the shape of two beams may be identical but the analysis and design are different depending on the size and configuration of the compression block. As such, one may conclude that:

The shape of the compression block area is what dictates and determines the analysis procedure of T–beams, not the shape of the beam. This is because the portion of the beam below the neutral axis is essentially cracked and largely inefficient in resisting bending moments; subsequently, it is neglected in the analysis and design. The procedure presented in Section 9.5, where the centroid of the compression block is utilized, can apply to any beam shape particularly if the compression block area is not rectangular.

Based on the above discussion, a host variety of beams with various shapes may be analyzed and designed as presented in this chapter as rectangular or T–shaped beams depending on the compression block. Figure 9.12 presents examples of these shapes.

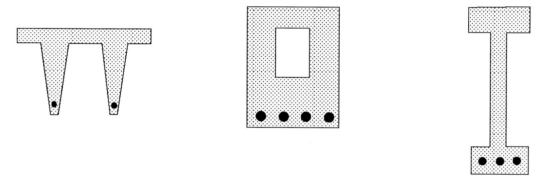

Figure 9.12: Examples of beam shapes that can be treated as T–beams.

Test your knowledge

1. What are the assumptions used in the analysis of reinforced concrete T–shaped beams in bending?

2. Define the spacing of T-beams in a floor system.

3. Define the tributary width of a T-beam in a slab or floor system.

4. Define the tributary area of a T-beam in a slab or floor system.

5. Define the effective flange width of a T–beam.

6. What dictates if a T–beam needs to be analyzed as a rectangular beam or as a T– beam?

7. Why is minimum steel reinforcement in T–beams provided for the web only and not the flange?

8. What purpose does the web serve in T–beams and what is the main factor for selection of b_w?

9. List the advantages versus disadvantages for using T- beams with close spacing.

10. Can one design a floor system with T-beam depth equal to slab thickness? Explain?

11. List the possible methods utilized of increasing the bending moment capacity ϕM_n of T–beam in a ribbed floor system.

12. Investigate the firms in your city that supply forms for ribbed concrete slabs and determine the standard spacings of T–beam webs and standard thickness of ribbed floors.

Perform the analysis for bending moment and compute the value of ϕM_n for the T–beams listed in Problems 13 to 24. One need not apply the ACI 318 code provisions for isolated T– beams. Use f'_c = 4,000 psi (28 MPa) and f_y = 60,000 psi (420 MPa).

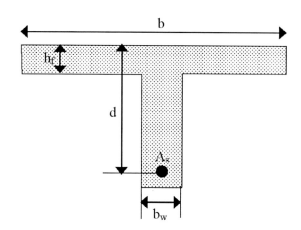

Problem	b(in)	b_w(in)	h_f(in)	A_s(in²)	d(in)
13	24	10	2.5	3#6	21
14	24	12	2.5	3#11	21
15	40	14	3	4#9	27
16	40	14	3	8#9	25
17	56	14	3	8#8	27
18	56	18	3	12#9	27
19	72	18	5	4#14	44
20	72	18	5	8#11	42
21	96	24	6	6#11	56
22	96	24	6	12#11	54
23	2 m	400 mm	100 mm	8 No. 25	1.2 m
24	3 m	0.5 m	120 mm	10 No. 28	1.5 m

Analyze the isolated T–beams shown below and determine the value of ϕM_n assuming that f'_c = 5,000 psi (35 MPa) and f_y = 60,000 psi (420 MPa).

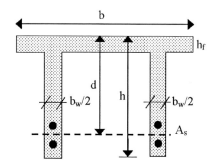

25. b = 36 in, h_f = 3.5 in, d = 25 in,
        b_w = 10 in, As = 4#10

26. b = 600 mm, h_f = 60 mm, d = 400 mm,
        b_w = 150 mm, As = 4 No. 25

Analyze the beams shown below and determine the value of ϕM_n assuming that f'_c = 4,000 psi (28 MPa) and f_y = 60,000 psi (420 MPa).

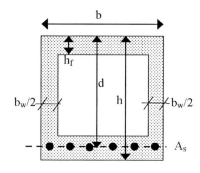

26.   b = 32 in, h_f = 6 in, d = 33 in, b_w = 8 in,
          A_s = 5#11

27.   b = 600 mm, h_f = 120 mm, d = 550 mm,
          b_w = 150 mm, A_s = 5 No. 28

Analyze the beams shown below and determine the value of ϕM_n assuming that f'_c = 4,000 psi and f_y = 60,000 psi.

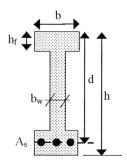

28.   b = 15 in, h_f = 5 in, d = 15 in, b_w = 5 in,
          A_s = 3#8

29.   b = 18 in, h_f = 6 in, d = 21 in, b_w = 6 in,
          A_s = 3#10

Analyze the beams shown below and determine the value of ϕM_n assuming that f'_c = 3,000 psi and f_y = 60,000 psi. Use ACI Code provisions for L-beams.

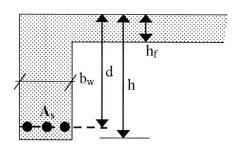

30. b_w = 12in, h_f = 4in, d = 17in, A_s = 3#7,
          Span = 20 ft

31. b_w = 14 in, h_f = 5 in, d = 2 2in, A_s = 3#9,
          Span = 25 ft

Analyze the beams shown below and determine the value of ϕM_n assuming that f'_c = 3,000 psi (20 MPa) and f_y = 60,000 psi (420 MPa) for the following problems.

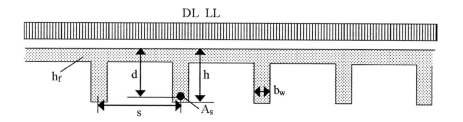

32. h_f = 4 in, h = 18 in, b_w = 10 in, A_s = 4#7, s = 6 ft, T–beam span = 20 ft, DL (excluding self wt) = 30 psf, LL = 80 psf

33. h_f = 3 in, h = 15 in, b_w = 8 in, A_s = 4#6, s = 5 ft, T–beam span = 15 ft, DL (excluding self wt) = 15 psf, LL = 50 psf

33. h_f = 5 in, h = 20 in, b_w = 12 in, A_s = 4#8, s = 8 ft, T–beam span = 25 ft, DL (excluding self wt) = 25 psf, LL = 75 psf

34. h_f = 6 in, h = 24 in, b_w = 14 in, A_s = 4#7, s = 8 ft, T–beam span = 28 ft, DL (excluding self wt) = 30 psf, LL = 120 psf

35. h_f = 6 in, h = 30 in, b_w = 18 in, A_s = 4#9, s = 10 ft, T–beam span = 30 ft, DL (excluding self wt) = 20 psf, LL = 120 psf

36. h_f = 150 mm, h = 800 mm, b_w = 500 mm, A_s = 4 No. 35, s = 3.5 m, T–beam span =11 m, DL (excluding self wt) = 1.5 kN/m^2, LL = 11.5 kN/m^2

37. h_f = 6 in, h = 6 in, b_w = 24 in, A_s = 4#9, s = 6 ft, T–beam span = 20 ft, DL (excluding self wt) = 15 psf, LL = 50 psf

38. h_f = 5 in, h = 5 in, b_w = 18 in, A_s = 3#10, s = 8 ft, T–beam span = 22 ft, DL (excluding self wt) = 10 psf, LL = 40 psf

For the following problems, design the T-beams as part of the floor system shown below determining the values of h, d and A_s as well as steel bar distribution. Use f'_c = 5,000 psi (35 MPa) and f_y = 60,000 psi (420 MPa).

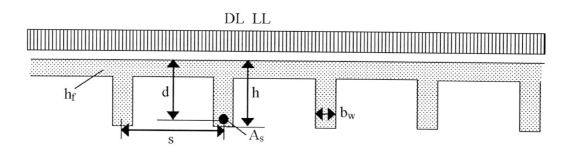

39. h_f = 4 in, T–beam span = 25 ft, DL (excluding self wt) = 30 psf, LL = 75 psf

40. h_f = 3 in, T–beam span = 18 ft, DL (excluding self wt) = 15 psf, LL = 50 psf

41. h_f = 5 in, T–beam span = 28 ft, DL (excluding self wt) = 35 psf, LL = 100 psf

42. h_f = 6 in, T–beam span = 30 ft, DL (excluding self wt) = 30 psf, LL = 120 psf

43. h_f = 6 in, T–beam span = 35 ft, DL (excluding self wt) = 30 psf, LL = 240 psf

44. h_f = 150 mm, T–beam span = 10 m, DL (excluding self wt) = 2 kN/m^2, LL = 6 kN/m^2

45. h_f = 150 mm, T–beam span = 12 m, DL(excluding self wt)= 2 kN/m^2, LL = 12 kN/m^2

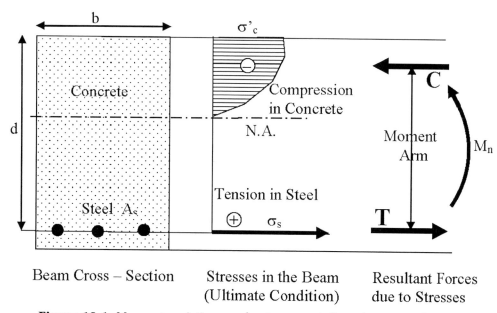

Chapter 10

ANALYSIS AND DESIGN OF DOUBLY REINFORCED CONCRETE BEAMS

10.1 The Need for Doubly Reinforced Concrete Beams

Bending resistance (M_n) of a typical reinforced concrete beams is provided by a couple of forces separated by a moment arm. These forces are tension in steel and compression in concrete. The bending moment resistance is the product of ultimate tension in steel (T) or ultimate compression in concrete (C) and the moment arm which is the perpendicular distance between the two forces (M_n = T or C x Moment Arm) as shown in Figure 10.1.

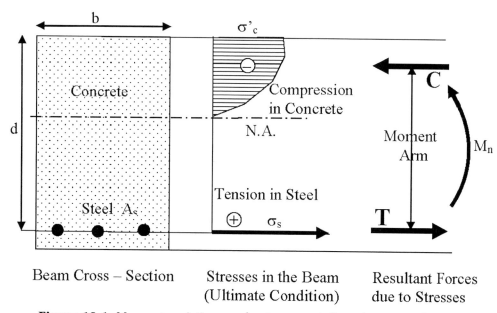

Figure 10.1: Moment resisting mechanism or reinforced concrete beams.

It is common for reinforced concrete beams to have constrained dimensions due to space restriction, such as floor height, or architectural features. Nevertheless, these beams must still be designed to resist high bending moments. To sustain such high usable bending moment capacity ($\phi M_n \geq M_u$), the designer may consider the following:

1. <u>Required Increase in Forces C and T to Increase M_n</u>: Since beam size is constrained, the bending moment arm is notably limited as well. With M_n = T or C × moment arm and that T = C for force equilibrium, the only viable way to increase the bending moment capacity is by increasing T and C. One method that may be used to increase beam bending capacity is to use a wider beam (increase b) which has a larger concrete area and can accommodate an increased steel area.

2. <u>Increase in Tension Steel Area A_s</u>: One may consider increasing the area of tension steel reinforcement A_s to increase T = $A_s.f_y$ the resultant tension force in steel. As T increases, a corresponding increase in C is required to maintain force equilibrium. Consequently, a larger area of concrete is utilized to generate a compression force, C, to counteract the tensile force in steel, T. However, there is a limit on increasing A_s where $\rho = A_s/bd$ must remain less than ρ_{max} or ρ_{maxp} dictated by ACI 318 Code to maintain a gradual or ductile mode of failure in bending. This ACI 318 Code limitation is imposed to ensure that available steel strength is less than available concrete strength. Thus, the failure mode remains gradual as triggered by steel yielding.

3. <u>Maximum Bending Moment Capacity of a Singly Reinforced Concrete Beam</u>: With tension reinforcement only, beam maximum bending capacity is determined as follows:

$$(M_n)_{maxsingly} = (R_n)_{max} \times bd^2$$

Where:

$(M_n)_{maxsingly}$ = maximum bending moment capacity for a singly reinforced concrete beam (with tension reinforcement only).

$(R_n)_{max}$ = design aid factor corresponding to the maximum code allowed reinforcement ratio for singly reinforced beams ρ_{max} or ρ_{maxp} (Chapter 7 and Appendix A).

b and d are the width and effective depth of the beam.

4. <u>Sufficiency of Singly Reinforced Beams</u>: One needs to check if the magnified applied bending moment, M_u, can be supported by a singly reinforced concrete designed within code limitation ($\rho < \rho_{max}$ or ρ_{maxp}). With that, there is no need to use a doubly reinforced concrete beam.

5. <u>Increasing Concrete Compressive Strength f'_c and/or Increasing Steel Yield Strength f_y</u>: If the discussion of 2 – 4 above results in $\phi M_n < M_u$, then one may consider increasing material strength of concrete (f'_c) to increase the resultant compression force C and/or steel (f_y) to increase the resultant tension force T to increase M_n of the beam. Force equilibrium (C = T) must be preserved with any material property change. Also, and reinforcement ratio, ρ, must remain within code imposed limits.

6. <u>Increasing the Area of Tension Steel and Using Compression Steel</u>: Further increases of C and T to increase the beam bending capacity can be accomplished by additional steel. Extra steel can be used to supplement the tension force (T) as well as the compression force (C) in the beam. Increased area of tension steel beyond ρ_{max} or ρ_{maxp} must be accompanied by added steel in the compression area of concrete to assist the concrete in resisting compression. These steel bars within the compression area of concrete will be subjected to compression and are referred to as compression steel. With that, equilibrium is maintained and the beam ductile failure mode is preserved as per ACI 318 Code. A reinforced concrete beam with both tension and compression steel reinforcement is referred to as doubly reinforced concrete beam (Figure 10.2) as opposed to a singly reinforced concrete beam containing tension steel only.

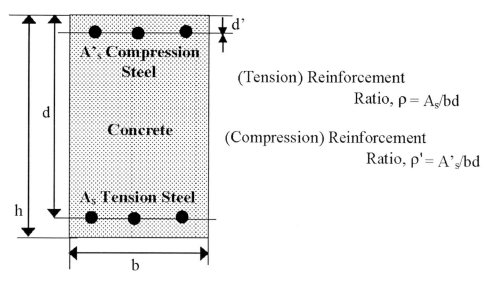

(Tension) Reinforcement
Ratio, $\rho = A_s/bd$

(Compression) Reinforcement
Ratio, $\rho' = A'_s/bd$

Figure 10.2: Schematics and notation for doubly reinforced concrete beams.

10.2 Effectiveness of Compression Steel in Increasing Bending Moment Capacity

There are two cases where compression steel is not significantly effective in increasing the bending moment capacity of reinforced concrete beams:

1. Doubly reinforced concrete beams with $\rho < \rho_{max}$ or ρ_{maxp}: Using tension reinforcement less than the maximum allowed by ACI 318 Code for singly reinforced beams is indicative of less than full utilization of beam capacity. This is due to the lack of available tension force. With force equilibrium (C = T), the available concrete compression area is not fully utilized as it can accommodate up to $\rho = \rho_{max}$. Since concrete is more than adequate for compression, extra compression capacity due to compression steel is neither useful nor necessary. Any increase of available compression force capacity beyond the value of tension force, T, is left unused. Typically, doubly reinforced beams with $\rho < \rho_{maxp}$ have a marginal increase in bending moment capacity (ϕM_n) due to the presence of compression steel and may be analyzed as singly reinforced.

2. Compression steel used as hanger bars: It is common that steel bars are used in the concrete compression area to tie the stirrups (shear reinforcement) where they are called hanger bars. In addition to tying the stirrups, they reduce the compression bearing stress in concrete (Figure 10.3). Tension in the stirrups causes a bearing force on the concrete. Bearing is concentrated compression. With the hanger bars situated directly at the hook, stirrup tension force is transmitted to the hanger bar which in turn transmits it to the concrete. Thus, the bearing area responsible for resisting stirrup tension force us increased from the underbelly area of stirrup hook to the area below hanger bar. Such dramatic increase in bearing area results in a significant decrease of bearing stress, thereby reducing the potential of concrete distress due to bearing. The area of hanger bars is typically 10 to 20% of the tension steel area. Their effectiveness in increasing bending moment capacity, M_n, is limited and they are often neglected in calculations where the beam is assumed to be singly reinforced.

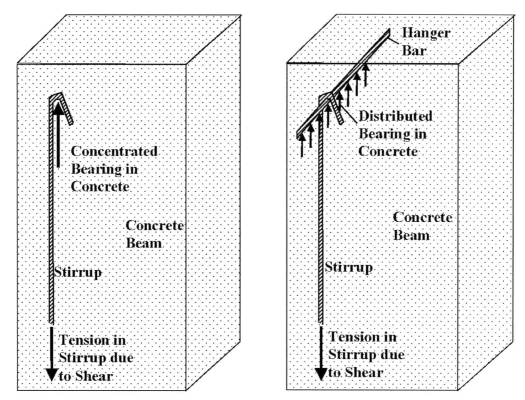

Figure 10.3: Hanger bar effect in reducing concrete bearing stress due to stirrup tension.

10.3 Analysis of Doubly Reinforced Concrete Beams

The goal of reinforced concrete beam analysis is to determine the value of ϕM_n and to compare this value to M_u to ascertain if the beam is adequate for moment resistance. It is generally assumed that analysis is for beams with known dimensions (b, d and d'), reinforcement (A_s and A'_s), material properties (f'_c and f_y) as well as loading conditions (loads, span and support types). Such analysis procedure is discussed as follows:

1.   Initially the value of magnified design bending moment acting on the beam (M_u) including self weight needs to be determined.

2.   Further, it is required to check if analysis as singly reinforced concrete is sufficient by checking if one of the following conditions is met:

     a. $A'_s < 0.20A_s$ Compression steel used as hanger bars

     b. $\rho = A_s/bd < \rho_{maxp}$ Compression steel is not needed since concrete is
                            adequate for compression

3.   If neither of the conditions of step 2 is met, then the beams needs to be analyzed as doubly reinforced.

4.   For analysis simplification, the doubly reinforced beam section is hypothetically divided into two overlapping beam sections (Figure 10.4). The first section (Beam 1) contains the concrete and a portion of tension steel A_{s1} adequate to generate a tension force to counteract the compression in concrete. The second beam (Beam 2) contains the remainder of tension steel (A_{s2}) as well as compression steel without concrete. A_{s2} is presumed adequate to generate a tension force to counteract the compression in compression steel A'_s. This is shown below in Figure 10.4.

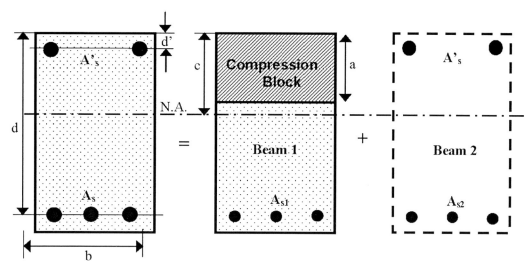

Figure 10.4: Hypothetical division of doubly reinforced beams.

As the tension steel is split into two portions, one with beam 1 and the other with beam 2, one needs to remember that A_s is unchanged and can be represented as:

$$A_s = A_{s1} + A_{s2}$$

Similar to singly reinforced beams, ultimate or failure condition of doubly reinforced concrete beams is defined by concrete reaching its crushing strain ($\varepsilon_c = 0.003$). Furthermore, ACI 318 Code dictates conditions upon the tensile strain in steel at ultimate condition so that beam failure is gradual. The strain in tension steel (ε_t) upon failure must be greater than 0.004 (preferably greater than 0.005). Since $\varepsilon_t = 0.004$ is generally greater $\varepsilon_y = f_y/E_s$, then, the stress in tension steel upon failure is f_y. This provision for tension steel strain is general for all tension controlled bending members. However, there are no code requirements or conditions regarding the strain or stress in compression steel at ultimate or failure condition of doubly reinforced concrete beams. Consequently, one can draft the following equilibrium equations for Beam1 and Beam 2 in Figure 10.4:

Equilibrium Equation of Beam 1 $A_{s1} \cdot f_y = 0.85 f'_c \cdot a \cdot b$

Equilibrium Equation of Beam 2 $A_{s2} \cdot f_y = A'_s \cdot f'_s$

Where the stress in A_{s2} is also f_y since it is a part of tension steel and f'_s is the stress in compression steel.

5.   To determine the value of strain and stress in compression steel at failure or ultimate condition, one needs to use strain analysis. Beam 2 does not contain concrete; it cannot be used to determine strain distribution in the section at failure. The backbone of such distribution is the crushing strain in concrete (0.003). Figure 10.5 shows the strains in Beam 1.

The strain distribution of Beam 1 (Figure 10.5) is overlapped or superimposed on Beam 2 to determine the strain and stress in compression steel as shown in Figure 10.6. The strain and stress in A_{s2} is the same as the strain and stress in A_{s1} shown in Figure 10.5 as they are parts of the tension steel in the beam A_s.

The strain and stress in compression steel A'_s are then determined based on Figure 10.6:

$$\varepsilon'_s = (0.003/c) \times (c - d')$$

$$f'_s = \varepsilon'_s \cdot E_s \le f_y \qquad\qquad (10.1)$$

Where:

E$_s$ = modulus of elasticity of steel, and

f$_y$ = yield strength of steel.

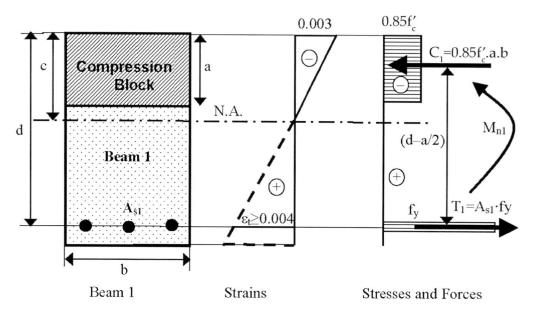

Figure 10.5: Strains stresses forces and bending moment capacity of Beam 1.

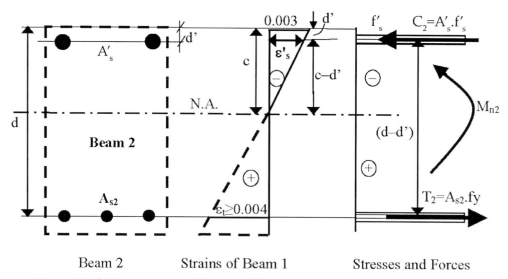

Figure 10.6: Strains stresses forces and bending moment capacity of beam 2.

Based on Figure 10.5, the bending moment capacity of Beam 1 is determined:

$$M_{n1} = A_{s1} \cdot f_y \cdot (d - a/2) = 0.85 f'_c \cdot a \cdot b \cdot (d - a/2)$$

The depth of Whitney's compression block (a) is determined based on Beam 1 as:

$$a = A_{s1} \cdot f_y / 0.85 f'_c \cdot b$$

The bending moment capacity of Beam 2 can be determined based on Figure 10.6:

$$M_{n2} = A_{s2} \cdot f_y \cdot (d - d') = A_s' \cdot f'_s \cdot (d - d')$$

6. As a first step in the analysis of a doubly reinforced concrete beam, the strain in compression steel is assumed to equal or exceed the yield strain of steel. As a result the stress in compression steel would be the yield stress (f_y) as shown below:

$$\text{Assume } \varepsilon'_s \geq \varepsilon_y = f_y/E_s \Longrightarrow f'_s = \varepsilon'_s \cdot E_s = f_y$$

Important Note: Due to concrete creep under compression (Chapter 2), strains increase significantly in the beam compression area including the strains in compression steel. Experiments have shown the strains in compression steel are especially high and the stress is consequently equal to yield stress (f_y) due to concrete creep. Creep essentially results in increase in strains and reduction of stresses or forces in concrete and, consequently, increase in strains and increase in stresses and forces in compression steel.

7. Consequently, utilizing the equilibrium equation of Beam 2, it can be concluded that:

$$A_{s2} \cdot f_y = A'_s \cdot f_y \Longrightarrow A_{s2} = A'_s \tag{10.2}$$

8. Thus, the values of A_{s1} and a can be determined as follows:

$$A_s = A_{s1} + A_{s2} \text{ and } A_{s2} = A'_s \Longrightarrow A_{s1} = A_s - A'_s$$

$$A_{s1} \cdot f_y = 0.85 f'_c \cdot a \cdot b \text{ and } A_{s1} = A_s - A'_s$$

$$\Longrightarrow a = (A_s - A'_s) \cdot f_y/(0.85 f'_c \cdot b) \tag{10.3}$$

9. The validity of the assumption made in step 6 ($f'_s = f_y$) is then checked. As $c = a/\beta_1$, Equation 10.1 may be rewritten as:

$$f'_s = (0.003/c) \times (c - d') \times E_s \leq f_y \tag{10.1}$$

10. If $f'_s = f_y$, then the assumption of step 6 is correct and one can proceed to calculate M_n for the whole beam; otherwise one needs to go to step 13:

$$M_n = 0.85 f'_c \cdot a \cdot b \cdot (d - a/2) + A_s' \cdot f_s' \cdot (d - d') \tag{10.4}$$

Beam 1 is treated similar to a comparable singly reinforced concrete beam subjected to all code limitations. To determine the value of ϕ, one needs to use the strain diagram of Beam 1. Tensile strain $\varepsilon_t = (0.003/c) \times (d - c)$ is first calculated. If $\varepsilon_t < 0.004$, then the beam is not within code and a more sophisticated analysis is warranted. If $\varepsilon_t \geq 0.004$, then Beam 1 conforms to ACI 318 Code. The following equation, from Chapter 5, can be used to determine the value of ϕ. As discussed earlier, it is recommended that ε_t exceeds 0.005 for better beam bending behavior and so that $\phi = 0.9$:

$$\phi = 0.48 + 83\varepsilon_t \leq 0.9 \tag{5.7}$$

11. In lieu of investigating strains to verify code compliance of Beam 1, one may check the area of tension steel A_s against the maximum area allowed for doubly reinforced concrete beams. Maximum allowed tension steel area in a doubly reinforced concrete beam may be determined as follows:

$$A_{smax,doubly} = A_{smaxbeam 1} + A_{smaxbeam 2} = A_{smaxsigly} + A_{s2}$$

$$A_{smax,doubly} = A_{smaxsigly} + A_{s2} = \rho_{maxsigly} \cdot b.d + A'_s \cdot f'_s/f_y$$

$$A_{smax,doubly} = 0.36 f'c \cdot \beta_1 \cdot b \cdot d/f_y + A'_s \cdot f'_s/f_y \tag{10.5}$$

$$A_{smax,doublypreferred} = 0.32 f'c \cdot \beta_1 \cdot b \cdot d/f_y + A'_s \cdot f'_s/f_y \tag{10.6}$$

Where $\rho_{max,singly}$ and $\rho_{max,singlypreferred}$ in Equations 10.5 and 10.6 are obtained from Chapter 6 for ε_t of 0.004 and 0.005, respectively.

12. Use the value of ϕ obtained in step 11 to determine the value of ϕM_n and draw one of the following conclusions:

If $\phi M_n \geq M_u$ ===> OK   Beam is satisfactory

If $\phi M_n < M_u$ ===> NG   Beam is inadequate

13. If the assumption in step 6 failed verification in step 9, then the stress in compression steel is less than the yield stress. The analysis of doubly reinforced concrete beam should then proceed based on the global equilibrium equation for the doubly reinforced beam:

Tension = Compression

or

Force in Tension Steel = Force in Concrete + Force in Compression Steel

$A_s \cdot f_y = 0.85 f'_c \cdot a \cdot b + A'_s \cdot f'_s$ ===> This equation may be rewritten as

$A_s \cdot f_y = 0.85 f'_c \cdot (\beta_1 c) \cdot b + A'_s \cdot (0.003/c) \cdot (c - d') \cdot E_s$ ===>

$$(0.85 f'_c \cdot b \cdot \beta_1) . c^2 + (0.003 A'_s \cdot E_s - A_s \cdot f_y) . c - 0.003 A'_s \cdot d' \cdot E_s = 0 \qquad \textbf{(10.7)}$$

Equation 10.7 is the global equilibrium equation for the doubly reinforced concrete beam. It is a quadratic equation with c as the unknown and can easily be solved to determine the value of c:

$c = \{A_s \cdot f_y - 0.003 A'_s \cdot E_s + [(0.003 A'_s \cdot E_s - A_s \cdot f_y)^2 + 4 \times 0.85 f'_c \cdot b \cdot \beta_1 \times 0.003 A'_s \cdot d' \cdot E_s]^{1/2}\}/(2 \times 0.85 f'_c \cdot b \cdot \beta_1)$

Simplifying the previous equation and assuming <u>units of kips and inches</u>:

$$c = \{A_s \cdot f_y - 87 A'_s + [(87 A'_s - A_s \cdot f_y)^2 + 296 f'_c \cdot A'_s \cdot b \cdot d' \cdot \beta_1]^{1/2}\}/(1.7 f'_c \cdot b \cdot \beta_1) \qquad \textbf{(10.8)}$$

14. After determining the value of c based on equation 10.8, the section compliance with ACI 318 Code is verified. The strain in tension steel is computed based on the strain diagram of bam 1 (Figure 10.5) and compared with code limits:

$\varepsilon_t = (0.003/c) \times (d - c)$

If $\varepsilon_t > 0.004$ (preferably > 0.005) ===> Section complies OK

If $\varepsilon_t < 0.004$ ===> Section does not comply NG

15. If the section does not comply with code limits, then further and more sophisticated analysis no covered in this textbook may be required.

16. If the beam complies with the code limits, then the stress in compression steel can be determined and the bending moment capacity of the beam may be computed as follows using Equations 10.1 and 10.4 rewritten below:

$f'_s = (0.003/c) \times (c - d') \times E_s$

$M_n = M_{n1} + M_{n2} = 0.85 f'_c \cdot b \cdot \beta_1 \cdot c \cdot (d - \beta_1 \cdot c/2) + A'_s \cdot f'_s \cdot (d - d')$

17. One needs to determine the value of ϕ based on the value of ε_t in step 14:

If $\varepsilon_t \geq 0.005$ then $\phi = 0.9$ ===> More efficient beam design is achieved

If $0.004 \leq \varepsilon_t < 0.005$ ===> $\phi = 0.48 + 83 \varepsilon_t \leq 0.9$ and beam is within code limits

18. Use the value of ϕ obtained in step 17 to determine the value of ϕM_n and draw one of the following conclusions:

If $\phi M_n \geq M_u$ ===> OK   Beam is satisfactory

If $\phi M_n < M_u$ ===> NG   Beam is inadequate

Important Note: There are two maximum reinforcement ratios for singly reinforced concrete beams, ρ_{max} and ρ_{maxp}. ρ_{max} occurs when $\varepsilon_t = 0.004$ and $\phi = 0.81$ and ρ_{maxp} occurs when $\varepsilon_t = 0.005$ and $\phi = 0.9$. It is strongly recommended that the designer uses ρ_{maxp} as the maximum allowed reinforcement ratio in singly as well as doubly reinforced beams (Equation 10.6) since it represents more enhanced and efficient design and better utilization of steel area as discussed earlier.

Example 10.1

Determine the reduced bending moment capacity (ϕM_n) of the doubly reinforced concrete beam shown below if $f'_c = 4,000$ psi and $f_y = 60,000$ psi.

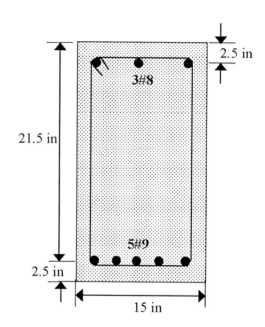

Solution

First check if the beam requires analysis as doubly reinforced concrete beam. To investigate this possibility, ρ needs to be calculated:

A_s = 5#9 bars = 5 in², A'_s = 3#8 = 2.34 in², b = 15 in, d = 21.5 in, β_1 = 0.85

$\rho = A_s/bd = 5/(15\times21.5) = 0.0155 < \rho_{maxp} = 0.0181$ (Table A.19)

Thus, compression steel is not significantly effective in enhancing the bending moment capacity of the beam. However, the beam in Example 10.1 will be analyzed as both a singly reinforced first and a doubly reinforced second for comparison purposes.

Analysis as Singly Reinforced

Based on ρ calculated above and Table A.19, R_n = 802.8 psi.
Also, since $\rho < \rho_{maxp}$ ===> $\phi = 0.9$

Thus, $\phi M_n = 0.9\times802.8\times15\times(21.5)^2/12,000 = 417.5$ k.ft

Analysis as Doubly Reinforced

Assume that compression steel yields ($f'_s = f_y$) at ultimate or failure condition of the beam. To verify this assumption, "a" needs to be calculated using Equation 10.3:

$a = (A_s - A'_s).f_y/(0.85f'_c \cdot b) = (5 - 2.37)\times60/(0.85\times4\times15) = 3.09$ in

and $c = a/\beta_1 = 3.09/0.85 = 3.64$ in

Furthermore, based on Equation 10.1:

$$f'_s = (0.003/c) \times (c - d') \times E_s = (0.003/3.64) \times (3.64 - 2.5) \times 29,000 = 27.2 \text{ ksi}$$

$$f'_s = 27.2 \text{ ksi} \leq f_y = 60 \text{ ksi} ===> \text{ the assumption of } f'_s = f_y \text{ is invalid}$$

Since it was determined that compression steel does not yield, then the global equilibrium equation needs to be used (Equation 10.7). Equation 10.8 contains the solution for global equilibrium equation as shown below:

$$c = \{A_s \cdot f_y - 87A'_s + [(87A'_s - A_s \cdot f_y)^2 + 296f'_c \cdot A'_s \cdot b \cdot d' \cdot \beta_1]^{1/2}\}/(1.7f'_c \cdot b \cdot \beta_1)$$

Thus, $c = \{5 \times 60 - 87 \times 2.37 + [(87 \times 2.37 - 5 \times 60)^2 + 296 \times 4 \times 2.37 \times 15 \times 2.5 \times 0.85]^{1/2}\}/(1.7 \times 4 \times 15 \times 0.85) = 4.70 \text{ in}$

It can now be investigated if the section conforms to ACI 318 Code:

$$\varepsilon_t = (0.003/c) \times (d - c) = (0.003/3.83) \times (21.5 - 3.83) = 0.0138 > 0.005$$

Thus, the section conforms to ACI 318 Code and $\phi = 0.9$. Next, Equations 10.1 and 10.4 need to be utilized:

$$f'_s = (0.003/c) \times (c - d') \times E_s = (0.003/4.70) \times (4.70 - 2.5) \times 29,000 = 40.7 \text{ ksi}$$

$$M_n = M_{n1} + M_{n2} = 0.85f'_c \cdot b \cdot \beta_1 \cdot c \cdot (d - \beta_1 \cdot c/2) + A'_s \cdot f'_s \cdot (d - d')$$

$$M_n = 0.85 \times 4 \times 15 \times 0.85 \times 4.70 \times (21.5 - 0.85 \times 4.70/2) + 2.37 \times 40.7 \times (21.5 - 2.5)$$

$$M_n = 3,970 + 1,825 = 5,795 \text{ k.in} = 482.9 \text{ k.ft}$$

$$===> \phi M_n = 0.9 \times 483.9 = 434.6 \text{ k.ft}$$

===> 434.6 k.ft obtained using doubly reinforced beam analysis is only about 4% larger than 417.5 k.ft obtained from the analysis of beam as singly reinforced.

Example 10.2

Determine the reduced bending moment capacity (ϕM_n) of the doubly reinforced concrete beam shown below if $f'_c = 4,000 \text{ psi}$ and $f_y = 60,000 \text{ psi}$.

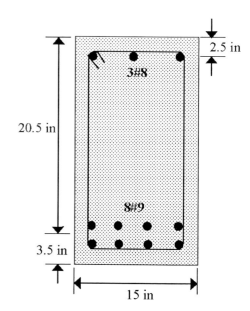

Solution

Check if the beam requires analysis as doubly reinforced concrete beam:

$A_s = 8\#9 \text{ bars} = 8 \text{ in}^2$, $A'_s = 3\#8 = 2.37 \text{ in}^2$, $b = 15 \text{ in}$, $d = 20.5 \text{ in}$, $\beta_1 = 0.85$

$\rho = A_s/bd = 8/(15 \times 20.5) = 0.026 > \rho_{max} = 2.04\%$ (Table A.9)

Thus, analysis as doubly reinforced is warranted.

Assume that compression steel yields at ultimate or failure condition of the beam. To verify this assumption, "a" needs to be calculated based on Equation 10.3:

$$a = (A_s - A'_s) \cdot f_y / (0.85 f'_c \cdot b) = (8 - 2.37) \times 60 / (0.85 \times 4 \times 15) = 6.64 \text{ in}$$

and $c = a/\beta_1 = 6.64/0.85 = 7.81 \text{ in}$

Furthermore, based on Equation 10.1:

$$f'_s = (0.003/c) \times (c - d') \times E_s = (0.003/7.83) \times (7.83 - 2.5) \times 29,000 \approx 60 \text{ ksi}$$

Thus, the initial assumption is correct and compression steel has yielded.

It can now be investigated if the section conforms to ACI 318 Code:

$$\varepsilon_t = (0.003/c) \times (d - c) = (0.003/7.81) \times (20.5 - 7.83) = 0.0049 < 0.005$$

Thus, the section conforms to ACI 318 Code and $\phi = 0.885$. Next, Equation 10.4 can be utilized to determine M_n:

$$M_n = M_{n1} + M_{n2} = 0.85 f'_c \cdot b \cdot \beta_1 \cdot c \cdot (d - \beta_1 \cdot c /2) + A'_s \cdot f'_s \cdot (d - d')$$

$$M_n = 0.85 \times 4 \times 15 \times 0.85 \times 7.81 \times (20.5 - 0.85 \times 7.81/2) + 2.37 \times 60 \times (20.5 - 2.5)$$

$$M_n = 5,815 + 2,549 = 8,363 \text{ k.in} = 696.9 \text{ k.ft}$$

$$\Longrightarrow \phi M_n = 0.885 \times 696.3 = 616.8 \text{ k.ft}$$

Example 10.3

Determine the reduced bending moment capacity (ϕM_n) of the doubly reinforced concrete beam shown below if $f'_c = 4,000 \text{ psi}$ and $f_y = 60,000 \text{ psi}$.

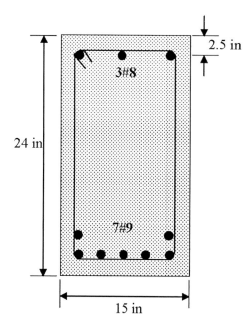

Solution

Check if the beam requires analysis as doubly reinforced concrete beam:

$$A_s = 7\#9 \text{ bars} = 7 \text{ in}^2, \quad A'_s = 3\#8 = 2.34 \text{ in}^2, \quad b = 15 \text{ in}, \quad \beta_1 = 0.85$$

Effective depth "d" needs to be determined via the weighted average between the lower and the upper layer of tension steel. Since bar size is the same fort tension steel, then:

d = (effective depth for lower bars × No. of lower bars + effective depth of upper bars x No. of upper bars)/total number of bars

d = [(h − cover − stirrup diameter − ½ the diameter of lower bars) x No. of lower bars + (h − cover − stirrup diameter − diameter of lower bars − space between upper and lower bars − ½ the diameter of upper bars) × No. of upper bars]/total No. of bars

$d = \{[24 − 1.5 − 3/8 − (9/8)/2]×5 + [24 − 1.5 − 3/8 − (9/8) − (9/8) − (9/8)/2]×2\}/7$

$d = [(21.56)×5 + (20.44)×2]/7 = 21.2$ in

$\rho = A_s/bd = 7/(15×21.2) = 0.022 > \rho_{max} = 2.06\%$ (Table A.9)

Thus, analysis as doubly reinforced is warranted.

Assume that compression steel yields. Then:

$a = (A_s − A'_s)\cdot f_y/(0.85f'_c \cdot b) = (7 − 2.34)×60/(0.85×4×15) = 5.48$ in

and $c = a/\beta_1 = 5.48/0.85 = 6.45$ in

Utilizing Equation 10.1:

$f'_s = (0.003/c)×(c − d')×E_s = (0.003/6.45)×(6.45 − 2.5)×29,000 = 53.3$ ksi

$f'_s = 53.3$ ksi $\leq f_y = 60$ ksi ===> the value of f'_s is invalid

Use the global equilibrium equation for the beam:

$c = \{A_s\cdot f_y − 87A'_s + [(87A'_s − A_s\cdot f_y)^2 + 296f'_c\cdot A'_s\cdot b\cdot d'\cdot\beta_1]^{1/2}\}/(1.7f'_c\cdot b\cdot\beta_1)$

Thus, c = {7×60 − 87×2.34 + [(87×2.34 − 7×60)² + 296×4×2.34×15×2.5×0.85]^{1/2}}/ (1.7×4×15×0.85) = 6.73 in

It can now be investigated if the section conforms to ACI 318 Code:

$\varepsilon_t = (0.003/c)×(d − c) = (0.003/6.73)×(21.2 − 6.73) = 0.0065 > 0.005$

Section conforms to ACI 318 Code and $\phi = 0.9$.

$f'_s = (0.003/c)×(c − d')×E_s = (0.003/6.73)×(6.73 − 2.5)×29,000 = 54.7$ ksi

$M_n = M_{n1} + M_{n2} = 0.85f'_c\cdot b\cdot\beta_1\cdot c\cdot(d − \beta_1\cdot c/2) + A'_s\cdot f'_s\cdot(d − d')$

$M_n = 0.85×4×15×0.85×6.73×(21.2 − 0.85×6.73/2) + 2.34×54.7×(21.2 − 2.5)$

$M_n = 5,351 + 2,394 = 7,745$ k.in = 645.4 k.ft

===> $\phi M_n = 0.9×645.4 = 580.9$ k.ft

Important Note: Some designers are comfortable assuming a stress in compression steel of f_y for analysis of doubly reinforced concrete beam without referring to the strain distribution. This is due to concrete creep and is supported by experiments.

Example 10.4

Determine the reduced bending moment capacity (ϕM_n) of the doubly reinforced concrete beam shown below if

$f'_c = 30$ MPa and $f_y = 420$ MPa

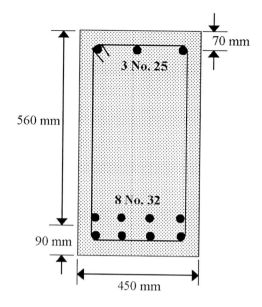

Solution

Check if the beam requires analysis as doubly reinforced concrete beam:

$A_s = 8$ No. 28 = 6,430 mm², $A'_s = 3$ No. 25 = 1,470 mm², b = 450 mm, d = 560 mm, $\beta_1 = 0.85$

$\rho = A_s/bd = 6,430/(450 \times 560) = 0.0255 > \rho_{max} = 2.04\%$ (Table A.9)

Thus, analysis as doubly reinforced is warranted.

Assume that compression steel yields at ultimate or failure condition of the beam. To verify this assumption, "a" needs to be calculated based on Equation 10.3:

$a = (A_s - A'_s) \cdot f_y/(0.85 f'_c.b) = (6,430 - 1,470) \times 420/(0.85 \times 30 \times 450) = 182$ mm

and $c = a/\beta_1 = 181/0.85 = 214$ mm

Furthermore, based on Equation 10.1:

$f'_s = (0.003/c) \times (c - d') \times E_s = (0.003/214) \times (214 - 70) \times 200 \times 10^3$

$f'_s = 404$ MPa ≈ 420 MPa ===> $f'_s = f_y = 420$ MPa where $E_s = 200$ GPa

Thus, the initial assumption is correct and compression steel has yielded.

It can now be investigated if the section conforms to ACI 318 Code:

$\varepsilon_t = (0.003/c) \times (d - c) = (0.003/214) \times (560 - 214) = 0.0049 < 0.005$

Thus, the section conforms to ACI 318 Code and $\phi = 0.885$. Next, Equation 10.4 can be utilized to determine M_n:

$M_n = M_{n1} + M_{n2} = 0.85 f'_c \cdot b \cdot \beta_1 \cdot c \cdot (d - \beta_1 \cdot c /2) + A'_s \cdot f'_s \cdot (d - d')$

$M_n = [0.85 \times 30 \times 450 \times 0.85 \times 214 \times (560 - 0.85 \times 214/2) + 1,470 \times 420 \times (560 - 70)]/10^6$

$M_n = 979 + 303 = 1,292$ kN.m where 10^6 is a conversion factor

===> $\phi M_n = 0.885 \times 1,292 = 1,240$ kN.m

10.4 Design of Doubly Reinforced Concrete Beams

The input data for design of a doubly reinforced concrete beam include beam size (b and h) as well as the design bending moment (M_u). Selection of materials (f_c' and f_y) is generally made by the designer with market availability in mind. Such selection is subject to adjustment based on calculations performed during the design process.

A first an essential step of the design of doubly reinforced beam is to investigate if the available beam section can resist the design bending moment as singly reinforced. Since h is given, the value of d can be estimated using Figures 6.2 and 6.3 and the following Equations:

$d \approx h - 2.5$ to 3 in (70 to 80 mm)   for one layer of tension steel

$d \approx h - 3.5$ to 4 in (90 to 100 mm)   for two layers of tension steel

The designer can choose if one or two layers should be used. Modifications follow after steel selection is made. Equipped with d, b, f_c' and f_y, the designer can determine the maximum bending moment capacity of the proposed beam section as singly reinforced. To achieve that, one needs to use the maximum preferred reinforcement ratio for singly reinforced beam or ρ_{maxp} as explained earlier. Based on ρ_{maxp}, one can determine the value of R_{nmax} using Appendix A. The maximum reduced bending moment capacity of the beam section as singly reinforced can be calculated as follows:

$$\phi M_{nmax,psingly} = 0.9 R_{nmax} \cdot b \cdot d^2$$

Where:

R_{nmax} is based on ρ_{maxp}, f_c' and f_y obtained from the tables in Appendix A, and b and d are beam section width and effective depth.

If $M_u < \phi M_{nmaxpsingly}$, then the beam can be designed as singly reinforced using previous chapters. However, if $M_u > \phi M_{nmaxpsingly}$, then the beam needs to be designed as a doubly reinforced beam. Such design would be carried out similar to the analysis discussed in Section 10.3. The design of Beam 1 is performed first followed by the design of Beam 2. Beam 1 is considered to contain the maximum reinforcing steel (ρ_{maxp}) for a singly reinforced beam. Beam 1 contains concrete and a part of tension steel. With maximum code allowed steel area in Beam1, one achieves maximum utilization of the concrete area and, consequently, employment of beam 1 would be maximized/optimized. Therefore:

$$A_{s1} = \rho_{maxp} \cdot b \cdot d \tag{10.9}$$

Where:

ρ_{maxp} = the maximum preferred steel area for singly reinforced beams and can be obtained from Appendix A based on f_c' and f_y.

With the selection of reinforcement ratio for Beam 1 as the maximum code preferred ratio for singly reinforced beams, then $\phi = 0.9$. The bending moment capacity of Beam 1 can be computed as follows:

$$M_{n1} = R_{nmaxp} \cdot bd^2$$

Where R_{nmaxp} is the value of R_n that corresponds to $\rho = \rho_{maxp}$. Beam 2 is composed of steel without concrete. The design of Beam 2 entails determining the area of compression steel (A_s') and remaining area of tension steel (A_{s2}). First, the residual bending moment to be resisted by beam 2 (M_{n2}) needs to be computed:

$$M_{n2} = M_u/\phi - M_{n1} = M_u/0.9 - R_{nmaxp} \cdot bd^2$$

The remaining area of tension steel can be determined as a derivative of Equation 10.10:

$$A_{s2} = M_{n2}/[f_y \cdot (d - d')] = [M_u/0.9 - R_{nmaxp} \cdot bd^2] / [f_y \cdot (d - d')] \qquad (10.10)$$

Prior to determining the value of A'_s, one needs to calculate the stress in compression steel. With $A_{s1} = A_{smaxpsingly}$, then, in Beam 1, the strain in tension steel at ultimate condition is 0.005 while the crushing strain in concrete is 0.003 and the depth of compression zone can be computed:

$$c = (3/8) \cdot d$$

And the stress in compression steel can be determined using Equation 10.1. Consequently, A'_s can be calculated:

$$A'_s = M_{n2}/[f'_s \cdot (d - d')]$$

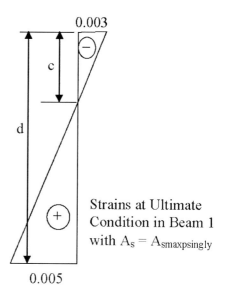

Strains at Ultimate Condition in Beam 1 with $A_s = A_{smaxpsingly}$

With steel areas for the doubly reinforced concrete beam known ($A_s = A_{s1} + A_{s2}$ and A'_s), the beam can be detailed and analyzed (Section 10.3) for design soundness and efficiency. Note that the above shown strain distribution in Beam 1 is adopted for the entire doubly reinforced concrete beam as explained earlier (Sections 10.2 1nd 10.3).

Example 10.5

Design a reinforced concrete beam with section dimensions limited to b = 12 in and h = 15 inch if it is subjected to a design bending moment M_u = 235 k.ft

f'_c = 5,000 psi and f_y = 60,000 psi.

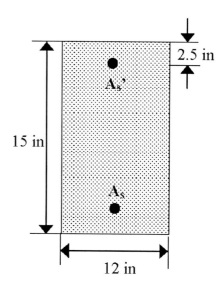

Solution

Check if the applied moment, M_u, can be supported with a singly reinforced beam.

$\phi M_{nmaxp,singly} = 0.9R_{nmaxp} \cdot b.d^2 = 0.9 \times 1,085.8 \times 12 \times (11)^2/12,000 = 0.9 \times 131 = 118$ k.ft
Where: $d \approx 15 - 4 = 11$ in assuming two layers of steel
$R_{nmaxp} = 1,085.8$ psi from Table A.20

Thus, the applied bending moment M_u cannot be resisted by a singly reinforced beam with the prescribed dimensions. One needs to determine the residual moment that needs To be determined by beams, M_{n2}:

$M_{n2} = M_u/\phi - M_{n1} = M_u/0.9 - M_{n1} = 235/0.9 - 131 = 130$ k.ft

The values of A_s and $A_{s'}$ can be determined as follows:

$A_s = \rho_{maxp,singly} \cdot b \cdot d + M_{n2}/[f_y \cdot (d - d')]$

$A_s = 0.0213 \times 12 \times 11 + 130 \times 12/[60 \times (11 - 2.5)] = 5.87$ in²

Where $\rho_{maxp,singly}$ is obtained from Table A.20 and $d' = 2.5$ in is assumed

$A_{s'} = M_{n2}/[f_{s'} \cdot (d - d')] = 130 \times 12/[34 \times (11 - 2.5)] = 5.40$ in²

Where $f_{s'} = (0.003/c) \times (c - d') \times E_s \leq f_y$ and $c = 3d/8 = 4.1$ in

$f_{s'} = (0.003/4.1) \times (4.1 - 2.5) \times 29{,}000 = 34$ ksi $< fy = 60$ ksi OK

Use 8#8 for tension steel ($A_s = 6.24$ in² > 5.87 in²) and for compression steel ($A_{s'} = 6.24$ in² > 5.40 in²)

The section can thus be detailed as shown.

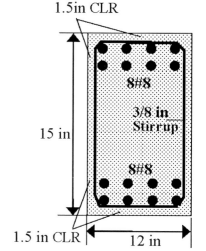

Based on Table A.12, the required minimum width for 4#8 bars in one layer is 11.3 inch which is less than the provided width of 12 in. It is important to realize that the area of $A_{s'}$ is close to A_s where the stresses in tension and compression steel are based on Bernouli's theory in bending. Since concrete is not elastic, then Bernouli's theory does not apply. Furthermore, the effect of creep is not taken into account. Nevertheless, one should follow ACI 318 Code in structural design in spite of possible flaws.

The beam designed for Example 10.4 needs to be analyzed. One of the analysis steps is included herein:

d = 15 in – 1.5 in(clear cover) – 3/8 in(stirrup) – 1 in(diameter of bars in lower layer) – ½ in(half of the space between the two bar layers) = 11.6 in

The resulting d = 11.6 in is greater than the assumer d of 11.0 inch indicating that the initial assumption was conservative but realistic. The remainder of analysis can be done similar to Examples 10.1 through 10.4.

Example 10.6

Design a reinforced concrete beam with section dimensions limited to b = 18 in and h = 35 inch if it is subjected to a design bending moment M_u = 1,300 kft f'_c = 4,000 psi and f_y = 60,000 psi.

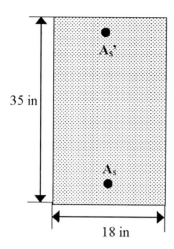

35 in

18 in

Solution

Check if the applied moment, M_u, can be supported with a singly reinforced beam.

$\phi M_{nmaxp,singly}$=0.9R_{nmaxp}·b·d² = 0.9×912.6×18×(31)²/12,000=0.9×1,315=1,183 ft.k

Where: d ≈ 35 – 4 = 31 in assuming two layers of steel

R_{nmaxp} = 912.6 psi from Table A.19

Thus, the applied bending moment M_u cannot be resisted by a singly reinforced beam with the prescribed dimensions. One needs to determine the residual moment that needs To be determined by beams, M_{n2}:

$M_{n2} = M_u/\phi − M_{n1} = M_u/0.9 − M_{n1}$ = 1,300/0.9 – 1,315 = 129 k.ft

The values of A_s and $A_{s'}$ can be determined as follows:

$A_s = \rho_{maxp,singly}$·b·d + $M_{n2}/[f_y·(d − d')]$
A_s = 0.0181×18×31+129×12/[60×(31–2.5)] = 11.0 in²

Where $\rho_{maxpsingly}$ is from Table A.19 and d' = 2.5 in is assumed.

$A_s' = M_{n2}/[f'_s·(d − d')]$ = 129×12/[60×(31 − 2.5)] = 0.91 in²

Where f'_s = (0.003/c)×(c − d')×E_s ≤ f_y and c = 3d/8 = 11.6 in

f'_s = (0.003/11.6)×(31 − 2.5)×29,000 = 213 ksi > fy = 60 ksi

NG ===> f's = fy = 60 ksi

Use 10#10 for tension steel (A_s = 12.86 in² > 11.9 in²)

Use 2#9 for compression steel (A_s' = 2.57 in² > 0.91 in²)

Then, Section analysis can be done as explained earlier.

The section can thus be detailed as shown. Note that although ACI 318 Code does not require skin reinforcement for this section, such reinforcement is used as a precautionary measure.

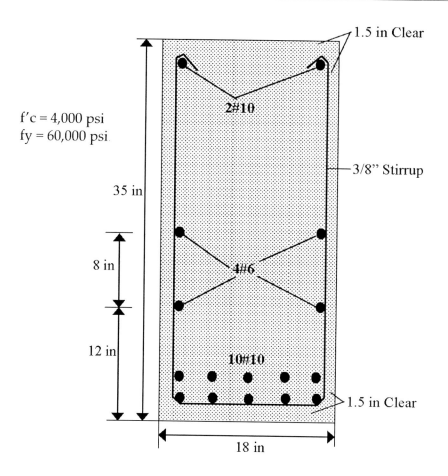

f'c = 4,000 psi
fy = 60,000 psi

Example 10.7

Design a reinforced concrete beam with section dimensions limited to b = 300 mm and h = 400 mm if it is subjected to a design bending moment $M_u = 350$ kN.m $f'_c = 35$ MPa psi and $f_y = 420$ MPa.

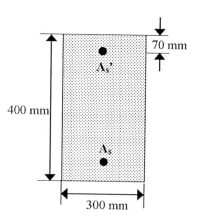

Solution

Check if the applied moment, M_u, can be supported with a singly reinforced beam.

$\phi M_{nmaxp,singly} = 0.9R_{nmaxp}.b.d^2 = 0.9 \times 7.54 \times 300 \times (300)^2/10^6 = 0.9 \times 204 = 183$ kN.m

Where: $d \approx 400 - 100 = 300$ mm assuming two layers of steel

$R_{nmaxp} = 7.54$ MPa from Table A.20

Thus, the applied bending moment M_u cannot be resisted by a singly reinforced beam with the prescribed dimensions. One needs to determine the residual moment that needs To be determined by beams, M_{n2}:

$M_{n2} = M_u/\phi - M_{n1} = M_u/0.9 - M_{n1} = 350/0.9 - 204 = 185$ kN.m

The values of A_s and $A_{s'}$ can be determined as follows:

$A_s = \rho_{maxp,singly} \cdot b \cdot d + M_{n2}/[f_y \cdot (d - d')] =$

$A_s = 0.0213 \times 300 \times 300 + 185 \times 10^6/[420 \times (300 - 70)] = 3,830 \text{ mm}^2$

Where $\rho_{maxp,singly}$ is obtained from Table A.20 and $d' = 70$ mm is assumed

$A_{s'} = M_{n2}/[f'_s \cdot (d - d')] = 185 \times 10^6/[228 \times (300 - 70)] = 3,530 \text{ mm}^2$

Where $f'_s = (0.003/c) \times (c - d') \times E_s \leq f_y$ and $c = 3d/8 = 113$ mm

$\qquad f'_s = (0.003/113) \times (113 - 70) \times 200 \times 10^3 = 230 \text{ MPa} < fy = 420 \text{ MPa OK}$

Use 8 No. 25 for tension steel ($A_s = 3930 \text{ mm}^2 > 3,830 \text{ mm}^2$) and for compression steel ($A_{s'} = 3,930 \text{ mm}^2 > 3,530 \text{ in}^2$)

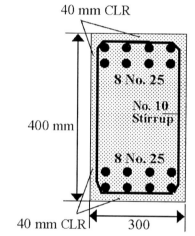

The section can thus be detailed as shown.

One can verify that, the required minimum width for 4 No. 25 bars in one layer is about 280 mm which is less than the provided width of 300 mm. It is important to realize that the area of A_s' is close to A_s where the stresses in tension and compression steel are based on Bernouli's theory in bending that does not take creep into account.

The beam designed for Example 1074 needs to be analyzed. One of the analysis steps is included herein:

d = 400 mm – 40 mm(clear cover) – 10 mm(stirrup) – 25 mm(diameter of bars
in lower layer) – 13 mm(half the space between the two bar layers) = 312 mm

The resulting d = 312 mm in is greater than the assumer d of 300 mm inch indicating that the initial assumption was conservative but realistic. The remainder of analysis can be done similar to Examples 10.1 through 10.4.

Test your knowledge

1.   Why is the stress in compression steel most probably equal to f_y at ultimate conditions in doubly reinforced concrete beams?

2.   What are the methods for increasing the bending moment capacity of a reinforced concrete beam with limited depth.

3.   What are the two cases when compression steel is not effective in increasing the bending moment capacity of reinforced concrete beams.

4.   Why is ρ_{maxp} used for the design of beam1 in doubly reinforced concrete beam design?

5.   Perform the analysis for bending moment and compute the value of ϕM_n for the doubly reinforced concrete beams listed in the following:

Problem	b(in)	d(in)	d'(in)	A_s(in²)	A_s'(in)	f'c(psi)	fy(ksi)
5	14	17	2.5	5#7	2#5	2,500	60
6	16	21	2.5	5#8	2#6	3,000	60
7	15	21	2.5	6#9	4#7	3,000	60
8	18	31	2.5	12#9	4#9	4,000	60
9	24	35	3	14#9	4#9	4,000	60
10	20	21	3	8#10	3#11	4,500	60
11	18	26	3	14#9	6#10	5,000	60
12	24	43	4	18#11	18#10	5,500	80
13	20	30	3	12#11	8#11	6,000	60
14	24	55	5	12#14	10#11	6,000	80
15	300 mm	520 mm	80 mm	5 NO. 25	2 NO. 20	30 MPa	420 MPa
16	600 mm	520 mm	80 mm	15 NO. 28	4 NO. 28	30 MPa	420 MPa

Design for bending moment and compute the value of ϕM_n for the doubly reinforced concrete beams listed in the following:

Problem	b(in)	h(in)	f'c(psi)	fy(ksi)	M_u(k.ft)
17	12	18	2,500	60	120
18	14	20	3,000	60	250
19	15	15	3,000	60	170
20	16	18	4,000	60	220
21	16	36	4,000	60	1,500
22	18	24	4,500	60	800
23	20	20	5,000	60	700
24	20	24	5,500	80	1,000
25	24	36	6,000	60	2,500
26	24	60	6,000	80	8,000
27	350 mm	450 mm	30 MPa	420 MPa	400 kN.m
28	300 mm	450 mm	25 MPa	420 MPa	180 kN.m
29	600 mm	900 mm	42 MPa	420 MPa	3,500 kN.m

Chapter 11

SHEAR BEHAVIOR OF REINFORCED CONCRETE BEAMS

11.1 Introduction to Beam Shear Stresses

Shear forces are tangential to the beam cross–section. Deformations or failure due to shear can be envisioned as a separation of adjacent beam sections from each other along a tangential plane or line. Shearing forces on a beam are shown in Figure 11.1.

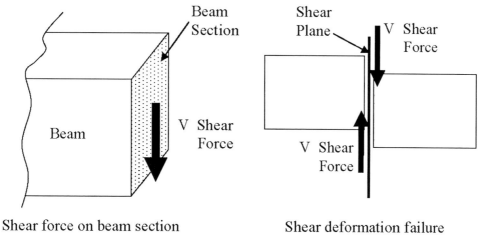

Shear force on beam section   Shear deformation failure

Figure 11.1: Effect of shear on beams.

The effect of shear force V on beam section is not uniform. Shear stress varies according to the location within the beam cross–section due to variation in shear deformations of the section. Based on Bernoulli's theorem in solid mechanics for beams, the following equation (Bernoulli's shear stress) equation may be utilized for determination of shear stress in a beam subjected to shearing force:

$$f_v = VQ/Ib$$

Where:

f_v = shear stress at point P of the section.

V = external shear force affecting the beam.

Q = first moment of area A which is bound by a horizontal line (perpendicular to shear force) at point P. Q = A·$\bar{y}$ is the general equation for Q.

A = area bound by a horizontal line (line perpendicular to shear force) at point P.

$\bar{y}$ = distance of center of gravity of area A to the neutral axis of beam section.

I = moment of inertia of the beam about its neutral axis, and

b = width of beam cross–section at point P.

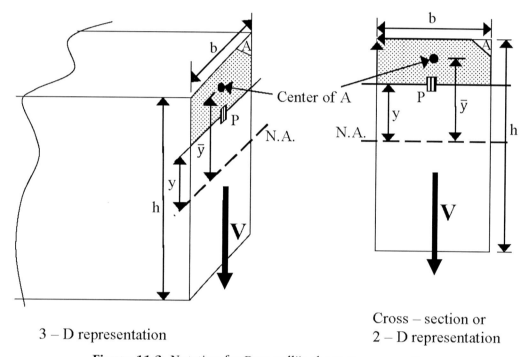

3 – D representation   Cross – section or
                                         2 – D representation

Figure 11.2: Notation for Bernoulli's shear stress equation.

The terms or notation for Bernoulli's shear stress equation are exhibited in Figure 11.2. One may conclude that shear stresses are equal at a straight line perpendicular to the shear force or horizontal line if V is vertical. In Figure 11.2, all the cross–sectional points along the horizontal line passing through point P have equal shear stresses due to force V. Shear stresses in a rectangular beam (height = h and width = b) subjected to an external shearing force, V, can be determined by the following equation that is a deduction from Bernoulli's shear stress equation:

$f_v = V·(h/2 − y)·b·[h/2 − (h/2 − y)/2]/[(b·h^3/12)·b]$

$f_v = 6V·(h^2/2 − y^2)/(b·h^3)$

Where y is the distance from the neutral axis to point P at which shear stress is f_v. The value of shear stress at the top and bottom of section (y = h/2 and y = −h/2) is zero. Maximum shear stress for a rectangular, based on the above equation, is at the N.A., where y = 0, is 150% of the average shear stress in a rectangular section. Average shear stress is determined by dividing the external shear stress by cross–sectional area:

$f_{vav} = V/bh$

$f_{vmax} = 1.5V/bh$ at N.A. where y = 0

Figure 11.3 illustrates the distribution of shear stresses in a beam with rectangular section subjected to external shear force based on the discussion above.

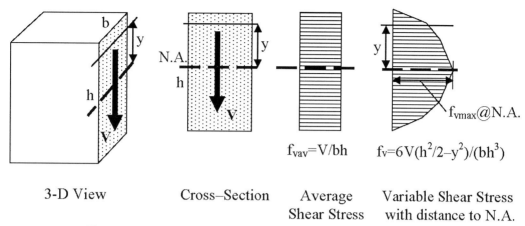

$$f_{vav}=V/bh \qquad f_v=6V(h^2/2-y^2)/(bh^3)$$

3-D View	Cross–Section	Average Shear Stress	Variable Shear Stress with distance to N.A.

Figure 11.3: Variation of shear stress in a rectangular section.

The above described shear stresses are tangential to the beam's cross–section and parallel to the path of external shear force. Nevertheless, shear stresses always exist in pairs to preserve rotational equilibrium. For further discussion, a beam stress element at the N.A. where shear stresses are maximum is selected as shown in Figure 11.4. f_{v1} stresses are the action and reaction due to the external shear force. They are equal in magnitude and opposite in direction (one is directed towards the positive y axis direction and the other is towards the negative y axis direction). Force equilibrium is achieved. Both shear stresses f_{v1} rotate clockwise about the center of the element (point O) shown in Figure 11.4. Thus, moment equilibrium is not preserved. Analysis reveals the presence of two additional shear stresses, f_{v2}, acting on the same element to satisfy moment equilibrium. f_{v2} stresses are equal in magnitude to f_{v1} stresses but cause a counter clockwise rotation that is in the opposite direction to the rotation caused by f_{v1}, thereby, preserving moment equilibrium.

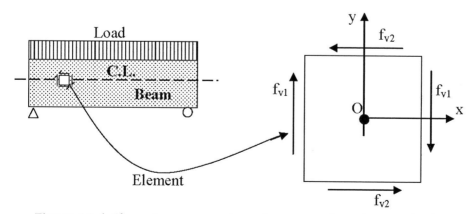

Figure 11.4: Shear stresses on a stress element at a beam's neutral axis.

Every beam stress element is typically subjected to axial and shear stresses due to bending and shear. Such stress situation of each beam element with infinitesimal dimensions resembles a plane stress situation. Analysis of similar elements is focused on finding the principal stresses or maximum tension stress and maximum compression stress. A review of Chapter 4 concludes that no bending stresses are present at the N.A. Therefore, the stress situation at any beam element in Figure 11.4 at the N.A. includes shear stresses only termed pure shear situation. This facilitates the determination of principal stresses, σ_1 and σ_2 maximum tension stress and maximum compression stress, respectively, also termed first and second principal stress. Based on one's background in solid mechanics, it can be determined that principal stresses for an element in pure shear are at 45° angles from axes x and y as shown in Figure 11.5. σ_1, maximum tensile stress and σ_2, maximum compression stress can be expressed as follow (assuming positive sign convention for tension):

$$\sigma_1 = -\sigma_2 = f_v$$

Where: $f_v = |f_{v1}| = |f_{v2}|$

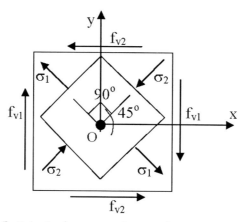

Figure 11.5: Principal stresses due to shear stresses at the N.A.
(No bending stresses)

11.2 Cracking Behavior of Concrete Beams Due to Shear and Moment

To study cracking behavior of reinforced concrete beams subjected to moment and shear, a typical simply supported beam subjected to uniformly distributed load is utilized as displayed in Figure 11.6. The shear force diagram shows that the magnitude of shear force is maximum at either support while shear force at midspan is zero. Therefore, the effect of tensile stresses caused by shear is more pronounced near the support. Assuming that concrete resistance to tension is negligible, the effect of shear is 45° cracks or diagonal cracks that are perpendicular to σ_1 first principal stress. Such cracks are larger, wider, and more frequent near the support due to increased tensile stresses caused by shear as schematically shown in Figure 11.6.

To understand the interaction between bending and shear, one needs to analyze the crack pattern due to bending moment. Therefore, the same typical beam of Figure 11.6 is used. As discussed in Chapter 4, flexural stresses are maximum at the top and bottom of beam section and zero at the N.A. Tension stresses caused by bending are concentrated at the lower portion of the beam termed the tension zone, and cause concrete cracking. Assuming the presence of adequate steel reinforcement within the beam and utilizing the principles of Chapter 4, one may recognize the crack pattern due to bending in the tension zone of the beam. Depth, width and frequency (intensity) of cracks increase as the magnitude of bending moment (and consequently tensile stresses) increases. Distributed load causes a maximum bending moment at beam center line while the value of bending moment at the support is zero. Thus, different from shear cracking, cracks in beam tension zone due to bending are more pronounced near midspan and less significant near the supports as attributed to bending moment changes along the beam (Figure 11.7).

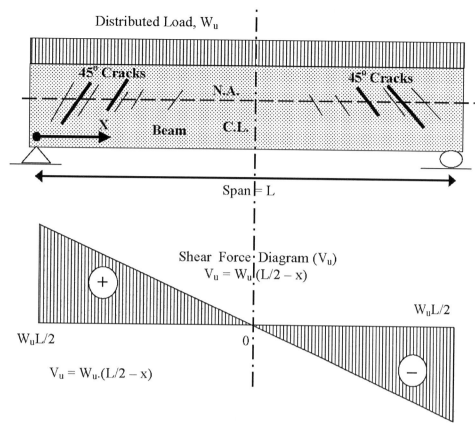

Figure 11.6: Schematic of cracks due to shear in a typical reinforced concrete beam.

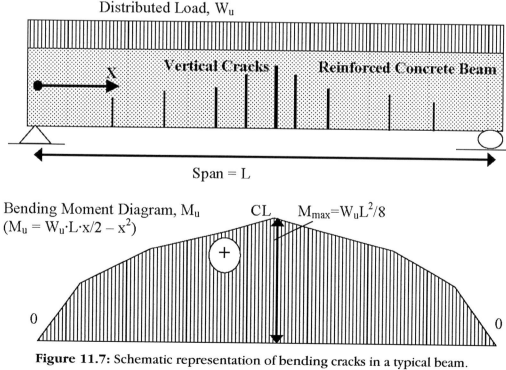

Figure 11.7: Schematic representation of bending cracks in a typical beam.

Figure 11.8 superimposes the cracks due to shear on the cracks due to bending moment for the typical beam selected herewith. While bending moment is maximum at midspan and zero at the supports, shear is maximum at the supports and zero at midspan. The critical locations for the combined action of shear and moment in a beam are at distance of "x_c" from each support face where shear forces are high while bending moments remain fairly high. Moment equation is parabolic and the reduction in value is gradual within short distances from midspan. The bending moment at x_c causes tension cracking in concrete (Phase 2 of behavior in Chapter 4). These tension cracks combine with the 45° or diagonal cracks caused by shear to generate such critical failure surfaces (Figure 11.8). The value of x_c is a function of loading, steel reinforcement as well as span/depth ratio.

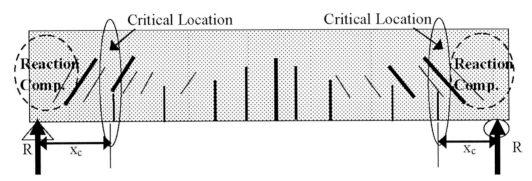

Figure 11.8: Beam cracking due to bending moment and shear.

The locations at distance x_c from support face are critical for beam failure under shear. This is in addition to the locations directly at the face of support where shear is maximum while bending moment is zero. However, support reactions cause compression stresses within the beam that counteract the effect of tension caused by shear forces at supports. This moderates shear force effect near the supports as illustrated in Figure 11.8. Consequently, the sections at support faces are typically less critical than the sections at distance x_c from supports. The configuration of crack surfaces due to shear and bending moment combine to generate a critical failure situation (Figure 11.9). This critical failure surface results from shear cracking at the N.A. and bending tension cracking below the N.A.

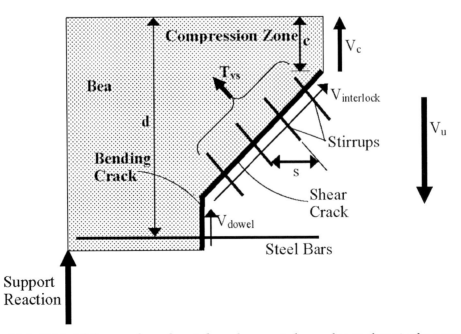

Figure 11.9: Critical failure surface of a reinforced concrete beam due to shear in the presence of bending tension cracks.

Furthermore, Figure 11.10 illustrates another failure scenario due primarily to shear. This failure condition is caused by shear cracks that extend throughout the whole depth of the beam (except through the concrete compression zone) without the presence of bending cracks.

Although there are differences between the aforementioned failure modes, the fundamentals of both modes are similar. The primary crack is a 45^0 crack caused by tension shear. The concrete compression zone remains intact. Stirrups (steel bars intercepting the 45^0 crack) are also present for both failure modes.

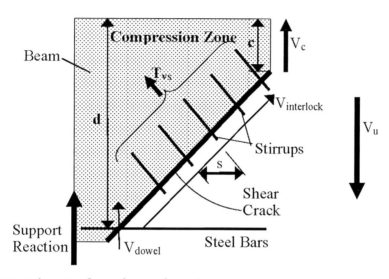

Figure 11.10: Failure Surface of a reinforced concrete beam due to shear in the absence of bending tension cracks.

In accordance with the strength design method, one needs to study the behavior at ultimate failure condition where the nominal shear resistance, ϕV_n, must be greater than the ultimate magnified external shear force, V_u. The forces contributing to beam resistance to external shear acting at the critical section upon ultimate failure conditions are:

1. Shear resistance of the uncreacked concrete compression zone, V_c.

2. Shear resistance of any shear reinforcement (stirrups) intercepting the shear crack, T_{vs}.

3. Aggregate interlock $V_{interlock}$.

4. Steel dowel action V_{dowel}.

Aggregate interlock shear resistance, $V_{interlock}$, is generated by the interlocking friction of the two sides of crack while crack width is small. As crack width increases and relative deformations between the two sides of the crack amplify, $V_{interlock}$ decreases in value significantly. As such, ACI 318 Code does not consider aggregate interlock as a reliable source of beam shear resistance. Also, shear resistance due to steel dowel action is caused by tension steel bars ramming or pressing against the adjacent concrete surface due to the effect of shear deformation. Such action may cause splitting within the concrete surrounding tension steel bars. Therefore, ACI 318 Code does not consider steel dowel action as a component in beam shear resistance.

Consequently, based on ACI 318 Code, shear resistance of concrete beams comes from two sources: the concrete compression zone and shear steel reinforcement (stirrups). The sum of these shear resistance is used for beam shear design.

11.3 ACI Code Shear Design Equation for Beams

For shear analysis and/or design of reinforced concrete beams, the failure surfaces/sections due to shear either with or without flexural tension cracks (Figures 11.9 and 11.10) may be simplified as shown in Figure 11.11.

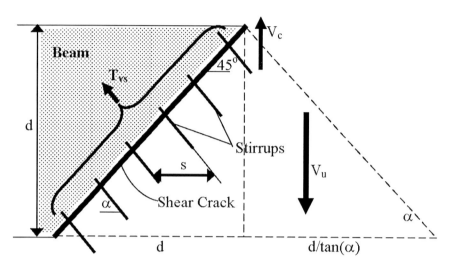

Figure 11.11: Simplified failure surface due to shear.

The following are code equations for shear resistance of reinforced concrete beams.

Concrete resistance to shear (mainly contributed by the compression zone):

1. For beams not subjected to axial forces:

$$V_c = [1.9\sqrt{f_c'} + 2500\rho(V_u\cdot d/M_u)]\cdot b_w\cdot d \le 3.5\sqrt{f_c'}\,b_w\cdot d \tag{11.1}$$

The simplified form of the above equations may also be used:

$$V_c = 2\sqrt{f_c'}\cdot b_w\cdot d \tag{11.2}$$

2. For beams subjected to axial compression:

$$V_c = [1.9\sqrt{f_c'} + 2500\rho(V_u\cdot d/M_m)]\cdot b_w\cdot d \le 3.5\sqrt{f_c'}\cdot b_w\cdot d\cdot\sqrt{1 + N_u/500A_g} \tag{11.3}$$

The following simplified equation may also be used:

$$V_c = 2\sqrt{f_c'}\cdot(1 + N_u/2000A_g)\cdot b_w\cdot d \tag{11.4}$$

3. For beams subjected to axial tension:

$$V_c = 2\sqrt{f_c'}\cdot b_w\cdot d\cdot(1 - N_u/500A_g) \tag{11.5}$$

The following simplified equation can also be used:

$$V_c = 0 \tag{11.6}$$

Where:

V_c = concrete shear resistance (lb).

f_c' = design compressive strength of concrete (psi).

b_w = width of rectangular concrete beam or web width of T-beam (in).

d = effective depth of concrete beam (in).

h = total depth or thickness of concrete beam (in).

A_s = area of tension steel reinforcing bars (in²).

$\rho = A_s/(b_w{\cdot}d)$ = tension steel reinforcement ratio.

M_u = ultimate or magnified bending moment (lb.in).

V_u = ultimate or magnified shear force (lb).

N_u = ultimate or magnified axial force (+ for compression and – for tension) in lb.

$A_g = b_w{\cdot}h$ = gross area of beam (in²).

$M_m = M_u - N_u{\cdot}(4h - d)/8$ (lb.in)

Equations 11.1 through 11.6 are valid only with U.S. units of lb in and psi. For practice problems with SI units, one needs to convert to U.S. prior to utilizing these equations.

The second expressions contained in Equations 11.1 and 11.3 following ≤ sign are for the upper bound (not to exceed) values of V_c. For light weight concrete, one may use Equations 11.1 through 11.6 with $\sqrt{f'_c}$ multiplied by 0.85 for sand light weight concrete and 0.75 for all light weight concrete. In this text, Equations 11.2, 11.4 and 11.6 will be utilized for determining concrete resistance to shear forces.

Equations 11.1 through 11.6 are based on average shear stress resistance of concrete over the entire effective beam area (b.d). Shear stress resistance is multiplied by effective beam area to obtain shear force resistance of concrete. Furthermore, code expressions for concrete shear resistance are all functions of $\sqrt{f'_c}$ In previous chapters, it was explained that concrete tensile strength is also a function of $\sqrt{f'_c}$ This denotes the close correlation between shear strength and tensile strength of concrete. Such correlation is understandable since shear forces generate tensile stresses responsible for diagonal cracking in concrete beams as explained earlier in this chapter.

Concrete shear resistance is due to the uncracked concrete compression zone with width and depth of b and c, respectively. Yet, $c = a/\beta_1$ and $a = A_s f_y/0.85 f'_c b$ at failure condition. Thus, tension steel reinforcement A_s dictates the size of the concrete compression zone. Designers commonly use a reinforcement ratio, $\rho = A_s/bd$, of at least 1.2% for this cause.

Steel shear reinforcement resistance to shear is due to the steel bars crossing the diagonal 45° degree crack called stirrups as shown in Figure 11.11. The number of bars with a horizontal spacing of "s" intercepting the diagonal crack may be determined using the following equation:

No. of bars intercepting diagonal shear crack = [d + d/tan(α)]/s

Ultimate Tension Force in Diagonal Bars or Stirrups = T_{sv} can then be computed:

$T_{sv} = \{[d + d/\tan(\alpha)]/s\}{\cdot}a_v{\cdot}f_y$

Shear Resistance of Diagonal Bars or Stirrups = V_s can, thus be derived as follows:

$V_s = T_{sv}{\cdot}\sin(\alpha) = \{[d + d/\tan(\alpha)]/s\}{\cdot}a_v{\cdot}f_y{\cdot}\sin(\alpha)$

$$V_s = [1 + \cos(\alpha)]{\cdot}a_v{\cdot}f_y{\cdot}d/s \qquad (11.7)$$

With α = 90° as in the case of vertical stirrups:

$$V_s = a_v{\cdot}f_y{\cdot}d/s \qquad (11.8)$$

Where:

T_{vs} = ultimate tension force in shear reinforcement (stirrups or diagonal bars).

V_s = ultimate shear resistance of shear reinforcement.

α = angle of inclination of shear reinforcement with beam center line.

f_y = yield strength of shear reinforcement.

s = horizontal spacing of shear reinforcement.

a_v = cross–sectional area of shear reinforcement intercepted by diagonal or 45° crack.

Figures 11.12 and 11.13 illustrate the computation of shear reinforcement area, a_v. Figure 11.12 is a three dimensional representations where cracks are planes instead of lines. The diagonal shear crack is plane ABCD. A beam section intersecting this diagonal crack is EFGH and shown in a dashed line. Shear reinforcement in Figure 11.12 is vertical stirrups at spacing = s. Longitudinal reinforcement is not shown to reduce congestion. In Figure 11.13, the diagonal shear crack ABCD projects as horizontal line IJ within the beam cross–section. This horizontal line intercepts each vertical stirrup at two points. As a result, the following general equation for a_v is used:

$$a_v = n \cdot a_s \tag{11.9}$$

Where:

a_v = area of stirrup effective for shear reinforcing.

n = number of vertical branches per each stirrup.

a_s = stirrup steel bar cross–sectional area.

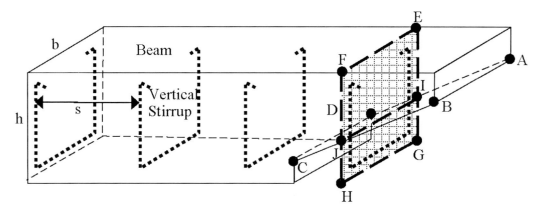

Figure 11.12: Intersection of shear cracking and vertical stirrups.

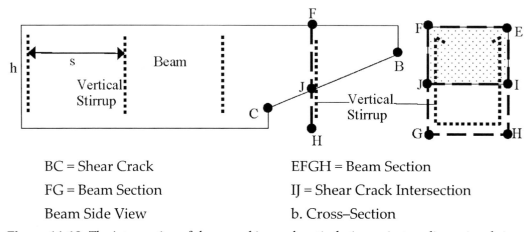

BC = Shear Crack   EFGH = Beam Section

FG = Beam Section   IJ = Shear Crack Intersection

Beam Side View   b. Cross–Section

Figure 11.13: The intersection of shear cracking and vertical stirrups in two dimensional view.

#3 ($^3/_8$ inch diameter, $a_s = 0.11$ in^2) steel bar is the most commonly used for stirrups. #4 ($^1/_2$ in diameter, $a_s = 0.20$ in^2) bar is also used. #5 ($^5/_8$ in diameter, $a_s = 0.31$ in^2) is occasionally used. For SI units No. 8, 10 and 12 (8, 10 and 12 mm) are typical stirrup sizes. U–shaped stirrups, with two branches (n = 2) are the most common. Figure 11.14 shows other shapes of stirrups with different number of branches. Since stirrups are also used to tie longitudinal bars, their shape is sometimes alternated to facilitate such function as shown in Figure 11.14 with n = 1. The shape in Figure 11.14 with n = 3 can be considered as two stirrups in the same cross–section. Utilizing Equation 11.9 can yield $a_v = a_{v1} + a_{v2}$ where a_{v1} and a_{v2} are for the first and second stirrup, respectively. If both stirrups are of the same size, then Equation 11.9 can be used with n = total number of branches for both stirrups.

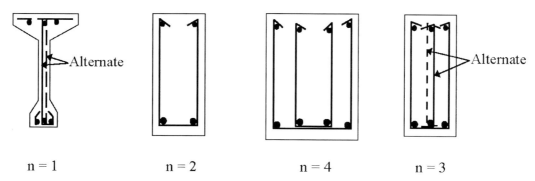

| n = 1 | n = 2 | n = 4 | n = 3 |

Figure 11.14: Stirrups with different shapes.

Anchor requirements for stirrups are also governed by ACI 318 Code. Coverage for this topic is included in Chapter 13 that deals with bond and development length.

11.4 The Truss Mechanism and The Arch Mechanism

The Truss Mechanism

The crack pattern due to bending and shear in Figure 11.8 may be simplified as shown in Figure 11.15. Cracks are shown in dashed lines. Longitudinal steel bars as well as stirrups are shown in thick solid lines representing tension forces. Compression in the concrete is shown as a top chord for the cracked beam. Also, the diagonal portions of the concrete beam between neighboring shear cracks are shown as compression elements of the beam. These portions are subjected to compression as their direction is parallel to the direction of σ_2, the second principal stress (maximum compression stress) due to shear as shown in Section 11.2 and Figure 11.5. Based on the crack pattern and the presence of steel reinforcement in the beam, one may redraw the beam represented by a truss containing tension and compression elements as shown in Figure 11.15 where tension members are shown in a solid line while compression members are shown in a dashed line. The truss mechanism is a theory of the method or mechanism by which reinforced concrete beams support gravity load.

The Arch Mechanism

The arch mechanism is another theory for reinforced concrete beam support of gravity load. In this mechanism, the concrete within the beam is considered cracked due to tension except for two regions, the compression zone and the area above support reactions (as explained in Section 11.2). The tension steel bars can then be viewed as a tension member or tie preventing horizontal movement at the support that may cause instability. In the arch mechanism, the action of stirrups is not considered. The arch mechanism is schematically illustrated in Figure 11.16.

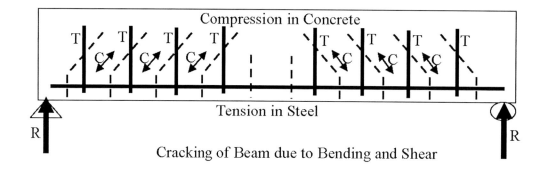

Cracking of Beam due to Bending and Shear

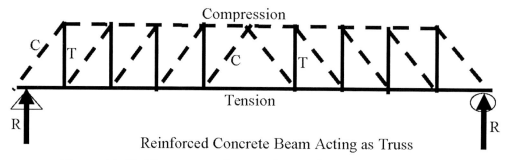

Reinforced Concrete Beam Acting as Truss

Figure 11.15: The truss mechanism for reinforced concrete beams.

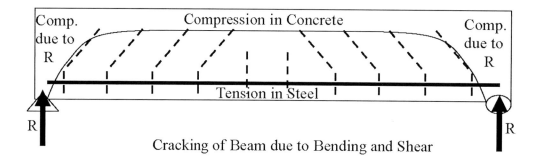

Cracking of Beam due to Bending and Shear

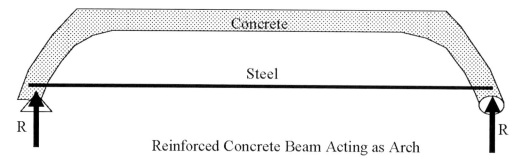

Reinforced Concrete Beam Acting as Arch

Figure 11.16: The arch mechanism for reinforced concrete beams.

Test your knowledge

1. Beam design for bending moment is design of beam cross–section; shear design, however, is for a longitudinal segment of the beam. Explain the reasons for these difference.

2. Where would one locate the most critical section for shear in a simply supported beam loaded with a uniformly distributed load?

3.   How is shear failure in a reinforced concrete beam characterized?

4.   What the most efficient stirrup angle with respect to beam center line and why?

5.   Explain the concept of stirrup number of branches?

6.   What are the major differences between the truss mechanism and the arch mechanism?

SHEAR DESIGN OF REINFORCED CONCRETE BEAMS

12.1 ACI 318 Code Provisions for Shear Design

ACI 318 Code has the following provisions (limitations) for shear design:

Governing Equations

The governing equations for shear design of reinforced concrete beams are:

$$\phi V_n \geq V_u \tag{12.1}$$

$$V_n = V_c + V_s \tag{12.2}$$

Where:

ϕ = the capacity reduction factor for shear design = 0.75

V_n = nominal beam shear resistance based on code assumptions and provisions.

V_u = factored applied (design) shear force acting on the beam.

V_c = concrete resistance to shear.

V_s = shear resistance of steel reinforcement.

Materials

Concrete: In calculating concrete shear resistance, the value of $\sqrt{f'_c}$ may not be taken greater than 100 psi (0.7 MPa). This precaution is implemented so than one does not overestimate high strength concrete's shear resistance. This precaution does not limit the value of f'_c but it imposes a limit of $\sqrt{f'_c}$ utilization for shear resistance.

Steel: Experiments have revealed a direct correlation between steel yield strength and the width of beam shear cracks. Thus, yield strength of stirrups or other steel bars used for steel reinforcement may not exceed 60 ksi (420 MPa). This provision is employed to reduce the width of diagonal 45° cracks caused by shear stresses.

<u>Special Case:</u> If welded wire fabric is used as shear reinforcement, then f_y may be increased to 80 ksi (560 MPa).

Maximum Factored Shear Force, V_{umax}

Figure 12.1 shows the factored shear force (V_u) diagram for a reinforced concrete beam subjected to a uniformly distributed load. The support reaction causes compression that counteracts the tension caused by shear. Because of this compression, cracking due to shear stresses does not occur in the beam sections directly above the support. The location of first possible shear crack is at the face of support as confirmed by experiments. Consequently, the portion of V_u diagram affecting beam sections above the support may be disregarded. Also, the factored shear force for this first possible shear crack occurs at about d (beam's effective depth) from support face of support which is the maximum factored shear force affecting the beam, V_{umax} as illustrated in Figures 12.2.

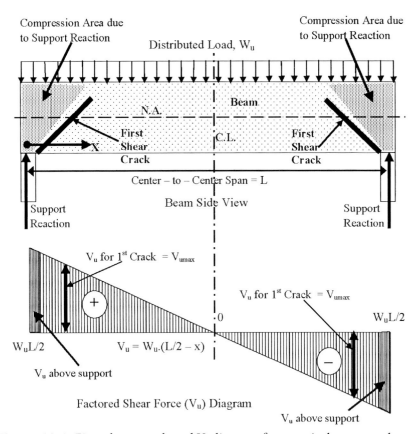

Figure 12.1: First shear crack and V_u diagram for a typical concrete beam.

In Figure 12.2, C = compressive force in concrete, T = tensile force in steel, V_c = shear resistance of the concrete compression zone, $\Sigma A_v f_y$ = collective tension force in the vertical stirrups, d = beam effective depth, and V_{umax} = maximum factored shear force for the beam shown in Figure 12.1. First (potential) shear crack commences at the face of support. The applied load within dimension d from the face of support is transferred directly to the support by compression. Consequently, the location for determining the value of V_{umax} for design is at distance d from support face as shown in Figure 12.2. This value of V_{umax} influences all beam sections within distance d from support face.

Considering the aforementioned discussion, ACI 318 Code allows disregarding the portion of shear force diagram above the support. Furthermore, ACI 318 Code allows that the calculated factored shear force at distance d from the face of support be used in beam design as maximum factored shear force, V_{umax}. Based on these code provisions, the shear diagram in Figure 12.1 is modified and presented in Figure 12.3.

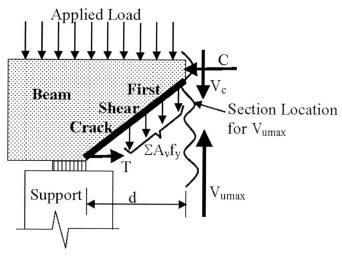

Figure 12.2: Location of critical section for calculating V_{umax} in a typical concrete beam.

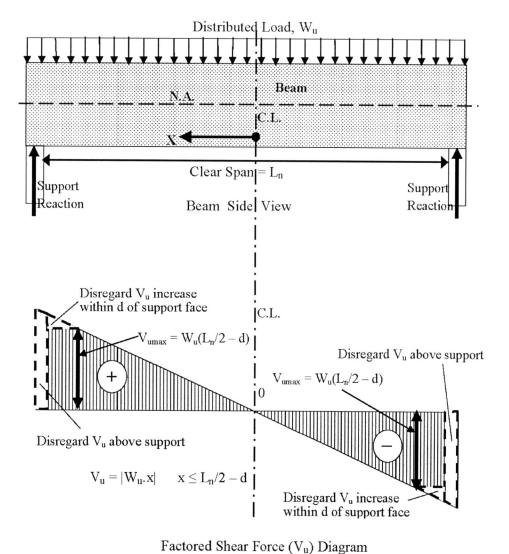

Factored Shear Force (V_u) Diagram

Figure 12.3: V_u diagram with ACI 318 Code provisions.

Exceptions

There are exceptions to the code provisions for disregarding portions of the factored shear force (V_u) diagram as shown in Figure 12.4 and illustrated as follows:

a.   Support reaction causes tension rather than compression within the concrete beam,

b.   The load is applied near the beam bottom portion. Thus, external loading causes tension within the beam,

c.   A concentrated force is applied to the beam within a distance less than d from the face of support including corbels and brackets.

For the exception cases above, design for unreduced factored shear force is required for all beam sections. Shear stresses within the beam sections above the support also require analysis and investigation.

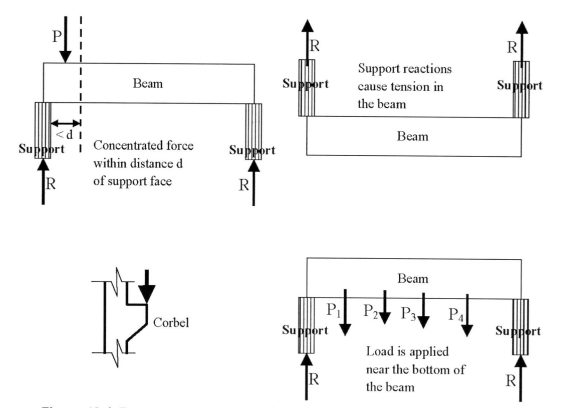

Figure 12.4: Exceptions to ACI 318 Code provisions of V_u diagram of Figure 12.3.

Shear Reinforcement Requirement

Based on ACI Code, shear reinforcement is required if the following equation applies:

$$V_u > \phi V_c/2$$

This code provision assumes that plain concrete can comfortably support an applied shear smaller than $\phi V_c/2$, and thus shear reinforcement is not required. However, designers are apt to provide minimum shear reinforcement in beams in spite of V_u being less than $\phi V_c/2$. Such precautionary measure is for safety against unpredicted shear cracking due to unusual or abnormal loading conditions. ACI Code does not require shear reinforcement for concrete slabs or footings if $V_u \leq \phi V_c$. In practice, the design for such members does not typically employ shear reinforcement due to potential construction complexities.

Maximum Spacing of Shear Reinforcement

Although, there are several types of shear reinforcement within the ACI Code, the following discussions in this chapter focus on vertical stirrups. For beam shear design, increasing stirrup spacing indicates reduction of shear reinforcement and vice versa. Larger spacing leads to fewer stirrups or less shear steel reinforcement and vice versa. Based on Equations 12.1 and 12.2, the amount of shear force required to be resisted by steel reinforcement can be determined:

$$V_s = V_u/\phi - V_c \tag{12.3}$$

There are limits or lower bounds for shear reinforcement or upper bounds for stirrup spacing (s) according to the ACI 318 Code depending on the value of V_u in order to reduce diagonal cracking caused by shear. These limits are listed below:

$V_s \leq 4\sqrt{f_c'} \cdot b_w d \implies s \leq d/2$ or 24 in whichever is smaller

$V_s > 4\sqrt{f_c'} \cdot b_w d \implies s \leq d/4$ or 12 in whichever is smaller

Where b_w is beam (web) width and d is its effective depth. As the horizontal projection of a typical 45° or diagonal shear crack is approximately equal to d, the code provision of $s \leq d/2$ is instated to insure that each shear crack is intercepted by at least one stirrup. The code provision of $s \leq d/4$ protects against potential extensive cracking at high values of beam shear forces. The above equations may be rewritten using the simplified expression for concrete shear resistance (Equation 11.2) as well as Equations 12.1 and 12.2 as:

$V_u \leq 6\phi\sqrt{f_c'} \cdot b_w d \implies s \leq d/2$ or 24 in whichever is smaller

$V_s > 6\phi\sqrt{f_c'} \cdot b_w d \implies s \leq d/4$ or 12 in whichever is smaller

Maximum Shear Force for Reinforced Concrete Beams

Shear forces applied to concrete beams are resisted has two components, concrete shear resistance and steel reinforcement shear resistance. While it appears that steel and concrete are cooperating in resisting shear, steel simply redirects and redistributes shear stresses to concrete that ultimately transfers all such forces to the support via the truss and/or the arch mechanisms explained in Chapter 11. Experiments have shown that reinforced concrete beams under the effect of excessively high shear forces experience extensive cracking in spite of the amount of provided steel reinforcement. ACI 318 Code imposes a limit upon the shear force that may be resisted by shear reinforcement in concrete beams as in the following equations:

$V_s \leq 8\sqrt{f_c'} \cdot b_w d \implies$ Beam may be designed to resist the applied shear force

$V_s > 8\sqrt{f_c'} \cdot b_w d \implies$ Beam section is inadequate to resist the applied shear force

Among the remedies for the case of beam inadequacy to resist shear are enlarging beam cross sectional dimensions and increasing concrete design compressive strength (f_c'). The above equations may be rewritten in terms of V_u using the simplified concrete shear resistance of Equation 11.7:

$V_u \leq 10\phi\sqrt{f_c'} \cdot b_w d \implies$ Beam may be designed to resist the applied shear force

$V_u > 10\phi\sqrt{f_c'} \cdot b_w d \implies$ Beam section is inadequate to resist the applied shear force

Based on the discussion above, one may summarize code requirements for concrete beam shear reinforcement:

$$V_u \leq \phi V_c/2 \implies \text{No shear reinforcement required} \tag{12.4}$$

$$\phi V_c/2 < V_u \leq 6\phi\sqrt{f_c'}\cdot b_w\cdot d \Longrightarrow s < d/2 \text{ or } 24 \text{ in } (600 \text{ mm}) \text{ whichever is smaller} \qquad (12.5)$$

$$6\phi\sqrt{f_c'}\cdot b_w\cdot d < V_u \leq 10\phi\sqrt{f_c'}\cdot b_w\cdot d \Longrightarrow s < d/4 \text{ or } 12 \text{ in } (300 \text{ mm}) \text{ whichever is smaller} \qquad (12.6)$$

$$V_u > 10\phi\sqrt{f_c'}\cdot b_w\cdot d \Longrightarrow \text{Beam section inadequate for the applied shear force} \qquad (12.7)$$

Minimum Area of Shear Reinforcement

The discussion in Section 6.4 of Chapter 6 regarding achieving a gradual failure mode in bending also applies to shear failure. The failure scenario of reinforced concrete in bending or shear to achieve a gradual failure mode is as follows:

a.   Tension cracking of concrete.

b.   Steel yielding accompanied by crack expansion and deformation increase.

c.   Compression crushing of concrete.

Flexural cracking is vertical while shear cracking is diagonal. Flexural steel is horizontal crossing flexural cracks at a 90–degree angle, resulting in high efficiency. The main type of shear reinforcement is vertical stirrups that cross shear cracks at 45–degree angle achieving an acceptable level of efficiency.

Step b of the aforementioned behavior scenario is pivotal for achieving a gradual failure mode. Steel yield is responsible for the gradual failure characteristic of reinforced concrete. In order to accomplish step b, steel reinforcement is required to support the load or stress previously supported by concrete prior to cracking. Thus, minimum reinforcement is required by ACI 318 Code for bending as well as shear so that sudden failure does not occur due to concrete tension cracking. Load increase following concrete tension cracking results in increased steel deformations and yielding, thereby, achieving gradual failure. Code required minimum shear reinforcement consists of provisions for maximum stirrup spacing and minimum shear reinforcement area. Maximum spacing of stirrups as a function of factored shear force has been covered in previous parts of this chapter. Minimum area of shear reinforcement based on ACI Code can be determined based on the following equation:

$$A_{vmin} = 0.75\sqrt{f_c'}\cdot b_w\cdot s/f_y \quad \text{or} \quad 50 b_w\cdot s/f_y \text{ whichever is larger} \qquad (12.8)$$

Consequent to tension yielding of steel, the concrete compression zone (located at crack tip in bending and shear) experiences crushing at 0.003 strain, resulting in failure as defined by ACI 318 Code. There are strong similarities between reinforced concrete behavior in bending and shear, revealing the uniqueness of the theory of reinforced concrete.

12.2 Design Principles

For one to understand the principles of shear design, it is useful to start by reiterating the principles of flexural design. Flexural design is essentially for determining the cross sectional configuration of reinforced concrete beams. In brevity, the outcome of beam flexural design is width, b, effective depth, d, and reinforcement ratio, ρ. Consequently, beam design for bending is design for an individual cross–section based on M_u and material properties (Figure 12.5). First, beam cross–sectional dimensions (b and d) are selected and typically remain constant for the entire beam. Since bending moment along a typical concrete beam is variable, it is customary that a segment of the beam is selected. M_{umax} for the selected segment is identified. This segment is, then, designed based on M_{umax}. With beam dimensions already chosen, segment design is focused on determining the value of ρ and steel bar configuration. This design is utilized throughout the entire segment.

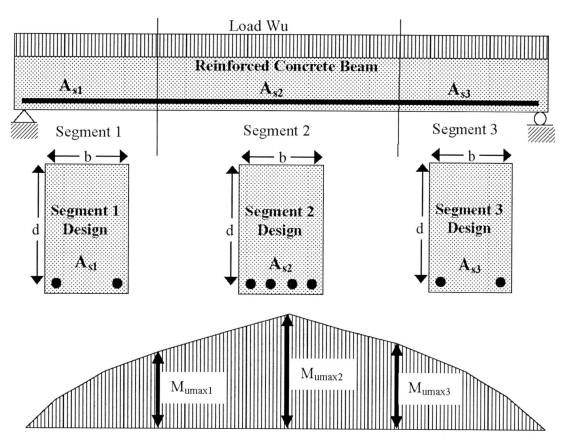

Factored Bending Moment (M_u) Diagram

Figure 12.5: Beam design for bending moment.

With the beam divided into several segment, several flexural cross–sectional designs are performed (Figure 12.5). Beam division into segments needs to be practical so that construction is not cumbersome. Also, tension steel bar adjustment or bar termination must be in accordance to ACI Code provisions as discussed in Chapter 13.

Shear design focuses upon determining stirrup configuration and spacing. In flexural design, section dimensions typically remain constant throughout the beam. Similarly, in shear design stirrup configuration is most likely to remain constant throughout the beam. The most common stirrup used for shear reinforcement of concrete beams is #3 (No. 10) U–shaped stirrup. It has two branches or legs (n = 2) and the cross–sectional area of each branch (a_s) is 0.11 in² (80 mm² for No. 10) resulting in a shear reinforcement area (A_v) of 0.22 in² (160 mm² for No. 10). In flexural design, the area of tension steel reinforcement (A_s) varies along the span of the beam dependent upon the value of M_u. In shear design, stirrup spacing (s) varies along the beam dependent upon the value of V_u. As in flexural design, the beam is divided into segments. For each segment, the value of maximum applied (magnified) shear force (V_{umax}) is determined and used for shear design utilized throughout the entire segment.

Beam division into segments needs to be practical. Segment design must follow ACI Code provisions outlined in this chapter. Figure 12.6 presents further illustrations.

Beam division into segments for bending does not normally match with beam division into segments for shear. Essentially, they should not match as each is based on different design principles or procedures. As an example, flexural design of simply supported beams generally considers the entire beam as one segment and utilizing A_s based on M_{umax} throughout the beam. However, it is typical that shear design of simply supported beams requires division into three segments; two symmetrical segments adjacent to the supports and a central segment.

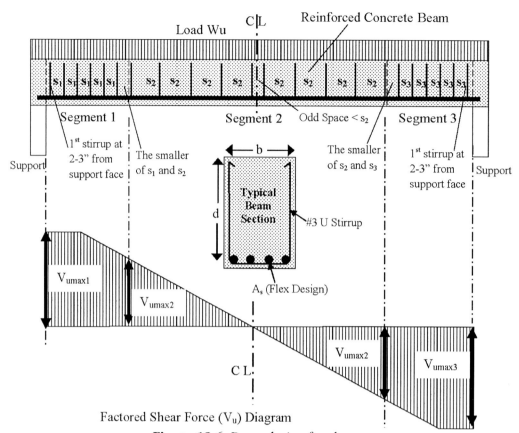

Figure 12.6: Beam design for shear.

12.3 Important Beam Shear Design Considerations

There are important design considerations for reinforced concrete beams under the effect of shear forces contained in Figure 12.6. They are aimed at achieving an efficient and practical (construction suitable) design. Some of such considerations are:

1. The beam portions above the supports do not contain stirrups. As explained earlier, shear stress effect upon the beam sections above the support is negligible.

2. Stirrup distribution throughout the beam commences by locating the first stirrup typically within 2 to 3 inches from the face of support. As such, the first potential diagonal crack at support face can be intercepted.

3. In spite of the low value of applied shear force V_u near midspan, it is prudent to use stirrups throughout the entire length of the beam to account for any unforeseen increased in shear force as explained in later sections of this chapter.

4. As stirrups are spaced evenly within a segment, an overlap into the adjacent segment frequently occurs. A spacing of two stirrups is likely to be directly at the division line between segments. For this spacing, the smaller value of stirrup spacing of the two adjacent segments is used. This solution equates to using more stirrups and increased beam shear resistance.

5. Stirrup distribution or location starts from both support faces towards the centerline of the beam. The resulting stirrup distribution is symmetrical about beam centerline. Accordingly, an odd space may result at beam centerline. Although odd stirrup spaces are not uncommon in beam shear design, one can attempt stirrup redistribution or beam redesign to avoid them and produce a more efficient or practical design.

12.4 Stirrup Design for a Segment

Based on the previous discussion in this chapter, the following steps can be used for segment shear design:

1. For design of beam segment for shear, beam cross–sectional dimensions and segment maximum design shear force V_{umax} are required. Beam cross–section is based on flexural design. Segment V_{umax} is determined from the beam factored shear force diagram obtained via structural analysis.

2. Assume stirrup configuration to determine the values of a_s and A_v. #3 U–shaped stirrups ($a_s = 0.11$ in² and $A_v = 0.22$ in²) are the most common due to their shaping/bending ease (for SI units common stirrups are No. 10 with $a_s = 80$ mm² and $A_v = 160$ mm²). Using multiple #3 U–shaped stirrups at the same location to increase A_v is also common. Larger diameter stirrups are more difficult to shape/bend into the desired configuration. Generally, #3 U–shaped stirrups are assumed. Based on beam shear design and particularly design stirrup spacing, this assumption may be modified.

3. Use $V_u = V_{umax}$ for the segment and utilize Equations 12.4 though 12.7 to:

   a. Determine if shear reinforcement is required based on Equation 12.4. In this textbook, however, the author recommends the use of shear reinforcement throughout the entire beam for safety and practicality.

   b. Determine if the beam section adequate for resistance of applied shear force V_u based on Equation 12.7.

   c. Compute maximum stirrup spacing based on Equations 12.5 and 12.6.

4. Determine the value of V_s based on Equation 12.4. Furthermore, determine the required stirrup spacing (s) for resistance to applied shear based on Equation 11.13 (Chapter 11) rewritten as follows:

$$s = a_v \cdot f_y \cdot d / V_s \qquad (12.9)$$

5. Based on steps 3 and 4, the stirrup spacing for the segment is selected. The selected value should be smaller or equal to the value of spacing from step 3 to satisfy code cracking requirements and smaller or equal to the value of spacing from step 4 to satisfy code cracking requirements for resistance the applied shear. Furthermore, the selected stirrup spacing (s) should be in multiples of 2 or 3 inches (100 mm) for construction suitable beam design. With vertical stirrups, beam shear design is unaffected by shear force direction or sign. Thus, antisymmetric factored shear force diagram may be considered symmetrical. The design is performed for half of the beam and mirrored to the other half.

6. Confirm that the assumed stirrups in step 2 meet ACI Code requirements for A_{vmin} based on Equation 12.8 utilizing the selected spacing in step 5.

Following shear design of each segment, stirrups are distributed within the beam in a manner that satisfies stirrup spacing for each segment and the design considerations discussed in Sections 12.2 and 12.3.

Based on Chapter 11 and the previous parts of this chapter, vertical stirrups are suitable for shear design with no effect for the sign of applied shear force V_u (positive or negative). Vertical stirrups are at 45° angle with either critical shear crack or plane due to positive or negative shear force (Figure 12.2).

Example 12.1

Select the required stirrup spacing for beam design for the beam section shown for the values of V_u listed. Assume $f'_c = 3,000$ psi and $f_y = 60,000$ psi

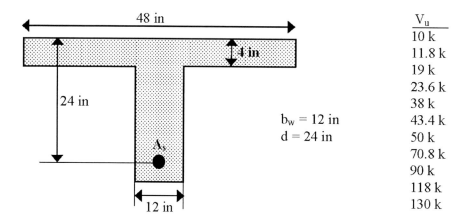

	V_u
	10 k
	11.8 k
	19 k
	23.6 k
	38 k
	43.4 k
	50 k
	70.8 k
	90 k
	118 k
	130 k

$b_w = 12$ in
$d = 24$ in

Solution

The following table that summarizes ACI 318 Code cracking requirements is useful for beam shear design:

V_u less than	$\phi\sqrt{f'_c}\,b_w d$	$6\phi\sqrt{f'_c}\,b_w d$	$10\phi\sqrt{f'_c}\,b_w d$
	11.8 k	70.8 k	118 k
Code required s_{max}	No stirrups Needed	d/2 or 24 in	d/4 or 12 in
	—	12 in	6 in

For the table of stirrup spacing design for the beam in following page:

1.   Assume that #3 U shaped stirrups will be used resulting in $A_v = 0.22$ in².

2.   Compute the values of V_c and ϕV_c: $V_c = 2\sqrt{f'_c}\times12\times24 = 31,500$ lb $= 31.5$ k

$$\phi V_c = 0.75\times2\sqrt{f'_c}\times12\times24 = 23,600 \text{ lb} = 23.6 \text{ k}$$

V_u (k)	s_{max} required by code (in) table above	Stirrup spacing for shear resistance Equation 12.10 (in)	Design stirrup spacing (in)	A_{vmin} required by code Equation (12.8)
10	No shear reinforcement required	$V_s = V_u/\phi - V_c < 0$ ===> Vs = 0 k ===> No stirrups needed	Use min shear reinforcement $s = s_{max} = 12$in	$A_{vmin} = 0.12$ in² $A_v = 0.22$ in² > A_{vmin} ===> OK
11.8	12 in	$V_s = V_u/\phi - V_c < 0$ ===> Vs = 0 k ===> No stirrups needed	Use s = 12 in	$A_{vmin} = 0.12$ in² $A_v = 0.22$ in² > A_{vmin} ===> OK

19	12 in	$V_s = V_u/\phi - V_c < 0$ $===> Vs = 0$ k $===>$ No stirrups needed	Use s = 12 in	$A_{vmin} = 0.12$ in^2 $A_v = 0.22$ in$^2 > A_{vmin}$ $===>$ OK
23.6	12 in	$V_s = V_u/\phi - V_c$ $===> V_s = 0$ k $===>$ No stirrups needed	Use s = 12 in	$A_{vmin} = 0.12$ in^2 $A_v = 0.22$ in$^2 > A_{vmin}$ $===>$ OK
38	12 in	$V_s = V_u/\phi - V_c$ $V_s = 19.2$ k $s = 0.22\times60\times24/19.2$ $s = 16.5$ in	Use s = 12 in	$A_{vmin} = 0.12$ in^2 $A_v = 0.22$ in$^2 > A_{vmin}$ $===>$ OK
43.4	12 in	$V_s = V_u/\phi - V_c$ $V_s = 26.4$ k $s = 0.22\times60\times24/26.4$ $s = 12$ in	Use s =12 in	$A_{vmin} = 0.12$ in^2 $A_v = 0.22$ in$^2 > A_{vmin}$ $===>$ OK
50	12 in	$V_s = V_u/\phi - V_c$ $V_s = 35.2$ k $s = 0.22\times60\times24/35.2$ $s = 9.2$ in	Use s = 9 in	$A_{vmin} = 0.09$ in^2 $A_v = 0.22$ in$^2 > A_{vmin}$ $===>$ OK
70.8	6 in	$V_s = V_u/\phi - V_c$ $V_s = 62.9$ k $s = 0.22\times60\times24/62.9$ $s = 5.1$ in	Use s = 5 in	$A_{vmin} = 0.05$ in^2 $A_v = 0.22$ in$^2 > A_{vmin}$ $===>$ OK
90	6in	$V_s = V_u/\phi - V_c$ $V_s = 88.5$ k $s = 0.22\times60\times24/88.5$ $s = 3.6$ in	Use s = 3 in	$A_{vmin} = 0.03$ in^2 $A_v = 0.22$ in$^2 > A_{vmin}$ $===>$ OK
118	6 in	$V_s = V_u/\phi - V_c$ $V_s = 125.8$ k $s = 0.22\times60\times24/125.8$ $s = 2.5$ in	Use s = 2.5 in	$A_{vmin} = 0.02$ in^2 $A_v = 0.22$ in$^2 > A_{vmin}$ $===>$ OK
130	$V_u > 10\phi(f'_c)^{1/2}b_wd = 118$ k $===>$		Per code, cross–section is inadequate for resistance of applied shear force (V_u) $===>$ Larger cross–section is required	

The solution for this example can be summarized in the following table:

V_u (k)	Stirrup Design	Code Required s_{max} Cracking Criteria
10	#3 U stirrups @ 12 in o.c.	N/A
$11.8 = \phi V_c/2$	#3 U stirrups @ 12 in o.c.	12 in
19	#3 U stirrups @ 12 in o.c.	12 in
$23.6 = \phi V_c$	#3 U stirrups @ 12 in o.c.	12 in
38	#3 U stirrups @ 12 in o.c.	12 in
$43.4 = \phi(V_c + A_vf_yd/s_{max})$	#3 U stirrups @ 12 in o.c.	12 in

50	#3 U stirrups @ 9 in o.c.	12 in
70.8	#3 U stirrups @ 5 in o.c.	6 in
90	#3 U stirrups @ 3 in o.c.	6 in
118	#3 U stirrups @ 2.5 in o.c.	6 in
130	Cross – section size need to be increased for shear resistance	

Of note is that for $V_u \leq \phi(V_c + A_v f_y d/s_{max})$, the required shear reinforcement spacing based on the ACI 318 code is $s_{max} = 12$ in (generally $d/2$ or 24 in whichever is smaller). As such, this value of V_u is important for beam division into segments as explained later in this chapter.

Example 12.2

Select the required stirrup spacing for design of the beam section shown for the values of V_u listed. Also, draw a diagram showing the variation of s with V_u.

Use $f'_c = 4,000$ psi and $f_y = 60,000$ psi.

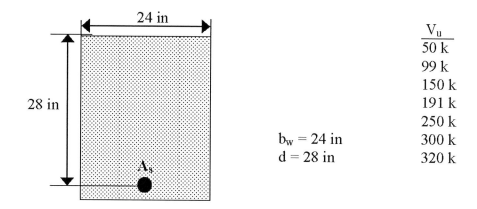

$$\frac{V_u}{}$$
50 k
99 k
150 k
191 k
250 k
300 k
320 k

$b_w = 24$ in
$d = 28$ in

Solution

ACI 318 Code requirements for cracking combined with the recommendation of always using minimum shear reinforcement in beams regardless of V_u are summarized in the following design aid table:

V_u less than	$6\phi\sqrt{f'_c}\,b_w d$	$10\phi\sqrt{f'_c}\,b_w d$
	191 k	318 k
Code required s_{max}	d/2 or 24 in	d/4 or 12 in
	14 in	7 in

Assume #4–U shaped stirrups with $A_v = 0.40$ in².

$V_c = 2\sqrt{f'_c} \times 24 \times 28 = 85,000$ lb $= 85.0$ k

$\phi V_c = 0.75 \times 2\sqrt{f'_c} \times 24 \times 28 = 63,800$ lb $= 63.8$ k

Based on the above design aid table, s_{max} = 14 in
For $V_u = \phi(V_c + A_v \cdot f_y \cdot d / s_{max})$, the required spacing for code cracking criteria is equal to the required spacing for resistance to applied shear
$\phi(V_c + A_v \cdot f_y \cdot d / s_{max}) = 0.75 \times (85.0 + 0.40 \times 60 \times 28/14) = 99,800$ lb = 99.8 k

Example 12.2 solution is tabulated and graphically presented in the following page.

V_u (k)	s_{max} by code (in)	$V_s = V_u/\phi - V_c$ (k)	$s = a_v \cdot f_y \cdot d / V_s$ (in)	s used for design (in)	A_{vmin} required Equation (12.8)
50	No stirrups	$V_s < 0$	Not Applicable	14	0.27 in² < A_v OK
99	14	$V_s = 47$	13.9	14	0.27 in² < A_v OK
150	14	$V_s = 115$	5.8	5	0.10 in² < A_v OK
191	14	$V_s = 170$	3.9	3	0.06 in² < A_v OK
250	7	$V_s = 248$	2.7	2.5	0.04 in² < A_v OK
300	7	$V_s = 315$	2.1	2	0.04 in² < A_v OK
320	Not allowed by code – Need a larger cross–section				

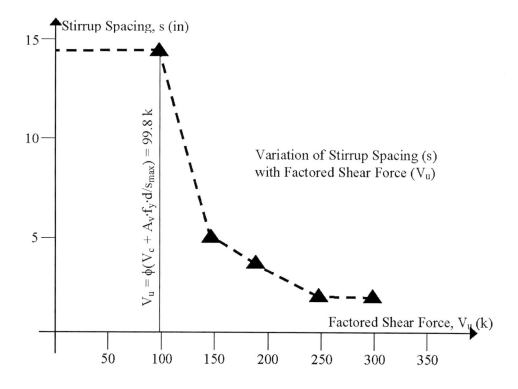

As previously stated, for shear design, the beam is divided into segments and each segment is designed individually (Section 12.2 and Figure 12.6). Based on Example 12.2, V_{umax} for the segment at beam center line should be equal or slightly lower than $\phi(V_c + A_v f_y d/s_{max})$. Such central segment requires stirrup spacing of s_{max} (d/2 or 24 in whichever is smaller) resulting in an efficient design. Beam division into segments and distribution of stirrups for beam shear design are discussed in the following examples.

Example 12.3

Design the beam shown below for shear including stirrup configuration, spacing and distribution using beam section of Example 12.1, $f'_c = 3,000$ psi and $f_y = 60,000$ psi.

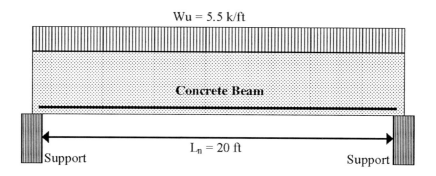

Solution

<u>First</u> draw the magnified shear force diagram as shown below using the beam clear span L_n in lieu of the center–to–center span since shear forces above the support are neglected per ACI Code.

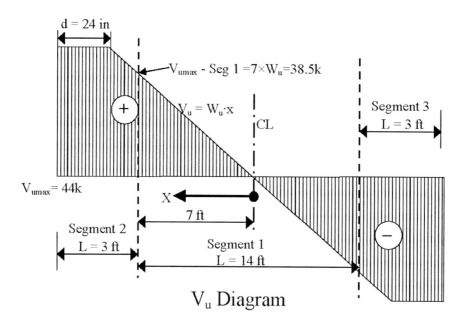

V_u Diagram

Second, a design aid table is composed as shown below is needed similar to Examples 12.1 and 12.2. In this table, #3 U shaped stirrups are assumed ($A_v = 0.22$ in²).

V_u less than	$6\phi\sqrt{f'_c}b_w d$	$10\phi\sqrt{f'_c}b_w d$	$\phi(V_c + A_v \cdot f_y \cdot d/s_{max})$
	70.8 k	118 k	43.4 k
Code required s_{max}	d/2 or 24 in	d/4 or 12 in	d/2 or 24 in
	12 in	6 in	12 in
Code required A_{vmin}	0.12 in²	0.03 in²	0.12 in²

For the above shown table, the following values were utilized:

$$- V_c = 2\sqrt{f_c'} \times 12 \times 24/1,000 = 31.5 \text{ k} \quad \phi V_c = 0.75 \times 2\sqrt{f_c'} \times 12 \times 24/1,000 = 23.6 \text{ k}$$

$- s_{max} = 12$ in based on ACI Code requirements, thus,

$$\phi(V_c + A_v \cdot f_y \cdot d/s_{max}) = 0.75 \times (31.5 + 0.22 \times 60 \times 24/12)/1,000 = 43.4 \text{ k}$$

– Minimum shear reinforcement: #3–U shaped stirrups ($A_v = 0.22$ in^2 > A_{vmin}) are adequate for the spacings in the table. Code required A_{vmin} decreases with decreased stirrup spacing. For $s = 12$ in, $A_{vmin} < A_v$ for #3–U stirrups = 0.22 in^2; thus, #3–U shaped stirrups are adequate for spacings of 12 in or less.

Third, divide the beam efficiently into segments so that suitable stirrup distribution can be achieved. A proposed segment division for Example 12.3 is shown in the previous page. For each segment, V_{umax} is determined; based on which the required stirrup spacing for resistance to applied shear is calculated. Consequently, a practical value for spacing for each segment is selected. Such value is required to be smaller than code dictated cracking requirement (included in the design aid table) and the requirement for resistance of applied shear. As such, both requirements would be satisfied.

Segment 1

$- V_{umax} = 5.5 \times 7 = 38.5 \text{ k} < 43.4 \text{ k}$ ===> Based on the design aid table, $s = 12$ in

satisfies code cracking requirement and resistance to applied shear (Example 12.2)

Thus, for segment 1, use #3–U stirrups @12 in O.C.

Segments 2&3

$- V_{umax} = 44 \text{ k}$ ===> $V_{smax} = 44/0.75 - 31.5 = 27.2 \text{ k}$

– Required stirrup spacing for resistance of applied shear:

$$s = A_v f_y d/V_{smax} = 0.22 \times 60 \times 24/27.2 = 11.6 \text{ in}$$

– Code required maximum stirrup spacing for cracking requirements for $V_u < 70.8$ k is 12 in based on the design aid table.

– A practical value for stirrup spacing can be 10 in which satisfies both code requirements of cracking and resistance to applied shear.

Fourth, a practical stirrup distribution within the beam is designed. Stirrups are distributed within the beam starting from the first stirrup (2 – 3 in from support face) moving towards the center of the beam, Due to the symmetry of V_u diagram for this example, stirrup distribution is also symmetrical. This distribution is required to be practical and construction efficient. Shown below are three different stirrup distributions; each can be considered a possible solution to Example 12.3.

Solution No. 1

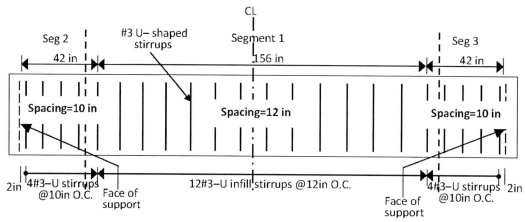

Total No. of stirrups = 22

Beam side view

Solution No. 2

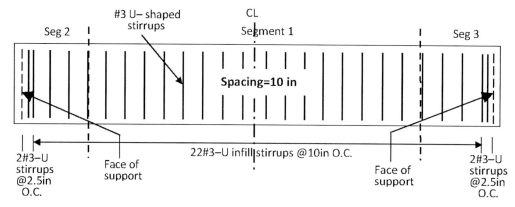

Total No. of stirrups = 26

Beam side view

Solution No. 3

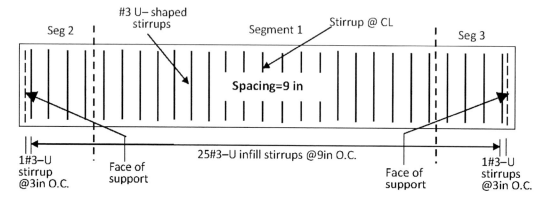

Total No. of stirrups = 27

Beam side view

Example 12.4

Design the beam shown below for shear including stirrup configuration, spacing and distribution for beam cross–section shown below.

Assume $f'_c = 4,000$ psi and $f_y = 60,000$ psi.

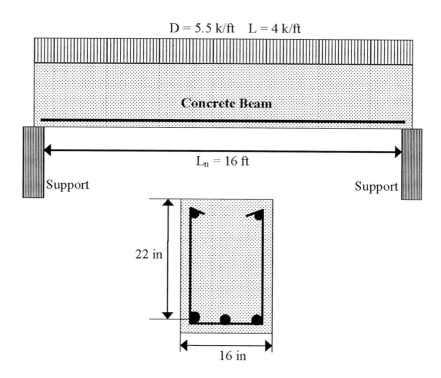

D = 5.5 k/ft   L = 4 k/ft

Concrete Beam

$L_n = 16$ ft

Support   Support

22 in

16 in

Solution

First draw the magnified shear force diagram as shown below:

$W_u = 1.2 \times 5.5 + 1.6 \times 4 = 6.6 + 6.4 = 13$ k/ft

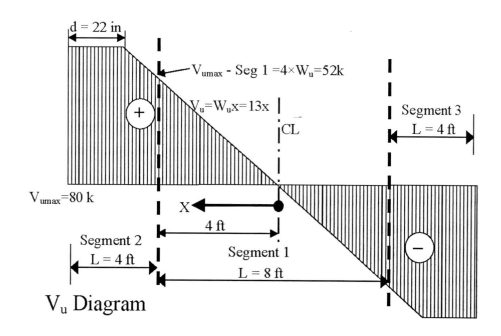

d = 22 in

V_{umax} - Seg 1 = $4 \times W_u$ = 52k

$V_u = W_u x = 13x$

CL

Segment 3
L = 4 ft

V_{umax} = 80 k

X

4 ft

Segment 2
L = 4 ft

Segment 1
L = 8 ft

V_u Diagram

Second, a design aid table as shown below needs to be assembled. In this table, #3 U shaped stirrups are assumed ($A_v = 0.22$ in²).

V_u less than	$6\phi\sqrt{f'_c}b_wd$	$10\phi\sqrt{f'_c}b_wd$	$\phi(V_c + A_vf_yd/s_{max})$
	100.2 k	167 k	55.2 k
Code required s_{max}	d/2 or 24 in	d/4 or 12 in	d/2 or 24 in
	10 in	5 in	10 in
Code required A_{vmin}	0.13 in²	0.07 in²	0.13 in²

For the above shown table, the following values were utilized:

– $V_c = 2\sqrt{f'_c}\times16\times22/1{,}000 = 44.5$ k $\phi V_c = 0.75\times2\sqrt{f'_c}\times16\times22/1{,}000 = 33.4$ k

– Although code requires d/2 = 11 in for $V_u \le 6\phi\sqrt{f'_c}b_wd$, 10 in is selected to simplify stirrup distribution.

– Similarly, code requires d/4 = 5.5 in for $6\phi\sqrt{f'_c}b_wd \le V_u \ge 6\phi\sqrt{f'_c}b_wd$, but 5 in is used.

– $s_{max} = 10$ in, thus

$$\phi(V_c + A_v\cdot f_y\cdot d/s_{max}) = 0.75\times(44.5 + 0.22\times60\times22/10)/1{,}000 = 55.2 \text{ k}$$

– Minimum shear reinforcement: #3–U shaped stirrups ($A_v = 0.22$ in² > A_{vmin}) are adequate for the spacings in the table. $A_{vmin} = 0.13$ in² for $s_{max} = 10$ in. With #3–U shaped stirrups, $A_v = 0.22$ in² > 0.13 in². Thus #3–U shaped stirrups are adequate for spacings of 10 in or less since A_{vmin} decreases with s.

Third, divide the beam efficiently into three segments as shown in the previous page. For each segment, a practical value for stirrup spacing is selected which is smaller than both code dictated spacing for cracking requirement and the spacing required for resistance of applied shear.

Segment 1

– $V_{umax} = 13\times4 = 52$ k ⟹ Use s = 10 in

Segment 1 design: #3–U stirrups @10 in O.C.

Segments 2&3

– $V_{umax} = 80$ k ⟹ $V_{smax} = 80/0.75 – 44.5 = 62.2$ k

– Required stirrup spacing for resistance of applied shear:

$$s = A_vf_yd/V_{smax} = 0.22\times60\times22/62.2 = 4.6 \text{ in}$$

– Code required max. stirrup spacing for $V_u < 167$ k is 5 in from the design aid table.

– A practical value for stirrup spacing can be 4 in which satisfies both code requirements of cracking and resistance to applied shear.

Segments 2&3 design: #3–U stirrups @4 in O.C.

Stirrup Distribution

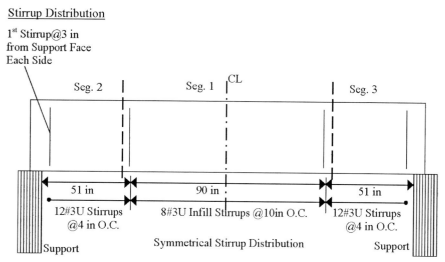

1st Stirrup@3 in
from Support Face
Each Side

Seg. 2 | Seg. 1 |CL Seg. 3

51 in 90 in 51 in

12#3U Stirrups
@4 in O.C.

8#3U Infill Stirrups @10in O.C.

12#3U Stirrups
@4 in O.C.

Support

Symmetrical Stirrup Distribution

Support

Total No. of Stirrups = 34

Example 12.5

Redo Example 12.4 taking into account distribution of live loads on the beam to achieve maximum applied shear.

Solution

For the various beam load cases, the entire dead load must always be applied.

First Load Case: dead load and live load on the entire span

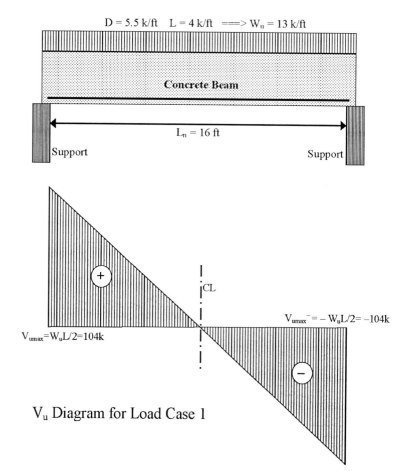

$D = 5.5$ k/ft $L = 4$ k/ft   ===> $W_u = 13$ k/ft

Concrete Beam

$L_n = 16$ ft

Support   Support

+

|CL

−

$V_{umax} = W_u L/2 = 104$k

$V_{umax}^- = -W_u L/2 = -104$k

V_u Diagram for Load Case 1

<u>Second Load Case</u>: dead load on the entire span and live load on the left half of the span.

Load factors 1.2 and 1.6 for dead and live load, respectively, are used for computation of ultimate load W_u. For the left half of the beam, the value of W_u is the same as the first load case. For the right half of the beam, $W_u = 1.2D + 1.6 \times 0 = 6.6$ k/ft.

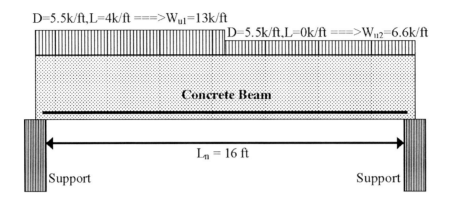

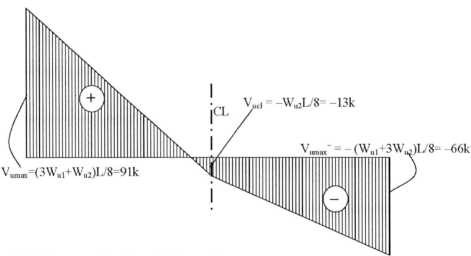

V_u Diagram for Load Case 2

<u>Third Load Case</u>: dead load on the entire span and live load on the right half of the span

Using load factors 1.2 and 1.6 for dead and live load, respectively, the following loading condition results.

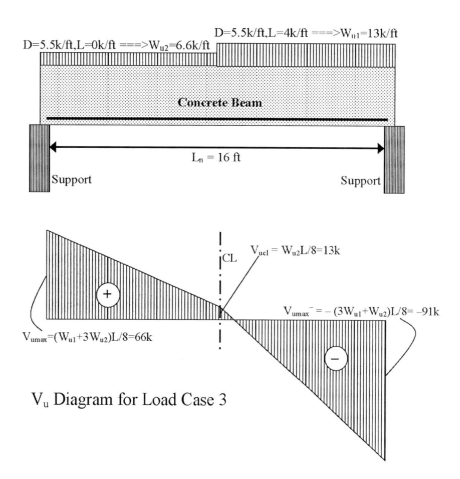

V_u Diagram for Load Case 3

As noted, load cases 2 and 3 have resulted in a non–zero shear at center line. Thus, it is prudent to consider redistribution of live load as stirrups may be required at beam CL. Designers typically utilize minimum shear reinforcement in beams for the entire span length.

The shear force envelope can be used to facilitate shear design taking into account all three load cases. It is produced by superimposing the shear force diagrams from the different load cases on each other as shown below:

Vertical stirrups have the same effect in shear resistance to positive or negative shear force V_u. Thus, half of the shear force envelope is used for stirrup design; then the design is mirrored to the other half. First, ACI 318 Code provisions are applied to the shear force envelope as shown in the following figure.

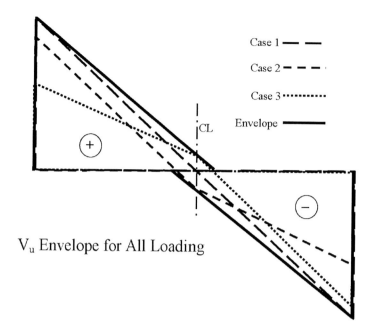

V_u Envelope for All Loading

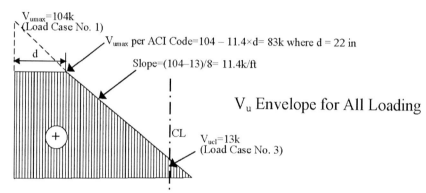

V_u Envelope for All Loading

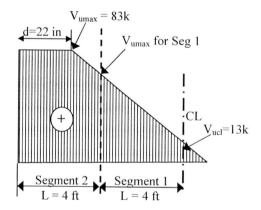

Beam Division into Segments

The design aid table of Example 12.4 can be reused for this example.

Segments 1

− V_{umax} = 58.6 k ⟹ V_{smax} = 58.6/0.75 − 44.5 = 33.6 k

− Required stirrup spacing for resistance of applied shear:

$$s = A_v f_y d / V_{smax} = 0.22 \times 60 \times 22 / 33.6 = 8.6 \text{ in}$$

– Maximum stirrup spacing is 10 in based on the design aid table ($V_u < 100.2$ k).

– A practical value for stirrup spacing of 8 in is selected which satisfies cracking and resistance to applied shear requirements.

Segment 1 design: #3–U stirrups @8 in O.C.

Segments 2

– V_{umax} = 83 k ===> V_{smax} = 83/0.75 – 44.5 = 66.2 k

– Required stirrup spacing for resistance of applied shear:

$s = A_v f_y d / V_{smax} = 0.22 \times 60 \times 22 / 66.2 = 4.4$ in

– Maximum stirrup spacing is 5 in based on the design aid table ($V_u < 167$ k).

– A practical value for stirrup spacing of 4 in is selected as 4 in which satisfies cracking and resistance to applied shear requirements.

Segment 2 design: #3–U stirrups @4 in O.C.

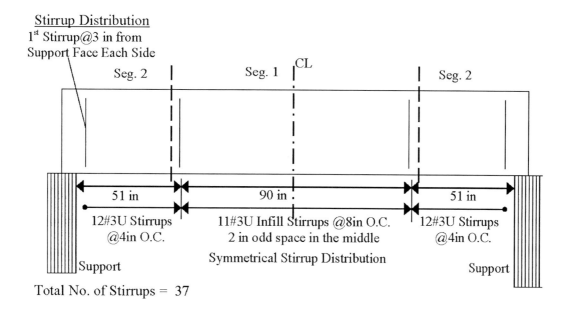

The utilization of live load redistribution results in a more conservative shear design for the middle part of the beam where stirrup spacing was reduced from 10 in to 8 in. Such design is prudent. However, the above distribution contains an odd space. A good student practice is to redistribute the stirrups above to eliminate the odd space. ACI Code does not require distribution of live load within a span to achieve maximum effect. The required distribution of live load is between spans within a structure or building.

Example 12.6

Design the beam shown below for shear including stirrup configuration, spacing and distribution. Beam cross–section is identical to the beam in Examples 12.4 and 12.5. The dead load is distributed while the live load is concentrated and not applied continuously or in tandem. Use f'_c = 3,000 psi and f_y = 60,000 psi.

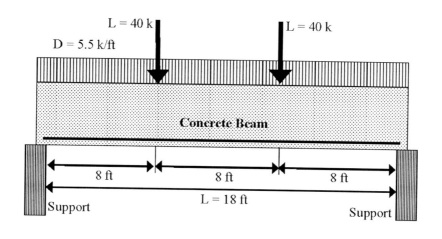

Solution

Use load combination: U = 1.2D + 1.6L. Dead load must be always applied to the beam in its entirety.
<u>First Load Case</u>: dead load on the entire span and both concentrated live loads.

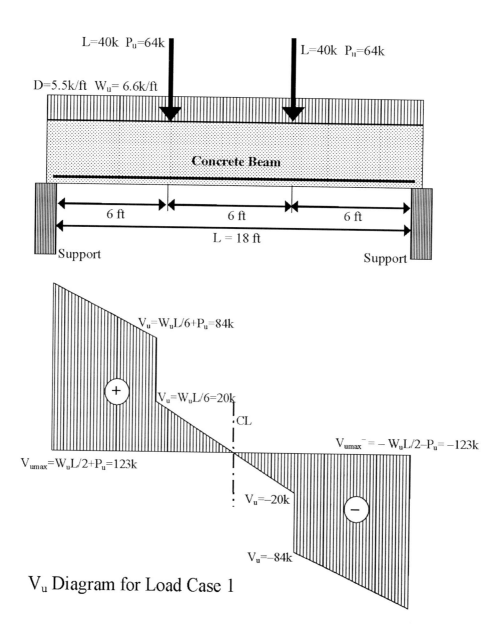

V_u Diagram for Load Case 1

<u>Second Load Case</u>: dead load on the entire span and live load on the left side.

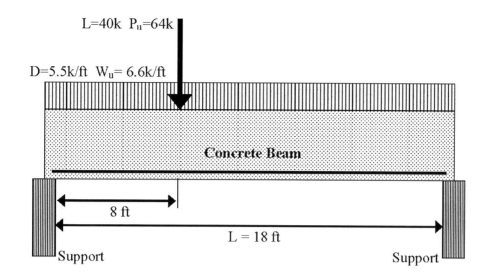

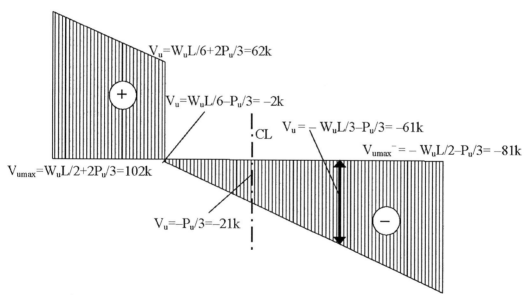

V_u Diagram for Load Case 2

<u>Third Load Case</u>: dead load on the entire span and live load on the right side.

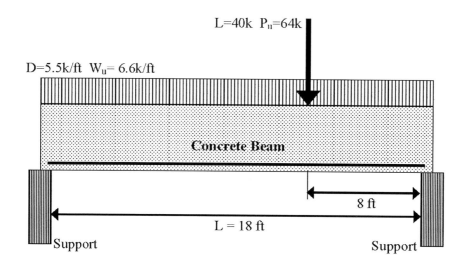

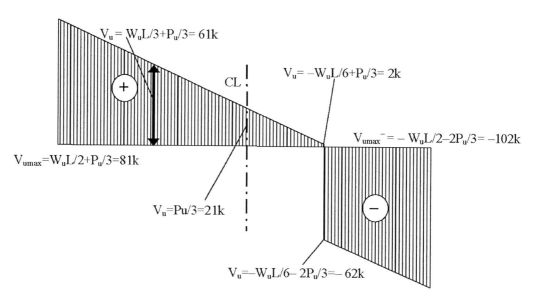

V_u Diagram for Load Case 3

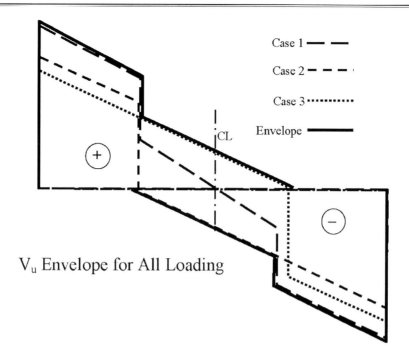

V_u Envelope for All Loading

The shear force envelope is utilized for design similar to Example 12.5. Also, half of the shear force envelope is used and the design is mirrored to the other half. The shear force envelope with ACI 318 Code provisions is shown below.

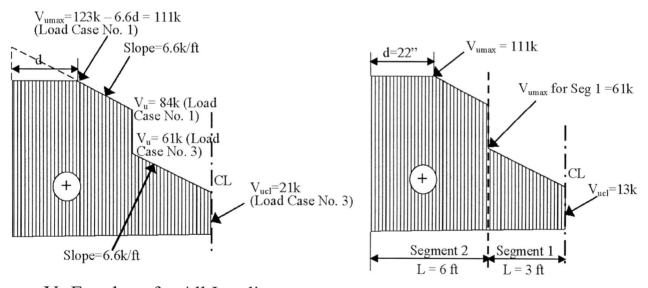

V_u Envelope for All Loading   Beam Division into Segments

Using the same design procedure for Examples 12.3 through 12.5:

Segments 1 – #3–U stirrups @6in O.C.

Segments 2 – #3–U stirrups @2.5in O.C.

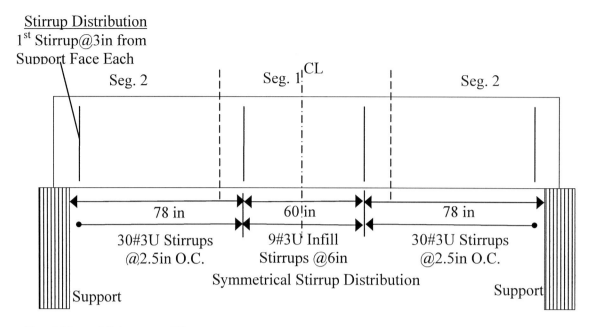

Stirrup Distribution
1st Stirrup@3in from Support Face Each

Total No. of Stirrups = 71

Each concentrated force is applied to an area with measurable dimensions and not to a point as mathematically modeled. Thus, the stirrups at 2.5 in spacing are extended well beyond the location of the concentrated loads for safety reasons against local shear forces due to these loads. A fourth load case with only dead load (U = 1.4D) is not considered in Example 12.6 since it does not generally govern.

Example 12.7

Design the beam shown below for shear including stirrup configuration, spacing and distribution. Use f'_c = 4,000 psi and f_y = 60,000 psi.

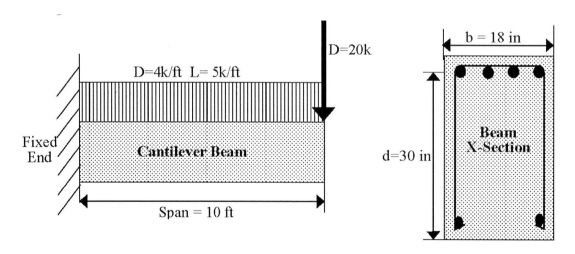

Solution

W_u = 1.2×4 + 1.6×5 = 12.8 k/ft P_u = 1.2×20 = 24 k

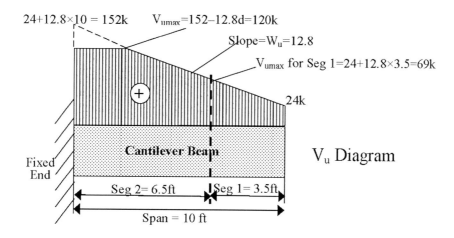

The design aid table for this example is shown in the next page. #3 U shaped stirrups are assumed (A_v = 0.22 in²).

V_u less than	$6\phi\sqrt{f_c'}b_wd$	$10\phi\sqrt{f_c'}b_wd$	$\phi(V_c + A_v \cdot f_y \cdot d/s_{max})$
	153.7 k	256.1 k	76.0 k
Code required s_{max}	d/2 or 24 in	d/4 or 12 in	d/2 or 24 in
	12 in	6 in	12 in
Code required A_{vmin}	0.18 in²	0.09 in²	0.18 in²

For the above shown table, the following values were utilized:

– $V_c = 2\sqrt{f_c'} \times 18 \times 30/1,000 = 68.3$ k $\phi V_c = 0.75 \times 2\sqrt{f_c'} \times 18 \times 30/1,000 = 51.2$ k

– Although code requires d/2 = 15 in for $V_u \leq 6\phi\sqrt{f_c'}b_wd$, 12 in is selected so that #3 U stirrups can be used. With s = 15 in, $A_{vmin} > 0.22$ in² and #3 U stirrups are not adequate.

– Also, code requires d/4 = 7.5 inch for $6\phi\sqrt{f_c'}b_wd < V_u \geq 6\phi\sqrt{f_c'}b_wd$, but 6 in is selected to facilitate construction.

– s_{max} = 12 in, thus

$$\phi(V_c + A_v \cdot f_y \cdot d/s_{max}) = 0.75 \times (68.3 + 0.22 \times 60 \times 30/12)/1,000 = 76.0 \text{ k}$$

– Minimum shear reinforcement: #3–U shaped stirrups (A_v = 0.22 in² > A_{vmin}) are adequate for the spacings in the table. Thus, #3–U shaped stirrups are adequate for spacings of 12 in or less.

Divide the beam efficiently into two segments. For each segment, stirrup spacing is a practical value selected smaller than both code cracking requirement and requirement of resistance of applied shear.

Segment 1

– V_{umax} = 69 k $\Longrightarrow$ V_{smax} = 69/0.75 – 68.3 = 23.7 k

– Required stirrup spacing for resistance of applied shear:

$s = A_v f_y d/V_{smax} = 0.22 \times 60 \times 30/23.7 = 17$ in

– Use s = 12 in to satisfy cracking and resistance to applied shear requirements

    Segment 1 design: #3–U stirrups @12 in O.C.

Segments 2

– $V_{umax} = 120$ k $\Longrightarrow$ $V_{smax} = 120/0.75 - 68.3 = 91.7$ k

– Required stirrup spacing for resistance of applied shear:

$\quad$ s = $A_v f_y d/V_{smax}$ = 0.22×60×30/91.7 = 4.3 in

– Use s = 4 in $\quad$ to satisfy cracking and resistance to applied shear requirements

$\quad\quad$ Segment 2 design: #3–U stirrups @4 in O.C.

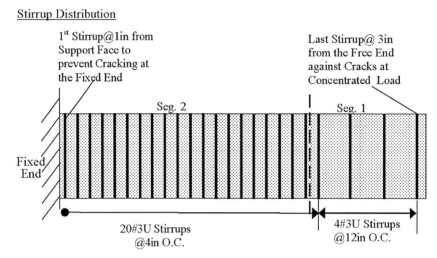

Total number of stirrups = 24

Example 12.8

Design the beam shown below for shear including stirrup configuration, spacing and distribution using beam section shown below, $f'_c = 30$ MPa and $f_y = 420$ MPa.

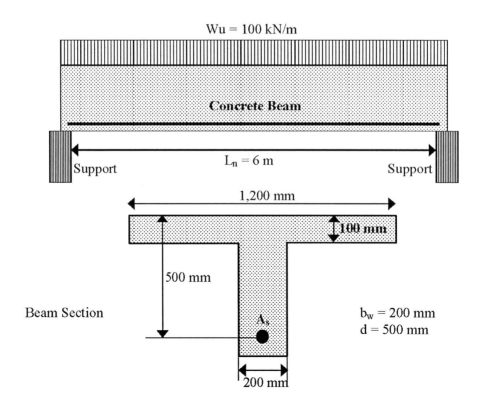

Solution

Draw the magnified shear force diagram as shown below using the beam clear span L_n in lieu of the center–to–center span since shear forces above the support are neglected per ACI Code.

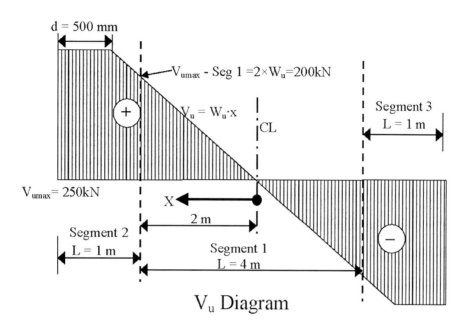

$$V_u \text{ Diagram}$$

Construct a design aid table is composed as shown below. In this table, No. 10–U shaped stirrups are assumed ($A_v = 160$ mm^2).

V_u less than	$6\phi\sqrt{f'_c}\,b_w d$	$10\phi\sqrt{f'_c}\,b_w d$	$\phi(V_c + A_v \cdot f_y \cdot d/s_{max})$
	205 kN	340 kN	168 kN
Code required s_{max}	d/2 or 600 mm	d/4 or 300 mm	d/2 or 24 mm
	250 mm	125 mm	250 mm
Code required A_{vmin}	40 mm^2	20 mm^2	40 mm^2

For the above shown table, the following values were utilized:

$\sqrt{f'_c} \times b_w \times d = \sqrt{30} \times 0.083 \times 200 \times 500/1{,}000 = 45$ kN where 0.083 is a conversion factor

$- V_c = 2\sqrt{f'_c} \times b_w \times d = 90$ kN $\phi V_c = 0.75 \times 90 = 68$ kN

$- 6\phi\sqrt{f'_c} \times b_w \times d = 205$ kN $10\phi\sqrt{f'_c} \times b_w \times d = 340$ kN

$- s_{max} = 250$ mm based on ACI Code requirements, thus,

$\phi(V_c + A_v \cdot f_y \cdot d/s_{max}) = 0.75 \times (90 + 160 \times 420 \times 500/250/1000) = 168$ k

$-$ Minimum shear reinforcement: No. 10–U shaped stirrups ($A_v = 160$ mm$^2 > A_{vmin}$) are adequate for the spacings in the table.

Divide the beam efficiently into segments so that suitable stirrup distribution can be achieved. A proposed segment division for Example 12.8 is shown in the previous page.

Segment 1

– V_{umax} = 80×2 = 160 kN < 168 kN ===> Based on the design aid table, s = 250 mm

satisfies code cracking requirement and resistance to applied shear

Thus, for segment 1, use No. 10–U stirrups @250 mm O.C.

Segments 2&3

– V_{umax} = 250 kN ===> V_{smax} = 250/0.75 – 90 = 240 kN

– Required stirrup spacing for resistance of applied shear:

$s = A_v f_y d/V_{smax}$ = 160×420×500/240/1,000 = 140 mm

– Code required maximum stirrup spacing for cracking requirements for $V_u \leq$ 250 kN is 250 mm based on the design aid table.

– A practical value for stirrup spacing can be 120 mm which satisfies both code requirements of cracking and resistance to applied shear.

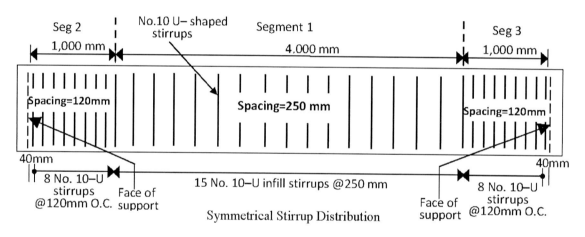

Total No. of stirrups = 33

Beam side view

12.5 Corbels and Shear Friction

Corbels are short cantilever members/beams attached to columns or walls. They are typically used in industrial buildings to support crane loads as shown in Figure 12.7.

The stress situation within a corbel is complicated and different than a conventional beam. Due to such complicated stress situation, corbel failure is via a near–vertical crack at the corbel–column or wall interface. Shear resistance of the corbel at the failure crack is provided by aggregate interlock. Aggregate interlock was considered not reliable in beam design for shear. However, for corbels aggregate interlock along with steel bars crossing the failure crack present a reliable method of shear resistance termed shear friction as explained in the following.

Steel bar reinforcement for a corbel is shown in Figure 12.8. Shear friction resistance at the failure crack for corbels is given by the following equation:

$$V_n = A_{vf} f_y \mu$$

(12.10)

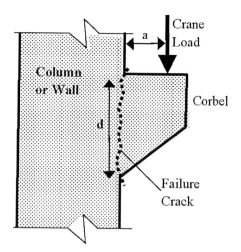

Figure 12.7: Corbel.

Where:

V_n = shear friction resistance at failure crack.

A_{vf} = area of shear friction reinforcement perpendicular to the failure crack.

f_y = yield strength of shear friction reinforcement.

μ = friction coefficient at failure crack:

$\mu = 1.4\lambda$ for monolithically cast concrete,

$\mu = 1.0\lambda$ for concrete placed against hardened concrete intentionally roughened to an amplitude of approximately ¼ inch, and

$\mu = 0.6\lambda$ for concrete placed against hardened concrete not intentionally roughened.

$\lambda = 1.0$ for normal–weight concrete, 0.85 for sand–lightweight concrete and 0.75 for lightweight concrete.

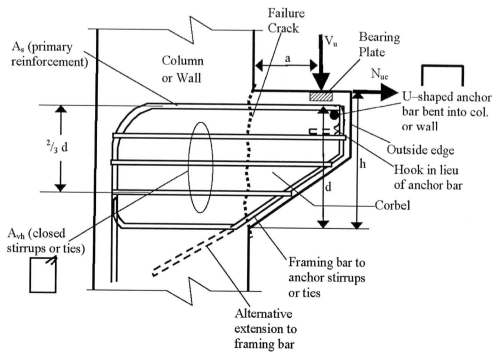

Figure 12.8: Corbel reinforcement detailing.

For the design method of Equation 12.10 to be accepted by ACI 318 Code, the following provisions/ conditions are required:

1. V_n shall not taken greater than $0.2f'_c A_c$ nor $V_n \le 800A_c$ in lb [$V_n \le 0.2f'_c A_c$ and $V_n \le 800A_c$ (lb)].

2.   Yield strength of shear friction reinforcement may not be larger than 60,000 psi ($f_y \le 60,000$ psi).

3.   a/d may not greater than 1.0 (a/d $\le$ 1.0).

4. N_{uc} may not larger than V_u ($N_{uc} \le V_u$).

5.   The depth of the outside edge of the corbel not less than d/2.

6. $\phi = 0.75$.

7. A_f = flexural reinforcement designed for the support face section to resist a bending moment $M_u = V_u a + N_{uc}(h - d)$

8. A_n = tension reinforcement designed to resist factored tension N_{uc} ($A_n = N_{uc}/\phi f_y$). For calculation of A_n, N_{uc} may not be less than $0.2V_u$ ($N_{uc} \ge 0.2V_u$). Thus, $A_{nmin} = 0.27V_u/f_y$. Tension is treated as a live load even if resulted from creep shrinkage or temperature changes.

9.   Primary reinforcement A_s shall be the greater of ($A_f + A_n$) and ($2A_{vf}/3 + A_n$), ($A_s \ge A_f + A_n$ and $A_s \ge 2A_{vf}/3 + A_n$).

10. A_{vf} shall consist of closed stirrups or ties parallel to A_s distributed evenly within the $^2/_3$d adjacent to A_s. Also, A_{vf} may not be less than half of A_s ($A_{vf} \ge A_s/2$).

11. $\rho = A_s/bd$ shall not be less than $0.04f'_c/f_y$ ($\rho = A_s/bd \ge 0.04f'_c/f_y$).

12. Transverse bar shall be at least the size of primary tension bars A_s.

13. A_s shall be anchored at the face of the corbel by:

     a.   Welding to a transverse bar of at least equal size. Weld to be designed to develop the specified yield strength of A_s (Figure 12.9), or

     b.   Bending back to form a horizontal loop.

14. Bearing area shall not project beyond the straight portion of primary tension bars nor beyond the interior face of or transverse anchor bars.

Where:

   V_n A_{vf} and f_y are as defined for Equation 12.10.

   A_c = area of concrete section resisting shear friction.

   a = distance of shear force from face of support (shear span).

   d = effective corbel depth at face of support.

   h = corbel thickness at face of support.

   V_u = factored shear force applied to corbel.

   N_{uc} = factored tension force applied to corbel.

   A_f = area of flexural reinforcement for corbel section at support face.

   A_n = area of reinforcement against axial tension for corbel section at support face.

   A_{nmin} = minimum code dictated area for A_n.

A_s = area of primary reinforcement for corbel section at support face.

$\rho = A_s/bd$ = reinforcement ratio for corbel section at support face.

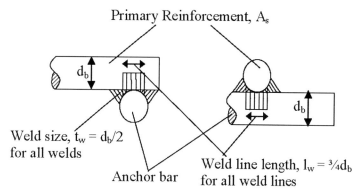

Figure 12.9: Weld detail of primary reinforcement and anchor bar at corbel outside edge.

Code requirement for upper cap on V_n (item 1 above) applies to normal weight concrete only. For lightweight concrete, V_n may not be taken greater than $(0.2 - 0.07a/d)f'_c b_w d$ nor $(800 - 280a/d)b_w d$ where b_w is the minimum width of corbel section at support face.

12.6 Deep Beams and the Strut–and–Tie Model

Deep beam are defined as beams with clear span–to–effective depth ratio less than four ($l_n/d \leq 4.0$). There are two methods of design for deep beams, a conventional method and a strut–and–tie model method.

12.6.1 Conventional Method for Deep Beam Shear Design

12.6.1.a. Loads are applied near the bottom of the beam

Examples of beams loads applied in this manner are shown in Figure 12.10. Such loading manner results in cracking and failure due to shear similar to conventional beams via 45° diagonal cracks. The design method for deep beams follows the same method described earlier for conventional beams. An important detailing criterion is that flexural reinforcement needs to be adequately anchored at the support by welding or hooks. Also, reinforcement at concentrated loads is required.

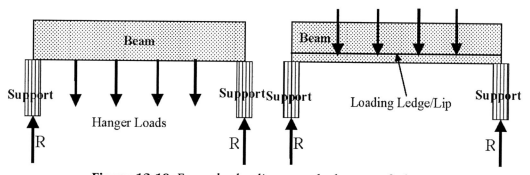

Figure 12.10: Examples loading near the bottom of a beam.

12.6.1.b. Loads are applied near the top of the beam

Beams are typically loaded near or at the top face. It has been noted that shear cracks in such beams is angled significantly less than 45° and are near vertical to some extent when beams are loaded at or near the top face (Figure 12.11). For this typical loading case, the provisions of deep beam design by ACI 318 Code shall apply. They are as follows:

1. Nominal shear capacity (V_n) of a deep beam may not exceed $10\sqrt{(f'_c)}\,b_w d$ ($V_n \leq 10\sqrt{(f'_c)}b_w d$).

2. Two layers of shear reinforcement shall be provided. One perpendicular to the beam span (A_v) and the other parallel to the beam span (A_{vh}). Perpendicular shear reinforcement area shall not be less than $0.0025 b_w s$ and its spacing (s) shall not be more than d/5 or 12 in. Parallel shear reinforcement area shall not be less than $0.0015\, b_w s_2$ and its spacing shall not be more than d/5 or 12 inch.

   Perpendicular shear reinforcement layer

   Area = $A_v \geq 0.0025 b_w s$ Spacing = $s \leq d/5$ and ≤ 12 inch

   Parallel shear reinforcement layer

   Area = $A_v \geq 0.0025 b_w s_2$ Spacing = $s_2 \leq d/5$ and ≤ 12 inch

   Where b_w = beam web width.

3. Shear and flexural reinforcement shall be adequately anchored by embedment, hooks or welding to assure full development of yield strength for these bars.

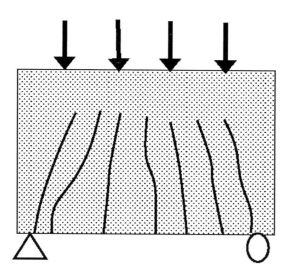

Figure 12.11: Cracking pattern of deep beams.

12.6.2 Introduction to the Strut – and – Tie Method for Deep Beam Shear Design

In the strut and tie method, resistance to external shear forces is considered provided by a concrete compression strut and steel tension tie. As such, three forces are joined at one point: external shear force in addition to concrete compressive force and a steel tensile force affecting a zero resultant force and member equilibrium. The concrete strut and the steel reinforcing bars need to be designed to resist the forces caused by the externally applied shear force. Examples of such conjuncture of forces are shown in Figure 12.12, 12.13 and 12.14. The design procedure using the strut and tie method is as follows:

1. Inclination angles of struts and the horizontal axis (θ) ranges between 70° and 50° and strut line assumptions shall be within this range.

2. Determine the forces in assumed struts (F_{us}) and ties (F_{ut}) based on force analysis under the effect of factored external forces.

3. The cross–sectional area of struts (A_c) shall be within the boundaries of loaded zone and perpendicular to the strut center–line.

4. The nodal zone is the junction of concurrence of forces within the deep beam. The compression force in nodal (F_{nu}) is equal to the force in the strut in the case of one strut joining at the nodal zone. It is equal to the resultant compression forces in all struts joining at the nodal zone for more than one strut. The area of the nodal zone (A_n) of a strut is typically equal to the area of the end of strut. The nodal zone may be extended along the boundaries of ties.

5. Check the adequacy of the strut, tie and nodal zone to resist the calculated forces.

6. All ACI 318 Code provisions for deep beams listed in section 12.6.1.b shall apply. Corbels may be designed using the strut and toe model for a/d ≤ 2.0 while deep beams ($l_n/d \leq 4.0$) may be designed by the strut and tie method for all loading methods (near the top or bottom of beam). However, the strut and tie method has been found more appropriate for deep beams loaded near the top.

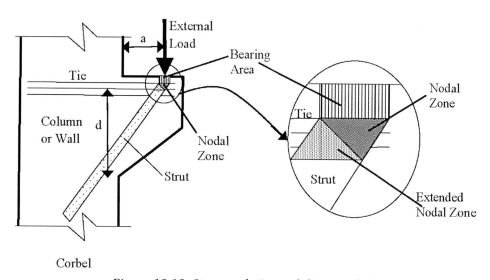

Figure 12.12: Strut–and–tie model in a corbel.

To investigate the adequacy of the strut, tie and nodal zone to resist forces caused by external loading, the following equations are used:

$$\phi F_n \geq F_u \qquad (12.11)$$

Where:

F_u = factored force in strut, tie or nodal zone due to external loading based on force analysis by the truss mechanism or strut and tie mechanism.

F_n = nominal strength of strut, tie or nodal zone due to external loading based on ACI 318 Equations for the strut and tie mechanism.

$\phi = 0.75$ = capacity reduction or resistance factor for the strut and toe mechanism.

$$F_{ns} = f_{cu} A_c \qquad (12.12)$$

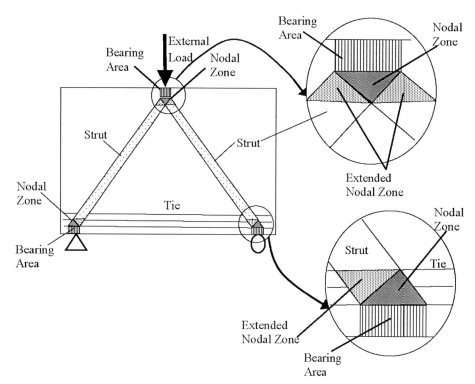

Figure 12.13: Strut–and–tie Model in a deep beam.

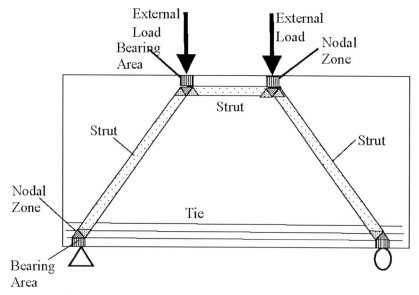

Figure 12.14: Strut–and–tie model in a deep beam.
(nodal zone configuration similar to Figures 12.12 and 12.13)

Where:

F_{ns} = nominal compressive strength of strut.

f_{cu} = effective compressive strength of the strut = the smaller of $0.85\beta_s f'_c$ and $0.85\beta_n f'_c$. The equations of β_s and β_n follows.

A_c = the cross–sectional area of the strut taken perpendicular to its centerline.

$$F_{nt} = A_{st}f_y \tag{12.13}$$

Where:

F_{nt} = nominal tensile strength of tie.

A_{st} = Area of steel reinforcement of the tie. Steel reinforcement shall be properly placed spaced and anchored. The axis of A_{st} shall coincide with the axis of the tie. Development length of the A_{st} shall be measured from edge of the extended nodal zone.

f_y = yield strength of A_{st}.

$$F_{nn} = f_{cu}A_n \qquad (12.14)$$

Where:

F_{nn} = nominal compressive strength of the nodal zone.

f_{cu} = $0.85\beta_n f'_c$ as follows.

A_n = the cross–sectional area of the nodal zone taken perpendicular to the force in the nodal zone centerline.

Where:

β_s = 1.0 for prismatic struts not passing through a tension zone.

β_s = 0.4 for struts passing through a tension zone.

β_s = 0.6 for all other cases.

Where:

β_n = 1.0 for nodal zones of struts containing no ties.

β_n = 0.8 for nodal zones anchoring one tie.

β_n = 0.6 for nodal zones anchoring two or more ties.

Test your knowledge

1. What are the limitations on concrete compressive strength (f'_c) and steel yield strength (f_y) for beam shear design?

2. For ACI 318 Code, what is the impetus for reducing the design shear force diagram at support by neglecting shear above the support and utilizing shear force at distance "d" from support face as maximum shear force?

3. List the exceptions to ACI 318 Code reductions of shear force diagram of question 2.

4. Why is it prudent to use stirrups at beam center line even if the value of factored shear force is zero?

5. Why is there a limit on shear resistance of steel reinforcement? Why is this limit a function of concrete shear resistance?

6. What is the reason for placing the first stirrup at 2 to 3 inch from support face?

7. Why are #3 U–shaped stirrups the most common for shear design in reinforced concrete?

8. What are the three criteria that need to be satisfied for shear design utilizing ACI 318 Code?

9. What is the purpose of the design aid table for beam design for shear?

10. What is the number of segments that is considered reasonable or construction suitable in shear design of simply supported beams?

11.   List the important conclusions that one may draw from the diagram of V_u vs. s included with Example 12.2.

12.   Can one use V_{umax} as V_u at "d" from face of support for shear design of cantilever beams and why?

13.   Define the shear force envelope and explain its role in shear design?

Design the following simply supported beams for shear assuming that the dead load includes beam self weight (N/A = not applicable):

Problem	Clear Span (ft)	Beam Size(in) b x d	f'_c & f_y (ksi)	Distributed Loads (k/ft) DL LL		Concentrated Load DL/LL P(k) & x(ft)		Concentrated Load DL/LL P(k) & x(ft)	
14	20	15 x 25.5	4 & 60	1.9	1.3	N/A		N/A	
15	10	12 x 15.5	3 & 60	4.5	2.8	N/A		N/A	
16	18	14 x 21.5	3.5 & 60	9.3	6.9	N/A		N/A	
17	30	20 x 32	4.5 & 60	7.5	4.5	N/A		N/A	
18	24	18 x 21	5 & 60	6.0	4.0	N/A		N/A	
19	20	48 x 7.0	4 & 60	5.0	3.5	N/A		N/A	
20	60	36 x 64	6 & 60	15	12	N/A		N/A	
21	14	12 x 17	2.5 & 40	1.6	0.8	N/A		N/A	
22	18	14 x 21.5	4 & 60	4.5	3.2	DL=20	x=9	N/A	
23	24	12 x 17	3 & 60	0.5	0.5	LL=60	x=12	N/A	
24	36	12 x 22	4 & 60	0.3	0.2	LL=60	x=12	LL=60	x=24
25	48	20 x 32	6 & 60	1.0	0.3	LL=80	x=16	LL=80	x=32
26	30	16 x 21	2.5 & 40	2.5	2.0	DL=10	x=10	N/A	
27	24	18 x 17	3 & 40	1.5	1.4	DL=8	x=8	N/A	
28	10	12 x 12	2.5 & 40	1.2	1.3	DL=12	x=4	N/A	

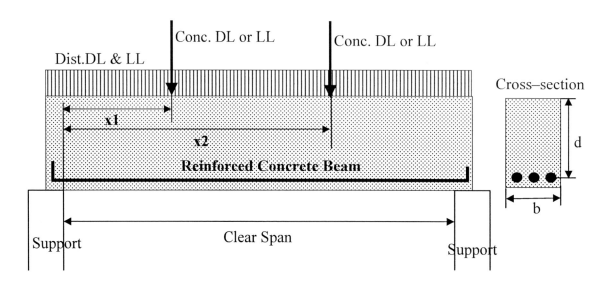

Design the following cantilevered beams for shear assuming that the dead load includes beam self weight (N/A = not applicable):

Problem	Clear Span (ft)	Beam Size(in) b x d	f'c&fy (ksi)	Distributed Loads (k/ft) DL LL		Concentrated Load DL/LL P(k) & x(ft)	Concentrated Load DL/LL P(k) & x(ft)
29	10	12 x 15	4 & 60	3.5	1.1	N/A	N/A
30	8	14 x 21	3 & 60	1.4	0.7	N/A	N/A
31	12	10 x 17	4.5 & 60	7.5	1.5	N/A	N/A
32	15	15 x 19	3.5 & 60	5.0	1.8	N/A	N/A
33	20	12 x 12	5 & 60	6.8	2.3	N/A	N/A
34	10	12 x 17	5 & 60	4.5	1.5	DL=7 x=10	N/A
35	18	16 x 20	6 & 60	5.5	2.1	DL=20 x=18	DL=20 x=12
36	14	12 x 17	2.5 & 40	1.6	0.8	N/A	N/A

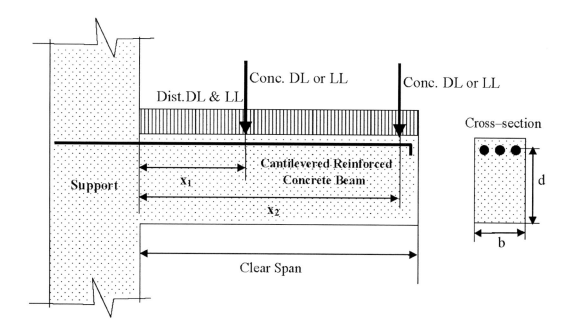

The student may use conversion factors for solving the above problems in SI units utilizing Example 12.8 in this chapter.

Chapter 13

BOND AND DEVELOPMENT LENGTH OF STEEL AND CONCRETE

13.1 Introduction

In the analysis of reinforced concrete beams in previous chapters, perfect bond between steel bars and concrete was assumed. To validate the analysis methods and results used for bending and shear, one needs to examine this initial assumption of perfect bond. This chapter discusses the specifics of bond between steel and concrete with special attention to the requirements of ACI 318 Code.

Bond of steel bars and concrete is of three distinct types:

1. Chemical bond or adhesion: the medium of cement paste is alkaline. pH of cement paste (portland cement + water) ranges between 12 and 14 indicative of its high alkalinity. The chemical reaction of cement paste and steel bars results in chemical bond or adhesion.

2. Physical bond or friction: as concrete shrinks, the cavities in which steel bars are located decrease in size. Thus, pressure is applied to steel bars resulting in enhanced friction that resists movement or slippage of bars with respect to the surrounding concrete.

3. Mechanical bond or anchorage: mechanical bond has been significantly enhanced with the utilization of deformed steel bars. Steel bars surface indentations or deformations create interlocking. Such interlocking results in effective resistance to relative movement or slippage between steel and concrete. For movement to occur, the steel bars' cavities are required to expand or deform to counterpart their surface deformation. As such expansion or deformation is not feasible, internal splitting pressure is applied to the concrete surrounding the steel bars resulting in greater resistance to bar slippage and significantly improved bond. Thus, bond failure is a splitting failure of concrete that occurs due to bar slippage or movement.

Experiments have revealed that any minute slippage causes chemical and physical bond to diminish appreciably. However, mechanical bond requires a larger relative movement or slippage for significant strength reduction. Thus, mechanical bond or anchorage is considered reliable while chemical and physical bond mechanisms are deemed undependable. ACI 318 Code design equations and provisions are based primarily on mechanical bond.

13.2 Splitting Failure of Concrete due to Bond/Anchorage Slippage

Figure 13.1 presents a block of concrete with an imbedded deformed steel bar subjected to an increasing tensile force. As the tensile force in the steel bar increases, bond stress of steel and concrete increases. Ultimately bond failure occurs. A steel bar within concrete occupies a cavity. This cavity expands in dimension or size and deforms in shape at slippage to match the surface deformations of steel bar. Expansion or deformation of the cavity results in radial tensile stresses in the concrete. Failure of concrete due to bond occurs by tensile splitting (Figure 13.1). Such failure occurs if the embedment or length of steel bar within concrete (l_d) is inadequate resulting in reduced concrete strength in splitting due to decreased area of split resisting surface.

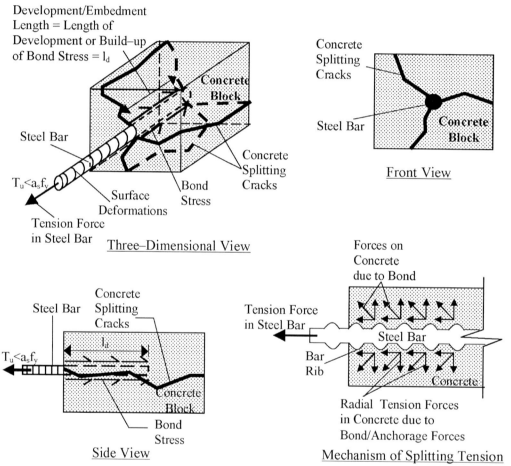

Figure 13.1: Splitting tension failure of concrete due to bond stresses of deformed steel bars caused by inadequate embedment length (l_d).

To prevent the potential for bond failure, steel bars must be properly embedded within concrete. With adequate embedment, increasing steel bar tension results in yielding prior to concrete splitting. Thereby, the steel bars can be properly functioning as concrete reinforcement. Figure 13.2 illustrates further examples of bond/anchorage failure in reinforced concrete.

Reinforced concrete presumably acts as a monolithic mass without slippage or relative deformations between its two components. As such, steel bars are considered to be securely embedded or lodged inside the concrete without significant movement. At ultimate condition, each steel bar is required to resist its ultimate tensile force, T_u, equal to its area, a_s, multiplied by its yield strength, f_y.

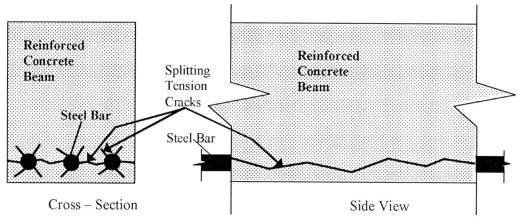

Figure 13.2: Splitting tension failure of a reinforced concrete beam due to bond/anchorage.

Consequently, the bond/anchorage stresses at the interface between each steel bar and concrete (bar surface area) must counteract and resist the steel bar ultimate tensile force so that it is fully effective unlike Figure 13.1. The length over which yield strength, f_y or the ultimate tensile force, $T_u = a_s f_y$, of steel bar is accumulated or developed via bond to concrete is termed the development length or l_d. This resultant force of bond stress within the development length resists or counteracts T_u. Adequate development length is required so that steel yield strength can develop and, consequently, steel bars would be fully effective for concrete reinforcement. Embedment length or development length as well as concrete cover of steel bars are required to satisfy ACI 318 Code for adequate safety against bond failure.

13.3 Introduction to Steel Bar Cut-off Points

To show the role of bond, development length and code compliance in reinforced concrete design, Figures 13.3 and 13.4 are introduced. Figure 13.3 illustrates that for a steel bar to be fully developed (bar tension stress = f_y), adequate anchorage needs to be provided on both sides of the cracked critical section (typically the section with maximum bending moment). The steel bar is subjected to counteracting bond stresses on either side of the crack. As bond forces are transferred to the steel bar, tension stress results in the bar at the cracked critical section. With increased loading, bond stresses increase. Consequently, with adequate embedment, bar tension stress reaches f_y and failure due to bond is prevented.

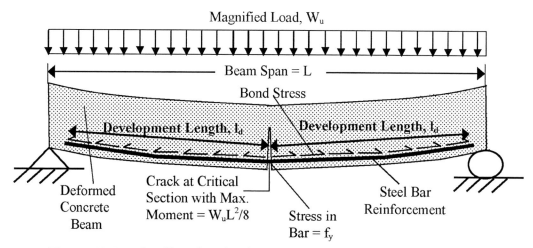

Figure 13.3: Role of bond in development of steel bar yield stress in tension

Figures 13.4 and 13.5 include a simply supported reinforced concrete beam with a span L subjected to a uniformly distributed magnified load W_u. At midspan (section 1), the factored bending moment $M_u = M_{umax} = W_u L^2/8$ (maximum factored moment applied to the beam) requires four reinforcing steel bars for adequate resistance (A_{smax}^+). Other cross–sections of the beam along the span are subjected to moments smaller than M_{umax}, thus, requiring less steel bar area for adequate resistance. At section 2, M_u requires only two steel bars. Section 2 is termed the theoretical cut–off point for the bars II and III as they are no longer needed for moment resistance. However, there are several code requirements for bar cut–off or termination in beams:

1. Each steel bars must extend beyond its theoretical cut–off point so that its actual bar cut–off point is at a distance of d (beam's effective depth) or $12d_b$ (where d_b is the diameter of terminated steel bar) whichever is greater. This precaution is implemented as a safety precaution against premature bond failure.

2. The distance between the point where a steel bar is considered fully effective and its actual cut–off point must be greater than the required embedment length or development length (l_d) by ACI 318 Code. At section 1, all steel bars are required to be fully effective for moment resistance. At sections 2, steel bars I and IV are required to be fully effective.

3. As potential shear–moment interaction can increase steel bar tension and result in premature shear failure (Figure 12.2), ACI 318 Code imposes other provisions for steel bar cut–off points including that tension flexural bars may not be terminated unless condition a, condition b OR condition c, and condition d are satisfied:

    a. $V_u \leq \frac{2}{3}(\phi V_n)$ at the location of bar cut–off point where V_u is the factored shear force and ϕV_n is nominal shear resistance of the beam section at the cut–off point.

    b. Extra shear reinforcement (in addition to shear and torsion requirements) is provided for at least ¾d beyond the cut–off point (on both sides) satisfying minimum area and maximum spacing provisions for shear reinforcement. Extra stirrup area, A_v, shall be not less than $60b_w s/f_y$ where d is beam effective depth, b_w is beam or web width and stirrup spacing, s, cannot be greater than $d/8\beta_d$, where β_d is the area of terminated bars to the area of total bars. See Chapters 11 and 12 for further illustrations.

    c. Terminating bars larger than #11 is not recommended. If #11 or smaller bars are used, then, at the cut–off point, continuing steel bar area (bars I and IV in Figure 13.5) must be at least twice the required area for resistance of bending moment and $V_u \leq \frac{3}{4}(\phi V_n)$.

    d. Furthermore, ACI 318 requires that the following equation be satisfied fr bar cut-off points. Thus, the size of steel bar that may be terminated is limited in order to conform to these equations, thereby achieving better bar detailing and improved beam structural behavior:

        $l_d \leq pM_n/V_u + l_a$

Where:

l_d = code required development length for the steel bar (as in Equation 13.1).

M_n = beam bending moment capacity at support face (Section A-A Figure 13.4).

V_u = magnified shear force at support face.

l_a = bar extension beyond support centerline (as illustrated in Figure 13.4). The value of l_a in this equation need not exceed d or $12d_b$, regardless of its actual value, where d is the effective beam depth and d_b is the diameter of steel bar.

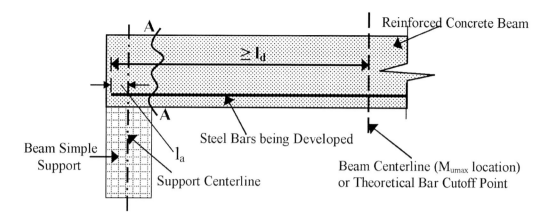

Figure 13.4: Illustration of ACI Code development length requirement at simple supports

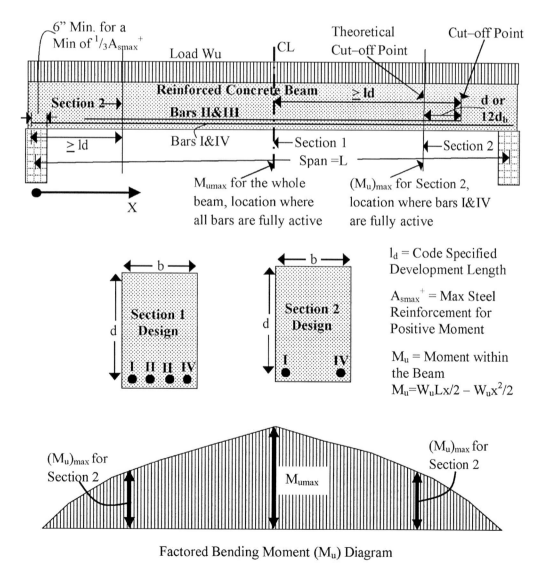

Factored Bending Moment (M_u) Diagram

Notes: Bars I&IV and II&III are drawn at two distinct horizontal planes in for illustration purposes. They are located at the same horizontal plane as shown in Section 1 detail.

Beam design and steel bar detailing are symmetrical about beam's center line.

Figure 13.5: Steel bar cut-off for a simply supported beam.

p = compression stress factor.

p = 1.3 if the steel bar extends into the support region subjected to compression stress (i.e. due to support reaction).

p = 1.0 for all other cases.

As such, for Figure 13.5, the diameter and/or area of bars I and IV may need to be adjusted in order to meet ACI 318 Code provisions (conditions a, b or c and d above) for bar cut–off points. Bases on condition (d) above, the permitted diameter steel bars terminating within a concrete support region subjected to compression is increased compared with other cases. ACI 318 Code does not require the equation in condition (d) for steel bars terminating with a hook at the support.

4.   Steel bars are, theoretically, no longer needed at the face of support since the bending moment value is zero. However, ACI 318 code requires that at least 1/3 and 1/4 of steel bars in simply supported beams and continuous beams continue at least 6in into the support, respectively.

Bars larger than #11 require special attention and are not generally terminated. The above ACI 318 code provisions are illustrated in Figures 13.4 and 13.5. As noted from the aforementioned discussion, detailing of steel bar cut–off is somewhat complex requiring extra and thorough effort of the design engineer. Generally, for simply supported beams, steel bars are not terminated within the beam. For simply supported beams subjected to a uniformly distributed load, the value of M_u within the middle third is nearly constant. As such, bar cutoff can only take place within a short distance near the support. Experience has shown that only limited savings results due to bar termination in simply supported beams. Furthermore, continuing all bars into the supports reduces construction complexity. This example is used to help the reader realize code requirements for bar cutoffs.

13.4 Code Required Bar Development Length in Tension

As stated earlier, the development length is the embedment length of a steel bar in concrete measured from the critical section or the point at which the steel bar in question is presumed fully active or can resist its yield strength. Embedment length is also termed development length since it is the length over which the steel bar develops it yield strength or full capacity. ACI 318 Code dictates that steel bar development length in tension satisfy the following equations and notes:

$$l_d = \frac{3}{40} \frac{f_y}{\lambda \sqrt{f'_c}} \frac{\psi_t \psi_e \psi_s}{\left(\frac{c_b + K_{tr}}{d_b}\right)} d_b \left(\frac{A_{srequired}}{A_{sprovided}}\right) \geq 12 \text{ in} \qquad (13.1)$$

Where:

l_d = ACI 318 Code required development length measured from the point of full steel bar utilization or critical section in inch. Critical sections include:

a. locations of maximum bending moment (M_{umax}).

b. theoretical bar cutoff points.

c. face of support.

d. points of inflection (bending moment changes signs).

f_y = yield strength of steel bars (psi). Increasing yield strength results in increased bar tension force that requires a larger embedment/development length.

f'_c = specified concrete compressive strength (psi). Increasing concrete strength enhances its resistance to tension splitting, thereby reducing the required development length. ACI 318 Code requires that $\sqrt{f'_c} \leq 100$ in Equation 13.1 regardless of its actual value. This is not a limit of f'_c but a limit on $\sqrt{f'_c}$ utilization for resistance of tension splitting due to bond.

ψ_t = top bar coefficient:

ψ_t = 1.3 with ≥ 12 in fresh concrete thickness below/underneath the steel bars due to the effect of bleed water of fresh concrete (Figure 13.6). Bleed water creates pockets underneath steel bars, thereby reducing their contact surface area with concrete.

ψ_t = 1.0 for all other cases.

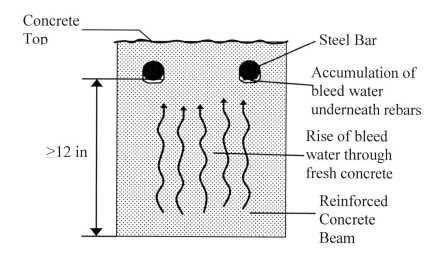

Figure 13.6: Illustration of the effect of bleed water on decreasing bond area of steel bars.

ψ_e = epoxy coating coefficient:

ψ_e = 1.5 for epoxy coated bars with cover (distance between bar centroid and nearest concrete surface) < $3d_b$ or clear distance/spacing between adjacent bars < $6d_b$.

ψ_e = 1.2 for all other epoxy coated bars.

ψ_e = 1.0 for non–coated (conventional) bars.

This factor increases the required bar development length since epoxy coating reduces the effect of mechanical anchorage.

The product of $\psi_t.\psi_e$ need not be greater than 1.7 as shown below:

$\psi_t.\psi_e \leq 1.7$ for use in Equation 13.1

ψ_s = steel bar size coefficient:

ψ_s = 0.8 for #6 and smaller size steel bars.

ψ_s = 1.0 for #7 and larger size steel bars.

The required development length is reduced for small size bars due to their large surface (bond) area compared with their cross–sectional (tension force) area compared with large size bars.

λ = lightweight concrete coefficient:

$\lambda = 0.75$ if lightweight concrete splitting tensile strength (f_{ct}) is unknown.

$\lambda = f_{ct}/6.7\sqrt{f'_c} \leq 1.0$ if splitting tensile strength (f_{ct}) is known.

$\lambda = 1.0$ for normal weight (conventional) concrete.

c_b = bar spacing and cover coefficient.

c_b = the smaller of c_1 and $c_2/2$ (Figure 13.7),

c_1 = concrete cover over the steel bar being developed = distance between bar centroid and the nearest concrete surface, and

c_2 = one–half of bar spacing for the steel bar being developed = distance between adjacent bar centroids/2.

The inclusion of c_1 and c_2 in the equation of l_d is to account for the effective concrete area surrounding each steel bar and affected by splitting tensile stresses caused by bar anchorage (Figure 13.2).

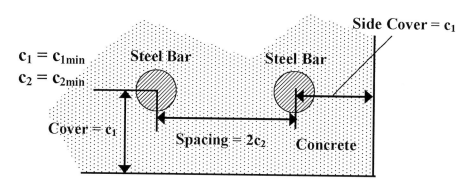

Figure 13.7: Concrete cover and rebar spacing for Equation 13.1.

K_{tr} = transverse reinforcement coefficient = $\dfrac{40A_{tr}}{sn}$

Transverse reinforcement includes stirrups and ties (closed stirrups see Chapter 15). Utilization of transverse reinforcement reduces the potential of concrete splitting tension failure caused by bond/anchorage stresses and increases bond strength to steel. As such, smaller development length is required. The terms in the equation of K_{tr} are as follows:

A_{tr} = area of transverse reinforcement = number of branches × bar area of each branch as explained in Chapter 11 for shear reinforcement.

s = spacing of transverse reinforcement (in) within the development length (l_d).

n = number of bars being developed along the plane of concrete splitting.

d_b = diameter of the steel bar being developed (in). Equation 13.1 is only applicable to deformed steel bars with standard deformations per ASTM A615 Specifications.

$A_{srequired}/A_{sprovided}$ = steel requirement coefficient

$A_{srequired}$ = steel bar area required for resistance of bending moment at the section from which development length is measured (in²).

$A_{provided}$ = steel bar area provided at the section from which development length is measured (in²).

Furthermore, ACI 318 Code states that $\dfrac{c_b + K_{tr}}{d_b} \leq 2.5$ for use in Equation 13.1 regardless of its actual value. This code limitation is a cap for the effect of transverse reinforcement and spacing on reducing development length, and not a limit for K_{tr} or c. The smaller of $\dfrac{c + K_{tr}}{d_b}$ and 2.5 is used in Equation 13.1. The designer may use a value of 1.0 for $\dfrac{c + K_{tr}}{d_b}$ for simplification even if transverse reinforcement is provided.

Special Case; Development Length for Welded Wire Fabric

For deformed welded wire fabric, Equation 13.1 is adjusted as follows:

$$l_d = \frac{3}{40} \frac{f_y}{\lambda \sqrt{f'_c}} \frac{\psi_t \psi_e \psi_s \psi_w}{\left(\dfrac{c_b + K_{tr}}{d_b}\right)} d_b \left(\frac{A_{srequired}}{A_{sprovided}}\right) \geq 8 \text{ in}$$

Where all the symbols of the above equation are identical to Equation 13.1 except:

ψ_w = welded wire fabric coefficient

ψ_w = the larger value of ψ_{w1} and ψ_{w2} if the cross wire condition (Figure 13.8) is satisfied and 1.0 if the cross wire condition is not satisfied,

$\psi_{w1} = (f_y - 35{,}000)/f_y \leq 1.0$

$\psi_{w2} = 5d_b/s \leq 1.0$ where d_b is wire diameter and s is wire spacing

Cross wire condition is that at least one cross wire within the development length with a distance not less than 2 inch from the critical section as shown in Figure 13.8.

ψ_e = epoxy coating factor

ψ_e = 1.0 if the cross wire condition above is satisfied and

ψ_e = same as in Equation 13.1 if the cross wire condition above is not satisfied.

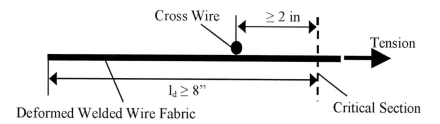

Figure 13.8: Cross wire condition for deformed welded wire fabric.

For plain welded wire fabric, two cross wires must be within the development length one of which not less than 2 in of the critical section (Figure 13.9). Furthermore, the development length of plain welded wire fabric may not be less than the value in the following equation:

$$l_d = 0.27 \frac{A_b}{s_w} \frac{f_y}{\lambda\sqrt{f'_c}} \left(\frac{A_{srequired}}{A_{sprovided}}\right) \geq 6 \text{ in}$$

Where all the symbols of the above equation are as illustrated earlier except:

A_b = area of individual wire of plain welded wire fabric (in²).

s_w = wire spacing of the welded wire fabric (in).

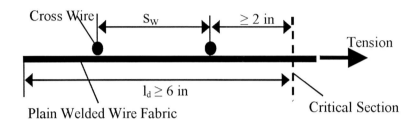

Figure 13.9: Cross wire condition for plain welded wire fabric.

Example 13.1

Determine the value of development length for the tension reinforcement in the beam section shown assuming: (a) conventional construction, (b) epoxy coated bars, and (c) light–weight concrete and epoxy coated bars. f'_c = 4,000 psi, f_y = 60,000 psi and $A_{srequired}/A_{sprovided}$ = 1.0

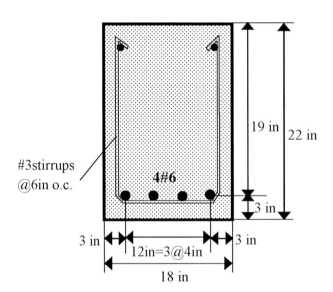

Solution

The variables of Equation 13.1 can be determined as shown below:

(a) f'_c = 4,000 psi f_y = 60,000 psi

  ψ_t = 1.0 for bottom bars ψ_e = 1.0 for conventional bars

  $\psi_t \cdot \psi_e$ = 1.0 ≤ 1.7 OK

  ψ_s = 0.8 for bars smaller than #7

$\lambda = 1.0$ for conventional concrete

c_1 = cover = 3 in c_2 = spacing/2 = 2 in

c_b = the smaller of c_1 and c_2 = 2 in

$K_{tr} = 40A_{tr}/(sn) = 40 \times 0.22/(6 \times 4) = 0.37$

$(c_b + K_{tr})/d_b = (2 + 0.37)/0.75 = 3.16 > 2.5$ N.G.

Then, one needs to use the maximum code value of $(c_b + K_{tr})/d_b = 2.5$

$$l_d = \frac{3}{40} \frac{f_y}{\lambda\sqrt{f'_c}} \frac{\psi_t\psi_e\psi_s}{\left(\dfrac{c_b + K_{tr}}{d_b}\right)} d_b \left(\frac{A_{srequired}}{A_{sprovided}}\right) \geq 12 \text{ in}$$

$$l_d = \frac{3}{40} \frac{60{,}000}{1.0 \times \sqrt{4{,}000}} \frac{1.0 \times 1.0 \times 0.8}{2.5} d_b = 22.8d_b = 17.1 \text{ in} \geq 12 \text{ in OK}$$

So for case (a) the development length for #6 bars is $22.8d_b$ or 17.1 in

(b)   All variables of Equation 13.1 are the same as case (a) except:

$\psi_e = 1.5$ for epoxy coated bars

$\psi_t \cdot \psi_e = 1.5 \leq 1.7$ OK

$$l_d = \frac{3}{40} \frac{60{,}000}{1.0 \times \sqrt{4{,}000}} \frac{1.0 \times 1.5 \times 0.8}{2.5} d_b = 34.2d_b = 25.6 \text{ in} \geq 12 \text{ in OK}$$

So for case (b) the development length for #6 bars is $34.2d_b$ or 25.6 in

(c)   All variables of Equation 13.1 are the same as case (b) except:

$\lambda = 0.75$ for light–weight concrete

$$l_d = \frac{3}{40} \frac{60{,}000}{0.75 \times \sqrt{4{,}000}} \frac{1.0 \times 1.5 \times 0.8}{2.5} d_b = 45.5d_b = 34.1 \text{ in} \geq 12 \text{ in OK}$$

So for case (c) the development length for #6 bars is $45.5d_b$ or 34.1 in

Example 13.2

Determine if the tension steel bars in the cantilever beam (3#9 bars) shown are adequately developed or anchored within the beam itself or within the supporting column. $f'_c = 5{,}000$ psi, $f_y = 60{,}000$ psi and $A_{srequired}/A_{sprovided} = 1.0$

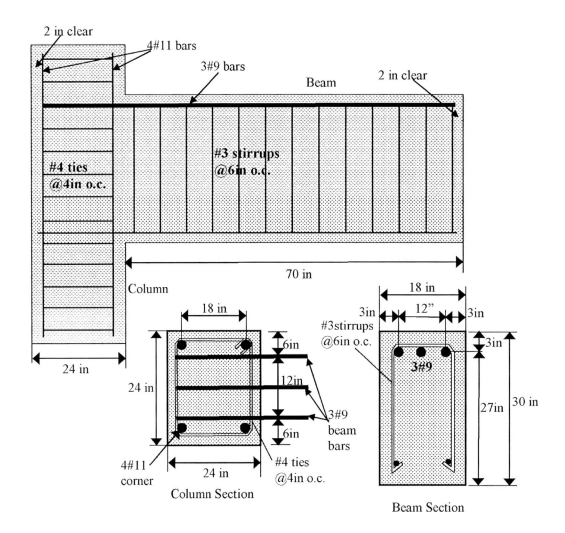

Solution

(a)   Development length within the beam:

$f'_c = 5,000$ psi $f_y = 60,000$ psi

$\psi_t = 1.3$ for top bars (≥ 12 in fresh concrete under the bars)

$\psi_e = 1.0$ for conventional bars $\psi_t \cdot \psi_e = 1.3 \leq 1.7$ OK

$\psi_s = 1.0$ for bars $\geq$ #7 $\lambda = 1.0$ for conventional concrete

$c_1 =$ cover $= 3$ in $c_2 =$ spacing/2 $= 3$ in

$c_b =$ the smaller of c_1 and $c_2 = 3$ in

$K_{tr} = 40A_{tr}/(sn) = 40 \times 0.22/(6 \times 3) = 0.49$

$(c_b + K_{tr})/d_b = (3 + 0.49)/1.125 = 3.10 > 2.5$ N.G.

Then, one needs to use the maximum code value of $(c_b + K_{tr})/d_b = 2.5$

$$l_d = \frac{3}{40} \frac{f_y}{\lambda\sqrt{f'_c}} \frac{\psi_t \psi_e \psi_s}{\left(\frac{c_b + K_{tr}}{d_b}\right)} d_b \left(\frac{A_{srequired}}{A_{sprovided}}\right) \geq 12 \text{ in}$$

$$l_d = \frac{3}{40} \frac{60,000}{1.0 \times \sqrt{5,000}} \frac{1.3 \times 1.0 \times 1.0}{2.5} d_b = 33.1 d_b = 37.2 \text{ in} \geq 12 \text{ in OK}$$

Within the cantilever beam, the required development length for the 3#9 bars is 37.2 in

Available development length within the beam = 70 − 2 = 68 in > 37.2 in OK

(b)   Development length within the column:

All variables of Equation 13.1 are the same as for the development length within the cantilever beam except for the following discussion:

ψ_t = 1.3 for top bars (potentially ≥ 12 in fresh concrete would be under the bars during column construction; this use ψ_t for top bars as a precaution)

c_1 = cover

> Top cover is very large since the concrete in the column is placed on top of the bars extending from the beam

> Side cover = (24 − 12)/2 = 6 in (see diagram with Example 13.2)

c_2 = spacing/2 = 3 inch remains the same as for case (a)

c_b = the smaller of c_1 and c_2 = 3 inch

Column vertical reinforcement acts as lateral reinforcement for the #9 bars extending from the beam   ===> A_{tr} = 4#11 = 6.25 in² s = 12 in

K_{tr} = 40A_{tr}/(sn) = 40×6.25/(12×3) = 6.9

$(c_b + K_{tr})/d_b$ = (3 + 6.9)/1.125 = 8.8 > 2.5 N.G.

Then, one needs to use the maximum code value of $(c_b + K_{tr})/d_b$ = 2.5

In spite of the large area of lateral reinforcement within the column, no effect on development length is due to ACI 318 Code cap that $(c_b + K_{tr})/d_b$ ≤ 2.5 in Equation 13.1.

Thus, all the variables of Equation 13.1 are the same for the development lengths within the column or beam, thus the required development length for both cases is the same which is 37.2 in

Available development length within the column = 24 − 2 = 22 in > 37.2 in NG

For beam column connections, as in Example 13.2, most designers tend to terminate beam bars with a hook within columns. Hooks will be discussed later in this chapter.

Tables 13.1 thru 13.3 (Tables A.30 thru A.32 in Appendix A) are design aid to provide the reader with reference values of tension development length assuming a typical situation with 2 in clear spacing between bars and #3@6 in o.c. stirrups for bars smaller than #11 and #4@6 in o.c. stirrups for #11 and larger bars. Cover of 2.5 in is also assumed. Other assumptions used are shown with each table.

Table 13.1: Tension development length for deformed steel bars (fy = 40,000 psi).

Assumptions		Bar Number								
		5	6	7	8	9	10	11	12	13
c_b		1.313	1.375	1.438	1.5	1.563	1.625	1.688	1.875	2.25
K_{tr}		0.293	0.293	0.293	0.293	0.293	0.293	0.293	0.533	0.533
$(c_b + K_{tr})/d_b$		2.5	2.224	1.978	1.793	1.65	1.535	1.441	1.376	1.237
f'_c (ksi)		Bottom Bars								
3	l_d/d_b	17.5	19.7	27.7	30.5	33.2	35.7	38.0	39.8	44.3
	l_d(in)	11.0	14.8	24.2	30.5	37.4	44.6	52.3	69.6	99.6
4	l_d/d_b	15.2	17.1	24.0	26.5	28.8	30.9	32.9	34.5	38.3
	l_d(in)	9.5	12.8	21.0	26.5	32.3	38.6	45.3	60.3	86.3
5	l_d/d_b	13.6	15.3	21.4	23.7	25.7	27.6	29.5	30.8	34.3
	l_d(in)	8.5	11.4	18.8	23.7	28.9	34.6	40.5	54.0	77.2
6	l_d/d_b	12.4	13.9	19.6	21.6	23.5	25.2	26.9	28.1	31.3
	l_d(in)	7.7	10.4	17.1	21.6	26.4	31.5	37.0	49.2	70.4
f'_c (ksi)		Top Bars								
3	l_d/d_b	22.8	25.6	36.0	39.7	43.2	46.4	49.4	51.7	57.6
	l_d(in)	14.2	19.2	31.5	39.7	48.6	58.0	68.0	90.5	129.5
4	l_d/d_b	19.7	22.2	31.2	34.4	37.4	40.2	42.8	44.8	49.8
	l_d(in)	12.3	16.6	27.3	34.4	42.1	50.2	58.9	78.4	112.2
5	l_d/d_b	17.6	19.8	27.9	30.8	33.4	35.9	38.3	40.1	44.6
	l_d(in)	11.0	14.9	24.4	30.8	37.6	44.9	52.6	70.1	100.3
6	l_d/d_b	16.1	18.1	25.5	28.1	30.5	32.8	34.9	36.6	40.7
	l_d(in)	10.1	13.6	22.3	28.1	34.3	41.0	48.1	64.0	91.6

Assumptions:

1. Clear spacing of steel bars = 2 in (2.25 in for #18 bars).

2. Steel bar cover = 2.5 in from center of steel bar.

3. Stirrup or tie diameter = #3 for bars ≤ #11 and #4 for #14 and #18 bars.

4. Stirrup or tie spacing = 6 in

5. Stirrup or tie yield strength = 60 ksi

6. No. of bars being developed = 5

Table 13.2: Tension development length for deformed steel bars (f_y = 60,000 psi)

Assumptions		Bar Number								
		5	6	7	8	9	10	11	14	18
c		1.313	1.375	1.438	1.5	1.563	1.625	1.688	1.875	2.25
K_{tr}		0.293	0.293	0.293	0.293	0.293	0.293	0.293	0.533	0.533
$(c+K_{tr})/d_b$		2.5	2.224	1.978	1.793	1.65	1.535	1.441	1.376	1.237
f'_c (ksi)		Bottom Bars								
3	l_d/d_b	26.3	29.5	41.5	45.8	49.8	53.5	57.0	59.7	66.4
	l_d(in)	16.4	22.2	36.3	45.8	56.0	66.9	78.4	104.5	149.4
4	l_d/d_b	22.8	25.6	36.0	39.7	43.1	46.4	49.4	51.7	57.5
	l_d(in)	14.2	19.2	31.5	39.7	48.5	58.0	67.9	90.5	129.4
5	l_d/d_b	20.4	22.9	32.2	35.5	38.6	41.5	44.2	46.2	51.4
	l_d(in)	12.7	17.2	28.2	35.5	43.4	51.8	60.7	80.9	115.8
6	l_d/d_b	18.6	20.9	29.4	32.4	35.2	37.9	40.3	42.2	47.0
	l_d(in)	11.6	15.7	25.7	32.4	39.6	47.3	55.4	73.9	105.7
f'_c (ksi)		Top Bars								
3	l_d/d_b	34.2	38.4	54.0	59.6	64.7	69.6	74.1	77.6	86.3
	l_d(in)	21.4	28.8	47.2	59.6	72.8	87.0	101.9	135.8	194.3
4	l_d/d_b	29.6	33.3	46.8	51.6	56.1	60.3	64.2	67.2	74.8
	l_d(in)	18.5	24.9	40.9	51.6	63.1	75.3	88.3	117.6	168.2
5	l_d/d_b	26.5	29.8	41.8	46.1	50.2	53.9	57.4	60.1	66.9
	l_d(in)	16.5	22.3	36.6	46.1	56.4	67.4	79.0	105.2	150.5
6	l_d/d_b	24.2	27.2	38.2	42.1	45.8	49.2	52.4	54.9	61.1
	l_d(in)	15.1	20.4	33.4	42.1	51.5	61.5	72.1	96.0	137.4

Assumptions:

1. Clear spacing of steel bars = 2 in (2.25 in for #18 bars).

2. Steel bar cover = 2.5 in from center of steel bar.

3. Stirrup or tie diameter = #3 for bars ≤ #11 and #4 for #14 and #18 bars.

4. Stirrup or tie spacing = 6 in

5. Stirrup or tie yield strength = 60 ksi

6. No. of bars being developed = 5

Table 13.3: Tension development length for deformed steel bars (f_y = 80,000 psi).

Assumptions		Bar Number								
		5	6	7	8	9	10	11	14	18
c		1.313	1.375	1.438	1.5	1.563	1.625	1.688	1.875	2.125
K_{tr}		0.293	0.293	0.293	0.293	0.293	0.293	0.293	0.533	0.533
$(c+K_{tr})/d_b$		2.5	2.224	1.978	1.793	1.65	1.535	1.441	1.376	1.181
f'_c (ksi)		Bottom Bars								
3	l_d/d_b	35.1	39.4	55.4	61.1	66.4	71.4	76.0	79.6	92.7
	l_d(in)	21.9	29.5	48.5	61.1	74.7	89.2	104.6	139.3	208.6
4	l_d/d_b	30.4	34.1	48.0	52.9	57.5	61.8	65.9	68.9	80.3
	l_d(in)	19.0	25.6	42.0	52.9	64.7	77.3	90.5	120.6	180.7
5	l_d/d_b	27.2	30.5	42.9	47.3	51.4	55.3	58.9	61.7	71.8
	l_d(in)	17.0	22.9	37.5	47.3	57.9	69.1	81.0	107.9	161.6
6	l_d/d_b	24.8	27.9	39.2	43.2	47.0	50.5	53.8	56.3	65.6
	l_d(in)	15.5	20.9	34.3	43.2	52.8	63.1	73.9	98.5	147.5
f'_c (ksi)		Top Bars								
3	l_d/d_b	45.6	51.2	72.0	79.4	86.3	92.8	98.9	103.5	120.5
	l_d(in)	28.5	38.4	63.0	79.4	97.1	116.0	135.9	181.1	271.2
4	l_d/d_b	39.5	44.4	62.3	68.8	74.8	80.4	85.6	89.6	104.4
	l_d(in)	24.7	33.3	54.6	68.8	84.1	100.5	117.7	156.8	234.9
5	l_d/d_b	35.3	39.7	55.8	61.5	66.9	71.9	76.6	80.2	93.4
	l_d(in)	22.1	29.8	48.8	61.5	75.2	89.8	105.3	140.3	210.1
6	l_d/d_b	32.2	36.2	50.9	56.2	61.0	65.6	69.9	73.2	85.2
	l_d(in)	20.1	27.2	44.5	56.2	68.7	82.0	96.1	128.0	191.8

Assumptions:

1. Clear spacing of steel bars = 2 in (2.25 in for #18 bars).

2. Steel bar cover = 2.5 in from center of steel bar.

3. Stirrup or tie diameter = #3 for bars ≤ #11 and #4 for #14 and #18 bars.

4. Stirrup or tie spacing = 6 in

5. Stirrup or tie yield strength = 60 ksi

6. No. of bars being developed = 5

13.5 Code Required Hooked Bar Development Length in Tension

ACI 318 Code permits using hooks to shorten the development length of steel bars in tension. It is required that hooks utilized for bond and development length meet the requirements of standard hooks. Two types of hooks are used for tension bar development, 90° hooks and 180° hooks. Figure 13.10 illustrates ACI Code requirements for standard hooks where d_b is hook bar diameter.

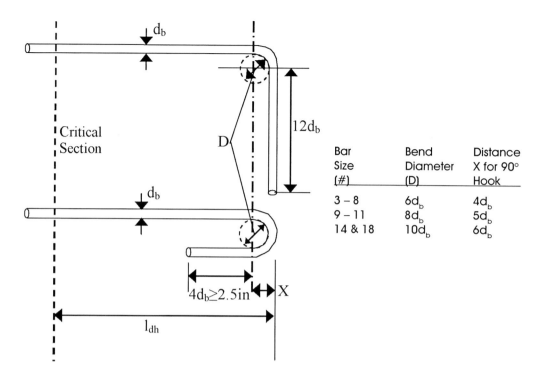

Figure 13.10: ACI 318 Code standard hook details for tension bars.

ACI code specifies that the development length required for steel bars in tension terminating with a standard hooks can be represented by the following equation:

$$l_{dh} = \frac{0.02\psi_e f_y}{\lambda \sqrt{f'_c}} \; d_b \left(\frac{A_{srequired}}{A_{sprovided}}\right) \prod_{i=1}^{3} M_i \geq 6 \text{ in or } 8d_b \text{ whichever is greater} \tag{13.2}$$

Where:

l_{dh} = hooked bar development length in tension in inch.

d_b = diameter of bar being developed in tension via a standard hook.

f'_c and f_y = concrete compressive strength and steel yield strength.

ψ_e = modification factor for epoxy coated bars. This factor is introduced since epoxy coating, while effective for corrosion protection, tends to reduce bond between hooked steel bars and concrete.

ψ_e = 1.2 for epoxy coated bars. ψ_e = 1.0 for all other cases.

λ = modification factor for light–weight concrete. This factor is utilized since light–weight concrete has inferior resistance to splitting failure caused by hooked bar bond failure or slippage.

λ = 0.75 for light–weight concrete. λ = 1.0 for all other cases.

M_1 = cover modification factor:

M_1 = 0.7   for #11 bars and smaller hooks with side cover ≥ 2.5 in and bar extension cover ≥ 2 inch for 90° hooks bar (Figure 13.11).
M_1 = 1.0   for all other cases.

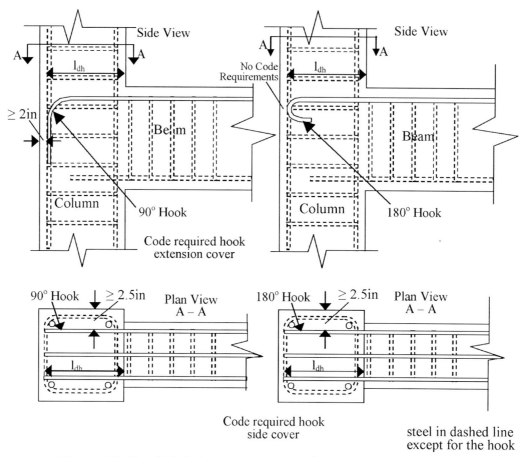

Figure 13.11: ACI 318 Code requirements for M_1 modification factor.

M_2 = lateral reinforcement modification factor for 90° hook not at discontinuous ends of members (see exception in Figure 13.14):

$M_2 = 0.8$ for #11 bars and smaller hooked bars enclosed within ties or stirrups perpendicular or parallel to the hooked bar being developed and spaced not greater than $3d_b$ with the first tie or stirrup within $2d_b$ of the bend exterior (Figure 13.12).

$M_2 = 1.0$ for all other cases including 180° hook.

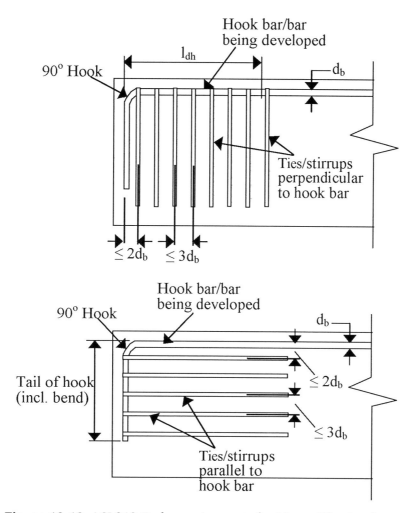

Figure 13.12: ACI 318 Code requirements for M_2 modification factor.

M_3 = lateral reinforcement modification factor for 180° hook not at discontinuous ends of members (see exception in Figure 13.14):

$M_3 = 0.8$ for #11 bars and smaller hooked bars enclosed within ties or stirrups perpendicular to the hooked bar being developed and spaced not greater than $3d_b$ with the first tie or stirrup within $2d_b$ of the bend exterior (Figure 13.13).

$M_3 = 1.0$ for all other cases including 90° hook.

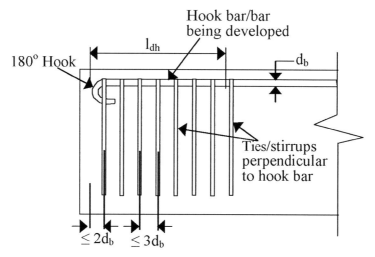

Figure 13.13: ACI 318 Code requirements for M_3 modification factor.

$A_{srequired}$ and $A_{provided}$ are as explained earlier for the case of straight bars.

Code Requirements for Lateral Reinforcement at Discontinuous Ends of Members and Exception for M_2 and M_3

ACI 318 Code requires the use of ties or stirrups for bars being developed with a standard hook at discontinuous ends of members with both side cover and top (or bottom) cover over the hook less than 2.5 in. Stirrup or tie shall used along the entire development length l_{dh} and shall not be spaced more than $3d_b$ with the first tie or stirrup not more than $2d_b$ from the outside of bend (Figure 13.4). Furthermore, factors M_2 and M_3 are not applicable.

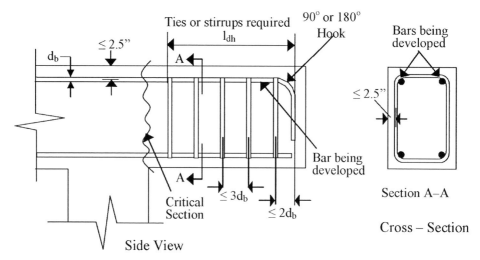

Figure 13.14: Illustration for code requirements for lateral reinforcement at discontinuous ends of members and exception for M_2 and M_3.

Example 13.3 (90° Hook)

Determine if the tension steel bars in the cantilever beam (3#9 bars) shown are adequately developed or anchored within the supporting column. $f'_c = 5,000$ psi, $f_y = 60,000$ psi and $A_{srequired}/A_{sprovided} = 1.0$.

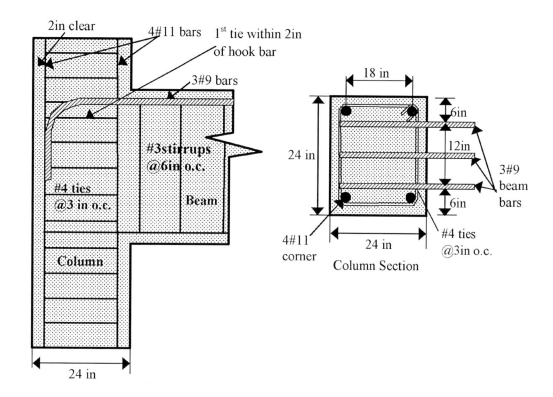

Solution

$$l_{dh} = \frac{0.02\psi_e f_y}{\lambda\sqrt{f'_c}} \, d_b \left(\frac{A_{srequired}}{A_{sprovided}}\right) \prod_{i=1}^{3} M_i \geq 6 \text{ in or } 8d_b \text{ whichever is greater}$$

$f'_c = 5,000 \text{ psi}$ $f_y = 60,000 \text{ psi}$

$\psi_e = 1.0$ for conventional bars

$\lambda = 1.0$ for conventional concrete

$A_{srequired}/A_{sprovided} = 1.0$

M_1: Cover modification factor for #11 and smaller bars

Bar size = #9 ≤ #11   OK

Side cover = 6 in ≥ 2.5 in OK $\Longrightarrow M_1 = 0.7$

Bar extension cover = 2 in ≥ 2 in OK

Exception to M_2 and M_3:

Hook is not at discontinuous member end but it terminates within a column.

One may reach this conclusion by comparing Figure 13.14 and the diagram with Example 13.3 ===> Exception to M2 and M3 is not applicable

M_2: lateral reinforcement modification factor for 90° hook ≤ #11 not at   discontinuous ends of members:

Not at discontinuous member end   OK

Bar size = #9 ≤ #11   OK

Enclosed within ties parallel or perpendicular to bar OK $\Longrightarrow M2 = 0.8$

Tie spacing = 3 in ≤ 3db = 3× 9/8 = 3.4 in   OK

First tie at 2 in from hook bar ≤ 2db = 2.25 in   OK

M3: lateral reinforcement modification factor for 180o hook ≤ #11 not at discontinuous ends of members:

Not applicable since Example 13.3 includes a 90o hook ⟹ M3 = 1.0

$$l_{dh} = \frac{0.02 \times 1.0 \times 60,000}{1.0 \times \sqrt{5,000}} \left(\frac{9}{8}\right) \times 1.0 \times 0.7 \times 0.8 \times 1.0 = 10.7 \text{ in}$$

$l_{dh} = 10.7 \text{ in} \geq 6 \text{ in or } 8 \times (9/8)$ whichever is greater = 9 in   ===> OK

$l_{dh} = 10.7 \text{ in}$ Available length = 24 − 2 = 22 in > 10.7 in   ===> OK

Beam 3#9 bars are adequately anchored within the column utilizing a 90o hook.

Example 13.4 (180° Hook)

Determine if the tension steel bars in the cantilever beam (3#9 bars) shown are adequately developed or anchored within the supporting column. f'_c = 5,000 psi, f_y = 60,000 psi and $A_{srequired}/A_{sprovided}$ = 1.0.

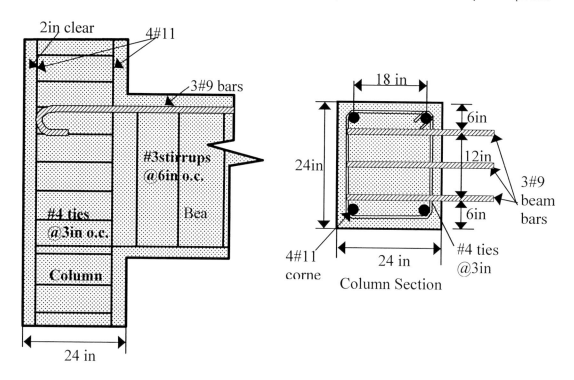

Solution

$$l_{dh} = \frac{0.02 \psi_e f_y}{\lambda \sqrt{f'_c}} \, d_b \left(\frac{A_{srequired}}{A_{sprovided}}\right) \prod_{i=1}^{3} M_i \geq 6 \text{ in or } 8d_b \text{ whichever is greater}$$

f'_c = 5,000 psi f_y = 60,000 psi

ψ_e = 1.0 for conventional bars

λ = 1.0 for conventional concrete

$A_{srequired}/A_{sprovided}$ = 1.0

M_1: Cover modification factor for #11 and smaller bars

Bar size = #9 ≤ #11 OK

Side cover = 6 in ≥ 2.5 in   OK $\Bigg\}$ ===> $M_1 = 0.7$

Bar extension cover = 2 in ≥ 2 in   OK

Exception to M_2 and M_3:

Hook is not at discontinuous end ===> Exception to M_2 and M_3 is not applicable as in Example 13.3

M_2: lateral reinforcement modification factor for 90° hook ≤ #11 not at discontinuous ends of members:

Not applicable since Example 13.4 includes a 180° hook ===> $M_2 = 1.0$

M_3: lateral reinforcement modification factor for 180° hook ≤ #11 not at discontinuous ends of members:

Not at discontinuous member end   OK

Bar size = #9 ≤ #11   OK $\Bigg\}$ ===> $M_3 = 1.0$

Enclosed within ties perpendicular to bar NG

$$l_{dh} = \frac{0.02 \times 1.0 \times 60{,}000}{1.0 \times \sqrt{5{,}000}} \left(\frac{9}{8}\right) \times 1.0 \times 0.7 \times 1.0 \times 1.0 = 13.4 \text{ in}$$

$l_{dh} = 13.4$ in ≥ 6 in or 8 × (9/8) whichever is greater = 9 in   ===> OK

$l_{dh} = 13.4$ in   Available length = 24 − 2 = 22 in > 13.7 in   ===> OK

Beam 3#9 bars are adequately anchored within the column utilizing 180° hook.

13.6 Tension Splices

Steel bars are typically manufactured in lengths of 20, 40 and 60 ft (10 m and 20 m). Bars are cut and bent to match reinforcement design of structural members. Generally, there are leftovers. It is a common practice that such leftovers are utilized. Bars are usually spliced or connected together to match the required reinforcement so that leftover or waste is minimized. Splicing or connecting steel bars is also needed for long members. ACI 318 Code permits using bar connections or splices in tension. Generally, bars should not be spliced near sections of maximum moments or maximum tensile stresses. Also, bar splicing should be staggered so that at each section only a portion (preferably less than 50%) of steel bars is spliced. This can alleviate any possible effect of splicing in creating weak sections or causing crowding or congestion of steel bars. Different types of tension splices are included in ACI Code:

1. Welded tension splices: welding of steel bars must be in accordance with the structural welding code (American Welding Society Standard D1.4) as ACI Code specifies. An approved welded splice is also required by ACI Code to have strength ≥ 125%×f_y of the spliced steel bar.

2. Mechanical tension splices: mechanical tension splices are typically proprietary devices for connection of steel bars. A common type consists of a sleeve fitted over the two steel bars being spliced that is filled with melted metallic filler. Another type utilized a sleeve fitted over bent or crimped bars. Tapered sleeves are also common. A mechanical splice is required by ACI Code to have strength ≥ 125%×f_y of the spliced steel bar.

3. Tension lap splices: in a lap splice, two steel bars are overlapped a distance sufficient to transfer the tension force from one bar to the other via bond. As such, the lapped steel bar tension force is transferred to the concrete that in turn transfers it to the other lapped steel bar. The following discussion and code provisions pertain to lap splices:

    a. Lap splices can cause extra radial tension stresses within the surrounding concrete that may result in premature concrete failure. Among the precautions against radial cracking due to bond stresses is the use of lateral reinforcement that confines the concrete against splitting tension cracks,

    b. ACI Code does not permit the use of tension lap splices for steel bar > #11 for the reasons explained in (a) above,

    c. Tension lap splices can be contact or noncontact splices (Figure 13.15). For contact splices, the two bars are tied with tie wire to maintain their position. For non contact splices, the distance between lapped bar centroids cannot exceed 1/5 the lap splice length or 6 in whichever is smaller,

    d. There are two types of lap splices in the ACI Code, type A and type B:

        (d.1) Type A lap splice: lap length = development length (l_d); where l_d is computed pursuant to Section 13.4 for the original beam section. ACI Code states that the designer may use lap splice type A provided that $A_{sprovided}/A_{srequired} \geq 2.0$ and that 50% or less of the section tension bars are being spliced at the same location.

        (d.2) Type B lap splice: lap length = 1.3 × development length (l_d); as l_d is per Section 13.4 for the original beam section. Type B lap splice is required by ACI Code if the requirements for Type A are not met.

It is prudent for the designer to: (a) not to use splices near maximum moment sections, (b) to use Type B lap splice, (c) not to splice more than 50% of the tension steel bars, and (d) to use lateral reinforcement in the concrete beams for the entire lap splice length.

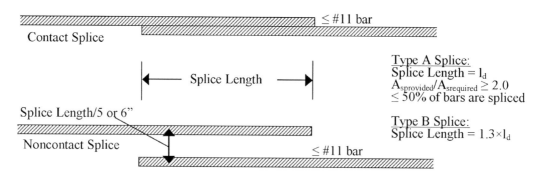

Figure 13.15: Code requirements for tension lap splices.

13.7 Bundled Bars

Parallel bars may be placed in contact so that they act as a unit termed bundled bars. (a) ACI allows the use of bundles of two, three and four bars. (b) Bundles of more than four bars are not allowed. (c) Bundles of > #11 bars are not permitted. (d) Use of lateral reinforcement where bars are bundled is required by ACI Code. (e) Termination of individual bars in a bundle must be staggered at least $40d_b$ where d_b is the individual diameter in the bundle. (f) Bundled bars can be stacked vertically or placed horizontally adjacent with no more than two bars high or wide as shown in Figure 13.1 advantages to using bundled bars. The contact area between a bundle and concrete is la? contact area of an individual bar of equal area. As such, bond to concrete of the bundle is impruv.. compared to bond to concrete of an individual bar of equal area. Thus, utilization of bundled would improve cracking behavior of reinforced concrete (Chapter 14).

Figure 13.16: Code allowed bundle bar arrangements.

Bundled bars may pose an added difficulty in construction especially with space limitations and when hooks are used. For design purposes, bundled bars are treated as a single bar of an equal area to the bundle and an equivalent diameter as explained below. Spacing and concrete cover are calculated on the basis of equivalent diameter of the bundle. Furthermore, development and splice lengths of bundled bars are computed based on the equations provided below:

$A_e = n \times a_s$

$d_e = \sqrt{na_s/\pi}$

$(l_d)_b = l_d$ for a single bar with $d_e \times b_f$

$(l_{dh})_b = l_{dh}$ for a single bar with $d_e \times b_f$

Bundle Splice Length = Splice Length for a single bar with $d_e \times b_f$

Where: A_e = area of bar equivalent to the bundle.

n = number of individual bars in the bundles (not to exceed 4).

a_s = area of individual bar in the bundle.

d_e = diameter of bar equivalent to the bundle.

$(l_d)_b$ = bundle development length.

$(l_{dh})_b$ = bundle hook development length.

b_f = bundle development length factor = 1.0 for two–bar bundle, 1.2 for three–bar bundle and 1.33 for four–bar bundle.

13.8 Development and Splices Compression

Development length is required for bars subjected to compression. Steel bar compressive load is transferred via bond along the embedded part as well as bearing at the end of the bar. As compression is the prevailing stress, no tension cracks are present. Consequently, bond of steel and concrete is enhanced. Hooks and end or mechanical anchorage are not effective for compression development. ACI Code has the following set of equations for determining the value of development length in compression:

$$
\left. \begin{aligned}
l_{dc} &= \{[0.02f_y/\sqrt{f'_c}] \times (A_{srequired}/A_{sprovided}) \times M_L\}d_b \\
l_{dc} &= (0.0003f_y) \times (A_{srequired}/A_{sprovided}) \times M_L \times d_b \\
l_{dc} &= 8 \text{ in}
\end{aligned} \right\} \quad \text{whichever is the largest} \quad (13.3)
$$

Where:

l_{dc} = compression development length.

λ = light–weight concrete modification factor.

λ = 0.75 for light – weight concrete.

λ = 1.0 for conventional concrete.

0.0003 factor carries the unit of in²/lb.

M_L = lateral reinforcement modification factor.

M_L = 0.75 if the bar is enclosed within spiral reinforcement (Chapter 15) with ≥ ¼ in diameter and ≤ 4 in pitch or within ≥ #4 ties spaced ≤ 4 in.

M_L = 1.0 for all other cases.

ACI Code also has the following set of equations for determining the length of lap splices in compression:

$$\text{Compression Lap Splice} = 0.0005f_y d_b M_{f'c} \geq 12M_{f'c} \qquad \text{for } f_y \leq 60,000 \text{ psi} \qquad \textbf{(13.4)}$$

$$\text{Compression Lap Splice} = (0.0009f_y - 24)d_b M_{f'c} \geq 12M_{f'c} \text{ for } f_y > 60,000 \text{ psi} \qquad \textbf{(13.5)}$$

Where:

0.0005 and 0.009 factors carry the unit of in²/lb,

$M_{f'c}$ = concrete compressive strength modification factor

$M_{f'c}$ = 1.3 if $f'_c \leq 3,000$ psi, and

$M_{f'c}$ = 1.0 for all other cases.

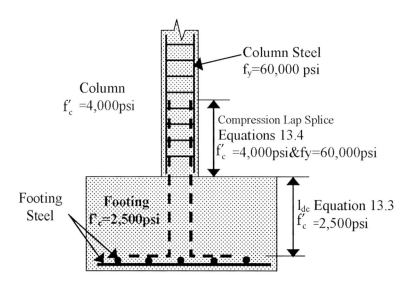

Figure 13.17: Column–footing dowels.

If two different bar sizes are lap spliced, then, Code equations for the larger size bar govern. #18 and #14 bars may not be spliced with #11 or smaller bars. A typical example for bar development/lap splicing in compression is at the connection of columns and footings. Dowel bars or dowels are bent bars that are placed on top of the footing steel bars. Upon footing concrete placement, the dowels extend out of the footing into the future location of the column. These dowels are spliced with the steel bars of the column. Consequently, the concrete column is placed. The dowels must extend into the footing to satisfy the compression development length of Equation 13.12 where the hooked end of dowels is not effective for compression development. The dowels must also extend into the column to satisfy the splice length of Equations 13.4 and 13.5. Figure 13.17 illustrates this typical example of steel bar compression development. For column–footing connections subjected to bending moments, extra precautions are needed so that steel bar development length/splices are adequate for tension and/or compression.

Compression mechanical or welded splices must have strength at least 125% the yield strength of the bar being spliced pursuant to ACI 318 Code. End bearing splices is another type of compression splices allowed by ACI Code. Steel bar end is terminated with a properly designed mechanical fitting that applies load to the surface of concrete as shown in Figure 13.18. Steel bars splices with end bearing splices must be enclosed within closed stirrups ties or spiral bar.

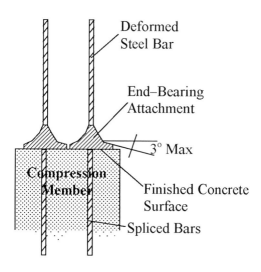

Figure 13.18: End–bearing compression splice.

13.9 Column Splices

Figure 13.19 presents two examples of column splices. As it is more convenient to splice columns at floor levels, from a structural integrity point of view, it is a more sound and prudent solution to splice columns within the middle third of column height. Within the middle third of column height, bending moments due to lateral loads (accidental or calculated) are minimal. Further analysis and discussion of column splices are presented in Chapter 16.

There are provisions by ACI Code for column splices due to their importance as listed below:

1. All types of splices including: lap splices, mechanical or welded splices and end bearing splices may be used for columns.

2. Clearly, column splices must satisfy all possible factored load combinations (see Chapter 3) for the structure.

3. Lap splices in compression:

   a. If the column, for all factored load combinations, is subjected to compression, then lap splices are designed as explained in Section 13.7 with the following:

b. Lap splice length may be multiplied by 0.83 but not less than 12 in if ties are provided throughout the lap splice length and tie bars cross sectional area is not less than 0.0015hs where h is member thickness and s is tie spacing.

c. Lap splice length may be multiplied by 0.75 but not less than 12 in if a spiral bar (Chapter 15) is provided throughout the lap splice length.

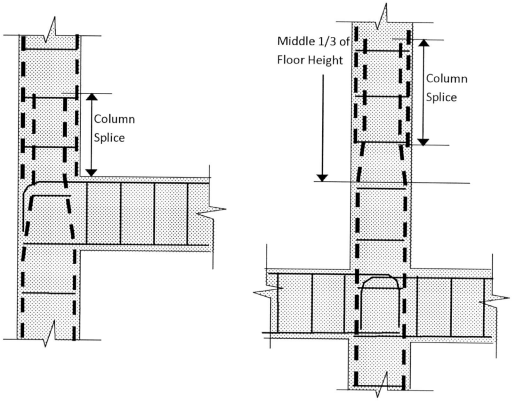

a.   Splice at floor level   b.   Splice within floor height

Figure 13.19: Column splice.

4. Lap splices in tension:

If steel bars within the column, for all factored load combinations, may be subjected to tension less than fy/2 and less than half of the steel bars are spliced at one location, then type A tension splice may be used.

For all other cases where column steel bars may be subjected to tension due to factored load combinations, type B lap splice is required.

5   Mechanical or welded splices must be designed to resist 125% of bar yield strength as is the case for conventional mechanical or welded splices for steel bars in compression.

6. End bearing splices are permitted for use in column splices provided that at least 25% of yield strength in tension of steel bars of each column side/face is continuously provided for stability of the splice. This can be accomplished by staggering the end bearing splices for each face/side or providing additional reinforcement within each face/side.

13.10 Development of Lateral Reinforcement

Stirrups and ties are typically subjected to tension and, thus, are required to be developed or anchored. ACI Code requires that stirrups and ties be extended to as close as permitted by code to the tension and compression surfaces of the concrete member. Such extension adds to their effectiveness in member reinforcement. Bond or anchorage failures have been noted in the past in stirrups and ties. Thus, ACI Code requires that stirrups and ties be hooked around a longitudinal reinforcement to achieve adequate anchorage. Furthermore, each bend in continuous portions of stirrups and ties shall enclose a longitudinal steel bar. ACI 318 Code does not specify 180 degree hooks for stirrups or ties. Steel bars > #8 may not be used for stirrups or ties per ACI Code. Three types of stirrup and tie hooks are listed by ACI Code as follows where d_b is stirrup or tie bar diameter:

 a.   135 degree hook plus $6b_d$ extension,

 b.   90 degree hook plus $6d_b$ extension for stirrups or ties of #5 bars or smaller, and

 c.   90 degree hook plus $12\, d_b$ extension for stirrups or ties of #6, #7 and #8 bars.

An additional code requirement applies to #6, #7 and #8 stirrups with yield strength greater than 40,000 psi as follows: Stirrup embedment, measured from member middepth (d/2) and outside end of the hook, shall be equal or greater than $0.014 d_b f_y / \sqrt{f'_c}$.

Pairs of U–shaped stirrups can be placed to form a closed loop are considered properly spliced/anchored if lap length $\geq 1.3 l_d$ where l_d is the development length for the stirrup bar. Furthermore, for members with h $\geq$ 18 in, splicing of U–shaped stirrup pairs is considered adequate if stirrup legs extend the full depth of member and $A_b f_y \leq 9,000$ lb where A_b and f_y are stirrup bar area and yield strength, respectively.

For steel bars bent to act as shear reinforcement, length measured from middepth (d/2) towards the beam compression region shall equal or exceed the bar development length taking into account required bar stress to satisfy shear strength requirements (multiply development length by bar stress under ultimate condition/bar yield strength f_y). Furthermore, such bars shall continue with longitudinal reinforcement within the beam tension region.

Figure 13.20 illustrates the approved stirrup and tie hooks by ACI Code. Figures 13.21 and 13.21 show the possible methods of anchoring stirrups and other shear reinforcement in beams as dictated by code.

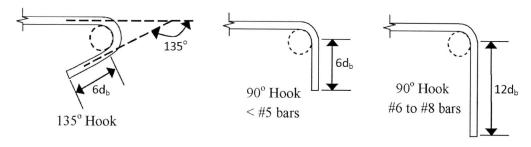

Figure 13.20: ACI standard stirrup and tie hooks.

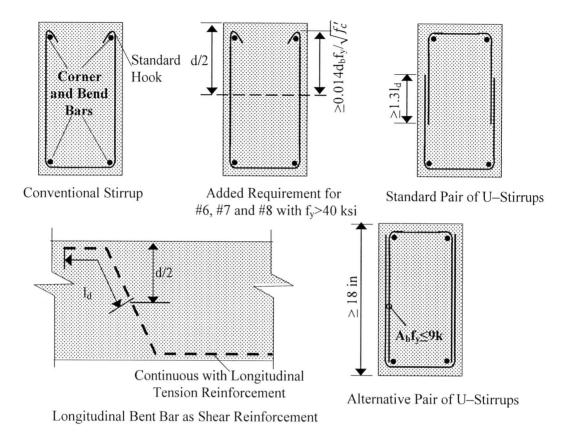

Figure 13.21: Development length for web reinforcement.

For deformed welded wire fabric acting as shear reinforcement, a standard hook around longitudinal steel is required. Alternative methods of anchorage are presented for plain welded wire fabric by ACI Code as shown in Figure 13.22. Each leg of U – shaped stirrup made of welded wire fabric is required to be properly anchored near the top of the beam via one of the methods of Figure 13.22.

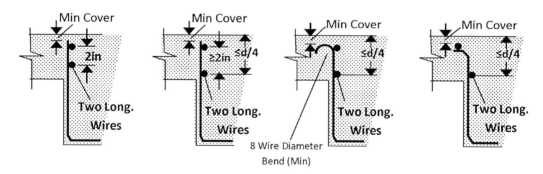

Figure 13.22: Development for web reinforcement of welded wire fabric.

Column ties require enclose all column longitudinal steel bars and anchorage to a longitudinal steels with two hooks of Figure 13.20 (Chapter 15). Spiral bars are used in concrete columns to confine the column inner part (termed column core) for an enhanced behavior/failure mode. They are continuous helical bars that wrap around the column core (Chapter 15) to achieve adequate confinement. ACI Code specifies that an extra 1.5 turns of the spiral bar at the very top and very bottom end of the column (at the footing or slab). Furthermore, splice lengths of $48d_b$, $72d_b$ and $72d_b$ are required for deformed, plain or epoxy coated spiral bars, respectively, where d_b is spiral bar diameter.

13.11 Mechanical Anchorage by Headed Bars in Tension

ACI Code allows the utilization of mechanical devices capable of developing the strength of steel bars without concrete damage as anchorage. Figure 13.23 illustrates a detail of a headed bar along with ACI Code requirements. At the anchor head, welding or threading is provided for connection of the deformed bar. Thus, an obstruction to bar deformation is introduced. ACI Code specifies that such obstruction length may not exceed $2d_b$. Also, the net bearing are of headed bar shall not be less than $4A_b$ where d_b and A_b are the diameter and area of the headed bar. Net bearing are is calculated as the inner or bearing surface area of bar head minus the area of obstructions.

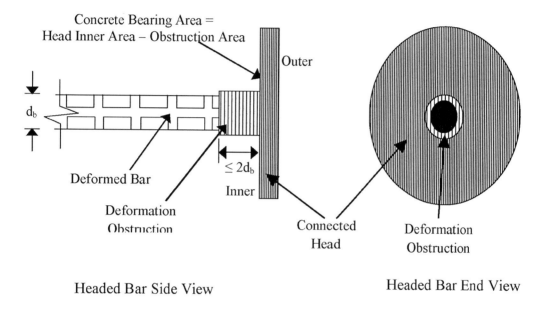

Figure 13.23: Schematics of headed bar.

Due to lack of experience with headed bars, ACI Code stipulates that they are not used for concrete with f'_c greater than 6,000 psi or steel with f_y greater than 60,000 psi. Headed bars may only be used for normal weight concrete per ACI Code. Clear cover and clear spacing of headed bars (measured from the bar not the head) may not be less than $2d_b$ and $4d_b$, respectively. As failure of headed bars by bond is localized and conical, lateral reinforcement (stirrups or ties) is not helpful for enhancing anchorage strength of headed bars. ACI Code requires that headed bars at discontinuous joints be extended through the entire joint for improved anchorage. ACI Code provides the following equation for development length of headed bars (Figure 13.24):

$$l_{dt} = \frac{0.016\psi_e f_y}{\sqrt{f'_c}} \, d_b \left(\frac{A_{srequired}}{A_{sprovided}}\right) \geq 6 \text{ in or } 8d_b \text{ whichever is greater} \qquad (13.6)$$

Where: l_{dt} = headed bar development length.

ψ_e = epoxy coated bar modification factor = 1.2 and 1.0 for epoxy coated and conventional bars, respectively.

f'_c and f_y = specified compressive strength of concrete and yield strength of steel,

d_b = diameter of headed bar.

$(A_{srequired}/A_{sprovided})$ = as explained earlier.

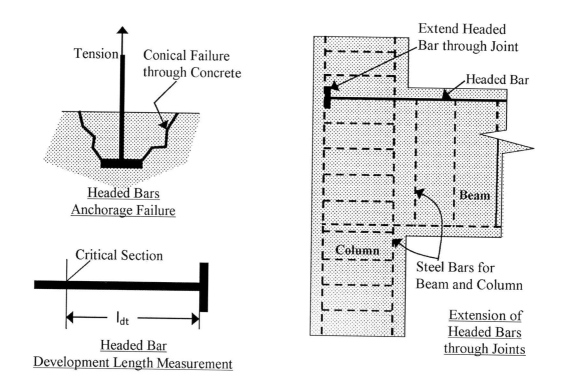

Figure 13.24: Headed bars

13.12 ACI Code Structural Integrity Provisions

Structural design of reinforced concrete members treats each element as a separate entity. However, a building structure responds to loading collectively. There are portions of ACI 318 Code that are dedicated to the structural integrity provisions of building structures. The essence of structural integrity for reinforced concrete structures is steel detailing. Among the important comments regarding structural integrity detailing are:

1. Avoid premature termination of steel bars. Code provisions for termination of steel bars are included in Section 13.3. Designers have typically extended steel bars throughout the entire beam to avoid bar termination. Bar termination does not represent a large material cost saving since it is accompanied by extra detailing work that may offset or even exceed such material safety.

2. If the designer desires to employ bar termination, then, all bars must extend beyond the point where they are no longer needed for resisting bending moment at least d, $12d_b$ or l/16 whichever is greater. Where d is beam's effective depth, d_b is bar diameter and l is beam span. Also, the provisions of Section 13.3 must be satisfied.

3. Negative tension reinforcement requires anchorage at the supports on both sides (into the beam and into the joint). Typically, with a straight bar extension into the beam and a standard hook into the support joint. Positive tension reinforcement requires simply extension into the support. Designers have also used hooks to terminate positive tension bars at joint as an extra precaution.

4. At least, $^1/_3$ of negative tension steel (minimum of two bars) is extended beyond the beam's inflection point through midspan. Also, at least $^1/_3$ of positive tension steel (minimum of two bars) is extended at least a foot into the support.

5. Negative steel splicing shall be close to midspan. Positive steel splicing shall be close to the supports. All splices shall be Class B Splices preferably enclosed within stirrups or ties.

6. For seismic joints, both tension and compression reinforcements are required to be anchored within the joint.

7. For perimeter beams, in addition to the above mentioned provisions, ACI Code requires the use of closed stirrups to safeguard against accidental torsion or flange tension splitting as shown in Figure 13.25. Pairs of U–stirrups are not permitted for perimeter beams.

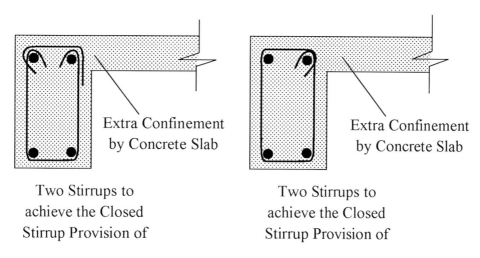

Figure 13.25: Closed stirrups for perimeter beams.

13.13 ACI Code Development Length Equations in SI Units

The following equations are replicas of previous equations in this chapter with unit conversion from U.S. customary units to SI units. Proper study requires examination of the fo llowing equations along with their counterparts. Each equation is numbered the same as its counterpart with the letter M added to facilitate reference.

Development length for straight bars in tension:

$$l_d = \frac{36}{40} \frac{f_y}{\lambda\sqrt{f_c'}} \frac{\psi_t\psi_e\psi_s}{\left(\frac{c_b + 25K_{tr}}{d_b}\right)} d_b \left(\frac{A_{srequired}}{A_{sprovided}}\right) \geq 300 \text{ mm} \qquad \text{(13.1M)}$$

Where:

l_d = ACI 318 Code required development length in tension in mm.

f_y = yield strength of steel bars (MPa).

f_c' = specified concrete compressive strength (MPa) with $\sqrt{f_c'} \leq 8.5$

ψ_t = top bar coefficient:

ψ_t = 1.3 with $\geq$ 300 mm fresh concrete below steel bars

ψ_t = 1.0 for all other cases.

ψ_e = epoxy coating coefficient:

ψ_e = 1.5 for epoxy coated bars with cover (distance between bar centroid and nearest concrete surface) < $3d_b$ or clear distance/spacing between adjacent bars < $6d_b$.

ψ_e = 1.2 for all other epoxy coated bars.

ψ_e = 1.0 for non–coated (conventional) bars.

The product of $\psi_t.\psi_e$ need not be greater than 1.7 as shown below:

$\psi_t.\psi_e \leq 1.7$ for use in Equation 13.1M

ψ_s = steel bar size coefficient:

ψ_s = 0.8 for No. 20 and smaller size steel bars.

ψ_s = 1.0 for 25 and larger size steel bars.

λ = lightweight concrete coefficient:

λ = 0.75 for lightweight concrete.

λ = 1.0 for conventional concrete.

c_b = bar spacing and cover coefficient.

c_b = the smaller of c_1 and $c_2/2$ in mm (Figure 13.7).

K_{tr} = transverse reinforcement coefficient = $\dfrac{1.6A_{tr}}{sn}$

A_{tr} = area of transverse reinforcement (mm²).

s = spacing of transverse reinforcement (mm).

n = number of bars being developed along the plane of concrete splitting.

d_b = diameter of the steel bar being developed (mm)

ACI requires that $(c + 25K_{tr})/d_b \leq 2.5$ for Equation 13.1M

ACI code require development length for steel bars in tension with a standard hooks:

$$l_{dh} = \frac{\psi_e f_y}{4\lambda\sqrt{f'_c}} \, d_b \left(\frac{A_{srequired}}{A_{sprovided}}\right) \prod_{i=1}^{3} M_i \geq 150 \text{ mm or } 8d_b \text{ whichever is greater} \qquad \textbf{(13.2M)}$$

Where:

l_{dh} = hooked bar development length in tension in mm.

d_b = diameter of bar (mm).

f'_c and f_y = concrete compressive strength and steel yield strength in MPa.

ψ_e = modification factor for epoxy coated bars

ψ_e = 1.2 for epoxy coated bars. ψ_e = 1.0 for all other cases.

λ = modification factor for light–weight concrete

λ = 0.75 for light–weight concrete. λ = 1.0 for all other cases.

M_1 = cover modification factor:

M_1 = 0.7   for No. 35 bars and smaller hooks with side cover $\geq$ 65 mm and bar extension cover $\geq$ 50 mm for 90° hooks bar (Figure 13.11).

M_1 = 1.0   for all other cases.

M_2 = lateral reinforcement modification factor for 90° hook:

M_2 = 0.8   for No. 35 bars and smaller hooked bars enclosed within perpendicular or parallel ties or stirrups (Figure 13.12).

M_2 = 1.0   for all other cases including 180° hook.

M_3 = lateral reinforcement modification factor for 180° hook:

$M_3 = 0.8$ for No. 35 bars and smaller hooked bars enclosed within perpendicular ties or stirrups (Figure 13.13).

$M_3 = 1.0$ for all other cases including 90° hook.

See Figure 13.14 for the exception to factors M2 and M3.

Example 13.5

Determine the value of development length for the tension reinforcement in the beam section shown assuming conventional steel bars with $f'_c = 30$ MPa, $f_y = 420$ MPa and $A_{srequired}/A_{sprovided} = 1.0$

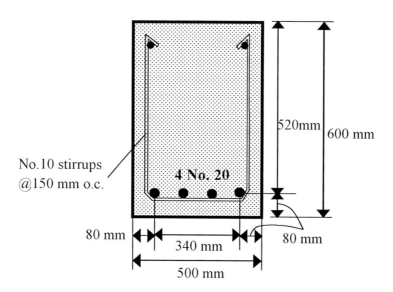

Solution

The variables of Equation 13.1M can be determined as shown below:

(d) $f'_c = 30$ MPa $f_y = 420$ MPa

$\psi_t = 1.0$ for bottom bars $\psi_e = 1.0$ for conventional bars

$\psi_t \cdot \psi_e = 1.0 \leq 1.7$ OK

$\psi_s = 0.8$ for bars smaller than No. 35

$\lambda = 1.0$ for conventional concrete

c_1 = cover = 80 mm c_2 = spacing/2 = 60 mm

c_b = the smaller of c_1 and c_2 = 60 mm

$K_{tr} = 1.6A_{tr}/(sn) = 1.6 \times 160/(150 \times 4) = 0.42$

$(c_b + 25K_{tr})/d_b = (60 + 25 \times 0.42)/20 = 3.5 > 2.5$ N.G.

Then, one needs to use the maximum code value of $(c_b + 25K_{tr})/d_b = 2.5$

$$l_d = \frac{36}{40} \frac{f_y}{\lambda \sqrt{f'_c}} \frac{\psi_t \psi_e \psi_s}{\left(\frac{c_b + 25K_{tr}}{d_b}\right)} d_b \left(\frac{A_{srequired}}{A_{sprovided}}\right) \geq 300 \text{ mm}$$

$$l_d = \frac{36}{40} \frac{420}{1.0 \times \sqrt{30}} \frac{1.0 \times 1.0 \times 0.8}{2.5} d_b = 22.1 d_b = 440 \text{ mm} \geq 300 \text{ mm} \quad \text{OK}$$

Example 13.6 (90o Hook)

Determine if the tension steel bars in the cantilever beam (3 No. 28 bars) shown are adequately developed or anchored within the supporting column. f'_c = 35 MPa, f_y = 420 MPa and $A_{srequired}/A_{sprovided}$ = 1.0.

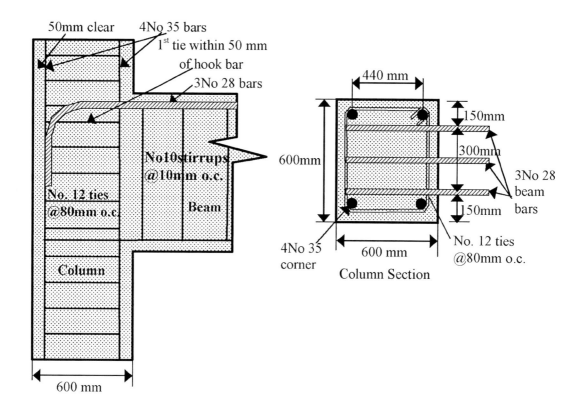

Solution

$$l_{dh} = \frac{\psi_e f_y}{4\lambda \sqrt{f'_c}} d_b \left(\frac{A_{srequired}}{A_{sprovided}}\right) \prod_{i=1}^{3} M_i \geq 150 \text{ mm or } 8d_b \text{ whichever is greater}$$

f'_c = 35 MPa psi f_y = 420 MPa

ψ_e = 1.0 for conventional bars

λ = 1.0 for conventional concrete

$A_{srequired}/A_{sprovided}$ = 1.0

M_1: Cover modification factor for #11 and smaller bars

        Bar size = No 28 ≤ No 35   OK

        Side cover = 150 mm ≥ 80 mm   OK $\Bigg\}$ ⟹ M_1 = 0.7

        Bar extension cover = 50 mm ≥ 50 mm   OK

Exception to M_2 and M_3:

Hook is not at discontinuous member end as in Example 13.3 ===> Exception to M_2 and M_3 is not applicable

M_2: lateral reinforcement modification factor for 90° hook ≤ No 35 not at discontinuous ends of members:

Not at discontinuous member end OK

Bar size = 28 ≤ No 35 OK

Enclosed within ties parallel or perpendicular to bar OK ===> $M_2 = 0.8$

Tie spacing = 80 mm ≤ $3d_b = 3×28 = 84$ mm OK

First tie at 50 mm from hook bar ≤ $2d_b = 56$ mm OK

M_3: Not applicable not 180° hook   ===> $M_3 = 1.0$

$$l_{dh} = \frac{1.0×420}{4×1.0×\sqrt{35}} (28) \times 1.0 \times 0.7 \times 0.8 \times 1.0 = 280 \text{ mm}$$

l_{dh} =280 mm ≥ 150 or 8 × (28) whichever is greater = 225 mm ===> OK

l_{dh} = 280 mm   Available length = 600 − 50 = 550 mm > 280 mm ===> OK

Beam 3 No. 28 bars are adequately anchored within the column utilizing a 90° hook.

Test your knowledge

1. What are three types of bond between steel bars and concrete?

2. What is the most reliable bond type of steel bars and concrete?

3. Describe the failure mode of concrete due to bond stresses?

4. Define the development length of steel bars in reinforced concrete?

5. Why does the equation of development length include f_y and $\sqrt{f'_c}$?

6. Generally, l_d/d_b is less in smaller diameter bars compared with larger diameter bars. Explain.

7. Why is lateral reinforcement important for computing the development length of steel bars in concrete?

8. Why are bar spacing and cover important for computing the development length?

9. Why is epoxy coating bar important for computing the development length?

10. Why is a larger development when embedded in light–weight concrete?

11. List the effective methods of reducing steel development in concrete structures?

12. What are the important points within a structure at which development length is required to be checked?

13. Why is bar spacing not important for computing hook bar development length?

14. Lateral reinforcement is beneficial for reducing the 90° hook development length. For 180° hooks, only lateral reinforcement perpendicular to the hook bar is effective. Explain.

Assuming straight deformed bar development length, determine the value of development length and type A and type B splice length and detail the cross–section for the beams illustrated below. Include with your solution any ACI Special Requirements:

No.	Element	b×h(inch)	f'_c, f_y(ksi)	Steel	Lat. Steel	A_{sreq}/A_{sprov}
15	Simply Supported Beam	14×20	4,60	3#8 1–layer	#3U–Stirrups @6in o.c.	0.82
16	Simply Supported Beam	12×18	3,40 Lightweight	3#9 1–layer	#3U–Stirrups @5in o.c.	0.76
17	Simply Supported Beam	15×24	5,60	4#10 1–layer Epoxy Coated	#3U–Stirrups @8in o.c.	0.92
18	Simply Supported Beam	10×15	4,60	3#5 1–layer	#3U–Stirrups @4in o.c.	0.95
19	Simply Supported Beam	18×36	4,60	8#10 2–layers Epoxy Coated	#4U–Stirrups @4in o.c.	0.90
20	Simply Supported Beam	14×24	2,60 Lightweight	6#7 2–layers	#3U–Stirrups @5in o.c.	0.89
21	Simply Supported Beam	16×18	4,60	4#6 1–layer	#3U–Stirrups @8in o.c.	0.50
22	Simply Supported Beam	10×16	3.5,60	two 2#7 bundles	#3U–Stirrups @4in o.c.	0.83
23	Simply Supported Beam	12×30	5,60	three 2#9 bundles	#4U–Stirrups @12in o.c.	0.90
24	Simply Supported Beam	14×27	4.3,60	three 2#8 bundles	#3U–Stirrups @10in o.c.	0.88
25	Cantilever Beam	20×24	3,60	4#8 1–layer	#3U–Stirrups @6in o.c.	0.85
26	Cantilever Beam	18×20	4,40	4#9 1–layer	#3U–Stirrups @4in o.c.	0.75
27	Cantilever Beam	18×20	4,40	4#9 1–layer	#3U–Stirrups @4in o.c.	0.70
28	Cantilever Beam	15×40	4,60	Four 2#9 bundles	#3U–Stirrups @15in o.c.	0.65
29	Concrete Slab	6in thick	3,60	#5@6in o.c.	#4@18in o.c. secondary	0.90
29	Concrete Slab	5in thick	4,60	#6@8in o.c. Epoxy Coated	#5@15in o.c. secondary	0.86
30	Concrete Slab	4in thick	3,60	#5@10in o.c.	#4@12in o.c. secondary	0.66
31	Concrete Slab	5in thick	2.5,60 Lightweight	#4@12in o.c.	#4@12in o.c. secondary	0.60
32	Concrete Slab	4.5in thick	3,80	WWF 6×6–W2×W2		0.90
33	Concrete Slab	4in thick	4,80	WWF 6×6–W4×W4		0.80
34	Concrete Slab	3.5in thick	5,80	WWF 6×6–W2.9×W2.9		0.77

35. Determine the required hook development length for the beam shown below and refer to ACI Code requirements if applicable:

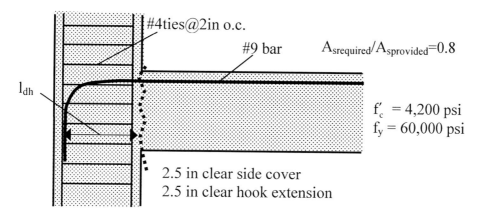

#4ties@2in o.c.

#9 bar

$A_{srequired}/A_{sprovided}=0.8$

l_{dh}

$f'_c = 4,200$ psi
$f_y = 60,000$ psi

2.5 in clear side cover
2.5 in clear hook extension

36. Repeat problem 35 if hook bar is #11.

37. Repeat problem 35 if hook bar is #7.

38. Repeat problem 35 if ties are #3@6in o.c.

39. Repeat problem 35 if clear cover is 3in.

40. Repeat problem 35 if hook bar is 180° hook is used.

41. Repeat problem 35 if hook bar is 180° hook is used and ties are vertical.

42. Repeat problem 35 if headed bar is used in place of hooked bar.

43. Repeat problem 36 if headed bar is used in place of hooked bar.

44. Repeat problem 37 if headed bar is used in place of hooked bar.

45. Repeat problem 38 if headed bar is used in place of hooked bar.

46. Repeat problem 39 if headed bar is used in place of hooked bar.

47. Determine the required development length for the main bars in footing shown below and refer to ACI Code requirements if applicable:

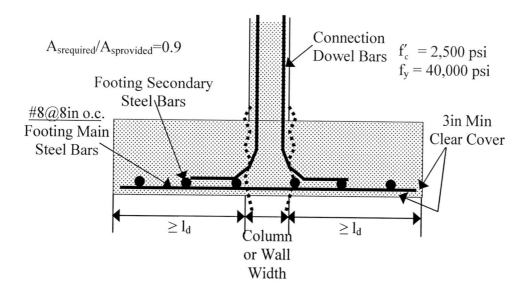

$A_{srequired}/A_{sprovided}=0.9$

Footing Secondary
Steel Bars

Connection
Dowel Bars

$f'_c = 2,500$ psi
$f_y = 40,000$ psi

#8@8in o.c.
Footing Main
Steel Bars

3in Min
Clear Cover

$\geq l_d$

$\geq l_d$

Column
or Wall
Width

48.   Repeat problem 47 if bars are #11@6in o.c.

49.   Repeat problem 47 if bars are #6@4in o.c.

50.   For problems 47, 48 and 49, conclude the minimum required footing width to satisfy ACI Code development length provisions.

51.   Determine the required straight bar and hooked development lengths for the cantilever beam shown below and refer to ACI Code requirements if applicable:

52.   Repeat problem 51 if $P_u = 0$.

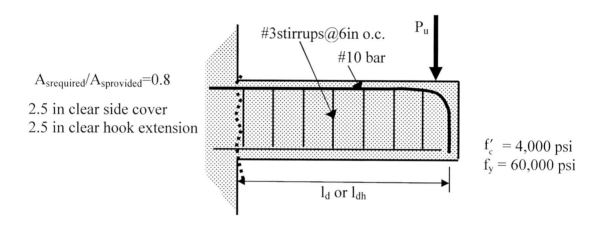

53.   Repeat problem 51 if stirrups are @3in o.c.

54.   Repeat problem 51 if steel bar is #7.

55.   Repeat problem 51 if $f'_c = 6,000$ psi.

56.   Detail the steel bars shown in the beam below taking into account the required development length or hook development length as well as ACI Code provisions:

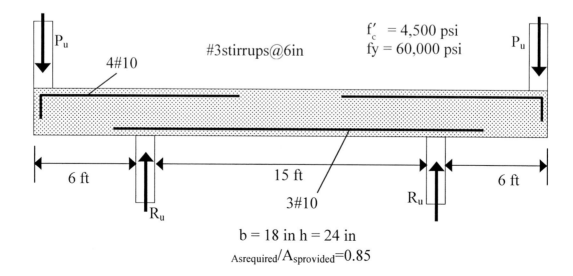

57. Repeat problem 56 if negative steel is 5#11 and positive steel is 3#11.

58. Repeat problem 56 if stirrups are #4@4in o.c.

59. Repeat problem 56 if cantilevers' length is 2 ft each.

60. Repeat problem 56 if span length is 6 ft.

61. Repeat problem 56 if positive and negative steels are 4#6 and 3#6.

62. Compute compression splice length and detail the column splice shown below:

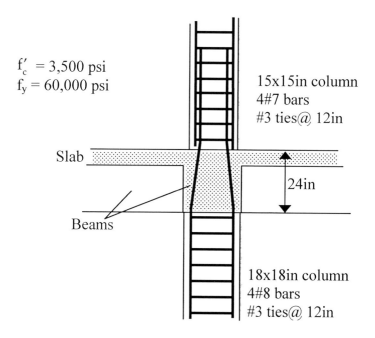

f'_c = 3,500 psi
f_y = 60,000 psi

15x15in column
4#7 bars
#3 ties@ 12in

Slab

24in

Beams

18x18in column
4#8 bars
#3 ties@ 12in

63. Repeat problem 62 if there are no beams but the slab is 8in thick.

64. Repeat problem 62 if the lower column is 16x24in with six bars and the upper column is 14x14in with four bars.

65. Repeat problem 62 if column are 24x24in with 8#11 bars and 18x18in with 6#9 bars.

66. Repeat problem 62 if column are 15x15in with 4#6 bars and 12x12in with 4#6 bars.

67. Repeat problem 62 if column are 20x24in with 8#9 bars and 16x12in with 8#7 bars.

68. Repeat problem 62 if column are 20x20in with 4#10 bars and 20x20in with 4#9 bars.

The student may use conversion factors for solving the above problems in SI units utilizing Examples 13.5 and 13.6 in this chapter.

Chapter 14

SERVICEABILITY OF REINFORCED CONCRETE BEAMS

14.1 The Concept of Serviceability

The previous chapters are focused on the ultimate conditions of reinforced concrete beams. However, a part of field performance of concrete beams is function under service conditions. Among the important defects that concrete beams, slabs and other elements may experience under service condition that could interfere with its function are:

a.   Excessive deflection:

Excessive deflections can result in misfits of as well as damage to non –structural elements such as partitions and window and door frames. This situation is exacerbated for deflections due to live load or moving live load. Floor vibration or wavering could further damage non–structural building elements. Excessive deflections are unsightly and can lead to distrust in the building structural integrity. The strength design method (LRFD) typically results in smaller element size compared with the elastic design method (ASD) that which can result in larger deflections.

b.   Excessive cracking:

Excessive cracking can result in detriment to steel–concrete bond. Furthermore, harmful materials may seep through large cracks resulting in steel bar corrosion. Cracks are unattractive and could also lead to occupant distrust in the building structural integrity.

c.   Discoloring:

Discoloring has two sources: concrete efflorescence and steel corrosion. Concrete efflorescence (white non–uniform flaky discoloring) is caused by leaching of calcium salts to the surface of concrete. This is due to excessive water content, non–conforming portland cement, or misuse of concrete additives such as fly ash or other pozzolans. Typically, efflorescence is not harmful to concrete strength or durability unless it persists for an extended time. Steel bar corrosion is another source of concrete surface discoloring (spotty rust spots). However, discoloring caused by steel corrosion is harmful to reinforced concrete as it is indicative of damage to steel bars and detriment to steel–concrete interfacial bond. Also, misuse of coloring agents could lead to unsightly spotting.

d.   Surface disintegration:

Surface disintegration can occur due to a number of reasons including honeycombing, raveling or pitting. Honeycombing occurs due to improper mix design and/or consolidation. It appears as an area of coarse aggregate without the binding cement paste or mortar. Raveling is caused by dislodging of coarse aggregate particles from the surface of concrete due to friction with vehicle tires in streets and highways or falling debris in a dam spillway as examples. It is also exacerbated by poor mix design. Pitting is a result of expansible chemical reaction between cement paste and incompatible aggregate particles typically near the surface of concrete slabs. Such surface disintegrations severely affect the aesthetic appearance of concrete.

Deflection and cracking are issues concerning structural design and member structural performance under service conditions. Discoloring and surface disintegration are primarily materials performance issues that relate to concrete mix design. This chapter deals with deflection and cracking of reinforced concrete beams and slabs.

14.2 Deflection

14.2.1 Transformed Uncracked Sections

The assumption of perfect bond entails that the strain in steel (ε_s) is equal to the strain in an adjacent concrete beam part (ε_c) provided that they are at equal distances from the neutral axis (N.A.). As explained in Chapter 4, prior to concrete cracking (phase 1 of behavior), concrete and steel may be considered within their respective elastic ranges where Hook's law remains applicable. Thus, the stresses in steel and adjacent concrete can be expressed as follows:

$$\sigma_s = E_s \cdot \varepsilon_s \qquad\qquad \sigma_c = E_c \cdot \varepsilon_c$$

Where E_s and E_c are the modulii of elasticity of steel and concrete, respectively.

Let: $\qquad \varepsilon = \varepsilon_s = \varepsilon_c$

The above equations may be rewritten as follows:

$$\sigma_s = n \cdot E_c \cdot \varepsilon \qquad\qquad \sigma_c = E_c \cdot \varepsilon$$

Where: $\qquad n = E_s / E_c = 7 - 9$ (14.1)

As such: $\qquad \sigma_s = n \cdot \sigma_c$

Based on the above equations, the force in steel may be expressed as shown:

$$F_s = \sigma_s \cdot A_s \qquad\qquad \text{where } A_s \text{ is the area of steel}$$

Hypothetically, if A_s was an adjacent concrete area subjected to a force F_c, then:

$$F_s / F_c = \sigma_s \cdot A_s / \sigma_c \cdot A_s = n$$

Consequently, an area of steel A_s is subjected to a stress or force n times if it was of concrete. In other words, an area of steel A_s is equivalent to an area of concrete of nA_s. Such is the concept of the transformed section. Steel is converted or transformed into an equivalent concrete area resulting in a homogeneous section entirely made of concrete. Utilizing transformed sections simplifies calculations of moments of inertia, stresses, strains as well as deflections of composite beams (made of several materials) including reinforced concrete beams. Figure 14.1 illustrates the concept of a transformed uncracked concrete beam section.

The rectangular beam cross–section in Figure 14.1(a) is converted into the transformed section (made entirely of concrete) in Figure 14.1(b). As explained earlier, the steel area, A_s, is converted into an equivalent concrete area, nA_s, where $n = E_s / E_c$. This magnification of steel area is used to fill the cavities previously occupied by steel bars, while the remainder, $(n-1)A_s$, is attached to the sides of the rectangular concrete section.

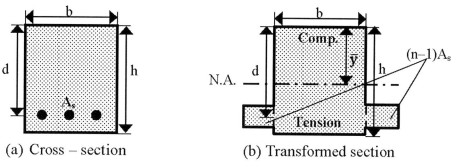

(a) Cross – section   (b) Transformed section

Figure 14.1: Transformed uncracked reinforced concrete beam section.

The moment of inertia of transformed uncracked concrete should first start with determining the location of neutral axis. The neutral axis (zero–strain axis) separates the tension area from the compression area in the uncracked section (see phase 1 of behavior in Chapter 4). The procedure for determining the neutral axis location for a beam in the elastic range follows. As shown in Figure 14.1(b), the neutral axis is located at a distance $\bar{y}$ from the top of the beam section (extreme compression strain location). The first moment of area (area × distance from the neutral axis) of the beam portion above the neutral axis is required to be equal to the first moment of area of the beam portion below the neutral axis as illustrated:

$$(b \cdot \bar{y})\bar{y}/2 = [b(h - \bar{y})] \cdot (h - \bar{y})/2 + (n - 1)A_s(d - \bar{y})$$

The above equation may be rewritten as:

$$b(h^2 - 2h\bar{y})/2 + (n - 1)A_s(d - \bar{y}) = 0$$

As a result $\bar{y}$ may be expressed as shown in the following equation:

$$\bar{y} = \frac{\dfrac{bh^2}{2} + (n - 1)dA_s}{bh + (n - 1)A_s}$$

One needs to note that the above equations apply to beams with rectangular sections. As the section remains in the elastic range, the location of N.A. remains constant and independent of the external loading or moment. Note that $\bar{y}$ in the above equation is a function of section geometry b, d, h and A_s as well as $n = E_s/E_c$ regardless of bending moment. The transformed moment of inertia of a reinforced concrete beam can be obtained by adding the moments of inertia of the beam portion above the neutral axis and the beam portions below the neutral axis. Since the moment of inertia of steel bars about their centroidal axis is negligible, it is typical ignored. The following equations serve as reminders for moments of inertia:

Moment of inertia of a rectangular section:

about its centerline axis = $bh^3/12$ CL

about its base axis = $bh^3/3$ BL

The parallel axis theorem for moments of inertia:

moment of inertia of a section about a non–centroidal axis = moment of inertia of the section about the parallel centroidal axis + section area × (distance between the axes)²

The above expression can be rewritten in an equation form as shown:

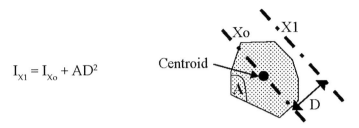

$$I_{X1} = I_{Xo} + AD^2$$

The transformed moment of inertia of a rectangular concrete beam can be obtained as follows:

$$I_{tr} = b\bar{y}^3/3 + b(h - \bar{y})^3/3 + n[\text{steel bar's moment of inertia about their centroidal axis} + A_s(d - \bar{y})^2]$$

As stated earlier, steel bars moment of inertia about their centroidal axis is typically equated to zero. As such, the above equation can be reduced:

$$I_{tr} = b\bar{y}^3/3 + b(h - \bar{y})^3/3 + nA_s(d - \bar{y})^2$$

Where:

I_{tr} = the transformed moment of inertia of the concrete beam.

b = width of concrete beam.

d = beam's effective depth.

h = beam's total depth or height.

$\bar{y}$ = distance of neutral axis from the extreme compression fiber.

A_s = area of steel bars.

n = ratio of steel and concrete modulii of elasticity as in Equation 14.1.

Example 14.1

Determine the values of n, $\bar{y}$ and I_{tr} for a rectangular reinforced concrete beam with the following properties: b = 18 in, d = 21 in, h = 24 in, A_s = 5#8 = 3.9 in², f'_c = 3,500 psi and f_y = 60,000 psi.

Solution

Concrete elastic modulus = E_c = 57,000$\sqrt{f'_c}$ = 57,000$\sqrt{3,500}$ = 3.37×10⁶ psi = 3.37×10³ ksi
(See Chapter 2)

Steel elastic modulus = E_s = 29×10⁶ psi = 29×10³ ksi
(See Chapter 2)

n = E_s/E_c = 29/3.37 = 8.6

$$\bar{y} = \frac{\dfrac{bh^2}{2} + (n - 1)dA_s}{bh + (n - 1)A_s} = \frac{\dfrac{18×(24)^2}{2} + (8.6 - 1)×21×3.9}{18×24 + (8.6 - 1)×3.9} = 12.6 \text{ in}$$

$$I_{tr} = b\bar{y}^3/3 + b(h - \bar{y})^3/3 + nA_s(d - \bar{y})^2 = 18×(12.6)^3/3 + 18×(24 - 12.6)^3/3 + 8.6×3.9×(21 - 12.6)^2$$

$$I_{tr} = 23,258 \text{ in}^4$$

14.2.2 Gross Moment of Inertia of Reinforced Concrete Beams

Prior to concrete cracking, contribution of reinforcing steel bars to reinforced concrete beam behavior under bending moments is somewhat limited due to the following reasons:

1. A relatively large uncracked area of concrete contributes to tension resistance, thus, steel bar role is overshadowed.

2. Bending moments on the beam prior to concrete tension cracking are small, and unreinforced concrete is generally adequate for resistance without a meaningful contribution of steel.

For these reasons, ACI 318 Code, consents to the utilization the gross moment of inertia for concrete beams pre–cracking in place of the transformed uncracked moment of inertia. The gross moment of inertia of a reinforced concrete beam is its moment of inertia assuming no steel reinforcement. For a rectangular concrete beam, the gross moment of inertia (I_g) can be expressed as:

$$I_g = bh^3/12 \tag{14.2}$$

Example 14.2

Determine the value of I_g for the reinforced concrete beam of Example 14.1 and the percent difference between I_{tr} and I_g.

Solution

$I_g = bh^3/12 = 18\times(24)^3/12 = 20{,}736 \text{ in}^4$

$(I_{tr} - I_g)/I_{tr} = (23{,}258 - 20{,}736)/23{,}258 = 10.8\%$

As such, utilization of I_g introduces about 10% error/reduction in the value of I_{tr}. Such approximation reduces the complexity of deflection calculations of reinforced concrete beams. Furthermore, replacing I_{tr} with I_g in deflection calculations is an approximation towards the safe side since it underestimates the value of moment of inertia and consequently overestimates the value of beam deflection resulting in a more prudent design from this standpoint.

14.2.3 Transformed Cracked Sections

Phase 2 of reinforced concrete beam behavior under bending (Chapter 4) assumes that concrete is cracked in tension while it remains, to some extent, elastic under compression. Furthermore, steel reinforcement is the only source of tension force in the section consequent to concrete cracking. Steel and concrete remain within their respective elastic ranges in tension and compression, respectively. Phase 2 of behavior is generally comparable to the service condition of concrete beams. The moment of inertia can be determined utilizing the transformed cracked section. As explained in section 14.2.1, the steel bar area is converted to an equivalent concrete area by multiplying with $n = E_s/E_c$. Below the neutral axis in a transformed cracked section, only steel bars are present while cracked concrete is neglected as it cannot resist stresses post cracking. Figure 14.2 illustrates the concept of a transformed cracked concrete beam section.

Figure 14.2(a) shows a typical rectangular concrete beam section. Figure 14.2(b) is the transformed cracked section. As explained earlier, the steel area, A_s, is converted into an equivalent concrete area, nA_s, where $n = E_s/E_c$. This magnified steel area is placed at the location of steel bars while the entire bottom portion of the beam is neglected due to cracking. The cavities previously occupied by steel bars are, thus, non–existent and cannot be filled contrary to the procedure in section 14.2.1. As such, nA_s is used in the transformed cracked section in place of the steel bars as shown in Figure 14.2(b).

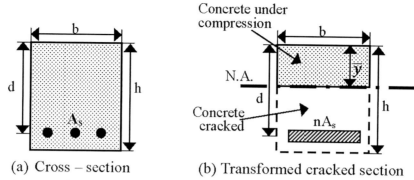

(a) Cross – section   (b) Transformed cracked section

Figure 14.2: Transformed cracked reinforced concrete beam section.

The transformed cracked section is considered elastic. Determination of the neutral axis location follows the same procedure outlined in Section 14.2.1. The neutral axis is located at distance $\bar{y}$ from the top of the beam section (Figure 14.2(b)). It limits the elastic concrete compression zone. The first moment of area (area × distance from the neutral axis) of the beam portion above the neutral axis is required to be equal to the first moment of area of the beam portion below the neutral axis:

$$(b \cdot \bar{y})\bar{y}/2 = nA_s(d - \bar{y})$$

The above equation may be rewritten as:

$$b\bar{y}^2/2 + nA_s\bar{y} - nA_s d = 0$$

As a result $\bar{y}$ for a rectangular beam may be expressed as shown in the following equation:

$$\bar{y} = \frac{-nA_s + \sqrt{(nA_s)^2 + 2(nA_s bd)}}{b} \tag{14.3}$$

As concrete and steel remain elastic in compression and tension, respectively, the location of N.A. in the cracked section remains constant and independent of the external loading or moment. Note that $\bar{y}$ in Equation 14.3 is a function of section geometry b, d, h, A_s and $n = E_s/E_c$ regardless of bending moment. Crack depth $(h - \bar{y})$ remains constant within the elastic range of steel and concrete. This is somewhat different from the theory presented in Chapter 4 for section behavior within the inelastic range since concrete is not considered fully elastic. As loading or stresses increase beyond the elastic range, crack depth increases due to variation of steel and concrete modulii of elasticity in the inelastic range.

The transformed cracked moment of inertia of a rectangular reinforced concrete beam can be obtained by adding the moments of inertia of the beam compression zone above the neutral axis to the moment of inertia of steel bars. The moment of inertia of steel bars about their centroidal axis is ignored:

$$I_{cr} = b\bar{y}^3/3 + nA_s(d - \bar{y})^2 \tag{14.4}$$

Where:

I_{cr} = the transformed moment of inertia of the cracked rectangular concrete beam.

b = width of concrete beam.

d = beam's effective depth.

$\bar{y}$ = depth of neutral axis as in Equation 14.3.

A_s = area of steel bars.

n = ratio of steel and concrete modulii of elasticity as in Equation 14.1.

Example 14.3

Determine the value of I_{cr} for the reinforced concrete beam of Example 14.1 and the percentage ratios I_{cr}/I_{tr} and I_{cr}/I_g.

Solution

$$\bar{y} = \frac{-nA_s + \sqrt{(nA_s)^2 + 2(nA_s bd)}}{b} = \frac{-8.6 \times 3.9 + \sqrt{(8.6 \times 3.9)^2 + 2(8.6 \times 3.9 \times 18 \times 21)}}{18} = 7.2 \text{ in}$$

$$I_{cr} = b\bar{y}^3/3 + nA_s(d - \bar{y})^2 = 18 \times (7.2)^3/3 + 8.6 \times 3.9(21 - 7.2)^2 = 8,627 \text{ in}^4$$

$$I_{cr}/I_{tr} = 8,627/23,258 = 39\% \qquad I_{cr}/I_g = 8,627/20,736 = 42\%$$

The cracked concrete tension zone is a significant portion of the concrete beam area that becomes inactive upon cracking. As such, cracking results in a significant reduction in beam area and moment of inertia for stress resistance and deflection opposition. Cracking of reinforced concrete beams results in rapid increase of stresses in concrete and steel as well as sudden increase in deflection (about 60% for Example 14.3).

14.2.4 Effective Moment of Inertia

Figure 14.3 illustrates the service condition of a reinforced concrete beam. Phase 2 of beam behavior in Chapter 4 can be considered analogues to service condition. At midspan where maximum bending moment occurs, the beam is cracked. At the crack location, bond of steel and concrete is nonexistent and tension stress in concrete is zero. Away from the crack, tension stress in concrete builds up since bond transfers stresses between steel and concrete. Eventually, the tensile strength of concrete is exceeded resulting in another crack in the beam. The frequency and depth of cracks lessens closer to support farther from midspan since bending moment and stresses in the beam decrease. However, there is a random element to concrete cracking due to nonuniformity of its properties. Concrete properties are variable since it is mostly made up of processed natural materials (water, sand, rock and cement) as explained in Chapter 2. It is also inaccurately assumed that plain concrete is an elastic material where its non–linear behavior results in variable crack depth with bending moment even under service conditions. The beam condition along the span is, thus, highly variable. Stresses in the beam vary depending on the bending moment and cracking condition.

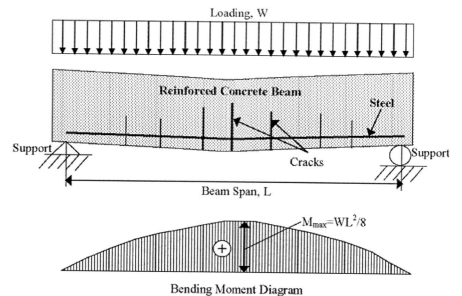

Bending Moment Diagram
Figure 14.3: Reinforced concrete beam under service conditions.

It is difficult to determine or select the moment of inertia of a concrete beam under service conditions for deflection calculation. Therefore, ACI 318 Code allows the designer to use an approximate moment of inertia for deflections of concrete beams termed the effective moment of inertia, I_e. If the beam is not cracked under service conditions, then the uncracked section moment of inertia may be used for deflection computation.

ACI 318 Code states that immediate (elastic) deflections can be computed utilizing the modulus of elasticity of concrete, E_c, and the effective moment of inertia, I_e, where:

$$I_e = \left(\frac{M_{cr}}{M_a}\right)^3 I_g + \left[1 - \left(\frac{M_{cr}}{M_a}\right)^3\right]I_{cr} \tag{14.5}$$

Where:

I_e = effective moment of inertia for computation of deflection.

M_{cr} = the bending moment that causes tension cracking in concrete determined as shown:

$$M_{cr} = \frac{f_r I_g}{y_t} \tag{14.6}$$

f_r = modulus of rupture of concrete as computed via Equation 14.6 multiplied by 1.0 for normal–weight concrete, 0.75 for all light–,weight concrete and 0.85 for sand light – weight concrete, as specified by ACI:

$$f_r(psi) = 7.5\sqrt{f'_c(psi)} \tag{14.7}$$

I_g = gross moment of inertia of the uncracked beam neglecting steel reinforcement.

y_t = distance between the neutral axis and extreme tension fiber in the uncracked section.

M_a = maximum bending moment within the member at deflection calculation.

At the onset of cracking, tension cracking stress = My/I as in Bernouli's theorem. Tension cracking stress = modulus of rupture = f_r (Chapter 2) expressed in Equation 14.7. M at the onset of cracking is the cracking moment, M_{cr}, as explained above (Equation 14.6). $y = y_t$ is the distance of the neutral axis in the pre–cracked beam and farthest location in the tension side of the beam where tension stress is maximum. I is the moment of inertia of the pre–cracked (uncracked) beam. As previously stated, ACI Code permits the use of I_g (gross moment of inertia) for the pre–cracked moment of inertia. As such, Equations 14.6 and 14.7 may be assembled. Some designers apt to use an approximate value of $I_e = 0.35I_g$ for reinforced concrete beams. Such approximation is not endorsed by ACI Code.

14.2.5 Immediate or Instantaneous Deflection of Concrete Beams

A reinforced concrete beam under service conditions may be considered elastic. As such, the traditional deflection equations of solid mechanics can be used for reinforced concrete with E_c and I_e for computing elastic deflection (termed immediate deflection). Such immediate deflection does not include long–term load effect caused by section inelastic or plastic deformations. Long–term deflections are discussed in Section 14.2.6.

Since ACI Code permits the use of elastic deflection equations for computing immediate deflections of reinforced concrete beams with E_c and I_e, a list of these equations is in Figures 14.4 and 14.5 duplicated in Figures A.33 and A.34 of Appendix A.

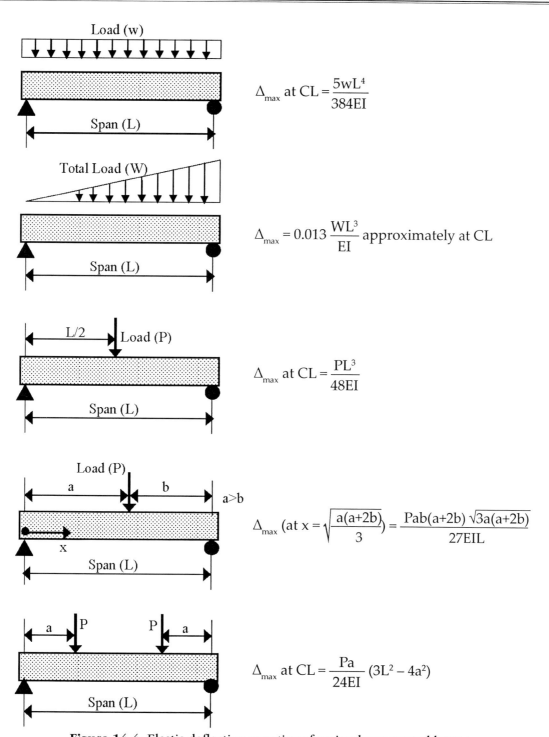

Figure 14.4: Elastic deflection equations for simply supported beams.

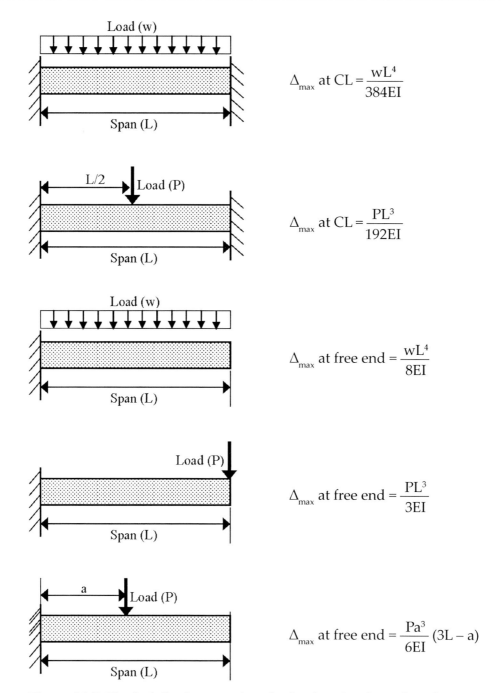

Figure 14.5: Elastic deflection equations for fixed–end and cantilever beams.

Example 14.4

Determine the immediate CL deflection of a concrete beam with a cross–section and materials as in Examples 14.1, 14.2 and 14.3 assuming normal–weight concrete and loading as shown below.

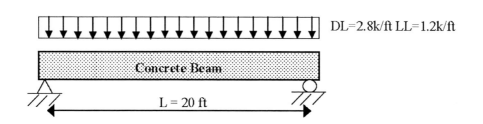

Solution

The solid mechanics elastic deflection equation that may be used for this example is:

$$\Delta = \frac{5wl^4}{384EI}$$

Immediate deflection due to dead load $= \Delta_{iDL} = \dfrac{5DL \cdot L^4}{384E_c I_e}$

DL = 2.8 k/ft = 2,800 lb/12 in = 233 lb/in

L = 20 ft = 240in

$E_c = 57,000 \sqrt{f'_c} = 57,000\sqrt{3,500} = 3.37 \times 10^6$ psi $= 3.37 \times 10^3$ ksi

$f_r = 7.5\sqrt{f'_c} = 7.5 \times \sqrt{3,500} = 444$ psi

$y_t = h/2 = 12$in

$I_g = 20,736$ in^4 (Example 14.2)

$$M_{cr} = \frac{f_r I_g}{y_t} = \frac{444 \times 20,736}{12} = 767,232 \text{ lb.in} = 63.9 \text{ k.ft}$$

$$M_a = \frac{DL \times L^2}{8} = \frac{233 \times (240)^2}{8} = 1,678,000 \text{ lb.in} = 139.8 \text{ k.ft}$$

$I_{cr} = 8,627$ in^4 (Example 14.3)

$$I_e = \left(\frac{M_{cr}}{M_a}\right)^3 I_g + \left[1 - \left(\frac{M_{cr}}{M_a}\right)^3\right]I_{cr} = \left(\frac{63.9}{139.8}\right)^3 20,736 + \left[1 - \left(\frac{63.9}{139.8}\right)^3\right]8,627$$

$I_e = 9,783$ in^4

$$\Delta_{iDL} = \frac{5 \times 233 \times (240)^4}{384 \times 3.37 \times 10^6 \times 9,783} = 0.31 \text{ in}$$

Immediate deflection due to total service load (dead load + live load) $= \Delta_{iTL} = \dfrac{5(DL+LL)L^4}{384E_c I_e}$

Total service load = TL = DL+LL = 4.0 k/ft = 4,000 lb/12 in = 333 lb/in

$$M_a = \frac{(DL+LL) \times L^2}{8} = \frac{333 \times (240)^2}{8} = 2,400,000 \text{ lb.in} = 200.0 \text{ k.ft}$$

$$I_e = \left(\frac{M_{cr}}{M_a}\right)^3 I_g + \left[1 - \left(\frac{M_{cr}}{M_a}\right)^3\right]I_{cr} = \left(\frac{63.9}{200.0}\right)^3 20,736 + \left[1 - \left(\frac{63.9}{200.0}\right)^3\right]8,627$$

$I_e = 9,021$ in^4

$$\Delta_{iTL} = \frac{5 \times 333 \times (240)^4}{384 \times 3.37 \times 10^6 \times 9,021} = 0.47 \text{ in}$$

Immediate deflection due to live load $= D_{iLL} = 0.47 - 0.31 = 0.16$ in

Discussion

The effective moment of inertia, I_e, used for deflection is dependent upon maximum service bending moment, M_a, which in turn is dependent upon loading. The loading condition of only dead load can be utilized for dead load deflection. However, live load may not be applied in the absence of dead load. Thus, to determine live load deflection, the deflection due to dead and live loads (total service load) is calculated, and then live load deflection may be computed by subtracting dead load deflection from dead and live load deflection as in Example 14.4. Thus, computing I_e is typically required multiple times for various loading conditions. To simplify deflection calculation, I_e can be determined only once for total service load and used for all loading cases, thus assuming that the extent of cracking within phase 2 of behavior (Chapter 4) has a limited effect on the cracked beam moment of inertia. This introduces a small deviation from ACI Code procedure. Another solution for Example 14.4 using the simplified procedure described above follows.

Simplified Solution for Example 14.4

The solid mechanics elastic deflection equation for this example is: $\Delta = \dfrac{5wl^4}{384EI}$

$E_c = 57{,}000\sqrt{f'_c} = 57{,}000\sqrt{3{,}500} = 3.37 \times 10^6 \text{ psi} = 3.37 \times 10^3 \text{ ksi}$

$f_r = 7.5\sqrt{f'_c} = 7.5\times\sqrt{3{,}500} = 444 \text{ ksi}$

$y_t = h/2 = 12\text{in}$

$I_g = 20{,}736 \text{ in}^4$ (Example 14.2)

$M_{cr} = \dfrac{f_r I_g}{y_t} = \dfrac{444\times20{,}736}{12} = 767{,}232 \text{ lb.in} = 63.9 \text{ k.ft}$

$I_{cr} = 8{,}627 \text{ in}^4$ (Example 14.3)

$M_{cr} = \dfrac{(DL + LL)\times L^2}{8} = \dfrac{333\times(240)^2}{8} = 767{,}232 \text{ lb.in} = 63.9 \text{ k.ft}$

DL+LL = 4 k/ft = 4,000 lb/12in = 333 lb/in

$I_e = \left(\dfrac{M_{cr}}{M_a}\right)^3 I_g + \left[1 - \left(\dfrac{M_{cr}}{M_a}\right)^3\right]I_{cr} = \left(\dfrac{63.9}{200.0}\right)^3 20{,}736 + \left[1 - \left(\dfrac{63.9}{200.0}\right)^3\right] 8{,}627$

$I_e = 9{,}021 \text{ in}^4$

L = 20 ft = 240 in

DL = 2.8 k/ft = 2,800 lb/12 in = 233 lb/in

Immediate deflection due to dead load $\Delta_{iDL} = \dfrac{5DL.L^4}{384E_c I_e}$

$\Delta_{iDL} = \dfrac{5\times233\times(240)^4}{384\times3.37\times10^6\times9{,}021} = 0.33 \text{ in}$

LL = 1.2 k/ft = 1,200 lb/12 in = 100 lb/in

Immediate deflection due to live load $= \Delta_{iLL} = \dfrac{5LL.L^4}{384E_c I_e}$

$\Delta_{iLL} = \dfrac{5\times100\times(240)^4}{384\times3.37\times10^6\times9{,}021} = 0.14 \text{ in}$

Immediate deflection due to DL and LL = 0.33 + 0.14 = 0.47 in

The following table compares the simplified methods of immediate deflection calculations:

	DL Deflection	LL Deflection	DL + LL Deflection
Detailed Method	0.31 in	0.16 in	0.47 in
Simplified Method	0.33 in	0.14 in	0.47 in

The above table reveals that the simplified method is within acceptable tolerance. It will be used throughout the remainder of this chapter.

Example 14.5

Determine the free–end immediate deflection of a one–way cantilevered concrete slab with 1–ft strip cross–section and material specifications of: $f'_c = 4,000$ psi and $f_y = 60,000$ psi.

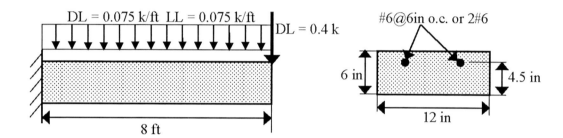

Solution

The solid mechanics elastic deflection equation that may be used for this example is:

For the distributed loads $\qquad \Delta_{max}$ at free end $= \dfrac{wL^4}{8EI}$

For the concentrated load $\qquad \Delta_{max}$ at free end $= \dfrac{PL^3}{3EI}$

Determine the effective moment of inertia of the slab strip cross–section:

$I_g = bh^3/12 = 12 \times (6)^3/12 = 216 \ in^4$

$E_c = 57,000\sqrt{f'_c} = 57,000\sqrt{4.000} = 3.6 \times 10^6 \ psi$

$n = E_s/E_c = 29 \times 10^6/3.6 \times 10^6 = 8.1$

$\bar{y} = \dfrac{-nA_s + \sqrt{(nA_s)^2 + 2(nA_s bd)}}{b} = \dfrac{-8.1 \times 0.88 + \sqrt{(8.1 \times 0.88)^2 + 2(8.1 \times 0.88 \times 12 \times 4.5)}}{12} = 1.8 \ in$

$I_{cr} = b\bar{y}^3/3 + nA_s(d - \bar{y})^2 = 12 \times (1.8)^3/3 + 8.1 \times 0.88 \times (4.5 - 1.8)^2 = 75.3 \ in^4$

$f_r = 7.5\sqrt{f'_c} = 7.5\sqrt{4,000} = 474 \ psi \qquad y_t = 3in$

$M_{cr} = \dfrac{f_r I_g}{y_t} = \dfrac{474 \times 216}{3}/12,000 = 2.8 \ k.ft$

$M_a = wL^2/2 + PL = (0.0.75 + 0.075) \times (8)^2/2 + 0.4 \times 8 = 8.0 \ k.ft$

$$I_e = \left(\frac{M_{cr}}{M_a}\right)^3 I_g + \left[1 - \left(\frac{M_{cr}}{M_a}\right)^3\right]I_{cr} = \left(\frac{2.8}{8.0}\right)^3 216 + \left[1 - \left(\frac{2.8}{8.0}\right)^3\right]75.3 = 81.3 \text{ in}^4$$

Distributed dead load immediate deflection $\Delta_{i1} = \dfrac{wL^4}{8EI} = \dfrac{\left(\frac{75}{12}\right)\times(8\times12)^4}{8\times3.6\times10^6\times81.3} = 0.23 \text{ in}$

Distributed live load immediate deflection $\Delta_{i2} = \dfrac{wL^4}{8EI} = \dfrac{\left(\frac{75}{12}\right)\times(8\times12)^4}{8\times3.6\times10^6\times81.3} = 0.23 \text{ in}$

Concentrated dead load immediate deflection $\Delta_{i3} = \dfrac{PL^3}{3EI} = \dfrac{400\times(8\times12)^3}{8\times3.6\times10^6\times81.3} = 0.40 \text{ in}$

Immediate total free–end deflection/total maximum deflection = 0.23 + 0.23 + 0.40 = 0.86 in

Example 14.6

Determine the maximum immediate deflection for the continuous beam span shown below with the cross–sections shown below and material specifications of f'_c = 4,500 psi and f_y = 60,000 psi.

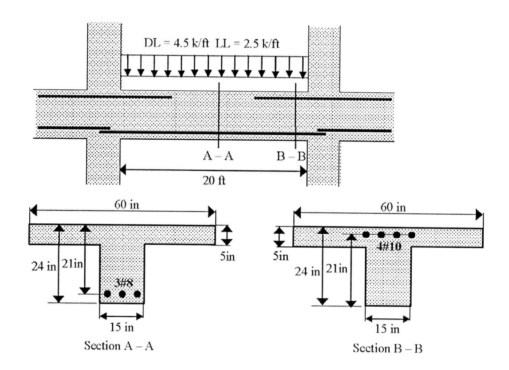

It is presumed that section A–A is a T–beam within a floor system. Its effective section (flange width, b = 60 in) is determined as illustrated in Chapter 9.

Solution

As the span shown in Example 14.6 is continuous on both sides, it is generally modeled as a fixed–end beam with a bending moment diagram as shown below:

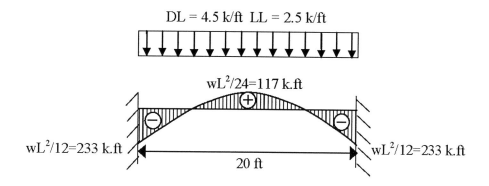

The solid mechanics elastic deflection equation for this example is: $\Delta = \dfrac{wl^4}{384EI}$

The continuous beam span shown in the problem statement has two distinct cross sections:

a T–beam subjected to positive bending moment (tension cracking below the N.A.) and a rectangular beam subjected to negative bending moment (tension cracking above the N.A. including the flange). Each needs to be handled separately.

Determine the effective moment of inertia of section A–A

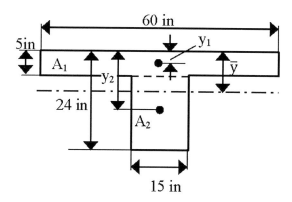

Gross moment of inertia, I_g:

Determine the location of section CL or $\bar{y}$:

$$\bar{y} = \frac{A_1 y_1 + A_2 y_2}{A_1 + A_2} = \frac{(5 \times 60) \times 2.5 + (15 \times 19) \times (5 + 9.5)}{(5 \times 60) + (15 \times 19)} = 8.3 \text{ in}$$

Gross moment of inertia, I_g, can be determined using the parallel axis theorem as explained earlier:

Area A_1: $I_{gA1} = 60 \times (5)^3/12 + (60 \times 5) \times (8.3 - 2.5)^2 = 10{,}717 \text{ in}^4$

Area A_2: $I_{gA2} = 15 \times (19)^3/12 + (15 \times 19) \times (14.5 - 8.3)^2 = 19{,}529 \text{ in}^4$

$I_g = 10{,}717 + 19{,}529 = 30{,}246 \text{ in}^4$

Cracking moment, M_{cr}:

$y_t = 24 - 8.3 = 15.7 \text{in}$

$f_r = 7.5\sqrt{4{,}500} = 503 \text{ psi}$

$M_{cr} = (503 \times 29{,}246/15.7)/12{,}000 = 80.8 \text{ k.ft}$

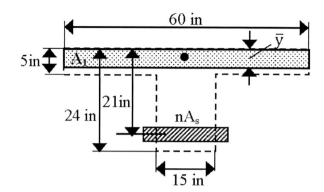

Cracked section moment of inertia, I_{cr}:

Determine the location of section CL or $\bar{y}$ of the cracked section:

$$n = 29\times10^6/57,000\sqrt{4,500} = 29\times10^6/3.8\times10^6 = 7.6$$

$$\bar{y} = \frac{-nA_s + \sqrt{(nA_s)^2 + 2(nA_sbd)}}{b}$$

$$\bar{y} = \frac{-7.6\times2.34 + \sqrt{(7.6\times2.34)^2 + 2(7.6\times2.34\times15\times21)}}{60} = 1.49 \text{ in}$$

The above equation assumes that the concrete elastic compression zone above neutral axis is within the flange resulting in the beam acting as a rectangular beam with a width equal to the flange width. The value of $\bar{y}$ computed with the above equation is less than 5 in, thus confirming the above assumption. Such assumption is logical for T–beams due to the large concrete flange area that is typically adequate for the required compression in the beam. If $\bar{y}$ was greater than the flange thickness, then, a similar solution procedure may be used based on area first moment as explained earlier.

Cracked moment of inertia, I_{cr}, can be determined using the parallel axis theorem:

Concrete area A_1: $I_{A1} = b\bar{y}^3/3 = 60\times(1.49)^3/3 = 60.2 \text{ in}^4$

Steel area nA_s: $I_{nAs} = nA_s(d - \bar{y})^2 = 7.6\times(2.34)\times(21 - 1.49)^2 = 6,769 \text{ in}^4$

$I_{cr} = 60.2 + 6,769 = 6,829 \text{ in}^4$

Maximum service moment for section A–A, $M_a = 117$ k.ft

$$I_e = \left(\frac{M_{cr}}{M_a}\right)^3 I_g + \left[1 - \left(\frac{M_{cr}}{M_a}\right)^3\right]I_{cr} = \left(\frac{80.8}{117}\right)^3 30,246 + \left[1 - \left(\frac{80.8}{117}\right)^3\right]6,829 = 14,541 \text{ in}^4$$

Determine the effective moment of inertia of section B–B

Section B–B is considered a rectangular section. It is subjected to a negative bending moment. The concrete flange is on the beam tension side and is, therefore, considered cracked and ineffective for bending moments.

Gross moment of inertia, I_g:

$I_g = bh^3/12 = 15\times(24)3/12 = 17,280 \text{ in}^4$

Cracking moment, M_{cr}:

$f_r = 503$ psi $y_t = 12$ in

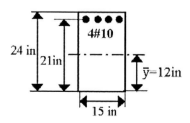

$$M_{cr} = \frac{f_r I_g}{y_t} = \frac{503\times17,280}{12}/12,000 = 60.4 \text{ k.ft}$$

Cracked moment of inertia:

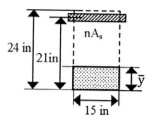

$$\bar{y} = \frac{-nA_s + \sqrt{(nA_s)^2 + 2(nA_sbd)}}{b}$$

$$\bar{y} = \frac{-7.6 \times 4.92 + \sqrt{(7.6 \times 4.92)^2 + 2(7.6 \times 4.92 \times 15 \times 21)}}{15} = 8.0 \text{ in}$$

$$I_{cr} = b\bar{y}^3/3 + nA_s(d - \bar{y})^2 = 15 \times (8.0)^3/3 + 7.6 \times 4.92 \times (21 - 8.0)^2 = 8{,}879 \text{ in}^4$$

Maximum service moment for section B–B, $M_a = 233$ k.ft

$$I_e = \left(\frac{M_{cr}}{M_a}\right)^3 I_g + \left[1 - \left(\frac{M_{cr}}{M_a}\right)^3\right] I_{cr} = \left(\frac{60.4}{233}\right)^3 17{,}280 + \left[1 - \left(\frac{60.4}{233}\right)^3\right] 8{,}879 = 9{,}025 \text{ in}^4$$

The continuous beam span in this example is subjected to positive as well as negative bending moments. The portions subjected to positive moments (Section A–A) have $I_e = 14{,}541$ in⁴ while the portions subjected to negative moments (Section B–B) have $I_e = 9{,}025$ in⁴. Judging from the bending moment diagram, about 40% of the span is subjected to negative moment while 60% is subjected to positive moment. Thus, I_e for the entire span may be determined using the weighted average as shown below:

$$I_e = 0.4 I_{eB-B} + 0.6 I_{eA-A} = 0.4 \times 9{,}025 + 0.6 \times 14{,}541 = 12{,}335 \text{ in}^4$$

$$\text{Dead load immediate deflection} = \Delta_{iDL} = \frac{DL \times L^4}{384 E_c I_e} = \frac{(4{,}500/12) \times (20 \times 12)^4}{384 \times 3.8 \times 10^6 \times 9{,}025} = 0.094 \text{ in}$$

$$\text{Live load immediate deflection} = \Delta_{iLL} = \frac{LL \times L^4}{384 E_c I_e} = \frac{(2{,}500/12) \times (20 \times 12)^4}{384 \times 3.8 \times 10^6 \times 9{,}025} = 0.052 \text{ in}$$

Total immediate deflection = 0.094 + 0.052 = 0.146 in

This deflection for a 20–ft span of a continuous beam is relatively small. The fixed–end support conditions cause deflection reduction by 80% compared with a simply supported beam (see Figures 14.4 and 14.5). The deflection of a simply supported beam loaded with a uniformly distributed load is $\frac{5wl^4}{384EI}$. However, the deflection of a fixed–end beam loaded with a uniformly distributed load is $\frac{wl^4}{384EI}$.

Furthermore, the portions of the fixed–end beam subjected to positive bending moment could have an increases moment of inertia due to slab/flange contribution which further reduces deflection.

To summarize, I_e equations for a continuous beam span are:

$$I_e = 0.6 I_{eM+} + 0.4 I_{eM-} \qquad \text{for two–end continuous beam}$$
$$I_e = 0.8 I_{eM+} + 0.2 I_{eM-} \qquad \text{for one–end continuous beam}$$

Where:

I_{eM+} = effective moment of inertia for sections subjected to positive moment.

I_{eM-} = effective moment of inertia for sections subjected to negative moment.

14.2.6 Long–Term Deflection of Concrete Beams

The deflection that occurs in a concrete beam or slab immediately or instantaneously as load is applied is the elastic deflection (discussed in Section 14.2.5). Strains and deflection increase as the load is sustained. Creep and shrinkage of concrete result in increasing compression deformations generally within the first few years of service life of concrete. Inelastic additional deflection caused by such deformation increase that is generally irrecoverable. This additional deflection is termed long term deflection due to sustained load in ACI Code. Other factors affect the long–term deflection of reinforced concrete including:

1.   Applied load or stress: increased load or stress obviously increases long–term deflection.

2.   Concrete compressive strength: increasing f'_c results in reduction of creep and shrinkage, consequently, reduction of long–term deflection.

3.   Age or maturity of concrete at the time of loading: if concrete is loaded at an early age, long term deflection significantly increases. With age, concrete hydration maturity improves causing an increase in strength and modulus of elasticity. With progress of hydration, more load resisting crystals and less air voids result which reduces creep and shrinkage.

4.   Compression reinforcement: use of compression reinforcement reduces the effect of concrete creep. Creep causes an increase in concrete compression strain and, consequently, increased deflection. Compression forces in steel bars increase due to such strain increase resulting in resistance to creep. As such, the effect of creep is moderated.

Pursuant to the ACI Code, following equations can be used for determining the value of long term deflection of reinforced concrete:

$$\lambda = \frac{\xi}{1+50\rho'}$$ (14.8)

Where:

λ = long–term deflection factor.

ξ = sustained load duration factor.

$\rho' = A'_s/bd$ = compression steel reinforcement ratio at midspan for simply supported and continuous beams and at support for cantilever beams.

ξ can be determined based on the duration of load utilizing Table 14.1 or Figure 14.6.

Table 14.1: Load Duration Factor (ξ)

Duration of sustained load	Value of load duration factor (ξ)
$\geq$ 5 years	2.0
12 months	1.4
6 months	1.2
3 months	1.0

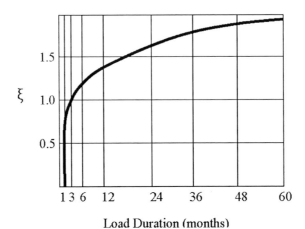

Load Duration (months)

Figure 14.6: Load duration factor (ξ).

The equations for long–term and total deflection due to a defined load:

$$\Delta_{LT} = \lambda \Delta_i \qquad (14.9)$$

$$\Delta_T = \Delta_i (1 + \lambda) \qquad (14.10)$$

Where:

$\Delta_i, \Delta_{LT}, \Delta_T$ = immediate, long–term and total (immediate + long–term) deflections, respectively, due to the load under consideration.

λ = long–term deflection factor based for the load under consideration (Equation 14.8).

Based on the aforementioned discussion, total service deflection of reinforced concrete element can be expressed as follows:

$$\Delta_{TL} = \Delta_{iDL}(1 + \lambda_\infty) + \Delta_{iLL} + \lambda_{SLL}\Delta_{iSLL} \qquad (14.11)$$

Where:

Δ_{TL} = total element service deflection.

Δ_{iDL} = immediate dead load deflection.

λ_∞ = long–term deflection factor for dead load (infinite duration in Equation 14.8).

Δ_{iLL} = immediate live load deflection.

Δ_{iSLL} = immediate deflection due to sustained live load. Part of live load is typically sustained for a period of time.

λ_{SLL} = long–term deflection factor for sustained live load determined via Equation 14.8.

For a typical reinforced concrete structure, a portion of the live load is sustained. For residential construction, about 30% may be considered sustained for few years. For storage facilities, about 70% of the live load is typically sustained. The duration and percentage of sustained live load are determined by the design engineer. The design engineer also needs to consider accuracy of deflection calculations. Several studies regarding the deflection of reinforced concrete beams have concluded that actual deflection is within about 30% of calculated deflection. This is a relatively high degree of variation. Generally, one needs to be on the safe side in deflection calculation of concrete structures for code compliance.

14.2.7 Code Requirements for Deflection of Reinforced Concrete Structures

Excessive deflections are detrimental. Non structural elements and attachments suffer defects (cracking, misalignment, detachment, etc) or may even malfunction. Appearance of sturdiness is jeopardized. Occupants or users frequently become concerned about structural integrity upon noticing excessive deflections. Such concerns can result in less desirability of the building or structure. Increased member thickness or depth results in increasing member moment of inertia and, consequently, reduced deflection. However, increasing member thickness increases member weight that results in increasing inertia forces affecting the structure (dead and seismic loads) and reducing usable space. Extra forces and space restrictions can complicate the design process. The designer needs to avoid excessive increases in member depths to reduce deflections for the above reasons. Optimization of structural design to reduce deflection including enhanced support conditions (e.g. fixed ends) and efficient structural element distribution can be utilized to reduce deflection. ACI Code places limits on permissible deflection of reinforced concrete element. There are two methods of complying with code requirements for deflection:

a. Method 1: elements (beams or slabs) that conform to the minimum thickness requirements, as explained in Chapter 6, are considered adequate. ACI Code specifies that deflections need not be computed unless the element in question supports sensitive partitions or attachments that may be damaged by excessive deflections.

b. Method 2: for elements that do not satisfy ACI Code minimum thickness requirements or are supporting deflection sensitive attachments, computed deflections using E_c and I_e must fall within the permissible limits specified in Table 14.2 or Table A.35.

Analysis of Table 14.2 reveals that ACI Code requirements regarding deflection are focused on roofs, floors and their supporting elements. Furthermore, focus is upon deflection effect on nonstructural attachment. ACI Code, however, states that the effect of deflection on structural elements must be carried out by structural analysis to determine extra forces or moments that are caused by deflection.

Table 14.2: Maximum permissible computed deflections

Type of member	Deflection to be considered	Deflection limitation
Flat roofs not supporting or attached to nonstructural elements likely to be damaged by large deflections	Immediate deflection due to live load, Δ_{iLL}	$\ell/180$*
Floors not supporting or attached to nonstructural elements likely to be damaged by large deflections	Immediate deflection due to live load, Δ_{iLL}	$\ell/360$
Roof or floor construction supporting or attached to nonstructural elements likely to be damaged by large deflections	That part of total deflection occurring after attachment of nonstructural elements , Δ_{ATT} (sum of the long–term deflection due to all sustained loads and the immediate deflection due to any additional live load) †	$\ell/480$‡
Roof or floor construction not supporting or attached to nonstructural elements likely to be damaged by large deflection		$\ell/240$§

Notes for Table 14.2:

*   Limit not intended to safeguard against ponding. Ponding should be checked by suitable calculations of deflection, including added deflections due to ponded water, and considering long–term effects of all sustained loads, camber, construction tolerances, and reliability of provisions for drainage.

†   Long – term deflection shall be determined per ACI Code provisions, but may be reduced by amount of deflection calculated to occur before attachment of nonstructural elements.

‡   Limit may be exceeded if adequate measures are taken to prevent damage to supported or attached elements.

§   Limit shall not be greater than tolerance provided for nonstructural elements. Limit may be exceeded if camber is provided so that total deflection minus camber does not exceed limit.

Where:

ℓ = member span defined as clear span + member depth not to exceed center–to–center span of simply supported and continuous beams and clear span of cantilevers.

Immediate deflection due to live load is as discussed earlier referred to as Δ_{iLL}.

Deflection that occurs after attachment of nonstructural members is the immediate deflection of live load + long term deflection of dead load + long term deflection of sustained live load and may be described in the following equation:

$$\Delta_{ATT} = \Delta_{iLL'} + \lambda_\infty \Delta_{iDL} + \lambda_{SLL}\Delta_{iSLL} \tag{14.12}$$

Where:

Δ_{ATT} = deflection after attachment of nonstructural elements.

Remaining notation is identical to Equation 14.11.

Ponding is accumulation of rain water on roofs if roof slope is not adequate to allow water to flow towards drains, and

Camber is negative deflection built in the concrete formwork so that it compensates for anticipated future deflection.

Example 14.7

Determine the CL total long term service deflection of the concrete beam in Example 14.4 with the added compression steel as shown if 50% of live load is sustained for 3 years and check if the beam is code compliant if it is:

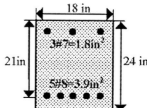

a.   flat roof beam without deflection sensitive attachment.

b.   floor beam without deflection sensitive attachment.

c.   roof or floor construction with deflection sensitive attachment.

Solution

From the solution of Example 14.4:

$-\Delta_{iDL} = 0.33$ in

$-\Delta_{iLL} = 0.14$ in $\rho' = A'_s/bd = 1.8/18\times21 = 0.0047$

$\xi_\infty = 2.0 \Longrightarrow \lambda_\infty = \xi_\infty/(1 + 50\rho') = 2/(1 + 50\times0.0047) = 1.62$

DL long term deflection $= \lambda_\infty \times \Delta_{iDL} = 1.62\times0.33 = 0.53$ in

Immediate deflection due to sustained live load $= 0.5\times0.14 = 0.07$ in

$\xi_{3yr} = 1.7 \Longrightarrow \lambda_{3yr} = \xi_{3yr}/(1 + 50\rho') = 1.7/(1 + 50\times0.0047) = 1.38$

SLL long term deflection $= \lambda_{3yr} \times \Delta_{iSLL} = 1.38\times0.07 = 0.10$ in

Total Deflection $= \Delta_{TL} = \Delta_{iDL}(1 + \lambda_\infty) + \Delta_{iLL} + \lambda_{SLL}\Delta_{iSLL}$

$\Delta_{TL} = (1 + 1.62)\times0.33 + 0.14 + 1.38\times0.07 = 1.10$ in

LL immediate deflection $= \Delta_{iLL} = 0.14$ in

Deflection after nonstructural element attachment $= \Delta_{ATT} = \Delta_{iLL'} + \lambda_\infty\Delta_{iDL} + \lambda_{SLL}\Delta_{iSLL}$

$\Delta_{ATT} = 0.14 + 1.62\times0.33 + 1.38\times0.07 = 0.77$ in

The following table summarizes compliance of Example 14.7 with code requirements for cases (a), (b) and (c):

Case	Code Requirement	Example 14.7
(a)	$\Delta_{iLL} \le \ell/180$ or $\ell/\Delta_{iLL} \ge 180$	$\ell/\Delta_{iLL} = 20\times12/0.14 \approx 1{,}700 > 180 \Longrightarrow$ OK
(b)	$\Delta_{iLL} \le \ell/360$ or $\ell/\Delta_{iLL} \ge 360$	$\ell/\Delta_{iLL} = 20\times12/0.14 \approx 1{,}700 > 360 \Longrightarrow$ OK
	$\Delta_{ATT} \le \ell/240$ or $\ell/\Delta_{ATT} \ge 240$	$\ell/\Delta_{ATT} = 20\times12/0.77 \approx 311 > 240 \Longrightarrow$ OK
(c)	$\Delta_{ATT} \le \ell/480$ or $\ell/\Delta_{ATT} \ge 480$	$\ell/\Delta_{ATT} = 20\times12/0.77 = 311 < 480 \Longrightarrow$ NG

It is noted that for floors without deflection sensitive attachments has two code requirements as shown above: one for Δ_{iLL} and the other for Δ_{ATT}.

Example 14.8

Determine the CL total and long term service deflection of the concrete slab in Example 14.5 if 25% of live load is sustained for 1 year and check if the slab is code compliant with $\rho' = 0$ with:

(d)   flat roof slab without deflection sensitive attachment.

(e)   floor slab without deflection sensitive attachment.

(f)   roof or floor construction with deflection sensitive attachment.

Solution

From the solution of Example 14.5:

$-\Delta_{iDL} = 0.63$ in

$-\Delta_{iLL} = 0.23$ in

$\xi_\infty = 2.0 \Longrightarrow \lambda_\infty = \xi_\infty/(1 + 50\rho') = 2/(1 + 50\times0) = 2.0$

DL long term deflection $= \lambda_\infty \times \Delta_{iDL} = 2.0\times0.63 = 1.26$ in

Immediate deflection due to sustained live load $= 0.25\times0.23 = 0.06$ in

$\xi_{1yr} = 1.4 \Longrightarrow \lambda_{1yr} = \xi_{1yr}/(1 + 50\rho') = 1.4/(1 + 50\times0) = 1.4$

SLL long term deflection $= \lambda_{1yr} \times \Delta_{iSLL} = 1.4\times0.06 = 0.08$ in

Total Def. $= \Delta_{TL} = \Delta_{iDL}(1 + \lambda_\infty) + \Delta_{iLL} + \lambda_{SLL}\Delta_{iSLL} = (1 + 2.0)\times0.63 + 0.23 + 1.4\times0.06 = 2.2$ in

LL immediate deflection $= \Delta_{iLL} = 0.23$ in

Deflection after nonstructural element attachment $= \Delta_{ATT} = \Delta_{iLL'} + \lambda_\infty\Delta_{iDL} + \lambda_{SLL}\Delta_{iSLL}$

$$\Delta_{ATT} = 0.23 + 2.0\times0.63 + 1.4\times0.06 = 1.57 \text{ in}$$

The following table summarizes compliance of Example 14.8 with code requirements for cases (a), (b) and (c):

Case	Code Requirement	Example 14.8	
(d)	$\Delta_{iLL} \le \ell/180$ or $\ell/\Delta_{iLL} \ge 180$	$\ell/\Delta_{iLL} = 8\times12/0.23 = 417 > 180$	$\Longrightarrow$ OK
(e)	$\Delta_{iLL} \le \ell/360$ or $\ell/\Delta_{iLL} \ge 360$	$\ell/\Delta_{iLL} = 8\times12/0.23 = 417 > 360$	$\Longrightarrow$ OK
	$\Delta_{ATT} \le \ell/240$ or $\ell/\Delta_{ATT} \ge 240$	$\ell/\Delta_{ATT} = 8\times12/1.57 = 61 < 240$	$\Longrightarrow$ NG
(f)	$\Delta_{ATT} \le \ell/480$ or $\ell/\Delta_{iLL} \ge 480$	$\ell/\Delta_{ATT} = 8\times12/1.57 = 61 < 480$	$\Longrightarrow$ NG

Important Note: Examples 14.4 and 14.7 represent a beam with a depth–to–span ratio of 24/(240) = 1/10 > 1/16 minimum code requirement for a simply supported beams (Table 6.1). Thus, deflections need not be computed unless the floor supports deflection sensitive attachments. Deflection computations are consistent with the above code requirement.

Examples 14.5 and 14.8 represent a slab with a depth–to–span ratio of 6/(96) = 1/16 < 1/8 minimum code requirement for a cantilevered slabs (Table 6.1). Thus, for all cases, deflections need to be computed and compared with code max permissible deflection (Table 14.2). Deflection computations are consistent with the above code requirement.

14.3 Cracking

14.3.1 Concrete Cracking Behavior

Cracks are unsightly and could invite harmful substances into contact with the steel bars and the inner portions of reinforced concrete elements. Corrosion or deterioration of steel and concrete may follow. Extensive cracking can result in the building dwellers losing confidence in its structural integrity as the case with excessive deflections.

Reinforced concrete is cracked at all stages of service life including the no–loading stage. Cracking develops due to the low tensile strength of concrete in addition to other factors (primarily shrinkage). Tensile stresses generally increase with increased loading. Since tensile strength of concrete is low, cracking is inevitable. Cracking relieves concrete of tension causing increased tension in steel bars. Cracking pattern depends upon the type of loading. Among the common types are flexural cracks, shear cracks, flexural–shear cracks as well as bond cracks. Among the factors that affect cracking of reinforced concrete are:

a.   Tensile stress in reinforcing steel bars: Research has determined that crack width is directly proportional to the tension stress in steel. With low or no tension stress under service conditions, only hairline cracks develop in concrete that may not be visible to the untrained. Reducing crack width may require increasing the area of steel reinforcement (A_s), thereby, decreasing the service stress in steel, f_s. The trend of using higher strength steel bars (f_y = 60 to 100 ksi) has prompted designers to utilize higher steel tensile stresses under service conditions resulting in increased cracking of reinforced concrete. Thus, design precautions are needed for crack treatment with high strength steel.

b.   Loading: With the sustained application of load or increased loading and with the continued shrinkage and creep deformations, initial tension hairline cracks expand in width and increase in number, thereby, exacerbating the cracking condition of the reinforced concrete element under consideration. Also, cyclic loading causes increased deformations in reinforced concrete resulting in a worsening crack situation,

c.   Concrete type: Higher–f'_c concretes are characterized by increased brittleness and decreased strain capacity (Chapter 2) compared with lower–f'_c concretes. Reinforced concrete elements made with higher strength concretes exhibit a smaller number of cracks of larger width (this situation is undesirable) especially at increased service loading. Well–designed steel reinforcement can improve this condition. Lower strength concretes may lead to better crack distribution within the element (larger number of cracks with small width). Proper concreting practice also enhances the cracking behavior of reinforced concrete structural elements. Such practice includes good mix design, proper consolidation and adequate curing.

d.   Steel reinforcement: Prior to cracking, the strains of steel and adjacent concrete are equal. As loading or tension increases, the strain capacity (rupture strain) of concrete is exceeded resulting in cracking. At the location of a crack, tension is completely supported by steel bars. Away from a crack, steel bars and concrete remain bonded. Via bond, steel transfers tensile stresses and forces to concrete. As the buildup of transferred forces exceeds concrete tensile strength, another crack develops. Between adjacent cracks, concrete remains subjected to tension due to bond forces. In a reinforced concrete element, the sum of crack widths is roughly equal to the tensile elongation of steel bars. Better distribution of the tension forces or stresses in concrete leads to improved cracking behavior manifested by a larger number of cracks of small width. Tensile stress distribution within concrete elements can be enhanced by increasing the surface area of steel bars transferring tension via bond to concrete. Consequently, a more uniform distribution of cracks is achieved. Common techniques for enhancing the cracking behavior of reinforced concrete include: (d.1) using deformed steel bars that distribute bond stresses along the entire bar rather than merely at the end hooks as for smooth bars (Chapter 1). Furthermore, (d.2) using smaller diameter bars to fulfill the required steel reinforcement area increases bar surface or bond area compared to larger diameter bars and generally leads to smaller crack widths. (d.3) The proper positioning of steel bars within concrete elements is also important. Steel bars are required to be located at potential cracks due to tension and/or shrinkage. Although theoretically not needed, designers typically utilize steel bars at corners, openings, top of walls, etc. to safeguard against potential cracking. Figure 14.7 illustrates some of the concepts in this paragraph.

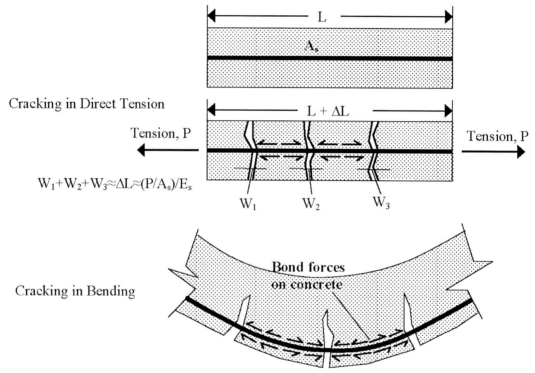

Figure 14.7: Cracks in reinforced concrete tension members and beams.

14.3.2 Crack Width Evaluation

A procedure for evaluating maximum tension crack width in beams subjected to bending is presented herewith. It is termed the Gergley–Lutz equation that is based on statistical data and analysis of a number of samples. This equation is presented below:

$$w_{max} = 0.076\beta f_s{}^3\sqrt{d_c A} \tag{14.13}$$

Where:

w_{max} = maximum flexural crack width in 0.001 in.

β = effective depth factor = $\dfrac{h - \bar{y}}{d - \bar{y}}$

Where $\bar{y}$ is determined as in Section 14.2.3 (Equation 14.3) for the elastic cracked section under service condition. Or the following approximate values may be used:

$\beta \approx 1.2$ for beams and $\beta \approx 1.35$ for slabs.

f_s = stress in steel bars under service conditions or service moment causing flexural cracks in ksi. For computation of f_s, the reinforced concrete section is considered cracked and elastic as explained in Section 14.2.3. The transformed cracked section is utilized for section stress determination. To assemble the transformed cracked section, steel bar area is magnified by factor n = E_s/E_c as in Section 14.2.3. As such, computation of f_s also requires multiplication by n since the stress is applied to a magnified steel area:

$$f_s = n \times (M_{service}/I_{cr}) \times y_{steelbars} = n \times (M_{service}/I_{cr}) \times (d - \bar{y})$$

Or the following simplified equation may be used:

$$f_s \approx 0.6f_y \text{ (ksi)}$$

d_c = concrete cover over steel bars measured from the extreme tension fiber to the centroid of steel bar closest to concrete outer surface (in).

A = the largest area of concrete beam centered around (concentric with or has the same centroid as) the tension steel bars divided by the number of tension steel bars (in²). The following equation may be used for A:

A = 2b(h – d)/number of bars (in²)

Where b is the beam's width at the tension side, h is total beam depth/height and d is beam's effective depth.

Figure 14.18 illustrates the above equations.

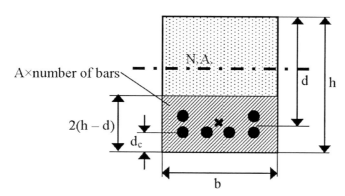

Figure 14.18: The terms of Equation 14.20.

Equation 14.13 indicates that crack width at steel location is a function of f_y the area of surrounding concrete as well as the number of bars. It also links crack width at steel location to crack width at extreme tension fiber location by incorporating factors β and d_c. Generally, β is considered the ratio of flexural crack width at extreme tension fiber to flexural crack width at steel location. Also, steel bar cover to the one–third degree, $d_c^{1/3}$, is proportional to local crack with at extreme tension fiber as noted in experimental research. The Gergley–Lutz equation has not been adopted by ACI Code.

14.3.3 Recommended Limits of Crack Width

A review of Equation 14.13 reveals that crack size reduction can be accomplished by controlling the values of d_c, f_y and A. Reducing the value of d_c can result in reduction of maximum flexural crack width. However, the designer must maintain a minimum concrete cover over steel reinforcing bars for corrosion protection per ACI Code as discussed in Chapter 6. Thus, only a minor crack controlling effect can be projected for d_c. Reasonably limiting the value of f_y is another approach for reducing crack width. Typically, f_y for steel bars for conventional structures ranges between 40 and 60 ksi. However, f_y can be as high as 120 ksi for high strength steel bars. To increase the bending moment capacity of reinforced concrete structural elements, designers have been utilizing high strength materials (steel and concrete) that may result in increased crack width. For elements with high f_y–steel bars, special reinforcement is generally designed to reduce cracking.

Proper bar detailing is generally very effective in reducing flexural crack width. Using smaller diameter steel bars to fulfill a required reinforcement area (A_s) involves the utilization of a larger number of steel bars, thereby reducing the value of A in Equation 14.17. Consequently, w_{max} is reduced. A general approach for crack reduction in reinforced concrete includes the utilization of properly spaced smaller diameter bars for beam reinforcement.

Table 14.3 (based on ACI Committee 224) contains the recommended maximum crack widths for different exposure conditions for protection of steel bars against corrosion. As cracking of concrete is unavoidable, a maximum crack width is even included for water retaining structures (waterproof coating and crackfill is used). The values within Table 14.3 (Table A.38) are not intended as rigid limits but as guidelines. Estimating crack width in reinforced concrete is a complex since it is affected by a number of variables including duration and history of loading and concrete creep and shrinkage. Equation 14.17 and Table 14.3 are approximate.

Table 14.3: Recommended limits of crack width for various exposure conditions.

Exposure Condition	Recommended Max. Crack Width
Dry air	0.016 inch
Moist air, soil	0.012 inch
Deicing chemicals	0.007 inch
Seawater and seawater spray	0.006 inch
Water retaining structures	0.004 inch

Example 14.9

Determine maximum crack width for the rectangular beam in Example14.7 and the exposure condition satisfied based on Table 14.3 with $f_y = 60$ ksi and assuming that tension reinforcement is: (a) 5#8 bars, and (b) 2#7 corner bars and 6#6 bars between the two corners.

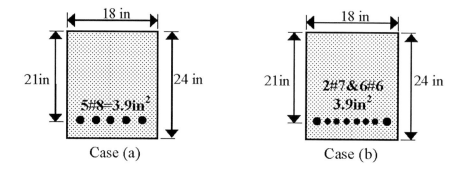

Case (a)   Case (b)

Solution

$$w_{max} = 0.076\beta f_s \sqrt[3]{d_c A}$$

$$\beta = 1.2$$

$$f_s = 0.6f_y = 0.6\times 60 = 36 \text{ ksi}$$

$d_c = 3$ in

A = 2b(h – d)/number of bars

For case (a)   A = 2×18×(24 – 21)/5 = 21.6 in²

w_{max} = 0.076×1.2×36³ $\sqrt{3×21.6}$ = 13×0.001 in = 0.013 in

Meets dry air exposure condition (Table 14.3)

For case (b)   A = 2×18×(24 – 21)/8 = 13.5 in²

w_{max} = 0.076×1.2×36³ $\sqrt{3×15.4}$ = 11×0.001 in = 0.011 in

Meets moist air/soil exposure condition (Table 14.3)

14.3.4 ACI Code Provisions for Crack Width

Due to the complexity of calculating crack width in reinforced concrete structures, ACI 318 Code includes only general guidelines for beams and one–way slabs not subjected to aggressive exposure (w_{max} = 0.016 inch). Special investigations and precautions for other cases are required. ACI Code specifies the maximum spacing of steel reinforcing bars so that the above crack width is not exceeded. Such specifications are better suited to one–way slabs and flanges of T–beams subjected to negative bending moments (see Chapter 9) than for conventional rectangular and T–beams under positive bending moments:

$$s = 540/f_s – 2.5C_c \quad \text{or} \quad s = 12(36/f_s) \quad \text{whichever is smaller} \tag{14.14}$$

Where:

s =   center–to–center spacing of flexural tension steel bars closes to the extreme tension face or the entire width of tension face for the case of single steel bar.

f_s =   stress in steel bars under service conditions or service moment causing flexural cracks in ksi (as explained earlier).

C_c =   clear cover of reinforcing steel bars nearest to the extreme tension face.

Example 14.10

Determine maximum crack width for the slab in Example 14.5 using Equation 14.13 and verify compliance with ACI Code provisions for crack width using Equation 14.14 with f_y = 60 ksi and assuming that tension reinforcement is: (a) #6@6in o.c, and (b) #9@12in o.c.

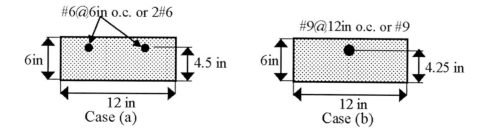

12 in
Case (a)   12 in
                                                                          Case (b)

Solution

w_{max} = 0.076βf_s³√d_cA

β = 1.35

$f_s = 0.6f_y = 0.6 \times 60 = 36$ ksi

$A = 2b(h - d)/$number of bars

<u>For case (a)</u> $d_c = 1.5$ in

$A = 2 \times 12 \times (6 - 4.5)/2 = 18$ in^2

$w_{max} = 0.076 \times 1.35 \times 36^3 \sqrt{15 \times 18} = 11 \times 0.001$ in $= 0.011$ in

Meets moist air/soil exposure condition (Table 14.3)

ACI Code provisions:

$s = 540/f_s - 2.5C_c$ or $s = 12(36/f_s)$ whichever is smaller

$C_c = 1.5 - 3/8 = 1.125$ in

$s = 540/36 - 2.5 \times 1.125 = 12.8$ in   or $s = 12(36/36) = 12.0$ in $\Longrightarrow s_{max} = 12$ in

Thus, with #6@6in o.c. fy = 60 ksi reinforcement, the slab meets ACI Code provisions and the predicted crack width is 0.011 in that meets ACI Committee 224 guidelines for moist air/soil exposure.

<u>For case (b)</u> $d_c = 1.75$ in

$A = 2 \times 12 \times (6 - 4.25) = 42$ in^2

$w_{max} = 0.076 \times 1.35 \times 36^3 \sqrt{1.75 \times 42} = 16 \times 0.001$ in $= 0.016$ in

Meets dry air exposure condition (Table 14.3)

ACI Code provisions: $s_{max} = 12$ in

Thus, with #9@12in o.c. fy = 60 ksi reinforcement, the slab meets ACI Code provisions and the predicted crack width is 0.016 in that meets ACI Committee 224 guidelines for dry air exposure.

14.4 Examples with SI Units

Examples 14.11 and 14.12 are examples with SI units utilizing the design procedures outlined in this chapter.

Example 14.11

Determine the values of immediate deflection for a uniformly loaded rectangular reinforced concrete beam with: b = 450 mm, d = 520 mm, h = 600 mm, A_s = 5 No. 25 = 2,500 mm^2, f'_c = 30 MPa, f_y = 420 MPa, span = 6 m and W = 80 kN/m.

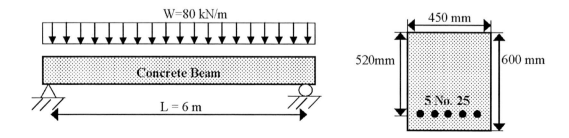

Solution

Gross moment of inertia:

$I_g = bh^3/12 = 450 \times (600)^3/12 = 8.1 \times 10^9$ mm^4

Cracked moment of inertia:

$E_c = 57{,}000\sqrt{f'_c} = 57{,}000\sqrt{30 \times 1{,}000/6.95} = 3.75 \times 10^6$ psi $= 26 \times 10^3$ MPa

$E_s = 29 \times 10^3$ ksi $= 200 \times 10^3$ MPa

$n = E_s/E_c = 200/26 = 7.7$

$\bar{y} = \dfrac{-nA_s + \sqrt{(nA_s)^2 + 2(nA_sbd)}}{b} = \dfrac{-7.7 \times 2{,}500 + \sqrt{(7.7 \times 2{,}500)^2 + 2(7.7 \times 2{,}500 \times 450 \times 600)}}{450} = 185$ mm

$I_{cr} = b\bar{y}^3/3 + nA_s(d - \bar{y})^2 = 450 \times (185)^3/3 + 7.7 \times 2{,}500 \times (520 - 185)^2 = 2.2 \times 10^9$ mm^4

Effective moment of inertia:

$f_r(\text{psi}) = 7.5\sqrt{f'_c(\text{psi})} = 7.5 \times \sqrt{30 \times 1{,}000/6.95} = 492$ psi $= 3.42$ MPa

$M_{cr} = \dfrac{f_r I_g}{y_t} = \dfrac{3.42 \times 8.1 \times 10^9}{600}/10^3 = 46.2$ kN.m

$M_a = 80 \times (6)^2/8 = 360$ kN.m

$I_e = \left(\dfrac{M_{cr}}{M_a}\right)^3 I_g + \left[1 - \left(\dfrac{M_{cr}}{M_a}\right)^3\right]I_{cr} = \{\left(\dfrac{46.2}{360}\right)^3 \times 8.1 + \left[1 - \left(\dfrac{46.2}{360}\right)^3\right] \times 2.2\} \times 10^9 = 2.2 \times 10^9$ mm^4

Deflection:

$\Delta = \dfrac{5wL^4}{384EI} = \dfrac{5 \times 80 \times (6)^4}{384 \times 26 \times 10^3 \times 2.2 \times 10^9}/10^{12} = 23.6$ mm

Span/Deflection $= 6 \times 10^3/23.6 = 254$

Example 14.12

Determine maximum crack width for the slab shown below using Equation 14.13 and verify compliance with ACI Code provisions for crack width using Equation 14.14 with $f_y = 420$ MPa and assuming tension reinforcement of: (a) No. 20@150mm o.c, and (b) No. 28@300mm o.c.

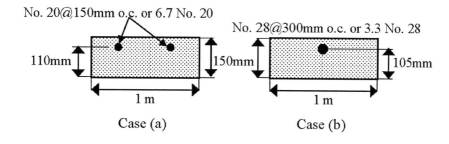

Case (a)   Case (b)

Solution

Crack Width – use Equation 14.13 with unit conversion: $w_{max} = \beta f_s \sqrt[3]{d_c A} \times 10^{-5}$

$\beta = 1.35$

$f_s = 0.6f_y = 0.6 \times 420 = 250$ MPa

$A = 2b(h - d)$/number of bars

<u>For case (a)</u>

    $d_c = 40$ mm

    $A = 2 \times 1,000 \times (150 - 110)/6.7 = 12,000$ mm^2

    $w_{max} = 1.35 \times 250 \sqrt[3]{40 \times 12,000} \times 10^{-5} = 0.26$ mm

    Meets moist air/soil exposure in Table 14.3 (crack width = 0.012 in = 0.31 mm)

<u>For case (b)</u>

    $d_c = 45$ mm

    $A = 2 \times 1,000 \times (150 - 105)/3.3 = 27,300$ mm^2

    $w_{max} = 1.35 \times 250 \sqrt[3]{45 \times 27,300} \times 10^{-5} = 0.36$ mm

    Meets dry air exposure in Table 14.3 (crack width = 0.016 in = 0.41 mm)

ACI Code provisions – use Equation 14.14 with nit conversion:

    $s = 10^5/f_s - 2.5C_c$ or $s = 7.5 \times 10^4/f_s$ whichever is smaller

<u>For case (a)</u>

    $C_c = 40 - 20/2 = 30$ mm

    $s = 10^5/250 - 2.5 \times 30 = 325$ mm   or $s = 7.5 \times 10^4/250 = 300$ mm $\Longrightarrow$ $s_{max} = 300$ mm

    ACI Code provisions: $s_{max} = 300$ mm

    Thus, with No. 20@150mm o.c. fy = 420 MPa reinforcement, the slab meets ACI Code provisions and the predicted crack width is 0.26 mm that meets ACI Committee 224 guidelines for moist air/soil exposure.

<u>For case (b)</u>

    $C_c = 45 - 28/2 = 31$ mm

    $s = 10^5/250 - 2.5 \times 31 = 323$ mm   or $s = 7.5 \times 10^4/250 = 300$ mm $\Longrightarrow$ $s_{max} = 300$ mm

    ACI Code provisions: $s_{max} = 300$ mms

    Thus, with No. 28@300mm o.c. fy = 420 MPa reinforcement, the slab meets ACI Code provisions and the predicted crack width is 0.36 mm that meets ACI Committee 224 guidelines for dry air exposure.

Test Your Knowledge

1.   List the various aspects of serviceability for reinforced concrete structures.

2.   Why is the area of steel bars multiplied by factor "n" for investigating service stresses in reinforced concrete sections?

3.   Explain the difference between the gross and the transformed moments of inertia.

4.   Is the cracked moment of inertia computed for a cracked transformed section? Explain.

5.   Define the effective moment of inertia.

6.   Define immediate and long term deflection.

7.   What are the factors affecting the amount of long term deflection?

8.   List the methods of reducing long term deflection in concrete.

9.   Why is compression steel considered very effective in reducing long term deflection?

10. ACI Code deflection criteria place upper limits on two types of deflection. Explain.

11. What is the relation between crack width in concrete beams and steel bar tensile extension?

12. What are the possible methods of limiting crack width in reinforced concrete beams?

Determine the value of maximum instantaneous and long term deflections and crack width and draw the cross–section detail for the simply supported beams described in the following table and figure. Investigate code compliance and recommendations for deflection and crack width:

No.	Span (ft)	f'$_c$ (ksi)	fy (ksi)	b (in)	h (in)	d (in)	A$_s$	A'$_s$	W$_D$ (k/ft)	W$_L$ (k/ft)	P$_D$ (kips)	P$_L$ (kips)	Sustained Load
10	20	4	60	15	20	17	5#9	N/A	2.5	1.4	0	15	0.5W$_l$ 2y 0.0P$_l$
11	18	3	60	14	24	21	6#7	2#6	1.6	1.1	10	12	0.5W$_l$ 2y 0.2P$_l$ 6mo
12	25	4.5	60	20	24	21	3#11	2#9	2.8	1.8	0	0	1.0W$_l$ 6mo
13	30	5	75	30	42	39	10#7 4#8 - 2 bundles		3.9	2.8	0	10	0.7W$_l$ 1y 0.0P$_l$
14	40	6	60	36	48	43	20#11-2 layers 4#9 – 2 bundles		8.5	7.0	5	20	0.2W$_l$ 6mo 0.5P$_l$ 6mo
15	22	3.5	60	18	22	19	7#9	2#9	3.5	2.5	0	0	1.0W$_l$ 6mo
16	35	4.8	60	24	36	31	8#7	2#7	4.0	3.0	0	0	0.5W$_l$ 2y

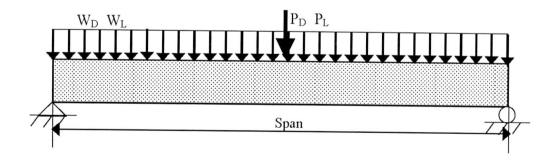

Determine the value of maximum instantaneous and long term deflections and crack width and draw the cross–section detail for the cantilever beams described in the following table and figure. Investigate code compliance and recommendations for deflection and crack width:

No.	Span (ft)	f'_c	f_y (ksi)	b	h	d (in)	A_s	A'_s	W_D	W_L (k/ft)	P_D	P_L (kips)	Sustained Load
17	9	4.5	60	16	20	17	4#9	2#5	2.3	1.5	6	0	0.5W_l 2y
18	6	4	60	12	18	15	4#6	2#5	1.5	1.2	3	0.5	0.5W_l 2y 0.0P_l
19	10	5	60	14	24	21	4#10 4#8 - 2 bundles		3.1	1.7	2	0	0.3W_l 4y
20	12	5.5	75	12	30	25.5	8#7 - 2 layers 4#8 - 2 bundles		3.5	2.2	2.5	1	0.7W_l 2y 0.0P_l
21	9	4	40	16	32	27.5	8#7-2 layers 4#7 – 2 bundles		2.5	1.4	1.5	1.2	0.2W_l 6mo 0.2P_l 6mo
22	8	3.8	60	10	22	19	6#6 - 2 layers		3	2	1	0.5	0.5W_l 1y 0.0P_l
23	15	6	80	20	36	31	7#9	2#9	4.0	2.0	2.0	1.5	0.5W_l 2y 0.0P_l

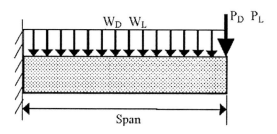

Determine deflection (immediate and long–term) and crack width and check code compliance for the following simply supported beams.

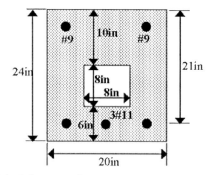

24. Other conditions same as No. 12

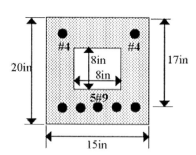

25. Other conditions same as No. 10

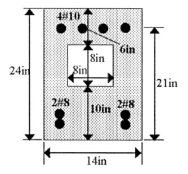

26. Other conditions same as No. 19

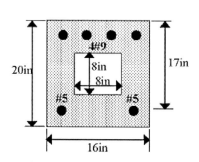

27. Other conditions same as No. 17

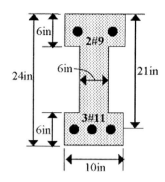

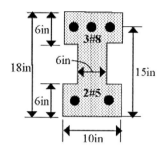

28. Other conditions same as No. 12   29. Other conditions same as No. 18

Determine the value of immediate and long term deflection for the beam shown in the following figure and check compliance with ACI Code:

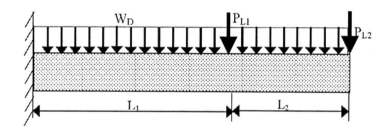

30. $L_1 = 12$ ft $L_2 = 8$ ft
$W_D = 1.5$ k/ft $P_{L1} = 8$ kips $P_{L2} = 10$ kips
$f'_c = 4,000$ psi $f_y = 60,000$ psi

31. $L_1 = 10$ ft $L_2 = 6$ ft
$W_D = 0.9$ k/ft $P_{L1} = 6$ kips $P_{L2} = 4$ kips
$f'_c = 3,000$ psi $f_y = 40,000$ psi

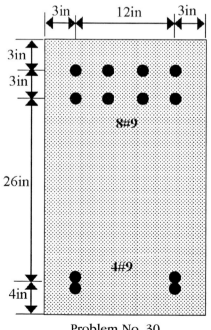

Problem No. 30

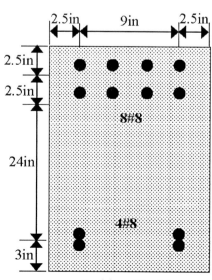

Problem No. 31

Compute the crack width for the beam sections shown and investigate compliance with ACI recommendations assuming that $f'_c = 3,500$ psi $f_y = 60$ ksi and $f_s = 0.6 f_y$

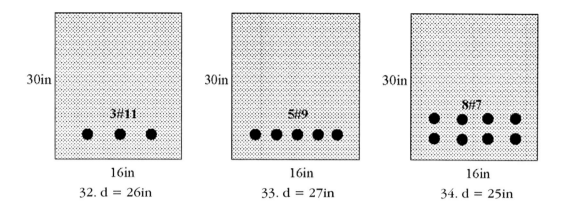

32. d = 26in   33. d = 27in   34. d = 25in

Compute the crack width for the beam sections shown and investigate compliance with ACI recommendations assuming that $f'_c = 4,000$ psi $f_y = 60$ ksi and $f_s = 0.6 f_y$

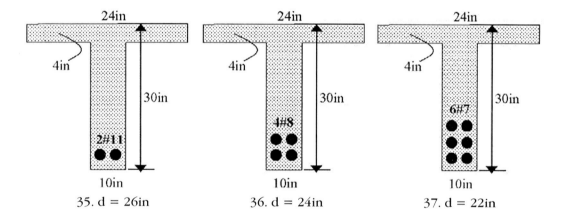

35. d = 26in   36. d = 24in   37. d = 22in

Determine the required bar spacing (s) for the concrete slabs below for ACI Code compliance:

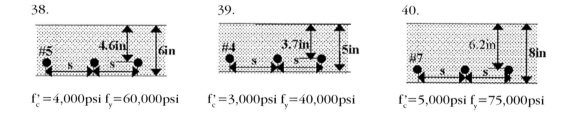

38.   39.   40.

$f'_c = 4,000$psi $f_y = 60,000$psi $f'_c = 3,000$psi $f_y = 40,000$psi $f'_c = 5,000$psi $f_y = 75,000$psi

Chapter 15

INTRODUCTION TO REINFORCED CONCRETE COLUMNS

15.1 Reinforced Concrete Columns

Columns are vertical members that receive loads from beams, girders or slabs and transfer these loads eventually to the footings which in turn convey such loads to the subgrade. Therefore, axial compression is the main stress mode in columns. With this mode of stress and since columns are typically long and slender, buckling becomes important. To simplify analysis and design, reinforced concrete columns are divided into three categories:

1. Short or compact columns subjected to concentric axial load (load is applied at column centroid). For these columns, buckling effect is minimal and generally neglected in analysis and design as discussed in this chapter.

2. Short (compact) columns subjected to axial force and bending moment. As with category (1), the effect of buckling is neglected as discussed in Chapter 16.

3. Slender columns. A major part of analysis and design of slender columns is the effect of bucking as explained in Chapter 17.

    Steel reinforcement for columns consists of two components:

1. Longitudinal.

2. Transverse.

    Longitudinal steel bars (generally parallel to the column centroidal axis) have several purposes including:

a. enhancing column load bearing capacity by supporting axial load.

b. resisting tensile stresses that may occur due to accidental eccentricity.

c. reducing the effect of concrete creep (as explained in Chapter 14).

d. providing confinement for the column core (the inner portion of column cross–section).

    Transverse reinforcement is steel bars that are perpendicular or near perpendicular to the column centroidal axis. They are configured or arranged to anchor to the longitudinal reinforcing bars. Transverse column reinforcement is used to tie or secure the longitudinal reinforcement during construction. Furthermore, transverse reinforcement works as lateral support for longitudinal reinforcing bars to avoid their buckling under axial compression load. Also, depending on its type and design, transverse reinforcement can provide considerable confinement for the column core which enhances column compressive load–deformation behavior. Columns are divided into two other categories based on the type of transverse reinforcement:

1. Tied columns containing a series of discrete closed ties wrapped (bent) around the longitudinal steel bars per ACI 318 Code. Such ties are used throughout the entire length of the column. Tie spacing (distance between adjacent ties) is constant along the column and must conform to code requirements. Tied columns are typically rectangular or square in cross–section.

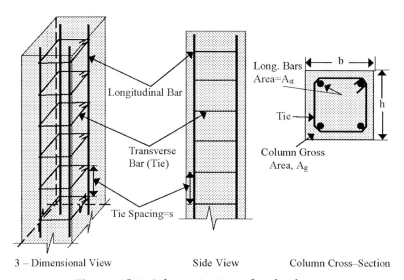

Figure 15.1: Schematic view of tied column.

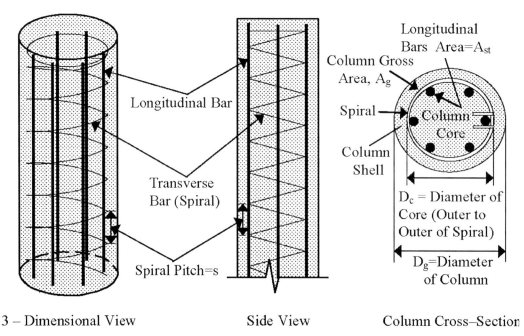

Figure 15.2: Schematic view of spiral column.

2. Spiral columns contain a continuous helical spiral bar that wraps around the longitudinal steel bars. The spiral bar provides confinement for the column core (the inner portion of the column). Such confinement would result in a significantly enhanced column behavior under compression as explained later in this chapter. Spiral pitch is the distance the spiral bar descends as it makes a full revolution around the column core. The column core is the inner portion of the column bound by the exterior surface of the spiral while the column shell is the exterior portion of the column beyond the exterior spiral bar surface. Spiral columns are generally circular in cross–section.

The subject matter of this chapter is column category 1 which is: short or compact columns subjected to concentric axial load (load is applied at column centroid). Chapter 16 deals with columns subjected to bending and axial forces. Chapter 17 deals with slender columns. The criteria used to classify columns as short or slender will be presented in Chapter 17.

15.2 Columns Axial Compression Capacity

Based on ACI 318 Code, axial compression capacity of concentrically loaded short/compact columns (column category 1) is determined using the following equation:

$$\phi P_n = \phi \alpha [0.85 f_c'(A_g - A_{st}) + A_{st} f_y]$$ (15.1)

Where:

ϕ = capacity reduction factor for columns under concentric compression load;
ϕ = 0.65 for tied columns and ϕ = 0.70 for spiral columns.

P_n = column nominal axial compressive force capacity taking into account code assumptions for column behavior under ultimate conditions.

α = minimum eccentricity capacity reduction factor;
α = 0.80 for tied columns and ϕ = 0.85 for spiral columns.

A_g = column gross cross–sectional area.

A_{st} = area of longitudinal steel bars of the column.

15.3 Discussion of Columns Axial Compression Capacity

Equation 15.1 for concentrically loaded short/compact columns (column category 1) implies the following assumptions:

1. Average compressive stress in concrete at ultimate condition is $0.85 f_c'$ which is the same assumption used for Whitney's compression block in beams (see Chapter 4).

2. Compressive stress in longitudinal steel bars at ultimate condition is f_y. As such, longitudinal steel bars yield prior or upon to reaching the column's ultimate condition.

Furthermore, the general ACI 318 Code assumptions for analysis of structural elements (primarily beams) are applicable to column analysis. These assumptions are restated in the following:

1. Ultimate or failure condition is defined by concrete reaching its compression crushing strain of 0.003.

2. Perfect bond of steel and concrete subsists. As such, the strains in longitudinal steel bars and adjacent concrete are equal.

3.   Concrete tensile strength = 0.

4.   Plane sections before deformation remain plane after deformation. This hypothesis applies mainly to beams. However, it is also applicable to columns since columns can be subjected to bending moments in addition to compression loading.

Additionally, ACI 318 Code makes particular assumptions regarding column behavior:

1.   Transverse reinforcement (ties or spiral) designed according to code provisions avert buckling of longitudinal bars under compression loading.

2.   Columns typically fail in a catastrophic manner since concrete crushing primarily governs the failure mode. In tied columns, failure starts by spalling of the concrete cover followed by buckling of longitudinal bars and crushing of the concrete core occurring rather swiftly. Consequent to shell spalling in spiral columns, the core and spiral bar can sustain further loading and crushing of the concrete core is delayed. Steel bars yield prior to concrete crushing. However, concrete crushing strain (0.003) is only slightly larger than steel yield strain (ε_y = 0.00209 for f_y = 60 ksi). As such, column gradual failure is not considerable compared with bending failure where ACI Code recommends a tensile strain $\geq$ 0.005. Columns are important structural elements. Major collapse of beams, slabs or building portions may occur upon failure of a supporting column. Thus, ACI 318 Code specifies a capacity reduction factor, ϕ, of 0.65 for tied columns to achieve a safe design. ACI Code specifies ϕ = 0.70 for spiral columns as they are designed to have a gradual failure manner (Section 15.5).

3.   ACI Code presumes the non–existence of a zero eccentricity column due to detailing and construction imperfections. Load eccentricities may increase the possibility of column buckling. Thus, all columns are designed for a minimum inherent eccentricity. This minimum eccentricity is specified as 0.10×column cross–sectional dimension in the plane of buckling in tied columns and 0.05×column diameter in spiral column. Factor "α" is used in Equation 15.1 as a reduction factor to account for accidental eccentricity. It is considered that spiral columns are of improved quality with lower inherent eccentricity and higher value of α compared with tied columns. The cross–sectional dimensions of tied column are called h and b and the cross–sectional diameter of spiral column is called D_g (Figures 15.1 and 15.2).

15.4 ACI Code Provisions for Columns

15.4.1 General Code Provisions

a. Column Reinforcement Ratio

Column reinforcement ratio may be determined based on the following equation:

$$\rho_g = A_{st}/A_g \qquad (15.2)$$

Where:

ρ_g = column reinforcement ratio.

A_{st} = cross–sectional area of longitudinal reinforcing bars.

A_g = column gross cross–sectional area.

ACI code requires that:

$1\% \leq \rho_g \leq 8\%$

ACI Code allows that ρ_g be reduced to 0.5% in low seismic risk area provided that Equation 15.1 is satisfied. Typically column reinforcement ratio ranges between 1% to about 4%. Reinforcement ratios greater than 4% present a construction difficulty as steel bars become congested within the column. Typically, designers adopt ρ_g between 2 and 3%. Furthermore, ACI Code requires that clear spacing between longitudinal steel bars not be less that $1.5d_b$ (where d_b is longitudinal bar diameter), 1.5 in nor 4/3×maximum aggregate size.

b. Minimum Number of Longitudinal Steel Bars

ACI code specifies that the minimum number of longitudinal steel bars in reinforced concrete columns as follows:

Column Cross – Section Shape	Minimum No. of Longitudinal Steel Bars
Triangular	3
Rectangular	4
Circular	6

15.4.2 Code Provisions for Tied Columns

a. Tie Diameter and Spacing

Tie bar diameter correlates to longitudinal bar diameter as stated below:

Longitudinal Bar Size	Tie Bar Size
≤ #10 (32 mm)	#3 (10 mm)
≥ #11 (35 mm)	#4 (12 mm)

Furthermore, #4 (12 mm) ties shall be used if the longitudinal steel bars of a column are bundled.

Tie vertical spacing, s (Figure 15.1), according to ACI code, shall be according to Equation 15.3:

$$s \leq \begin{cases} s_1 = 16 \times \text{longitudinal bar diameter} \\ s_2 = 48 \times \text{tie bar diameter} \\ s_3 = \text{least dimension of column cross–section} \end{cases} \tag{15.3}$$

As such, s_1, s_2 and s_3 are computed. Then, a value for s that is smaller than any of these values as well as construction suitable is selected. Typically, s is in multiples of 2, 3 or 5 inch. The highest tie may not farther than s/2 from the horizontal steel within the beam or slab at the top of column. The lowest tie may not be farther than s/2 from the top of footing, beam or slab at the bottom of column.

b. Tie Geometric Configuration

Tie bar bend diameter around longitudinal steel bars shall conform to ACI code. Each tie bar shall terminate (on both sides) with a hook around a longitudinal bar. Such hook could be 135° or 90° standard hook (Chapter 13) as shown in Figure 15.3. ACI code recommends that hooks along the length of the column alternate among longitudinal bars. Furthermore, the corner angle of a tie shall not exceed 135° (180° hooks are not allowed).

Column longitudinal bars are classified as laterally supported or laterally unsupported. A laterally supported bar is located at the corner of a tie. Bars at tie sides are not considered laterally supported due to incomplete shoring in two directions. Longitudinal bars are required to laterally supported or within 6 in (150 mm) clear distance of a laterally supported bar on both sides. Thus, laterally unsupported bars may not be adjacent to each other. Examples of ties acceptable by ACI 318 Code are shown in Figure 15.4.

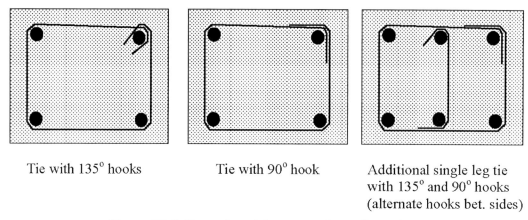

Tie with 135° hooks   Tie with 90° hook   Additional single leg tie
                                                         with 135° and 90° hooks
                                                         (alternate hooks bet. sides)

Figure 15.3: Examples of tie geometric configuration.

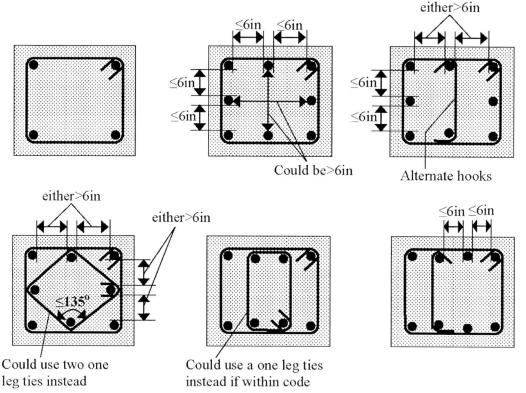

Figure 15.4: Examples of tie geometric configuration.

15.4.3 Code Provisions for Spiral Columns

a. Spiral Diameter and Pitch

ACI 318 Code specifies that:

1. Spiral bar diameter may not be less than #3 (10 mm).

2. Clear spacing of spiral bar loops may not be less than 1 in (25 mm) or greater than 3 in (75 mm).

3. Anchorage of the spiral bar shall be accomplished by 1-½ extra revolution (turn) at each end of the column at distance equal to the pitch from slab, beam, or footing reinforcement.

4. Lap splice length of uncoated deformed spiral bar may not be less than 12 in (300 mm) or $48d_b$ (d_b = spiral bar diameter) whichever is greater.

As such, spiral pitch (s) may not be less than $1\text{-}^3/_8$ in ($\approx$35 mm) or more than $3\text{-}^3/_8$ in ($\approx$85 mm) when #3 bar (10 mm) is used for the spiral. Also, spiral pitch may not be less than 1.5 in ($\approx$40 mm) or more than 3.5 in ($\approx$90 mm) when #4 bar (12 mm) is used for the spiral. Spacers may be used to secure the spiral in place and to ensure even spiral pitch along the column during construction.

Special case: If smooth or epoxy coated steel bars are used for the spiral, then, lap splice length may not be less than 12 in (300 mm) or $72d_b$.

b. Spiral Reinforcement Ratio

ACI 318 Code defines the spiral reinforcement ratio, ρ_s, as the volume of spiral bar in a full revolution divided by the volume of core in single pitch. As such, the following equation may be used:

$$\rho_s = \frac{\pi(D_c - d_b)a_s}{s(\pi D_c^2/4)}$$

Where:

ρ_s = spiral reinforcement ratio.

D_c = diameter of column core (outer–to–outer of spiral).

d_b = diameter of spiral bar.

a_s = cross – sectional area of spiral bar.

s = spiral pitch.

The above equation may be simplified as shown below:

$$\rho_s = 4a_s/sD_c \tag{15.4}$$

The spiral bar plays a special role in the behavior of spiral columns under compression load. It confines the core, thereby, allowing the concrete within the core to sustain higher compressive stresses and strains as the spiral bar undergoes yielding. Such function results in a gradual failure scenario as recommended for all reinforced concrete elements and structures. The failure scenario of a spirally reinforced column is typically as follows:

1. Entire column reaching a compressive strain of 0.003. At this strain, the average compressive stress in concrete is $0.85f_c'$ and the stress in longitudinal steel is f_y.

2. As load increases, strain in the column increases. The column shell (exterior portion of the column section beyond the outer of spiral), due to lack of confinement, spalls (cracks) off.

3. As load and strain in the column continue to increase, the column core continues to deform and resist the added load. This is accomplished with the aid of spiral confinement. Confinement results in extra concrete strength of four times the confining pressure. Further loading eventually leads to yielding of the spiral bar.

4. Yielding of spiral bar results in reduction of core confinement and, consequently, crushing of the column core.

Ductile/gradual failure of spiral columns is accomplished via yielding of the spiral bar prior to crushing of concrete as in bending and shear. For gradual failure to take place, the spiral bar effect upon the column core must exceed the lost contribution of shell spalling to column compression load capacity. This precaution is needed so that the spiral bar undergoes extensive yielding prior to column failure, thereby achieving the required gradual failure scenario. Without such requirement, the spiral column behavior includes a short step 3 (spiral yielding), thereby, not allowing gradual failure to take place (see previous page for spiral column failure scenario). Based on this, ACI 318 Code specifies a minimum spiral reinforcement ratio in columns:

$$\rho_{smin} = 0.45(A_g/A_c - 1) \cdot f'_c/f_y \qquad (15.5)$$

Where:

ρ_{smin} = code required minimum spiral reinforcement ratio.

A_g = gross cross − sectional area of the spiral column.

A_c = core area of the column.

f'_c = specified/design compressive strength of concrete.

f_y = yield strength of steel.

The derivation of Equation 15.5 is shown below:

Shell contribution to column capacity = $0.85f'_c \times (A_g - A_c)$

Yielding force in spiral bar = $a_s \cdot f_y$

Area of core subject to spiral yielding confining pressure in one pitch = $D_c \cdot s$

Numbers of spiral bar branches intercepted in one pitch = 2

Confining pressure on core due to spiral yielding = $2a_s \cdot f_y/D_c \cdot s$

Increase in concrete compressive strength due to lateral
confining pressure = 4×lateral confining pressure

Increase in core strength due to confining pressure = $4(2a_s \cdot f_y/D_c \cdot s) \cdot \pi D_c^2/4$

Based on the above discussion, the following equation needs to be satisfied for the spiral reinforcing bar to be effective:

$$4(2a_s \cdot f_y/D_c \cdot s) \cdot \pi D_c^2/4 \geq 0.85f'_c \times (A_g - A_c)$$

Substitute in the above equation the value of ρ_s from Equation 15.5:

$$2 \cdot \rho_s \cdot \pi D_c^2/4 \geq 0.85f'_c \times (A_g - A_c)$$

Since $A_c = \pi D_c^2/4$, then the above equation may be rewritten as:

$$\rho_s \cdot \pi D_c^2/4 \geq 0.425f'_c \times (A_g/A_c - 1) \quad \underline{OR} \quad \rho_{smin} \approx 0.45(A_g/A_c - 1) \cdot f'_c/f_y$$

As such, the equation for spiral pitch can be conservatively presented as follows:

$$s = 8a_s \cdot f_y/[f'_c \cdot D_c \cdot (A_g/A_c - 1)] \qquad (15.6)$$

Figure 15.5 can assist to understand the computation of spiral bar confining pressure. It presents a diagram of a spiral column sliced longitudinally. The area of column core within a spiral pitch is sD_c. Also, within a spiral pitch, s, the spiral bar intercept the concrete core in two locations. Hence, at yielding, the spiral bar causes a confining pressure of $a_s f_y/sD_c$ upon the concrete code.

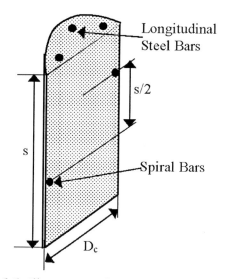

Figure 15.5: Illustration of spiral bar confining pressure.

Example 15.1

Determine the value of ϕP_n for the tied reinforced concrete column shown in the figure with $f'_c = 3,000$ psi and $f_y = 60,000$ psi.

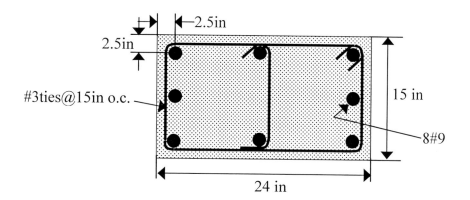

Solution

Check if column cross–section design is within code limits.

Reinforcement ratio, $\rho_g = 8/(15\times24) = 2.22\%$ within code limits of 1 to 8% ===> OK

Tie spacing:

$$s \le \left.\begin{array}{l} s_1 = 16 \times (9/8) = 18 \text{ in} \\ s_2 = 48 \times (3/8) = 18 \text{ in} \\ s_3 = 15 \text{ in} \end{array}\right\} s = 15 \text{ in is within code limits}$$

Tie bar diameter of #3 for longitudinal bars of #9 < #11 ===> OK

Clear spacing of bars:

For short side = [15 in − 2×2.5 in − 2×(9/8) in]/2 = 3.9 in < 6 in

===> No need for extra tie along the short side.

For long side = [24 in − 2×2.5 in − 2×(9/8) in]/2 = 8.4 in > 6 in

===> Extra tie is needed along the long side.

Thus, column cross–section design is within code. Use Equation15.1 for ϕP_n:

$\phi P_n = 0.65×0.80×[(15×24 − 8×1.00)×0.85×3 + 8×1.00×60]$

$\phi P_n = 716.4$ k

Example 15.2

Determine the value of ϕP_n for the spiral reinforced concrete column shown in the figure with $f'_c = 4{,}000$ psi and $f_y = 60{,}000$ psi.

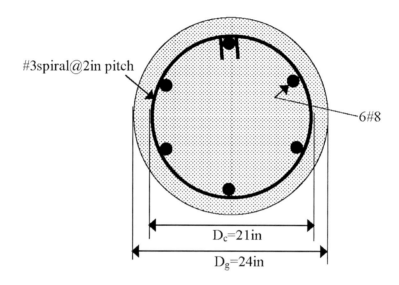

Solution

Check if column cross–section design is within code limits.

Reinforcement ratio, $\rho_g = 6×0.78/[\pi×(24)^2/4] = 1.03\%$ within code limits of 1 − 8% OK

Spiral pitch: $s = 8×0.11×60/\{4×21×[(24/21)^2 − 1]\} = 2.05$ in > 2 in ===> OK

Thus, column cross – section design is within code. Use Equation 15.1 for ϕP_n:

$\phi P_n = 0.70×0.85×\{[(\pi×(24)^2/4) − 6×0.78]×0.85×4 + 6×0.78×60\}$

$\phi P_n = 1{,}078$ k

15.5 Design of Reinforced Concrete Columns

15.5.1 Column Cross–Sectional Dimensions are Provided

Typically, column size is predetermined in the architectural plans. As such, material properties (f'_c and f_y) as well as reinforcement size and details are required for design of the column. f'_c and f_y are initially assumed based on material availability. Consequently, Equation 15.1 can be used to determine the value of A_{st}. If A_{st}/A_g is within code limits, then the designer may proceed with bar selection and detailing based on ACI Code criteria. Otherwise, modification of column cross–sectional size or material properties is needed. As design is an iterative process, it is recommended to investigate several column designs to select the most efficient based on cost and construction complexity.

Example 15.3

Design the reinforcement for a 15×15 inch tied column subjected to P_u = 430 k provided that f'_c = 3,000 psi and f_y = 60,000 psi.

Solution

Use Equation 15.1:

$$\phi P_n = 0.65 \times 0.80 \times [(15 \times 15 - A_{st}) \times 0.85 \times 3 + A_{st} \times 60] \geq 430 \text{ kips} \Longrightarrow A_{st} \geq 4.4 \text{ in}^2$$

Calculate ρ_g based on the computed value of A_{st}:

$$\rho_g = 4.4/15 \times 15 = 1.96\% \Longrightarrow \text{OK within ACI Code limits}$$

Steel bar selection: Options: 8#7 bars, 6#8 bars or 4#10 bars. As stated in Chapter 14, smaller diameter bars are generally advantageous for the cracking behavior of concrete compared with larger diameter bars. Thus, for this example 8#7 bars are selected.

Clear spacing of bars = $(15 - 5 - 2 \times 7/8)/2 = 4.1$ in $\Longrightarrow$ one tie is adequate for all bars

Use #3 ties since longitudinal bars are < #11. Tie spacing:

$$s \leq \begin{cases} s_1 = 16 \times (7/8) = 14 \text{ in} \\ s_2 = 48 \times (3/8) = 18 \text{ in} \\ s_3 = 15 \text{ in} \end{cases} \quad \text{use } s = 14 \text{ in}$$

Cross–section detail:

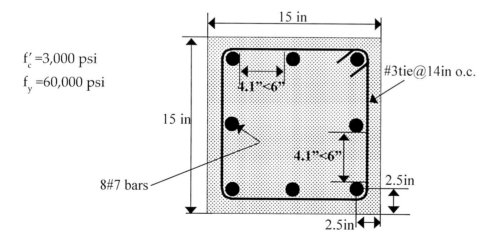

The actual column cross–sectional design is required to be verified using Equation 15.1:

$$\phi P_n = 0.65 \times 0.80 \times [(15 \times 15 - 4.88) \times 0.85 \times 3 + 4.88 \times 60] = 444 \text{ k} \geq 430 \text{ k} \Longrightarrow \text{OK}$$

15.5.2 Column Cross–Sectional Dimensions are not Provided

As the dimensions of column cross–section are not given, the designer typically assumes a reinforcement ratio (ρ_g) for the column within code limits. f'_c and f_y are chosen based on material availability. Equation 15.1 can, then, be used to determine the value of A_g. Consequently, a construction suitable section can be selected. Column design can, thus, proceed as described in Section 15.5.1. Several column designs are investigated to facilitate the selection of a suitable cross-section.

Example 15.4

Design a spiral column subjected to P_u = 600 k provided that f'_c = 4,000 psi and f_y = 60,000 psi.

Solution

Assume a reinforcement ratio for the column: ρ_g = 2%

Use Equation 15.1:

$$\phi P_n = 0.70 \times 0.85 \times [(A_g - 0.02 A_g) \times 0.85 \times 4 + 0.02 A_g \times 60] \geq 650 \text{ k}$$
$$\Longrightarrow A_g \geq 222 \text{ in}^2 \Longrightarrow \text{use 18 in–diameter section}$$
$$D_g = 18 \text{ in \& cover} = 1.5 \text{ in} \Longrightarrow D_c = 18 - 2 \times 1.5 = 15 \text{ in}$$
$$A_{st} \geq 0.02 \times \pi (18)^2 / 4 = 5.1 \text{ in}^2$$

Steel bar selection: Options 9#7 bars, 7#8 bars or 6#9 bars. For this example 8#8 bars are selected so that bar distribution is simplified as the central angle between two adjacent bars is 45°.

Use #3 spiral. Spiral pitch (Equation 15.6):

$$s = 8 \times 0.11 \times 60 / \{4 \times 15 \times [(18/15)^2 - 1]\} = 2 \text{ in}$$

Cross–section detail:

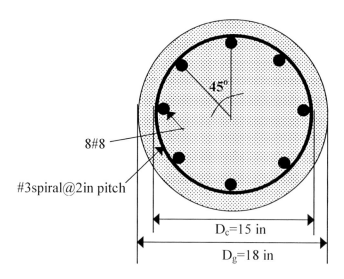

The actual column cross–sectional design is required to be verified using Equation 15.1:

$$\phi P_n = 0.70 \times 0.85 \times [(\pi \times (18)^2 / 4 - 6.24) \times 0.85 \times 4 + 6.24 \times 60] = 724 \text{ k} \geq 600 \text{ k} \Longrightarrow \text{OK}$$

Test your knowledge

1. As concrete can resist compression; what are the purposes for longitudinal reinforcement in columns?

2. What are the purposes for transverse reinforcement in columns?

3.   How is slow or gradual failure achieved in reinforced concrete columns?

4.   What is the difference in behavior between tied columns and spirally reinforced columns?

5.   What does factor α represent in the equation of ultimate compressive capacity?

6.   Why is minimum ACI–mandated accidental eccentricity smaller in spirally reinforced columns compared to tied columns?

7.   Why are the assumptions used for analysis of columns under ultimate conditions?

8.   What causes a steel bar within a column cross–section to be considered laterally supported?

9.   How does one ensure that failure of a spirally reinforced column is gradual with large and visible deformations?

10.   Describe the sequence of failure scenario of a properly designed spirally reinforced column?

Compute ϕP_n for the tied column sections listed below, design the ties and detail the column section assuming that $f_y = 60$ ksi:

No.	f'_c (ksi)	b (in)	h (in)	Steel Bars
11	2.5	12	12	4#6
12	3	15	18	6#7
13	3	14	24	8#8
14	4	20	20	8#9
15	4	18	24	6#10
16	4	12	12	4#6
17	4.5	16	16	6#7
18	4.5	15	20	6#10
19	5	20	30	10#10
20	6	24	24	12#11

Compute ϕP_n for the spirally reinforced column sections listed below, design the spiral bar and detail the column section:

No.	f'_c (ksi)	f_y (ksi)	D (in)	Steel Bars
21	3	60	18	6#6
22	3	60	24	8#7
23	4	60	20	8#8
24	4	60	24	8#9
25	4.5	60	30	8#10
26	4.5	75	30	8#11
27	5	60	18	6#7
28	5	75	24	6#10
29	6	75	24	10#10
30	6	80	30	12#11

Design the tied column sections for the conditions below listed below, design the ties and detail the column section assuming that f_y = 60 ksi:

No.	f'_c (ksi)	bxh (in)	ρ (%)	P_u (kips)
31	2.5	12x12		180
32	3	14x22		350
33	3		1.5	480
34	4		2	680
35	4	16x16		590
36	4	18x18		750
37	4.5	16x24		810
38	4.5		2.5	1,050
39	5		1.8	950
40	6		2.25	2,210

Design the spirally reinforced column sections for the conditions below listed below, design the spiral bar and detail the column section:

No.	f'_cxf_y (ksi)	D (in)	ρ (%)	P_u (kips)
41	4x60	18		570
42	4x60	24		880
43	4.5x60		2	730
44	4.5x60		2.2	980
45	4.5x75	24		1,200
46	5x60	32		2,200
47	5x75		1.9	1,600
48	5.5x60		2.5	2,700
49	6x75		1.8	1,150
50	6x80		2.25	2,200

Chapter 16

REINFORCED CONCRETE COLUMNS UNDER AXIAL FORCE AND BENDING MOMENT

16.1 Failure of Columns Subjected to Axial Force and Bending Moment

In this chapter, short or nonslender reinforced concrete columns under axial compression and bending are discussed. Effect of slenderness or buckling is presumed insignificant. The assumptions for analysis of reinforced concrete beams are the foundation of reinforced concrete theory. They are applicable to columns and are restated below:

1. Plane sections before loading (bending) remain plane after loading. Thus, strains in the section are linearly proportional to the distance from neutral axis.

2. Steel and concrete are perfectly bonded. Thus, strain in steel bars is equal to strain in neighboring concrete provided that they have the same distance from the neutral axis.

3. Ultimate or failure condition is defined by crushing of concrete. Concrete is crushed due to reaching the crushing strain in compression of 0.003.

4. Concrete tensile strength is zero.

5. Whitney's rectangular compression block is used for simplifying concrete compression stresses.

The axial compression force, P_n, and bending moment, M_n, on a column may be substituted with an eccentric compressive load. Eccentricity (e) can be computed as:

$$e = M_n/P_n \tag{16.1}$$

An illustration is shown in Figure 16.1. Based on the principles of solid mechanics, the axial force–bending moment and eccentric force are equivalent force systems with the same effect and one can substitute the other in structural analysis. In this chapter, focus is on columns subjected to uniaxial bending moments, M_n about the section centerline or axis of symmetry (Figure 16.1).

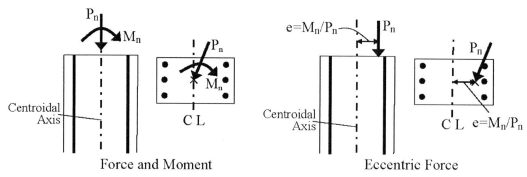

Figure 16.1: Equivalent force systems on a column.

It is important to note that ACI Code does not specify limits on tension or compression steel strains (ε_t and ε_s') at failure condition for columns. There are five categories of ultimate or failure condition of reinforced concrete columns under bending moment (M_n) and axial compression load (P_n). They are discussed in the following for a simplified rectangular column cross–section containing two layers of steel at the tension and compression faces as shown in Figure 16.2:

a.   Columns subjected to axial compression without bending moment (zero eccentricity): Cross–section strains and stresses at failure or ultimate condition are in Figure 16.2. As discussed in Chapter 15, ACI Code specifies that axial compression loading upon reinforced concrete columns includes an inherent minimal eccentricity due to design and construction imperfections. To simplify analysis, ACI Code permits the use of a reduction factor (α) for column axial load capacity if compression load eccentricity is $\leq$ the specified minimum eccentricity. This minimum eccentricity is listed as 0.1h and $0.05D_g$ where h is the cross section dimension in the plane of buckling and D_g is the gross section diameter for tied and spiral columns, respectively (Chapter 15).

b.   Columns subjected to axial compression load with small eccentricity: Section strains and stresses at failure/ultimate condition are shown in Figure 16.2. Strain in tension steel (ε_t) is less than 0.002. With this, steel does not experience yielding prior to failure by concrete crushing and compression is the dominant stress in the section. Thus, failure of columns with small eccentricity loading is somewhat catastrophic rather than the recommended gradual failure mode. Such types of column sections are termed by ACI Code compression controlled sections.

c.   Columns subjected to axial compression load with intermediate eccentricity: Failure strains and stresses are shown in Figure 16.2. Tension steel strain ranges between 0.002 and 0.005. Steel bars do not experience substantial yielding prior to concrete crushing. The failure mode of columns with intermediate eccentricity loading is to some extent gradual with moderate deformations and cracking. Such types of column sections are termed by ACI Code transition sections.

d.   Columns subjected to axial compression load with large eccentricity: Ultimate condition strains and stresses are shown in Figure 16.2. As noted, tension steel strain is greater than 0.005. Steel experiences significant yielding prior to concrete compression crushing. Thus, failure of columns with large eccentricity axial load is the required gradual failure mode accompanied with considerable deformations and cracking. Such types of column sections are termed by ACI Code tension controlled sections.

e.   Sections subjected to bending moment without axial force: Analysis is as discussed in Chapters 4 through 7 for beams.

Notes:

Axial load eccentricity is termed small, intermediate or large based on column's response to loading corresponding to section classification as compression controlled, transition or tension controlled. Classification of eccentricity or column's loading response is a function of column cross sectional dimensions, tension steel reinforcement ratio (ρ) and concrete and steel specified strengths, f'_c and f_y in addition to the applied load and moment. In Figure 16.2, the reader needs to distinguish between the centerline (CL) that is the section geometric centroid and the neutral axis (NA) with zero strains and stresses.

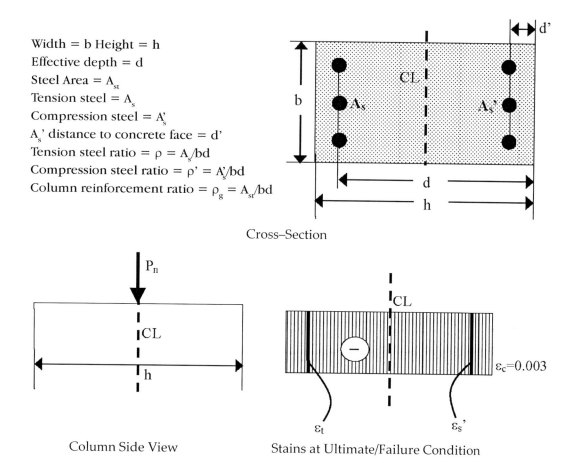

Width = b Height = h
Effective depth = d
Steel Area = A_{st}
Tension steel = A_s
Compression steel = A'_s
A'_s distance to concrete face = d'
Tension steel ratio = $\rho = A_s/bd$
Compression steel ratio = $\rho' = A'_s/bd$
Column reinforcement ratio = $\rho_g = A_{st}/bd$

Cross–Section

Column Side View

Stains at Ultimate/Failure Condition

(a) Column subjected to concentric loading.

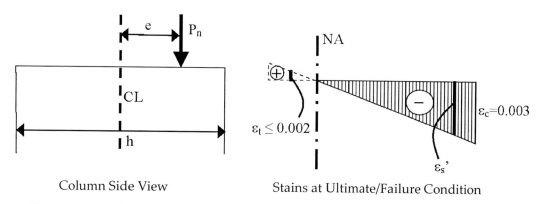

Column Side View

Stains at Ultimate/Failure Condition

(b) Column subjected to small eccentricity loading – compression controlled section.

Notes: ε_t = tension steel strain at failure/ultimate condition.

ε_s' = compression steel strain at failure/ultimate condition.

0.003 = concrete crushing strain at failure/ultimate condition.

Figure 16.2: Failure/ultimate strain conditions of columns.

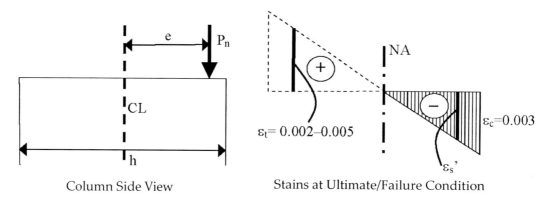

Column Side View   Stains at Ultimate/Failure Condition

(c) Column subjected to intermediate eccentricity loading – transition section.

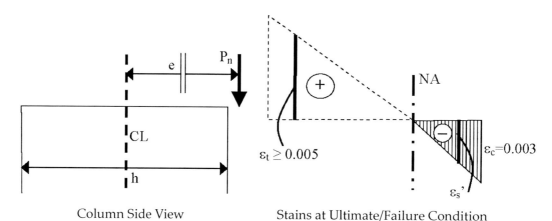

Column Side View   Stains at Ultimate/Failure Condition

(d) Column subjected to large eccentricity loading – tension controlled section.

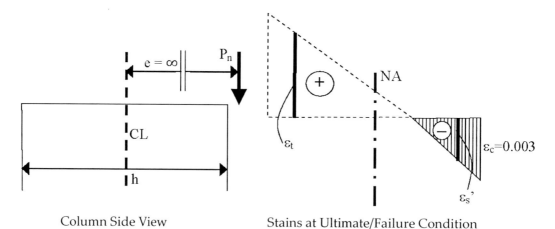

Column Side View   Stains at Ultimate/Failure Condition

(e) Column subjected to pure bending or axial force with ∞ eccentricity.

Figure 16.2 (continued): Failure/ultimate strain conditions of columns.

Note:

strain on the column tension side is shown in dashed line since concrete is cracked and tensile strain is hypothetical due to absence of material except at steel bar location.

The balanced failure condition of a column occurs if steel reaches yield strain, ε_y ($\varepsilon_y = f_y/E_s = 0.00209$ for $f_y = 60$ ksi), and concrete reaches crushing strain, 0.003, simultaneously. As such, for $f_y = 60$ ksi, balanced failure represents a borderline between compression controlled sections and transition sections.

It is important to note that Figure 16.2 applies only to column sections under uniaxial bending moment about section center line axis (CL). Tension and compression in the section are on opposite sides of the neutral axis (NA) parallel to CL. Section height or thickness (h) is the cross–sectional dimension in the plane of buckling perpendicular to the axis of bending moment or CL. Eccentricity (e) is also perpendicular to CL.

16.2 Strength Reduction Factor (φ) for Columns

As ε_t at ultimate condition (limit state) increases, a more gradual failure of the reinforced concrete element is achieved via extensive tension yielding of steel reinforcing bars. This results in enhanced failure safety. Such enhanced safety is reduced in transition and compression controlled sections. To safeguard against potential semi–catastrophic or catastrophic failure, ACI Code assigns a lower strength (capacity) reduction factor for transition and compression controlled sections compared with tension controlled sections as shown in Figure 16.3. ε_t is defined as the strain in extreme tension steel at ultimate condition. For sections with a single layer of tension steel bars, Figure 16.3 includes for $\varepsilon_t = 0.002$ and 0.005, the corresponding values of c/d where c is the depth of concrete compression zone and d is the effective depth of the concrete element. Also, spirally reinforced columns are assigned an increased value of φ as they are generally characterized by improved ultimate behavior or gradual failure mode.

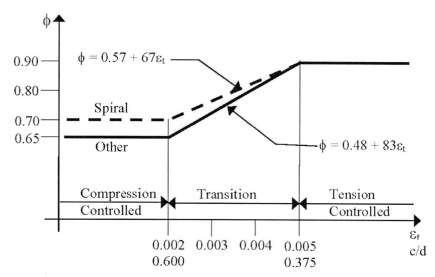

Figure 16.3: The relationship of strength reduction factor (φ) and strain in extreme tension steel (ε_t) and ratio of compression zone depth to effective depth in columns (c/d).

16.3 The Plastic Centroid

In Section 16.1, columns were classified based on the eccentricity of applied load (e). Ultimate condition under the effect of concentric loading with zero eccentricity causes uniform strains in the column cross–section equal to 0.003. Thus, compressive stresses in concrete and steel are $0.85f'_c$ and f_y, respectively. Eccentricity needs to be determined with respect to the above mentioned condition of concentric loading or the plastic centroid. The plastic centroid is defined as the location of resultant of compressive forces of concrete and steel under ultimate condition with $0.85f'_c$ and f_y stresses, respectively. Most column cross–sections are symmetrical. Thus, the plastic centroid coincides with the geometric centroid. However, column cross sections or steel reinforcing bars are at times not symmetrical designed to resist undirectional bending moments. For such cases, eccentricity needs to be evaluated with respect to the plastic centroid. Figure 16.4 illustrates two examples of unsymmetrical column cross–section with guidelines for determining the plastic centroids for each section.

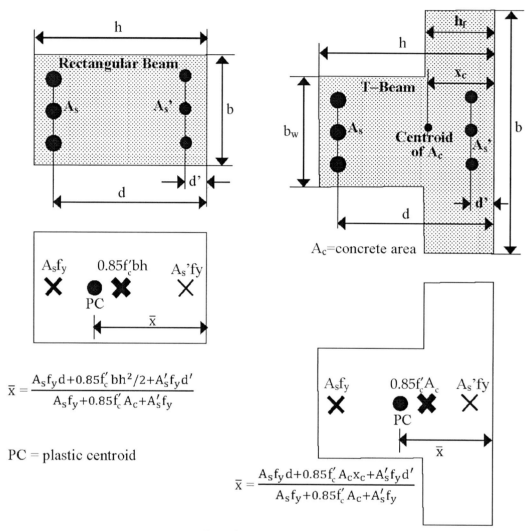

$$\bar{x} = \frac{A_s f_y d + 0.85 f_c' bh^2/2 + A_s' f_y d'}{A_s f_y + 0.85 f_c' A_c + A_s' f_y}$$

PC = plastic centroid

$$\bar{x} = \frac{A_s f_y d + 0.85 f_c' A_c x_c + A_s' f_y d'}{A_s f_y + 0.85 f_c' A_c + A_s' f_y}$$

Figure 16.4: Plastic centroid examples.

16.4 Analysis of Columns Subjected to Bending and Axial Compression

Based on the assumptions of Section 16.1, the strains in the column section can be determined followed by computing the corresponding stresses and forces. Consequently, P_n and M_n can be calculated. Figure 16.5 and Examples 16.1 and 16.2 illustrate this method of analysis.

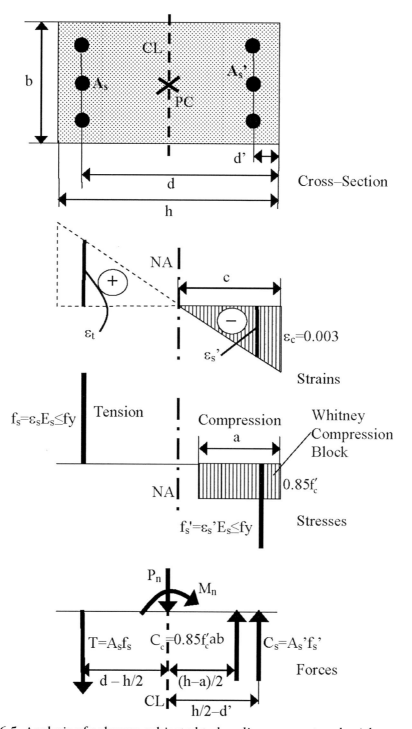

Figure 16.5: Analysis of columns subjected to bending moment and axial compression.

The notations in Figure 16.2 apply to Figure 16.5 with the following additional notations:

f'_c and f_y = specified compressive and yield strengths of concrete and steel.

c and a = depth of compression zone and block, respectively.

f_s = tension stress in tension steel at ultimate condition.

f'_s = compression stress in compression steel at ultimate condition.

T = resultant tension force in tension steel at ultimate condition.

C_c = resultant compression force in concrete at ultimate condition.

C_s = resultant compression force in compression steel at ultimate condition.

$d - h/2$ = moment arm for force in tension steel.

$h/2 - d'$ = moment arm for force in compression steel.

$(h - a)/2$ = moment arm for compression force in concrete.

Based on Figure 16.5, the nominal force P_n and bending moment M_n on a column section at ultimate/failure condition are:

$$P_n = C_c + C_s - T = 0.85f'_c ab + A'_s f'_s - A_s f_s \qquad (16.2)$$

$$M_n = 0.85f'_c ab(h - a/2) + A'_s f'_s(h/2 - d') + A_s f_s(d - h/2) \qquad (16.3)$$

Where the stresses in tension and compression steel may be expressed as follows:

$$f_s = (d - c) \times (0.003/c) \times E_s \le f_y \qquad (16.4)$$

$$f'_s = (c - d') \times (0.003/c) \times E_s \le f_y \qquad (16.5)$$

Where E_s is the modulus of elasticity of steel. With a given cross–section, Equations 16.2 and 16.3 may be used to recognize the loading condition of a column as explained below:

1. If the column nominal axial compressive force, P_n, is given, then, the value of c in Equation 16.2 may be determined via trial and error or iteration. Consequently, M_n can be computed using Equation 16.3 given the value of c.

2. If the column nominal axial compressive force, M_n, is given, then, the value of c in Equation 16.3 may be determined via trial and error or iteration. Consequently, P_n can be computed using Equation 16.2 given the value of c.

3. If the value of axial compression eccentricity, $e = M_n/P_n$ is given, then c can be determined via trial and error or iteration method utilizing Equation 16.6 given below. Then P_n and M_n may be determined using Equations 16.2 and 16.3.

$$e = [0.85f'_c ab(h - a/2) + A'_s f'_s(h/2 - d') + A_s f_s(d - h/2)]/ [0.85f'_c ab + A'_s f'_s - A_s f_s] \qquad (16.6)$$

Note:

The analysis method in this section remains focused on a simplified column cross–section with rectangular shape and two layers of steel at the tension and compression faces.

Example 16.1

Determine the values of M_n and $e = P_n/M_n$ for the tied reinforced concrete column section shown in the figure if $P_n = 750$ k with $f'_c = 4,000$ psi and $f_y = 60,000$ psi.

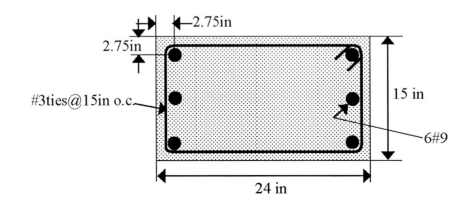

Solution

Using Equations 16.2, 16.4 and 16.5:

$$P_n = 0.85f'_c ab + A'_s f'_s - A_s f_s$$

or $\quad P_n = 0.72f'_c cb + A'_s \times \underbrace{(c - d') \times (0.003/c) \times E_s}_{\leq f_y} - A_s \times \underbrace{(d - c) \times (0.003/c) \times E_s}_{\leq f_y}$

In the above equation, the value of c (depth of compression zone) is the only unknown. Using trial and error resulted in c = 15.4 in and a = 13.1 in, $\varepsilon_t = 0.0011$ and $\varepsilon_s' = 0.0025$, and $f_s = 33$ ksi and $f'_s = 60$ ksi

Using Equations 16.3 with the above values:

$$M_n = 0.85f'_c ab(h - a/2) + A'_s f'_s (h/2 - d') + A_s f_s(d - h/2)$$
$$M_n = 0.85 \times 4 \times 13.1 \times 15 \times (24 - 13.1/2) + 3 \times 60 \times (24/2 - 2.75) + 3 \times 33 \times (21.25 - 24/2)$$
$$\Longrightarrow \quad M_n = 519 \text{ k.ft} \quad \Longrightarrow \quad e = M_n/P_n = 522 \times 12/750 = 8 \text{ in}$$
$$\varepsilon_t = 0.0011 \Longrightarrow \phi = 0.65 \text{ (Figure 16.3)}$$
$$\Longrightarrow \phi P_n = 488 \text{ k} \qquad \phi M_n = 337 \text{ k.ft}$$

Example 16.2

Determine the values of P_n and $e = P_n/M_n$ for the reinforced concrete column section in Example 16.1 if $M_n = 500$ k.ft with $f'_c = 4,000$ psi and $f_y = 60,000$ psi.

Solution

Using Equations 16.3, 16.4 and 16.5:

$$M_n = 0.85f'_c ab(h - a/2) + A'_s f_s'(h/2 - d') + A_s f_s(d - h/2)$$

The previous equation may be rewritten as follows:

$$M_n = 0.85f'_c ab(h - a/2) + A'_s \times \underbrace{(c - d') \times (0.003/c) \times E_s}_{\leq f_y} \times (h/2 - d') + A_s \times \underbrace{(d - c) \times (0.003/c) \times E_s}_{\leq f_y} \times (d - h/2)$$

In the above equation, the value of c (depth of compression zone) is the only unknown. Using trial and error resulted in c = 7.3 in and a = 6.2 in, $\varepsilon_t = 0.0057$ and $\varepsilon_s' = 0.0019$, and $f_s = 60$ ksi and $f'_s = 54$ ksi

Using Equations 16.3 with the above values:

$$P_n = 0.85f'_c ab + A'_s f'_s - A_s f_s$$
$$P_n = 0.85 \times 4 \times 6.2 \times 15 + 3 \times 54 - 3 \times 60$$
$$\Longrightarrow \quad P_n = 299 \text{ k} \Longrightarrow \quad e = M_n/P_n = 500 \times 12/299 = 20 \text{ in}$$
$$\varepsilon_t = 0.0057 \Longrightarrow \phi = 0.90 \text{ (Figure 16.3)}$$
$$\Longrightarrow \phi P_n = 269 \text{ k} \qquad \qquad \phi M_n = 449 \text{ k.ft}$$

Example 16.3

Determine the values of P_n and M_n for the reinforced concrete column section in Example 16.1 if e = $P_n/M_n = 12$ in with $f'_c = 4,000$ psi and $f_y = 60,000$ psi.

Solution

Using Equations 16.4, 16.5 and 16.6:

$$e = [0.85f_c'ab(h - a/2) + A_s'f_s'(h/2 - d') + A_sf_s(d - h/2)]/ [0.85f_c'ab + A_s'f_s' - A_sf_s]$$

$$f_s = (d - c) \times (0.003/c) \times E_s \leq f_y$$

$$f_s' = (c - d') \times (0.003/c) \times E_s \leq f_y$$

The value of c (depth of compression zone) is the only unknown. Using trial and error resulted in c = 13.0 in and a = 11.1 in, ε_t = 0.0019 and ε_s' = 0.0024, and f_s = 55 ksi and f_s' = 60 ksi

Using Equations 16.2 and 16.3 with the above values results in:

$$P_n = 575 \text{ k} \qquad\qquad M_n = 575 \text{ k.ft}$$

$$\varepsilon_t = 0.0019 \Longrightarrow \phi = 0.65 \text{ (Figure 16.3)}$$

$$\Longrightarrow \phi P_n = 373 \text{ k} \qquad\qquad \phi M_n = 373 \text{ k.ft}$$

Example 16.4

Determine the values of P_n and M_n for the reinforced concrete column section in Example 16.1 for the balanced failure condition with f_c' = 4,000 psi and f_y = 60,000 psi.

Solution

At the balanced condition, $\varepsilon_t = \varepsilon_y = f_y/E_s$ under failure condition where ε_t is strain in tension steel. Using the following equations:

$$\text{For the balanced condition} \Longrightarrow \varepsilon_t = (d - c) \times (0.003/c) = \varepsilon_y$$

$$\varepsilon_y = f_y/E_s = 60/29,000 = 0.00209 \Longrightarrow f_s = f_y = 60 \text{ ksi}$$

The value of c_b, depth of compression zone under balanced condition, can be computed:

$$c_b = \frac{3}{5.09}d = \frac{3}{5.09} \times 21.25 = 12.5 \text{ in}$$

Using the above values of c_b and $\varepsilon_t \Longrightarrow a_b = 10.6$ in, $\varepsilon_t = \varepsilon_s' = 0.0023$, and $f_s' = 60$ ksi

Using Equations 16.2 and 16.3 with the above values results in:

$$P_{nb} = 543 \text{ k} \qquad\qquad M_{nb} = 583 \text{ k.ft}$$

$$\varepsilon_t = 0.00209 \Longrightarrow \phi = 0.65 \text{ (Figure 16.3)}$$

$$\Longrightarrow \phi P_n = 353 \text{ k} \qquad\qquad \phi M_n = 377 \text{ k.ft}$$

For a given column section, generally, the value of P_u or/and M_u are determined based on structural analysis due to applied loading rather than P_n and/or M_n. The solution for such typical situations is to assume the value of ϕ. Utilizing the assumed ϕ, $P_n = P_u/\phi$ and/or $M_n = M_u/\phi$ can be calculated. The column section can then be analyzed to with P_n and/or M_n (Examples 16.1 thru 16.4) and the value of ϕ can be computed. The initial assumption of ϕ is afterward modified and the problem solution is repeated until the assumed and computed ϕ values converge.

The aforementioned analysis procedure applies to all column cross–sections with various shapes including circular as well as columns with multi–layers of steel reinforcement. Strains and stresses for all steel bars are determined as functions of c = depth of compression zone. The moment arms of all forces within the section at failure conditions are also expressed as functions of c. As such, the solution may become somewhat more sophisticated as graphically illustrated in Figure 16.6.

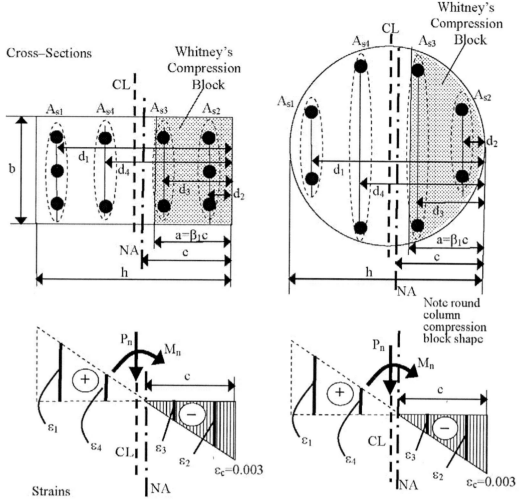

Figure 16.6: Analysis of columns of rectangular cross–section with multi steel bar layers and columns with circular cross–sections.

As shown in Figure 16.6, analysis of columns with round section subjected to axial compression and bending is particularly complicated. This is due to the circular layout of steel bars and the irregular shape of compression block. An approximate procedure is followed by designers at times. This procedure entails converting the circular column section into an equivalent rectangular section as shown in Figure 16.7. The resulting rectangular section is analyzed with relative simplicity as explained earlier.

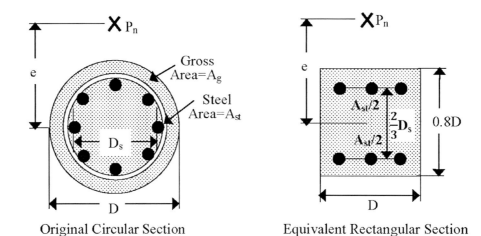

Original Circular Section   Equivalent Rectangular Section

Figure 16.7: Conversion of circular column section into equivalent rectangular section.

16.5 The Column Interaction Diagram for Bending and Axial Compression

Analysis of column cross–sections subjected to bending and axial compression is a complicated process requiring an iterative solution. In addition to computers software, designers have generally utilized a graphical solution termed the interaction diagram. Figure 16.8 represents a typical interaction diagram relating P_n to M_n for a given column cross–section. Each ultimate/failure loading condition/case within the interaction diagram is represented by a point of the line graph of Figure 16.8 with x, y and angular coordinates of M_n, P_n and e, respectively. All possible failure load cases (Figure 16.2) are included in the interaction diagram. Within the interaction diagram (the shaded area in Figure 16.8) represents loading cases with axial compression forces and/or bending moments less than the column section capacity (P_n and M_n). The area outside the interaction diagram is for loading cases with loading greater than column section capacity.

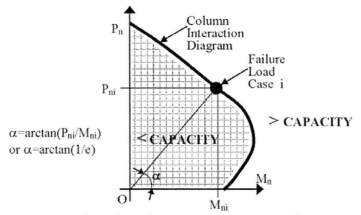

Figure 16.8: Conversion of circular column section into equivalent rectangular section.

With the availability of interaction diagram for a given section, analysis and design of such column section is simplified. ACI Code has developed a limited number of interaction diagrams that apply to a wide range of reinforced concrete column cross–sections. Examples of such interaction diagrams are in Figures 16.9 and 16.10. More interaction diagrams are provided in Appendix A. The following parameters are constant within each interaction diagram:

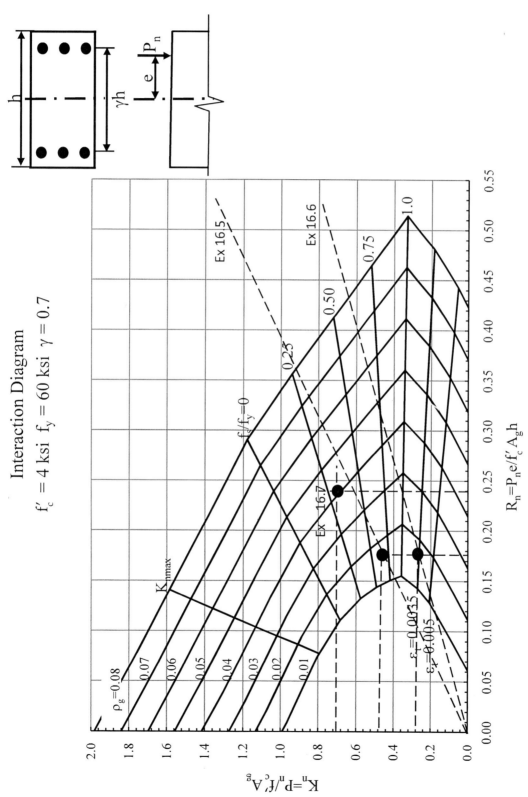

Figure 16.9: Example of column interaction diagram.

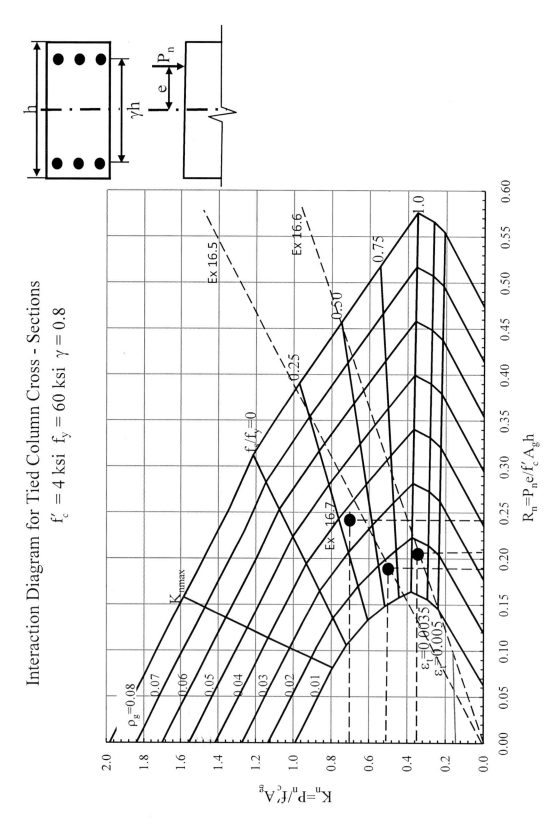

Interaction Diagram for Tied Column Cross - Sections

$f'_c = 4$ ksi $f_y = 60$ ksi $\gamma = 0.8$

$R_n = P_n e / f'_c A_g h$

$K_n = P_n / f'_c A_g$

Figure 16.10: Example of column interaction diagram.

1. Specified concrete and steel strengths (f'_c and f_y).

2. Shape of column cross–section (rectangular or circular).

3. Configuration of steel bar reinforcement. For rectangular sections, steel bars may be distributed on two faces of column section or on the entire perimeter of column section. For circular sections, steel bars are generally distributed on the entire perimeter.

4. Type of column: tied or spiral.

5. The ratio of distance between tension and compression steel bars to section height or entire thickness (γ).

Changing any of the above factors requires the utilization of a different interaction diagram. Two interaction diagrams can be used for a column section. Then, the results are linearly interpolated for the column in question as illustrated in the following examples. Each interaction diagram can be used for column section with variable:

1. Cross–sectional area = A_g as x and y coordinates are ratios of A_g.

2. Cross–sectional dimensions (b and h) as y coordinate is a ratio of h and x and y coordinates are ratios of A_g.

3. Reinforcement ratio, ρ_g. Each interaction diagram in Figures 16.9 and 16.10 contains a scale for ρ_g ranging from 1 to 8%.

The x and y coordinate axes in of column interaction diagrams are $R_n = P_n e/f'_c A_g h$ and $K_n = P_n/f'_c A_g$, respectively. Column interaction diagrams may be used for design or analysis to determine the values of P_n and/or $M_n = P_n e$ for a specific failure/ultimate loading condition. As such, column adequacy for supporting factored load and moment is assessed as shown in following examples.

To properly utilize column interaction diagrams, the following ACI Code provisions need to be verified:

1. The value of P_n or K_n may not exceed Equation 15.1 (Chapter 15) that incorporates factor α. A straight line is shown in each interaction diagram that depicts K_{nmax} for this reason.

2. The value of ϕ is determined based on the value of ε_t. Lines are shown within each interaction diagram that represent different values of f_s or ε_t. As such, ε_t is directly read from the line label or calculated as f_s/E_s. Consequently, ϕ may be determined as explained in Section 16.2.

3. Column reinforcement ration ρ_g is required to be within code limits (typically 1 to 8%) as explained in Chapter 15.

16.6 Analysis of Column Using the Interaction Diagram

For analysis problems, material properties (f'_c and f_y) and column cross–section details (b, h, d, d', A_s and A'_s) as well as column type (tied or spiral) are presumed given. Factored axial load (P_u) and factored bending moment (M_u) applied to the column are also given. They are determined as results of structural analysis. Analysis can determine the adequacy of column cross–section for resisting given axial force and moment. Analysis procedure can proceed according to the following steps:

1. Compute the value of reinforcement ratio $\rho_g = A_{st}/A_g$ for the column in question and confirm code compliance $1\% < \rho_g < 8\%$.

2. Determine the amount $h/e = P_u h/M_u$.

3. Select the appropriate interaction diagram for the column in question. If column properties do not match a single interaction diagram, then two or more diagrams can be used and the results are, then, interpolated. Interpolation can be utilized for the values of γ, f'_c and f_y. If two or more interaction diagrams need to be used, then steps 4 thru 6 are repeated for each diagram.

4. Draw a line from the origin of interaction diagram (0,0) at angle α from x axis where α = arctan(h/e).

5. On the line of step 4, utilizing the ρ_g scale, determine the point corresponding to ρ_g of the section in question computed in step 2. Based on this point's location, the values of ε_t can be determined using the design aid lines within the interaction diagram. Consequently, ϕ can be computed as explained earlier.

6. From the point of step 5, draw a horizontal line towards the y–axis to determine the value of K_n = $P_n/f'_c A_g \leq K_{nmax}$ and a vertical line towards the x–axis to determine the value of $R_n = M_n/f'_c A_g h$. Then determine the values of P_n and M_n.

7. By interpolation between the relevant interaction diagrams, ϕP_n and ϕM_n can be determined.

8. If $K_n > K_{nmax}$, then the section is inadequate since it does not conform to the ACI Code.

9. If $\phi M_n \geq M_u$ and $\phi P_n \geq P_u$ ===>   Section is adequate

    If $\phi M_n < M_u$ or $\phi P_n < P_u$ ===>   Section is inadequate

Example 16.5

Analyze the adequacy of the tied column cross–section of Example 16.1 under the effect of P_u and M_u of 450 k and 340 k.ft, respectively, with f'_c = 4,000 psi and f_y = 60,000 psi.

Solution

$\rho_g = A_{st}/A_g = 6/15 \times 24 = 1.67\%$ ===> as $1\% \leq \rho_g \leq 8\%$, section is within code OK

h/e = $P_u h/M_u$ = $450 \times 24/350 \times 12$ = 2.6 where 12 is a conversion factor from k.ft to k.in

The column section of Example 16.1 contains steel reinforcement is on two sides

$\gamma = (24 - 2 \times 2.75)/24 = 0.77$

f'_c = 4,000 psi and f_y = 60,000 psi

The interaction diagrams of Figures 16.9 and 16.10 (f'_c = 4,000 psi, f_y = 60,000 psi and steel on two sides for both diagrams whereas γ = 0.7 and 0.8, respectively) can be used and the results are interpolated for γ = 0.77

$\alpha = $ arctan(h/e) = 69°

A line from the origin has been drawn for this example in Figures 16.9 and 16.10. A point representing this example is also depicted on each interaction diagram.

From Figure 16.9   ===> $\varepsilon_t = f_s/E_s \approx 0.70 \times 60/29,000 = 0.0014$ ===> ϕ = 0.65

                    ===> $K_n = P_n/f'_c A_g = 0.48$ ===> $P_n = 0.48 \times 4 \times (24 \times 15) = 691$ k

                    ===> $R_n = M_n/f'_c A_g h = 0.175$

                    ===> $M_n = 0.175 \times 4 \times (24 \times 15) \times 24/12 = 504$ k.ft (12 is for conversion)

From Figure 16.10 $\Longrightarrow \varepsilon_t = f_s/E_s \approx 0.65\times60/29{,}000 = 0.0013 \Longrightarrow \phi = 0.65$

$\Longrightarrow K_n = P_n/f'_c A_g = 0.50 \Longrightarrow P_n = 0.50\times4\times(24\times15) = 720 \text{ k}$

$\Longrightarrow R_n = M_n/f'_c A_g h = 0.190$

$\Longrightarrow M_n = 0.190\times4\times(24\times15)\times24/12 = 547 \text{ k.ft (12 is for conversion)}$

For $\gamma = 0.77 \Longrightarrow P_n = 691 + (720 - 691)\times(0.77 - 0.7)/(0.8 - 0.7) = 711 \text{ k}$

For $\gamma = 0.77 \Longrightarrow M_n = 504 + (547 - 504)\times(0.77 - 0.7)/(0.8 - 0.7) = 534 \text{ k}$

$\phi P_n = 0.65\times711 = 462 \text{ k} \geq P_u = 450 \text{ k}$

$\phi M_n = 0.65\times 534 = 347 \text{ k.ft} \geq M_u = 340 \text{ k.ft} \Longrightarrow$ Section is adequate or OK

Compare the above obtained answers with the results of Example 16.1. Note that both examples represent identical situation with a manual solution for Example 16.1 and graphical solution for Example 16.5:

Example 16.1	Example 16.5
$\phi P_n = 488 \text{ k}$	$\phi P_n = 462 \text{ k}$
$\phi M_n = 337 \text{ k.ft}$	$\phi M_n = 347 \text{ k.ft}$

The difference between the two methods of solution is indicative of the approximate nature of graphical solutions. Engineers generally implement extra safety when using the interaction diagrams due to their approximate nature.

Example 16.6

Analyze the adequacy of the tied column cross–section of Example 16.1 under the effect of P_u and M_u of 350 k and 500 k.ft, respectively, with $f'_c = 4{,}000$ psi and $f_y = 60{,}000$ psi.

Solution

$\rho_g = A_{st}/A_g = 6/15\times24 = 1.67\% \Longrightarrow$ as $1\% \leq \rho_g \leq 8\%$, section is within code OK

$h/e = P_u h/M_u = 350\times24/500\times12 = 1.4$ where 12 is a conversion factor from k.ft to k.in

The column section of Example 16.1 contains steel reinforcement is on two sides

$\gamma = (24 - 2\times2.75)/24 = 0.77$

$f'_c = 4{,}000$ psi and $f_y = 60{,}000$ psi

The interaction diagrams of Figure 16.9 and 16.10 ($f'_c = 4{,}000$ psi, $f_y = 60{,}000$ psi and steel on two sides for both diagrams whereas $\gamma = 0.7, 0.8$, respectively) can be used and the results are interpolated for $\gamma = 0.77$

$\alpha = \arctan(h/e) = 54°$

A line from the origin has been drawn for this example in Figures 16.9 and 16.10. A point representing this example is also depicted on each interaction diagram.

From Figure 16.9

$\Longrightarrow$ the section is located between on the line of $\varepsilon_t = 0.0035$

$\Longrightarrow \phi = 0.48 + 83\varepsilon_t = 0.48 + 83\times0.0035 = 0.77$ (Figure 16.3)

===> $K_n = P_n/f'_c A_g = 0.28$ ===> $\phi P_n = 0.77 \times 0.28 \times 4 \times (24 \times 15) = 310$ k

===> $R_n = M_n/f'_c A_g h = 0.21$ ===> $\phi M_n = 0.77 \times 0.21 \times 4 \times (24 \times 15) \times 24/12 = 466$ k.ft

(12 is for conversion)

From Figure 16.10

===> the section is located between two lines: 1st line: $f_s = f_y$ line ($\varepsilon_t \approx 0.002$)
                                                       2nd line: $\varepsilon_t = 0.0035$

===> $\varepsilon_t \approx 0.0025$ ===> $\phi = 0.48 + 83\varepsilon_t = 0.48 + 83 \times 0.0025 = 0.69$ (Figure 16.3)

===> $K_n = P_n/f'_c A_g = 0.34$ ===> $\phi P_n = 0.73 \times 0.34 \times 4 \times (24 \times 15) = 357$ k

===> $R_n = M_n/f'_c A_g h = 0.21$ ===> $\phi M_n = 0.73 \times 0.21 \times 4 \times (24 \times 15) \times 24/12 = 442$ k.ft

(12 is for conversion)

For $\gamma = 0.77$ ===> $\phi P_n = 310 + (357 - 310) \times (0.77 - 0.7)/(0.8 - 0.7) = 342$ k

For $\gamma = 0.77$ ===> $\phi M_n = 466 + (442 - 442) \times (0.77 - 0.7)/(0.8 - 0.7) = 449$ k

$\phi P_n = 342$ k < $P_u = 350$ k

$\phi M_n = 449$ k.ft < $M_u = 500$ k.ft ===> Section is inadequate or NG

16.7 Column Section Design Using the Interaction Diagram

For design problems, material properties (f'_c and f_y) and column cross–section details (b, h, d, d', A_s and A_s') as well as column type (tied or spiral) are unknown parameters. Factored axial load (P_u) and factored bending moment (M_u) applied to the column are determined by structural analysis. Design can achieve a column section that can resist the applied load and bending moment and reasonable for construction. Typically, material properties are assumed based on regional material availability. Consequently, column section design can proceed as follows:

1.   Select the arrangement of steel bars and column type. Figure 16.11 contains several possible arrangements of steel bars along with their column type. Such selection is based on loading conditions (uniaxial or biaxial bending), column location (interior, exterior or corner column) and the need for using spiral columns. Column dimensions are at this time assumed.

2.   Conservatively assume $\phi = 0.65$ for tied columns and 0.70 for spiral columns,

3.   Compute the value of γ based on required cover over steel bars. Typically, γ ranges between (h – 5)/h and (h – 6)/h for two layer of steel bars (A_s and A_s') and ranges between (h – 8)/h and (h – 10)/h for four layer of steel bars.

4.   Select the appropriate interaction diagram for the column in question. If column properties do not match a single interaction diagram, then two or more diagrams can be used and the results are, then, interpolated. Interpolation can be utilized for the values of γ, f'_c as well as f_y. If two or more interaction diagrams need to be used, then step 4 is repeated for each diagram.

5.   Utilize the limit values of $P_n = P_u/\phi$ and $M_n = M_u/\phi$ for column design.

6.   Draw a horizontal line from $K_n = P_n/f'_c A_g \leq K_{nmax}$ on the y–axis to and a vertical line from $R_n = M_n/f'_c A_g h$ on the x–axis. The point of intersection of these two lines is used to determine the value of ρ_g required for the column section for resistance of P_u and M_u to determine the value of ρ_g. If $K_n > K_{nmax}$, then column section dimensions need to be adjusted/increased.

7.   By interpolation between the relevant interaction diagrams, ρ_g required can be determined. If the resulting ρ_g is within code continue to step 6; otherwise reassume column cross–section.

8.   The required column section is detailed including concrete cover and selection of steel bars.

9.   Perform analysis of the designed column section to confirm its adequacy for resisting P_u and M_u.

10.   If section design is not satisfactory for any reason (e.g. overdesign, inadequate for resisting applied load and/or moment, design is not convenient for construction, steel bar size is relatively large, etc), the design is revised (material properties as well as section dimensions can be reassumed) until a satisfactory design is achieved.

Example 16.7

Design 18×18 in tied column cross–section under the effect of axial force (P_u) and uniaxial bending moment (M_u) of 600 k and 300 k.ft, respectively, with f'_c = 4,000 psi and f_y = 60,000 psi.

Solution

Since this column is subjected to unidirectional bending moment, then, arrangement of steel bars on two sides/faces of column cross–section is reasonable.

$\gamma = (h-5)/h = (18-5)/18 = 0.72$

Thus, the interaction diagrams of Figures 16.9 and 16.10 can be used and the results are interpolated to obtain a suitable solution for the column in question.

$P_n = P_u/\phi = 600/0.65 = 923$ k

$M_n = M_u/\phi = 300/0.65 = 462$ ft.k

$K_n = P_n/f'_c A_g = 923/[4×(18)^2] = 0.71$

$R_n = M_n/f'_c A_g h = 462×12/[4×(18)^2×18] = 0.24$ where 12 in for conversion

Figure 16.9 ⟹ $\rho_g = 0.04$ ⟹ $\varepsilon_t = f_s/E_s = 0.25 f_y/E_s = 0.0007$ ⟹ $\phi = 0.65$

Figure 16.10 ⟹ $\rho_g = 0.037$ ⟹ $\varepsilon_t = f_s/E_s = 0.35 f_y/E_s = 0.0005$ ⟹ $\phi = 0.65$

For $\gamma = 0.77$ ⟹ $\rho_g = 0.04 + (0.037 - 0.04)×(0.77 - 0.7)/(0.8 - 0.7) = 0.038$

$A_{st} = 0.038×(18)^2 = 12.3$ in^2 ⟹ $A_s = A_s' = 12.3/2 = 6.2$ in^2

Use 10 #11 steel bars for the column with 5 #11 bars on each side.

Cross – section detail is as shown below:

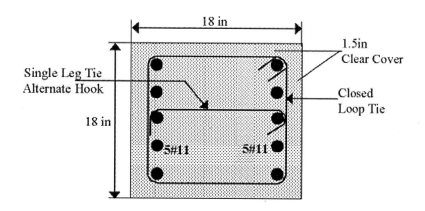

16.8 Additional Topics in Reinforced Concrete Columns

16.8.1 Shear in Columns

Columns are typically subjected to shear in addition to axial compression and bending moment. Concrete shear resistance of members subjected to axial compression has been discussed in Chapter 11. The ACI following equation described shear resistance of concrete in members subjected to axial compression:

$$V_c = 2(1 + \frac{N_u}{2000A_g})\sqrt{f_c'}b_w d \qquad (16.7)$$

Where:

    V_c = concrete shear resistance in columns.

    N_u = factored axial compression force on the column.

    f_c' = concrete design compressive strength.

    b_w = section width.

    d = section effective depth.

If factored applied shear force V_u exceeds $\phi V_c/2$, then closed stirrups are added to column ties for resistance to shear. Typically, the number of column ties is increased to accommodate both requirements (total number of ties = number of ties required for the column + number of stirrup/ties required for shear). As such, combined tie spacing is calculated by dividing the column length by the total number of ties. Furthermore, ACI Code modifies Equation 16.7 for application to circular columns (b_w = D and d = 0.8D) as follows:

$$V_c = 1.6(1 + \frac{N_u}{2000A_g})\sqrt{f_c'}D^2 \qquad (16.8)$$

Where D is the diameter of circular column cross–section.

16.8.2 Column Steel Bar Layout Design

The previous parts of this chapter have focused on columns subjected to uniaxial bending moments about the section centerline or principal axis deemed as the axis of bending. Steel bar distribution/ layout within the section reflects upon steel effectiveness for moment resistance, consequently, affecting column behavior significantly. Each interaction diagram is developed for a specific bar layout with respect to the bending axis (centerline). Various steel bar layout design are used for columns. Figures 16.11 and 16.12 include the following common steel bar layouts for tied and spiral columns, respectively.

Tied columns

a.   Steel bars distributed on the entire column perimeter

b.   Steel bars distributed on two column sides/faces parallel to the axis of bending

c.   Steel bars distributed on two column sides/faces perpendicular to the axis of bending

Spiral Columns

a. Spiral column with circular cross – section

b. Spiral column with square cross–section and circular core

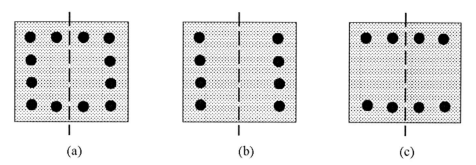

Figure 16.11: Common tied column reinforcement layout.

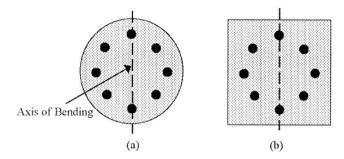

Figure 16.12: Common spiral column reinforcement layout.

16.8.3 Biaxial Bending

Figure 16.13 illustrates the contrast among column sections subjected to uniaxial bending about principal axis x, uniaxial bending about principal axis y and biaxial bending. It is important to realize that the axis of bending is perpendicular to force eccentricity. Furthermore, the neutral axes for bending bout x and y are parallel to x and y, respectively.

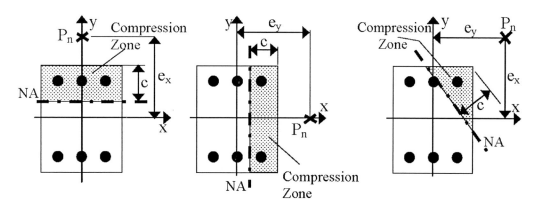

Figure 16.13: Uniaxial and biaxial bending of columns.

Analysis of column sections subjected to biaxial bending is iterative and complicated. Several methods are typically used for analysis for biaxial bending. Utilization of computer software and/ or design aid is typical. Furthermore, ACI Code introduces a conservative approximate solution for doubly symmetric sections based on the following equation:

$$\frac{1}{P_{ni}} = \frac{1}{P_{nx}} + \frac{1}{P_{ny}} - \frac{1}{P_o}$$ (16.9)

Where:

P_{ni} = section nominal axial load capacity with eccentricities e_x and e_y about x and y axes, respectively.

P_{nx} = section nominal axial load capacity with eccentricity e_x about x axis and zero eccentricity about y axis.

P_{ny} = section nominal axial load capacity with eccentricity e_y about x axis and zero eccentricity about x axis.

P_o = section nominal axial load capacity with zero eccentricity about both axes.

The value of ϕP_{ni} may not exceed $\phi \alpha P_n$ of Equation 15.1 in Chapter 15.

16.8.4 ACI Code Provisions for Column Splices

Figure 16.14 illustrates reinforcing steel bar arrangements for column splices. Typically, steel bars of the lower column portion are bent to fit within the higher column portion. Slope of bent portions of column bars shall not exceed 1/6 per ACI Code. If the required bend offset is greater than 3 in, then added dowels splicing bars are required in lieu of the bend. Column sizes normally change at floor levels due to load changing. Thus, column splices are generally used at floor levels as shown in Figure 16.15. However, column splices near floor midheight are more suitable for structural integrity since any bending moment within the column would be minimum. For splices at column midheight, tie spacing shall be s/2 throughout the entire splice length where s is code required tie spacing (Chapter 15). The lowest column tie shall be within s/2 from the top of concrete beam, slab or footing. The highest column tie shall be within s/2 from the steel bars in slab. If beam are provided on four sides, then the first tie shall be within 3 in from the beam bottom steel. Furthermore, special ties are required to be provided within 6 in of bar bends for splicing. These ties are required to resist 1.5 times the outward force caused by the bend (1.5×longitudinal bar yield force×bend slope).

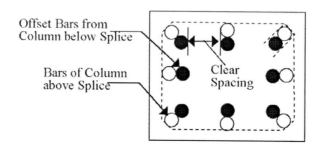

Offset Bars from
Column below Splice

Bars of Column
above Splice

Clear
Spacing

$\rho_g < 2 - 3\%$
To avoid bar congestion

Figure 16.14: Column splice steel bar arrangement.

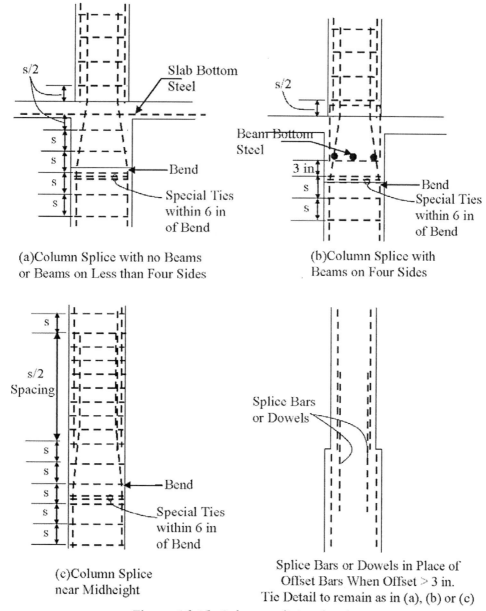

(a)Column Splice with no Beams or Beams on Less than Four Sides

(b)Column Splice with Beams on Four Sides

(c)Column Splice near Midheight

Splice Bars or Dowels in Place of Offset Bars When Offset > 3 in. Tie Detail to remain as in (a), (b) or (c)

Figure 16.15: Column splicing details.

Test your knowledge

1.   What is the difference between the analysis assumptions of beams and columns?

2.   What are the five categories of columns subjected to eccentric loading?

3.   Which category of the above five is not allowed by ACI Code?

4.   What are the three types of section from the standpoint of capacity reduction factor ϕ?

5.   Define the plastic centroid and point out its importance for columns under eccentric loading.

6.   What are the equations used for analysis of reinforced concrete columns subjected to eccentric loading?

7.   What is the solution basis of column analysis equations of Question 6, direct or iterative, and why?

8.   Define the interaction diagram. Why is it important for analysis of columns subjected to eccentric loading?

9.   What are the abscissa and coordinate of the interaction diagram used for a particular column?

10.   What are the abscissa and coordinate of a multi–use interaction diagram that can be used for numerous columns?

11.   What are the constants in a multi–use interaction diagram?

12.   What are the variables in a multi–use interaction diagram?

13.   What is the reason for the special ties in column splices required by ACI Code?

Draw the interaction diagram for the tied column sections shown below using a spreadsheet programming software:

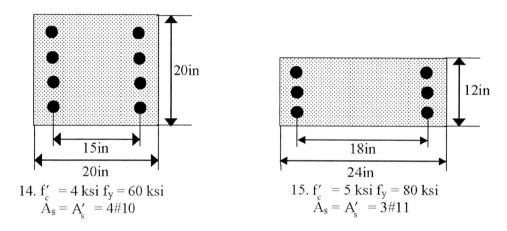

14. f'_c = 4 ksi f_y = 60 ksi
    $A_s = A'_s$ = 4#10

15. f'_c = 5 ksi f_y = 80 ksi
    $A_s = A'_s$ = 3#11

Draw the interaction diagram for the spirally reinforced column sections shown below using a spreadsheet programming software:

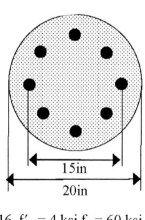 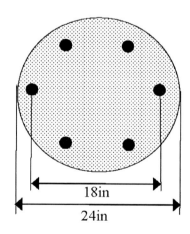

16. f'_c = 4 ksi f_y = 60 ksi
    A_s = 8#10

17. f'_c = 5 ksi f_y = 80 ksi
    A_s = 6#11

Use the interaction diagrams to determine the values of ϕP_n and ϕM_n for the tied column sections listed below under the effect of the listed load eccentricity e:

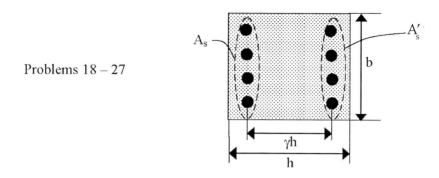

Problems 18 – 27

Problem	f'_c	f_y (ksi)	h	γh	b (in)	A_s	A'_s	e (in)
18	4	60	18	13	15	3#9	each	8
19	4.5	60	18	13	15	3#9	each	24
20	5.5	60	20	14	20	4#10	each	10
21	5	60	20	14	20	4#10	each	30
22	6	60	24	18	18	5#11	each	10
23	6.5	60	24	18	18	5#11	each	40
24	6	60	22	16	22	5#10	each	9
25	7	60	22	16	22	5#10	each	25
26	7.5	80	30	24	24	5#14	each	12
27	8	80	30	24	24	5#14	each	36

Use the interaction diagram to determine the values of ϕP_n and ϕM_n for the tied column sections listed below under the effect of the listed load eccentricity e:

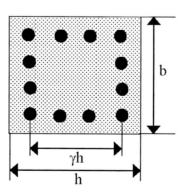

Problems 28 – 31

Problem	f'_c	f_y (ksi)	h	γh	b (in)	A_{st}	e (in)
28	4	60	18	13	15	8#8	8
29	4.5	60	18	13	15	8#8	24
30	5.5	60	20	14	20	10#10	10
31	5	60	20	14	20	10#10	30

Use the interaction diagram to determine the values of ϕP_n and ϕM_n for the spirally reinforced column sections listed below under the effect of the listed load eccentricity e:

Problems 32 – 41

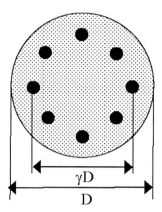

Problem	f'_c	f_y (ksi)	D	γD (in)	A_{st}	e (in)
32	4.3	60	18	13	6#8	8
33	4.3	60	18	13	6#8	23
34	5.1	60	20	14	7#10	10
35	5.1	60	20	14	7#10	30
36	6	60	24	18	8#11	12
37	6	60	24	18	8#11	36
38	6.5	60	22	16	6#10	9
39	6.5	60	22	16	6#10	28
40	8	80	30	24	7#14	12
41	8	80	30	24	7#14	30

Use the interaction diagram to determine the values of ϕP_n and ϕM_n for the spirally reinforced column sections listed below under the effect of the listed load eccentricity e:

Problems 42 – 45

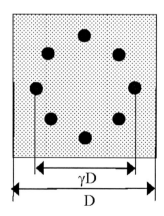

Problem	f'_c	f_y (ksi)	D	γD (in)	A_{st}	e (in)
42	4.5	60	18	13	6#8	10
43	4.8	60	18	13	6#8	20
44	5.5	60	20	14	7#10	10
45	5.6	60	20	14	7#10	30

Use the interaction diagram to design a tied column section for the values of P_n and e listed below so that ρ_g is between 1.0 and 2.0%:

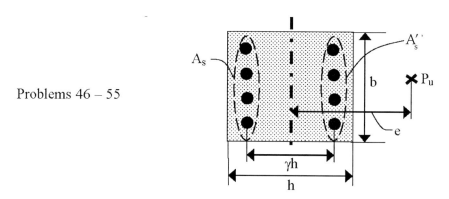

Problems 46 – 55

Problem	f'_c	f_y (ksi)	b/h	Pu (k)	e (in)
46	4.5	60	1	600	8
47	4.5	60	1	600	20
48	5	60	0.75	1,100	10
49	5	60	0.75	1,100	30
50	6.5	60	0.5	1,500	10
51	6.5	60	0.5	1,500	40
52	7	60	1	1,800	9
53	7	60	1	1,800	28
54	7.5	80	0.8	2,200	12
55	7.5	80	0.8	2,200	36

Use the interaction diagram to design a spirally reinforced column section for the values of P_n and e listed below so that ρ_g is between 1.5 and 2.5%:

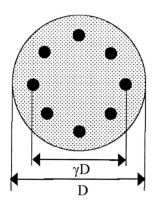

Problems 56 – 65

Problem	f'_c	f_y (ksi)	P_u (k)	e (in)
56	4.4	60	580	9
57	4.4	60	640	24
58	5.4	60	980	10
59	5.7	60	1,020	32
60	6.2	60	1,100	11
61	6.4	60	1,200	27
62	6	60	1,400	9
63	6	60	1,600	25
64	8	80	2,400	12
65	8	80	2,200	30

Use the interaction diagram to determine the values of ϕP_n and ϕM_n for the tied column sections listed below under the effect of the listed load eccentricities e_x and e_y:

Problems 66 – 69

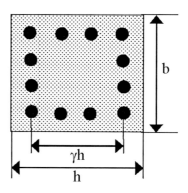

Problem	f'_c	f_y (ksi)	h	γh	b (in)	A_{st}	e_x	e_y (in)
66	4	60	18	13	15	8#8	8	8
67	4.5	60	18	13	15	8#8	24	15
68	5.5	60	20	14	20	10#9	10	10
69	5	60	20	14	20	10#9	36	24

Chapter 17

SLENDER REINFORCED CONCRETE COLUMNS

17.1 Supplementary Moment due to Column Buckling

In Chapters 15 and 16, short reinforced concrete columns under axial compression and bending were discussed. Effect of slenderness/buckling was presumed insignificant. With such presumption, column analysis is termed: "first order analysis." Buckling occurs due to imperfections of materials, column geometry and loading conditions. As discussed in Chapter 15, ACI Code implements factor α to account for imperfections of short or compact columns while using first order analysis. This factor accounts for accidental eccentricities of 0.10b or 0.10h in tied columns where b and h are column cross–sectional dimensions or 0.05D in spiral columns where D is column diameter.

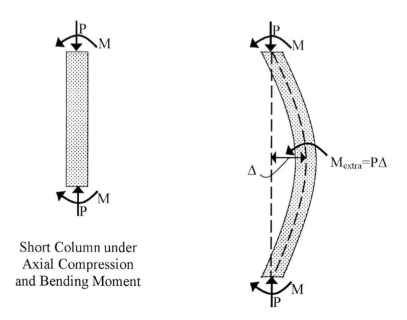

Short Column under
Axial Compression
and Bending Moment

Slender Column under Axial
Compression and Bending Moment with
Supplementary Deformation Moment

Figure 17.1: Behavior of columns under axial compression and bending moment.

With increasing column length or decreasing column section dimensions, the effects of the above mentioned imperfections are exacerbated. Such effects result in column center line deflection or deformations causing an added bending moment on the column section, thereby, decreasing its capacity or strength. To account for such supplementary deformation bending moment, a second order analysis is required. Figure 17.1 illustrates the phenomenon of supplementary bending moment.

17.2 Column Slenderness Ratio

Column slenderness ratio is used to evaluate buckling deformation effect on column capacity. Column slenderness ratio is the ratio of column effective length to radius of gyration ($k\ell_u/r$). Column effective length $k\ell_u$ is the length between points of zero moment or points of inflection. Column unsupported length, ℓ_u, is defined as the distance between points of lateral supports or, typically, floors. k is the effective length factor. Buckling analysis of columns is based on Euler's theory which uses a model column hinged at both ends and buckles/deforms in a shape of ½ sine wave as shown in Figure 17.2. For such model column, effective length = actual or unsupported length and k = 1.0. For columns with other support conditions, the values of k are also shown in Figure 17.2.

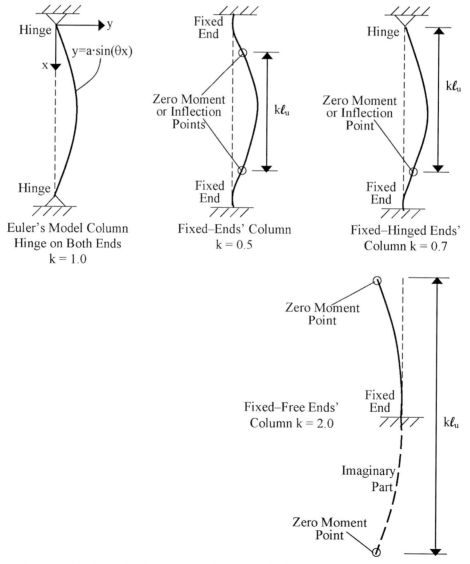

Figure 17.2: Effective length of columns with different support conditions.

Defining the axis of buckling is essential for column analysis. The axis of buckling is member cross–section axis of bending moment caused by buckling deformations. Thus, buckling deformations are perpendicular to the member cross–section axis of bending. The radius of gyration required for slenderness ratio can be computed as the square root of the ratio of moment of inertia about axis of buckling to section area. Slenderness ratio and radius of gyration equations can be mathematically expressed as follows:

$$\text{Slenderness Ratio} = k\ell_{ui}/r_i \tag{17.1}$$

$$\text{Radius of Gyration} = \sqrt{I_{ui}/A} \tag{17.2}$$

Where:

ℓ_{ui} = column unsupported length with respect to axis i.

k = effective length factor for buckling about axis i.

$k\ell_{ui}$ = column effective length for buckling about axis i.

I_i = moment of inertia of column cross–section about axis i.

A = column cross–sectional area.

Typically, every column is analyzed for buckling about its cross–section two principal axes (x and y) as shown in Figure 17.3.

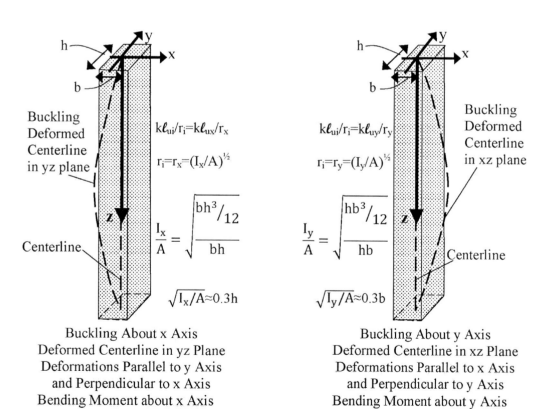

Buckling About x Axis
Deformed Centerline in yz Plane
Deformations Parallel to y Axis
and Perpendicular to x Axis
Bending Moment about x Axis

Buckling About y Axis
Deformed Centerline in xz Plane
Deformations Parallel to x Axis
and Perpendicular to y Axis
Bending Moment about y Axis

Figure 17.3: Buckling axes or planes for a typical rectangular column.

In Figure 17.3, the radius of gyration for rectangular/square columns is shown. For circular columns, the radius of gyration, r, may be taken as 0.25D where D is column diameter.

In order to study column buckling, end supports need to be evaluated (as hinged, fixed or free) for both buckling directions/axes to determine the value of $k\ell_u$ based on Figure 17.2. End supports need not match for both buckling directions. An end support can be fixed for buckling about one axis while hinged for buckling about the other axis. Following evaluation of end supports, the slenderness ratios about both axes, $k\ell_{ux}/r_x$ and $k\ell_{uy}/r_y$, are computed. Column design is based on the larger value of slenderness ratio representing a more critical buckling condition.

Commonly, a column is noted to be fully braced against buckling in the weak direction (the direction with smaller moment of inertia) by an abutting wall while subject to buckling in the strong direction. Also, a lateral support member can brace a column against buckling which may affect the value of unsupported length, ℓ_u, in the weak direction as illustrated in Figure 17.4.

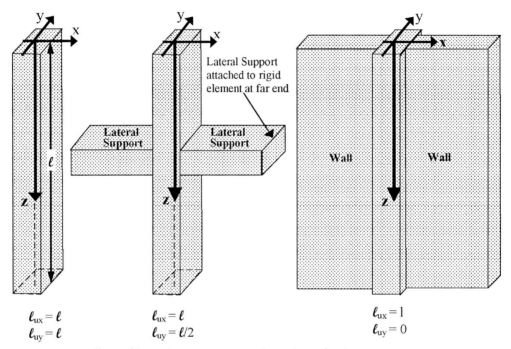

$\ell_{ux} = \ell$
$\ell_{uy} = \ell$

$\ell_{ux} = \ell$
$\ell_{uy} = \ell/2$

$\ell_{ux} = 1$
$\ell_{uy} = 0$

Figure 17.4: Effect of lateral support upon the value of column unsupported length ℓ_u.

17.3 Effect of Sidesway on Column Slenderness

Lateral displacement of column end points increases bucking susceptibility. Such phenomenon is termed sidesway. Sidesway results in an added bending moment via increasing the eccentricity of applied load (Figure 17.5). Thus, columns braced against sidesway are less prone to buckling than columns subjected to sidesway. The effect of sidesway upon column performance is considered in determining the effective length factor (k). As such, k is used as a measure for column susceptibility to buckling taking into account end conditions as well as sidesway as explained schematically in Figure 17.5 where examples of columns subjected to sidesway are included. The concept of single and double curvature is in Section 17.5.1.

There are several methods for reinforcing a column against sidesway as follows:

(a)   The column is a part of a building where lateral forces are resisted by a shear walls, braced frames or moment frames elsewhere within the building termed the lateral force resisting elements (or lateral force resisting system) where they resist and practically eliminate sidesway. Generally, building columns that are not part of the lateral force resisting system are not subjected to or designed for sidesway or lateral forces causing sidesway (e.g. wind and seismic forces).

(b)   The column is in a braced frame where sidesway is inhibited via the braces.

However, columns within moment frames are designed to resist lateral forces and consequential bending moments and potential sidesway. Figure 17.6 presents examples of columns where sidesway is inhibited or uninhibited. A building gridline is a line or row of columns or walls within the building.

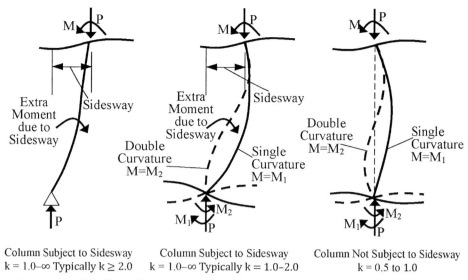

Figure 17.5: Effect of sidesway upon the value of effective length factor (k)

17.4 Use of Alignment Charts for Determination of k

The support conditions shown in Figure 17.2 are mathematical models. A hinge is defined as a support condition that allows rotation but not movement. A fixed end support does not permit rotation or movement. A member free end is allowed to rotate and move. Clearly, a beam–column joint provides some support for the column against rotation and displacement. The level or efficiency of such support is dependent upon the relative rigidity of beams with respect to column. Furthermore, extension of steel into the beam–column joint enhances the level of support significantly. A typical support does not represent a true hinge or a true fixed end but a blend of both. Such typical support presents an incomplete or partial column restraint against rotation and displacement. The degree of column end restraint can be determined via factor ψ. Factor ψ equation and rules specified by ACI Code are as follows:

$$\psi_{Ai} = \frac{\sum^{E_{cc}I_{ci}}/_{l_{uci}}}{\sum^{E_{cb}I_b}/_{l_b}} \tag{17.3}$$

Where:

ψ_{Ai} = column end restraint factor for joint A for buckling about axis i.

Σ represents the sum for all columns meeting at joint A including the column in question.

E_{cc} = modulus of elasticity of column concrete.

I_{ci} = moment of inertia of column section about axis i.

l_{uci} = unsupported length of column for buckling about axis i.

Σ is for all beams meeting at joint A and in the plane of buckling.

E_{cb} = modulus of elasticity of beam concrete.

I_b = moment of inertia of pertinent beam about its horizontal centroidal axis.

l_b = span length of beam between vertical supports.

– A typical column AB has four values of ψ:
 two for joint A for buckling about x and y, ψ_{Ax} and ψ_{Ay} respectively, and two for joint B for buckling about x and y, ψ_{Bx} and ψ_{By} respectively.

– If it is determined that a joint acts very closely to a fixed end, hinge or free end, the value of ψ may be considered as 0, 10 or ∞, respectively. A competent designer is capable of joint design as fixed or hinged where the above values of ψ apply.

For a typical column AB, consequent to the computing ψ_{Ax} and ψ_{Bx}, the value of k_x can be determined. Also, based on ψ_{Ay} with ψ_{By}, the value of k_y is determined. There are two possible procedures of determining k_x and k_y, graphically and analytically/mathematically. The graphical solution is via the alignment charts where the user connects the points of ψ_A and ψ_B by a straight line that intersects the k axis. The value of k is read at the point of intersection. As sidesway affects the value of k, there are two alignment chars introduced by ACI Code, one for non sway frames (k = 0.5 to 1.0) and the other is for sway frames (k = 1 to ∞) as shown in Figures 17.7 and 17.8.

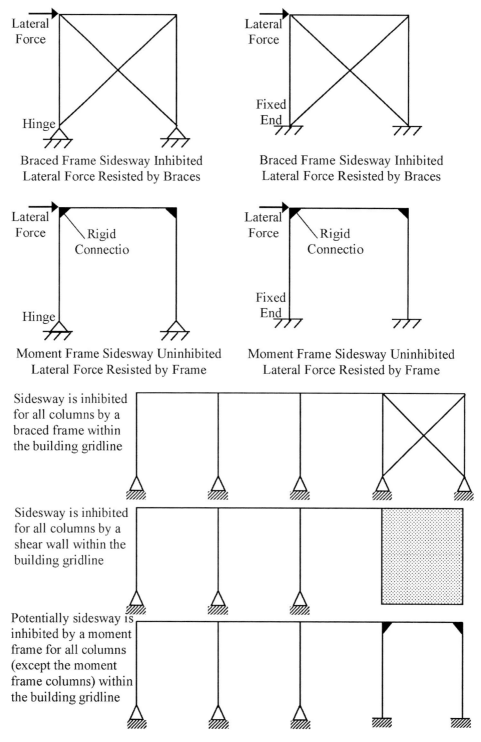

Figure 17.6:

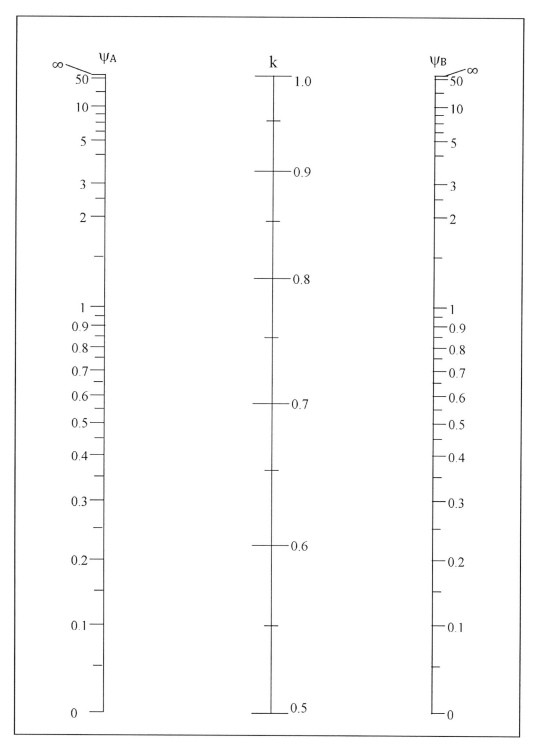

Figure 17.7: Alignment chart for sway frames.

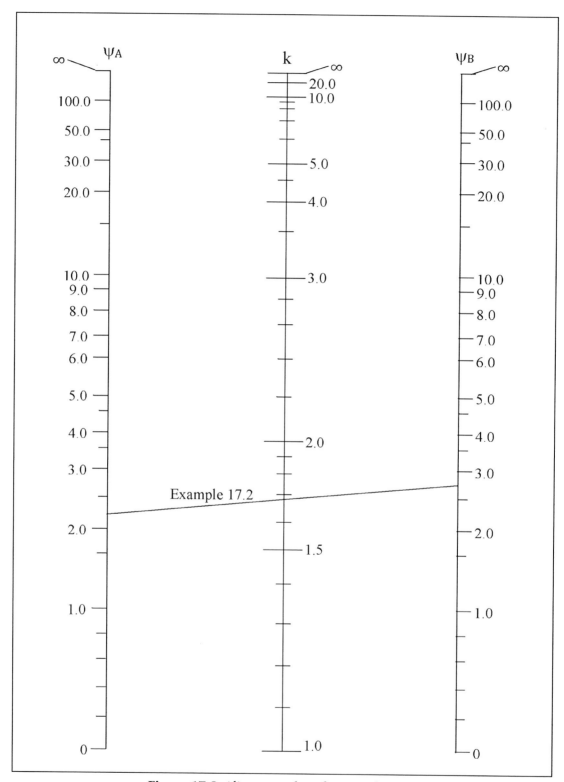

Figure 17.8: Alignment chart for sway frames.

The computation of ψ based on Equation 17.3 requires utilization of the moments of inertia of beams and columns. Beams and columns are likely cracked under ultimate or failure condition per the strength design. For computation of ψ, ACI Code requires the use of E_c (concrete elastic modulus), moments of inertia and section areas as shown below:

Elastic Modulus: $\qquad E_c = 57{,}000\sqrt{f_c'}$

Moments of Inertia:

   Beams $\qquad\qquad$ $I = 0.35I_g$

   Columns $\qquad\qquad$ $I = 0.70I_g$

   Walls – Uncracked $\qquad$ $I = 0.70I_g$ $\qquad$ Walls – Cracked $I = 0.35I_g$

   Slabs $\qquad\qquad$ $I = 0.25I_g$

Areas: $A = A_g$

Where I_g and A_g and the gross moment of inertia and gross section area (assuming no steel reinforcement) determined as illustrated in Chapter 14.

Figure 17.9 is introduced to further illustrate the computation of ψ. Joint A is the top joint of column C1 and the bottom joint of column C2. At joint A, four beams B1, B2, B3 and B4 connect to columns C1 and C2. There are two values of ψ for joint A, ψ_{Ax} for buckling about axis x (buckling in the yz plane) and ψ_{Ay} for buckling about axis y (buckling in the xz plane). For computation of ψ_{Ax}, only beams B1 and B2 are located in the buckling plane (plane yz) cause significant restraint of joint A. Also, for computation of ψ_{Ay}, only beams B3 and B4 are located in the buckling plane (plane xz) are responsible for restraint of joint A. The values of ψ_{Ax} and ψ_{Ay} are used for determination of k for both of columns C1 and C2. Based on the above discussion, the equations for ψ_{Ax} and ψ_{Ay} can be expressed as follows:

$$\psi_{Ax} = 2 \times (E_{cc}I_{gC1x}/\ell_{uC1x} + E_{cc}I_{gC2x}/\ell_{uC2x})/\ (E_{cb}I_{gB1}/\ell_{B1} + E_{cb}I_{gB2}/\ell_{B2})$$

$$\psi_{Ay} = 2 \times (E_{cc}I_{gC1y}/\ell_{uC1y} + E_{cc}I_{gC2y}/\ell_{uC2y})/\ (E_{cb}I_{gB3}/\ell_{B3} + E_{cb}I_{gB4}/\ell_{B4})$$

Where:

   E_{cc} and E_{cb} $\quad$ = as explained for Equation 17.3.

   I_{gC1x} and I_{gC1y} = gross moments of inertia of column C1 about axes x and y, respectively.

   I_{gC2x} and I_{gC2y} = gross moments of inertia of column C2 about axes x and y, respectively.

   ℓ_{uC1x} and ℓ_{uC1y} = unsupported length of column C1 for buckling about axes x (in yz plane) and y (in xz plane), respectively.

   ℓ_{uC2x} and ℓ_{uC2y} = unsupported length of column C2 for buckling about axes x (in yz plane) and y (in xz plane), respectively.

   I_{gB1}, I_{gB2}, I_{gB3} and I_{gB4} = gross moment of inertia of beams B1, B2, B3 and B4 about axes x_{B1}, x_{B2}, x_{B3} and x_{B4}, respectively.

   ℓ_{B1}, ℓ_{B2}, ℓ_{B3} and ℓ_{B4} = length of beams B1, B2, B3 and B4, respectively.

For joints with less than two columns or less than four beams, a zero replaces the terms pertaining to the mission column or beams.

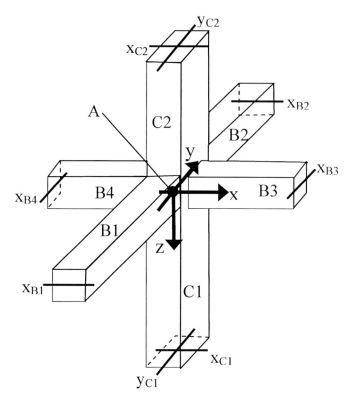

Figure 17.9: Joint A of beams and columns.

As stated earlier, $\psi = 0$, 10 and ∞ for fixed, hinged, and free joints, respectively. Generally, fixed joints are more common in moment frames than other types. Actual hinged joints design may require elaborate steel detailing. However, it is not common that reinforced concrete members are hinged in moment frames.

In lieu of using the interaction diagram to determine the value of k, it is typical for designers to use k = 1.0 for all columns in non–sway frames. For sway frames, designers often utilize values of 2 to 3 for k. Such approximation is to the safe side and avoids the complexity of alignment charts.

17.5 ACI Code Characterization of Columns

ACI Code includes two analysis procedures for slender columns in: (a) non–sway frames, and (b) sway frames. An approximate method is outlined in Section 17.3 for classification of frames as sway or non–sway. ACI Code recommends classification of each building story or level as a sway or non–sway story. Such classification is required for each building story in two perpendicular directions (typically termed x and y directions). Factor Q, termed story stability index, is introduced by ACI Code to be utilized for sway or non–sway classification. Factor Q can be computed using the following equation:

$$Q = \Sigma P_u \Delta_o / V_u \ell_c \qquad\qquad (17.4)$$

Where:

Q = story stability index.

Σ_{Pu} = summation of ultimate axial compression load for all the columns within the story in question.

Δ_o = relative lateral deflection between the top and bottom of the story in question.

V_u = lateral shear force affecting the story in question resulting from the lateral forces of all the stories above the story in question. Vu is of the same load combination (See Chapter 3) as ΣPu above. Generally, Equation 17.4 is applied to the most critical load combination.

ℓ_c = length of columns within the story in question between the center of joints.

Figure 17.10 presents a schematic illustration of the terms of Equation 17.4 for a three story building frame.

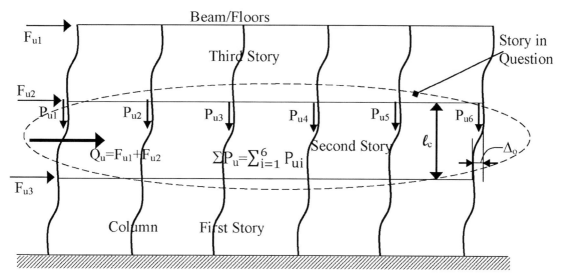

Figure 17.10: Schematic illustration of Equation 17.4 for a three story building frame.

The utilization of Q, story stability index, as a possible method to classify frames as sway or non–sway is outlined by ACI Code as follows:

$Q \leq 5$ Non–Sway Frame $Q > 5$ Sway Frame

With Q ≤ 5, ACI Code considers that sway effect need not be considered in column design as it is included with factor α (see Chapter 15). Each column that is not a part of the lateral force resisting system is typically considered of a non–sway frame by designers assuming adequate lateral support for the story or building. Columns utilized for resistance of lateral loads, such as in moment frames, are generally reviewed based on Equation 17.4 to assess if sway effect is required for column design per ACI Code. Slender columns require second order analysis as explained in Section 17.2. ACI Code criteria for assessing if a column is slender are as follows:

17.5.1 ACI Code Criterion for Slender Columns in Non–Sway Frames

Figure 17.11 illustrates the ultimate condition of a column under magnified load in a non sway frame. ACI Code states that:

$$\text{If } k\ell_u/r \leq 34 - 12(M_1/M_2) \Longrightarrow \text{Slenderness effect may be neglected} \qquad \textbf{(17.5)}$$
$$\text{Column is considered non–slender}$$

Where:

$k\ell_u/r$ = column slenderness ratio as explained earlier.

M_1, M_2 = factored column end moments where $|M_1| \leq |M_2|$

$M_1/M_2 > 0$ if column is in single curvature. In single curvature, each side of the column is continuously subjected to compression or tension with no stress relief resulting in a more critical condition as compared with double curvature.

$M_1/M_2 < 0$ if column is in double curvature. In double curvature, each side of the column is subjected to alternating compression and tension with an inflection point of zero moment/stress resulting in a less critical condition as compared with single curvature.

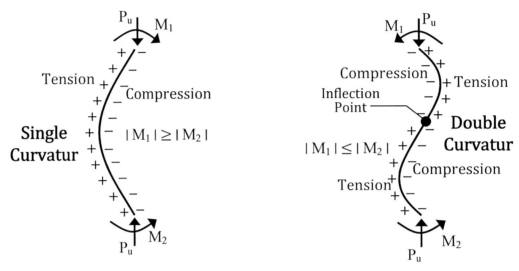

Figure 17.11: Single vs. double curvature in columns in non–sway frames.

17.5.2 ACI Code Criterion for Slender Columns in Sway Frames

For column in sway frames, ACI Code states that:

$$\text{If } k\ell_u/r \leq 22 \implies \text{Slenderness effect may be neglected} \qquad (17.6)$$
$$\text{Column is considered non–slender}$$

Where:

   $k\ell_u/r$ = column slenderness ratio as explained earlier.

17.6 Second Order Analysis of Columns by the Moment Magnifier Method

ACI Code specifies the use of the moment magnifier method for analysis of slender columns. This method increases/magnifies the value of factored (ultimate) bending moment affecting the column to take into account column slenderness. As such, for each load combination of concern, the column is designed for the factored load and magnified factored moment as explained in the following. Load combinations of concern include lateral load effect causing significant bending moment on the column in question.

17.6.1 The Moment Magnifier Method for Slender Columns in Non–Sway Frames

Column analysis assumes that material specifications (f_c' and f_y), column cross–sectional dimensions, and reinforcement are predetermined. The building structure is loaded with factored load. A first order elastic frame analysis is performed to determine the values of factored axial load (P_u) and factored end moments (M_1 and M_2 where $|M_1| \le |M_2|$) affecting the column under consideration. Since columns of non–sway frames are generally not subjected to lateral loads, the sources of M_1 and M_2 are typically dead load and live load. Equation 17.5 is utilized to determine if the column in question is slender or compact/short. For short or compact columns, analysis is performed as discussed in Chapter 16 under the effect of P_u and M_2 (since M_2 is the larger of the two end moments) using symmetrical reinforcement so that the column can resist bending moment M_2 or a smaller moment with alternating direction. If the column under discussion is determined to be slender based on Equation 17.5, then analysis is performed under the effect of P_u and M_c (in lieu of M_2). M_c is the magnified bending moment based on ACI Code provisions computed as follows:

$$M_c = \delta_{ns} M_2 \qquad \text{Non–Sway Magnified Moment}$$

Where:

$$\delta_{ns} = \frac{C_m}{1 - \dfrac{P_u}{0.75 P_c}} \ge 1.0 \qquad \text{Non-Sway Frames Magnification Factor} \qquad (17.7)$$

$$C_m = 0.6 + 0.4 M_1/M_2 \ge 0.4 \qquad \text{Buckling Mode Coefficient}$$

$$P_c = \pi EI/(k\ell_u)^2 \qquad \text{Euler's Buckling Load}$$

$$EI = 0.4 E_c I_g/(1 + \beta_{dns}) \qquad \text{Column Moment of Inertia}$$

β_{dns} = Creep Coefficient for Non–Sway Frames computed as follows:

β_{dns} = Maximum Factored Sustained Axial Load/Maximum Factored Axial Load for the same load combination

As shown in the aforementioned equations, ACI Code presents an estimate for the computation of P_c utilizing adjustment factors of 0.4 and β_{dns} and 0.75 included in the equations for EI and δ_{ns}, respectively. As an example, β_{dns} could be computed as $(1.2DL + 1.6SLL)/(1.2DL + 1.6LL)$ where DL, LL and SLL are dead load, live load and sustained live load.

ACI Code also requires a minimum value of M_2 for slender column analysis as in the following equation:

$$M_{2,min} = P_u(0.6in + 0.03h)$$

Where h is the column cross–sectional dimension in the plane of buckling (perpendicular to buckling/bending axis). Furthermore, P_u, M_1, M_2 and M_1/M_2 as in Section 17.5.1.

For design, columns' cross–sections are first assumed. Analysis is performed. The initial assumptions are then revised if column design is not satisfactory until a satisfactory design is achieved. The analysis method described in this section requires repetition in two perpendicular directions for buckling about x and y. In analysis of columns in non–sway frame, designers conservatively assume k = 1.0 and C_m = 1.0.

Example 17.1:

Compute the design moment and axial force for column AB shown below for each loading case considering that it is a part of a non–sway frame. Consider buckling is in the plane of the paper only with no potential for buckling in the perpendicular plane.

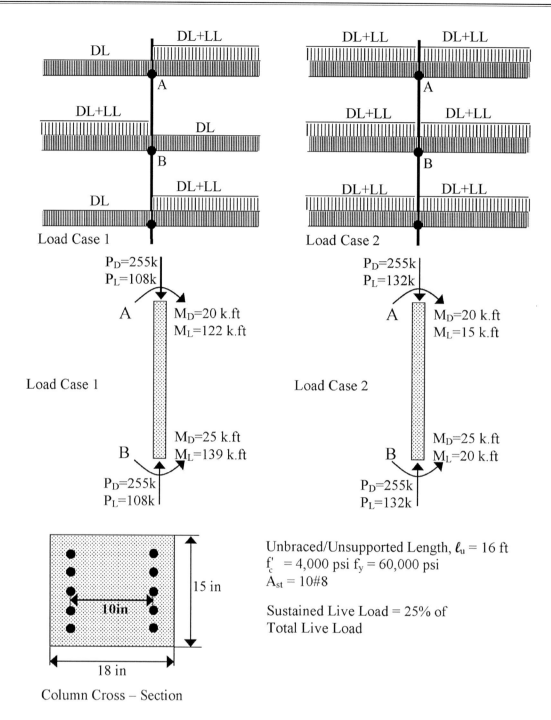

Column Cross – Section

Unbraced/Unsupported Length, ℓ_u = 16 ft
f_c' = 4,000 psi f_y = 60,000 psi
A_{st} = 10#8

Sustained Live Load = 25% of
Total Live Load

The figure accompanying problem 17.1 illustrates live load distribution near column AB to generate maximum live load moment. Load distribution at locations not near to column AB has limited effect. Spans or floors above the column location are assumed fully loaded with dead and live load to generate maximum axial compressive load. The frame of problem 17.1 is presumed to have rigid joints capable of transferring moments between columns and beams. However, the lateral load resisting system of the building does not include the frame containing column AB which only supports gravity loads.

Solution:

Factored Load and Moment

$P_u = 1.2P_D + 1.6P_L$ $M_u = 1.2M_D + 1.6M_L$

Load Case 1	Load Case 2
$P_u = 479$ k	$P_u = 517$ k
$M_1 = 219$ k.ft $M_2 = 252$ k.ft	$M_1 = 48$ k.ft $M_2 = 62$ k.ft

Moment Magnification for Non–Sway Frames

$M_{2min} = P_u(0.6in + 0.03h)$

Load Case 1

$M_{2min} = 479 \times (0.6 + 0.03 \times 15)/12 = 42$ k.ft

$M_2 = 252$ k.ft $> M_{2min} = 42$ k.ft $\Longrightarrow$ OK

Load Case 2

$M_{2min} = 517 \times (0.6 + 0.03 \times 15)/12 = 45$ k.ft

$M_2 = 62$ k.ft $> M_{2min} = 45$ k.ft $\Longrightarrow$ OK

$C_m = 0.6 + 0.4M_1/M_2 \geq 0.4$ and M_1/M_2

Load Case 1

$M_1/M_2 = 219/252 = 0.87 > 0$ Single Curvature

$C_m = 0.6 + 0.4 \times 0.87 = 0.95 \geq 0.4$ OK

Load Case 2

$M_1/M_2 = 48/62 = 0.77 > 0$ Single Curvature

$C_m = 0.6 + 0.4 \times 0.77 = 0.91 \geq 0.4$ OK

If $k\ell_u/r \leq 34 - 12(M_1/M_2)$ $\Longrightarrow$ Slenderness effect may be neglected
Column is considered non–slender

$k\ell_u/r = 1.0 \times (16 \times 12)/(0.3 \times 15) = 43$ applies to both load cases
Where k = 1.0 is a conservative estimate for column in non–sway frames.

Load Case 1

$34 - 12 \times 0.95 = 22.6$

Column is slender for Load Case 1

Load Case 2

$34 - 12 \times 0.91 = 23.1$

Column is slender for Load Case 2

β_{dns} = Maximum Factored Sustained Axial Load/Maximum Factored Axial Load for the same load combination

Load Case 1

$\beta_{dns} = 1.2 \times 255 + 1.6 \times 0.5 \times 108/479 = 0.75$

Load Case 2

$\beta_{dns} = 1.2 \times 255 + 1.6 \times 0.5 \times 132/517 = 0.72$

$$EI = \frac{0.4E_c I_g}{1 + \beta_{dns}} = \frac{0.4 \times 57,000\sqrt{4,000} \times 15 \times \frac{(18)^3}{12}}{1 + \beta_{dns}} = 10.5 \times 10^9/(1 + \beta_{dns}) \text{ lb.in}^2$$

Load Case 1

$EI = 10.5 \times 10^9/(1 + 0.75) = 6.01 \times 10^9$ lb.in^2

$P_c = = \pi EI/(k\ell_u)^2$

Load Case 2

$EI = 10.5 \times 10^9/(1 + 0.72) = 6.11 \times 10^9$ lb.in^2

Load Case 1

$P_c = \pi^2 \times 6.01 \times 10^9/(16 \times 12)^2/1,000 = 1,609$ k

Load Case 2

$P_c = \pi^2 \times 6.11 \times 10^9/(16 \times 12)^2/1,000 = 1,636$ k

$$\delta_{ns} = \frac{C_m}{1 - \dfrac{P_u}{0.75P_c}} \geq 1.0$$

Load Case 1

$\delta_{ns} = 0.95/[1 - 479/(0.75 \times 1,609)] = 1.58$

Load Case 2

$\delta_{ns} = 0.91/[1 - 517/(0.75 \times 1,609)] = 1.59$

Design Axial Compression Load and Bending Moment

Load Case 1	**Load Case 2**
$P_u = 479$ k	$P_u = 517$ k
$M_u = M_c = 1.58 \times 252 = 398$ k.ft	$M_u = M_c = 1.59 \times 62 = 99$ k.ft

Assume $C_m = 1.0$ as a design simplification, the design axial compression and bending are:

Load Case 1	**Load Case 2**
$\delta_{ns} = 1.0/[1 - 479/(0.75 \times 1,609)] = 1.66$	$\delta_{ns} = 1.0/[1 - 517/(0.75 \times 1,609)] = 1.75$
$P_u = 479$ k	$P_u = 517$ k
$M_u = M_c = 1.66 \times 252 = 418$ k.ft	$M_u = M_c = 1.75 \times 62 = 109$ k.ft

For each load case of this example, the methods of analysis and design of columns subjected to axial load and bending of Chapter 16 can be used.

17.6.2 The Moment Magnifier Method for Slender Columns in Sway Frames

Column analysis assumes that material specifications (f'_c and f_y), column cross–sectional dimensions and steel reinforcement are predetermined.

17.6.2.a Design for Lateral Load

A critical factored load combination containing lateral load is selected. The building structure is loaded with the selected factored load. Two first order elastic frame analyses are performed. The first analysis is under the effect of the non–sway (gravity load) portion of factored load. The second analysis is under the effect of the sway (wind or seismic load) portion of factored load. Total factored axial load (P_u) and total factored end moments (M_1 and M_2 where $|M_1| \leq |M_2|$) affecting the column under consideration are determined as the sum of results of both analyses. The end moments resulting from non–sway factored load portion (first analysis) are termed M_{1ns} and M_{2ns}. The end moments resulting from sway factored load portion (second analysis) are termed M_{1s} and M_{2s}. As such, M_{1ns}, M_{1s}, M_{2ns} and M_{2s} are the non–sway portion of M_1, sway portion of M_1, non–sway portion of M_2 and sway portion of M_2, respectively. ACI Code requires column analysis and design under the effect of P_u and the following factored end moments:

$$M_1 = M_{1ns} + \delta_s M_{1s} \tag{17.8}$$

$$M_2 = M_{2ns} + \delta_s M_{2s} \tag{17.9}$$

Where:

M_1, M_2 = factored column end moments due to the load combination under consideration, where $|M_1| \leq |M_2|$,

M_{1ns}, M_{1s}, M_{2ns} and M_{2s} are the non–sway portion of M_1, sway portion of M_1, non–sway portion of M_2 and sway portion of M_2, respectively, and

δ_s = sway frames magnification factor.

There are three methods included in the ACI Code for determination of δ_s described as follows:

Method 1: $\qquad \delta_s = \dfrac{1}{1-Q} \geq 1.0$

Where Q is based on Equation 17.4. ACI Code permits utilization of the above equation if the resulting value of δ_s is ≤ 1.5. Otherwise, Method 2 or Method 3 shall be used.

<u>Method 2:</u>

$$\delta_{ns} = \frac{1}{1 - \dfrac{\sum P_u}{0.75\sum P_c}} \geq 1.0 \qquad\qquad (17.10)$$

Where $\sum$ indicates summation for all the columns within the story under consideration. The method for determining P_c explained in Section 17.6.1 is valid except for the substitution of β_{dns} with β_{ds} determined as shown below:

β_{ds} = Creep Coefficient for Sway Frames computed as follows:

β_{ds} = Maximum Factored Sustained Story Shear/Maximum Factored Story Shear for the story of the column under consideration for same load combination

Commonly designers use $\beta_{ds} = 0$ as gravity load do not contribute significantly to story shear. However, for buildings on slopes where soil pressure can cause sidesway, a more sophisticated analysis is required.

<u>Method 3:</u> A third alternative for determining the value of δ_s is via a second order elastic analysis using the moments of inertia of Section 17.4. This method is generally the most accurate for determining the value of δ_s.

There has been a trend in ACI Code to limit the value of δ_s to 2.5. Story rigidity increase is required if $\delta_s > 2.5$. The story in question is considered a soft story with large deformations that can cause significant irreversible structural or cracking damage.

17.6.2.b Design for Gravity Load

Individual columns in sway frames may buckle under the effect of gravity load. ACI Code requires column analysis in sway frames under the effect of gravity load combinations without lateral loads using the method illustrated in Section 17.6.1. However, k is determined for the column as a part of a sway frame.

Analysis and design of columns in sway frames requires the consideration of various load combinations (Chapter 3) including only gravity load or combined gravity and lateral loads.

Example 17.2

Using the alignment charts, determine the value of effective length factor (k) for column AB shown below presuming that it is a part of a rigidly connected moment frame subjected to sidesway. Columns and beams are of the same concrete type with $f'_c = 4,000$ psi and $E_c = 3,600$ ksi.

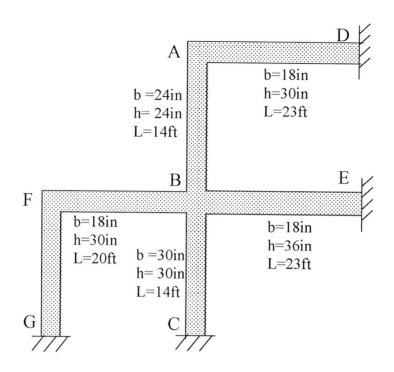

Solution

The frame is assumed to have rigidly connected members (columns and beams). Thus, no effect of hinges needs to be considered in the analysis and computation of ψ_A or ψ_B. Equation 17.3 can be simplified as $E_{cc} = E_{cb}$ since the concretes used for columns or beams have matching properties. For this problem solution, the following table is constructed:

Member	$I_g = bh^3/12$ (in⁴)	I/ℓ (in⁴/ft)
Column AB	27,648	1,975
Column BC	67,500	4,821
Beam AD	40,500	1,761
Beam BE	69,984	3,043
Beam BF	40,500	2,025

Based on Equation 17.3: $\psi_A = 2 \times \dfrac{1,975}{1,761} = 2.24$ $\psi_B = 2 \times \dfrac{1,975+4,821}{3,043+2,025} = 2.68$

Based on the alignment chart for sway frames, effective length factor = k = 1.68 for column AB as illustrated in Figure 17.8.

Example 17.3

Tow one–story frames subjected to gravity and lateral loads as shown. Using the portal frame method, determine the design axial forces and moments for the columns in the frame without using the moment magnification method of the ACI Code. Gravity dead load, W_D = 1.5 k/ft, gravity roof live load, W_{Lr} = 1.2 k/ft, lateral earthquake load, E = 12.0 kips and lateral wind load, W = 5.3 kips.

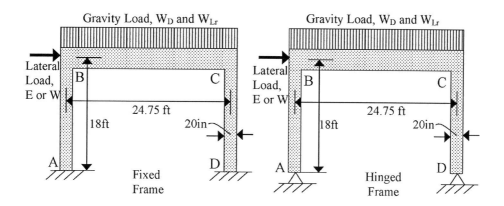

Solution

Using the portal frame method, the axial forces and bending moments due to gravity and lateral loads are as follows utilizing a sign convention generally used for frames. Analysis is performed twice for each frame; with gravity load acting alone first and with lateral load acting alone second. This procedure for determining ultimate load effect upon a structure is termed factored loads as explained in Chapter 5. Addition of factored load effect is consequently utilized as explained in the following.

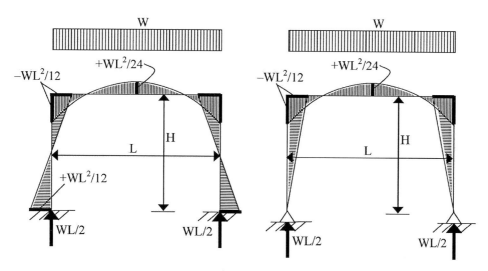

Moments and Axial Forces due to Gravity Loads

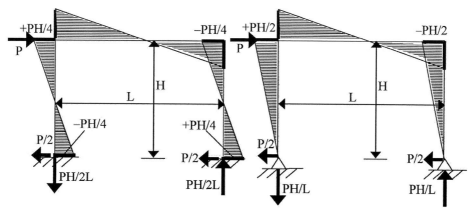

Moments and Axial Forces due to Lateral Loads

Based on the analysis shown above, the following tables can be assembled where the moments and axial forces in each frame column top joint (joints B or C) are shown using the same moment sign convention as in the moment diagrams and assuming that the negative sign for axial load indicates compression:

Load Effect	Dead Load Moment (k.ft)	Dead Load Axial Force (k)	Roof Live Load Moment (k.ft)	Roof Live Load Axial Force (k)	Seismic Load Moment (k.ft)	Seismic Load Axial Force (k)	Wind Load Moment (k.ft)	Wind Load Axial Force (k)
Fixed Frame	−76.6	−26.3	−61.3	−21.0	±54.0	±4.4	±23.9	±1.9
Hinged Frame	−76.6	−26.3	−61.3	−21.0	±108.0	±8.8	±47.7	±3.8

The load combinations of interest (see Chapter 3) for the frames of Example 17.3 are as follows where the non–existing loads are not included:

$U2 = 1.2D + 0.5Lr$

$U3 = 1.2D + 1.6Lr \pm 0.8W$

$U4 = 1.2D \pm 1.6W + 0.5Lr$

$U5 = 1.2D \pm 1.0E$

$U6 = 0.9D \pm 1.6W$

$U7 = 0.9D \pm 1.0E$

Considering the above load combinations, the following tables can be assembled for the fixed and hinged frames where column moments and forces at joints B or C are used. For the fixed frame, joints B or C are under the same design axial load and moment as joints A or D. In the hinged frame, joints B or C are under more critical axial load and moment than joints A or D in the hinged frame (see the diagrams for moments and axial forces in the previous page).

Moments and Forces at joints B or C in the Fixed Frame:

Load Comb.	U2	U3	U4	U5	U6	U7
(−) Moment (k.ft)	−122.6	−209.1	−160.8	−145.9	−107.2	−122.9
Axial Force (k)	−42.1	−66.7	−45.1	−36.0	−26.7	−28.1
(+) Moment (k.ft)	−122.6	−170.9	−84.3	−37.9	−30.7	−14.9
Axial Force (k)	−42.1	−63.6	−39.0	−27.2	−20.6	−29.3

Moments and Forces at joint B or C in the Hinged Frame:

Load Comb.	U2	U3	U4	U5	U6	U7
(–) Moment (k.ft)	–122.6	–228.2	–198.9	–199.9	–145.3	–176.9
Axial Force (k)	–42.1	–68.2	–48.1	–40.4	–29.8	–32.5
(+) Moment (k.ft)	–122.6	–151.8	–46.3	+16.1	+7.4	+39.1
Axial Force (k)	–42.1	–62.1	–36.0	–22.8	–17.6	–14.9

In the above table, (–) implies that the sign used for seismic load and wind is negative in the load combination in question and vice versa. Seismic and wind loads have alternating direction. Thus, each is applied twice for every related load combination as the tables above indicate. Without using the moment magnification method, it appears that U3 with (–) for the fixed frame is the most critical load combination. Similarly, for the hinged frame, U3 with (–) is critical. Furthermore, U7 with (+) is also critical since a reversal of moment sign or directio n requires care in steel bar placement. Thus, the design moments and axial forces for both frames are as follows: (a) for the fixed frame: $M_u = -209.1$ k.ft and $P_u = -66.7$ k, and (b) for the hinged frame: case 1: $M_u = -228.2$ k.ft and $P_u = -68.2$ k, and case 2: $M_u = +39.1$ k.ft and $P_u = -14.9$ k.

Note: most designers use columns with symmetrical reinforcement as a precaution against reversal of moment direction and for design simplification.

Example 17.4

Determine the design bending moments for the frames of Example 17.3 using the moment magnifier method of ACI Code assuming that column dimensions are width b = 12 in and height h = 20 in, k = 2.0 for the hinged frame and 1.5 for the fixed frame and $f'_c = 4,000$ psi.

Solution

First, one needs to determine if the columns in these frames are slender:

For the fixed frame: $k\ell_u/r = 1.5 \times 18 \times 12/(0.29 \times 20) = 56 > 22 \implies$ columns are slender

For the hinged frame: $k\ell_u/r = 2.0 \times 18 \times 12/(0.29 \times 20) = 74 > 22 \implies$ columns are slender

Where $0.29h = 0.29 \times 20$ in is the radius of gyration of the column in the plane of the paper as shown below:

$$r = (I/A)^{1/2} = ([bh^3/12]/(bh))^{1/2} = 0.29h$$

This example uses Equation 17.10 for calculating δ_s in Section 17.6.2. ΣP_c is for all the columns in the frame contributing to resistance of lateral loads in the direction of interest. For Example 17.3, there are two columns, for that ΣP_c calculation is as follows:

$$P_c = \pi EI/(k\ell_u)^2 \qquad \text{Euler's Buckling Load}$$

$$EI = \frac{0.4E_cI_g}{1+\beta_{ds}} \qquad \text{Column Moment of Inertia}$$

β_{ds} = Maximum Factored Sustained Story Shear/Maximum Factored Story Shear for the story of the column under consideration for same load combination

β_{ds} = 0 for this example since gravity load is not causing any lateral forces upon the frames in question.

For the fixed frame:

$$EI = \frac{0.4E_cI_g}{1+\beta_{ds}} = \frac{0.4 \times 57,000\sqrt{4,000} \times 12 \times (20)^3/12}{1} = 1.15 \times 10^{10} \text{ lb.in}^2$$

$$\Sigma P_c = 2\times\pi EI/(k\ell_u)^2 = 2\times\frac{\pi^2\times1.15\times10^{10}}{(1.5\times18\times12)^2}/1,000 = 2,161\text{ k}$$

$$\delta_s = \frac{1}{1-\dfrac{\Sigma P_u}{0.75\Sigma P_c}} = \frac{1}{1-\dfrac{\Sigma P_u}{1,621}} \geq 1.0$$

For the hinged frame:

$$EI = \frac{0.4E_cI_g}{1+\beta_{ds}} = \frac{0.4\times57,000\sqrt{4,000}\times12\times(20)^3/12}{1} = 1.15\times10^{10}\text{ lb.in}^2$$

$$\Sigma P_c = 2\times\pi EI/(k\ell_u)^2 = 2\times\frac{\pi^2\times1.15\times10^{10}}{(2.0\times18\times12)^2}/1,000 = 1,216\text{ k}$$

$$\delta_s = \frac{1}{1-\dfrac{\Sigma P_u}{0.75\Sigma P_c}} = \frac{1}{1-\dfrac{\Sigma P_u}{912}} \geq 1.0$$

The following tables can assist in calculating δ_s utilizing Method 2 in Section 17.6.2 for the sway frames magnification factor:

Magnified Axial Forces and δ_s in the Fixed Frame:

Load Comb.	U2	U3	U4	U5	U6	U7
ΣP_u (k)	84.2	133.4	90.2	76.0	53.4	56.2
Σ_s	1.05	1.09	1.06	1.05	1.03	1.04

Magnified Axial Forces and δ_s in the Hinged Frame:

Load Comb.	U2	U3	U4	U5	U6	U7 (–)	U7 (+)
ΣP_u (k)	84.2	136.4	96.2	80.8	59.6	65.0	29.8
δ_s	1.10	1.18	1.12	1.10	1.07	1.08	1.03

$k\ell_u/r = 74$ for the column in the hinged frame is somewhat high and resulted in a moment magnification factor of 1.18 which is also relatively high indicating possible large deformations for the columns under lateral loads. It is generally recommended that column slenderness ratio not exceed about 60 to 70. With the above tables, one can determine the design moments for the columns in the fixed and hinged frame:

For the fixed frame: it is clear that load combination U3 is the most critical one.

For load combination U3:

$M_{ns} = 1.2\times76.6 + 1.6\times61.3 = 190\text{ k.ft}$ $M_s = 0.8\times23.9 = 19.1\text{ k.ft}$

$M_u = M_{ns} + \delta_s M_s = 190 + 1.09\times19.1 = 211\text{ k.ft}$ tension at the column interior side.

$P_u = 66.7\text{ k}$ compression

For the hinged frame: there are two critical load combinations U3 and U7(+).

For load combination U3:

$M_{ns} = 1.2\times76.6 + 1.6\times61.3 = 190\text{ k.ft}$ $M_s = 0.8\times47.7 = 38.2\text{ k.ft}$

$M_u = M_{ns} + \delta_s M_s = 190 + 1.18 \times 38.2 = 235$ k.ft tension at the column interior side.

$P_u = 68.2$ k compression

For load combination U7(+):

$M_{ns} = -0.9 \times 76.6 = 68.9$ k.ft $M_s = +1.0 \times 108 = 38.2$ k.ft

$M_u = M_{ns} + \delta_s M_s = -68.9 + 1.03 \times 108.0 = 42.3$ k.ft tension at the column exterior.

$P_u = 14.9$ k compression

Design is required to take into account the above critical load combinations. It is clear from Examples 17.3 and 17.4 that using fixed–end frames is advantageous for reducing the effect of lateral loads.

17.7 Concluding Remarks

Beams connected to columns in sway frames require design for the magnified column moments determined as explained earlier in this chapter.

For column design in non–sway or sway frames, column dimensions are first assumed. Analysis and design are performed. Column detailing is utilized to achieve optimum results. Column dimensions are reassumed if the design is not satisfactory (column section is too large or too small for the applied load or moment).

Test your knowledge

1. What is the minimum assumed eccentricity for short columns by ACI Code?

2. What is the second order analysis of columns?

3. What is the ideal Euler's column based on which buckling theory was developed?

4. How is column susceptibility to buckling estimated or assessed?

5. How many planes of buckling exist for each column? What are these plane?

6. What is the difference between braced and unbraced frames?

7. List three methods of lateral bracing against sidesway in frames.

8. What are the principles of alignment charts?

9. What does y in the alignment charts represent?

10. What are the upper and lower bounds of k for braced and unbraced columns in the alignment charts?

11. What is the recommended ψ value in alignment charts for a fixed–end joint and for a hinged joint?

12. How many values of ψ are used for each column joint?

13. Why is column gross moment of inertia multiplied by 0.70 while beams gross moment of inertia is multiplied by 0.35 for use in the alignment charts?

14. What is the second order analysis method of column buckling adopted by ACI Code?

Determine the magnified design moment and axial force for the braced columns listed below where h is column dimension in the plane of buckling:

No.	M_{D1}-M_{D2} (k.ft)	M_{L1}-M_{L2} (k.ft)	P_D (k)	P_L (k)	Sustai-ned LL	kl_u (ft)	Cross Section b×h (in)	Curva-ture	f'c (psi)
15	110-150	60-92	260	140	0.25	14	12×12	single	3,000
16	130-175	110-160	370	209	0.30	16	12×14	double	3,000
17	180-210	140-180	450	320	0.15	18	15×15	single	3,500
18	190-200	80-108	430	140	0.20	18	12×16	single	4,000
19	210-260	80-100	500	120	0.40	20	14×18	single	4,000
20	120-140	40-75	320	100	0.10	20	12×16	double	4,500
21	70-80	30-35	220	80	0.20	16.5	10×12	single	2,500
22	180-190	160-180	700	410	0.35	24	24×30	single	5,000
23	130-150	70-90	330	180	0.0	17	12×16	double	4,200
24	110-140	50-75	260	140	0.20	15	12×12	single	3,000

Determine the values of kl_{ux}/r_x (buckling about x in the yz plane) and kl_{uy}/r_y (buckling about y in the xz plane) for column C1 utilizing the interaction diagram in Figure 17.8 for sway frames using the information in the following table and figures assuming that the same type of concrete (same f'_c) is used for columns and beams:

No.	C1 bxh(in)	C2 bxh(in)	C3 bxh(in)	B1 bxh(in)	B2 bxh(in)	B3 bxh(in)	B4 bxh(in)	B5 bxh(in)	B6 bxh(in)	B7 bxh(in)	B8 bxh(in)
25	12x12	12x12	15x15	12x20	12x20	12x18	12x18	14x20	14x20	12x18	12x18
26	12x15	12x12	15x15	12x24	12x24	12x16	12x16	14x24	14x24	12x16	12x16
27	14x14	12x12	16x16	12x18	12x20	14x18	14x18	12x18	12x20	14x18	14x18
28	14x18	12x12	16x20	14x20	14x20	14x18	14x18	14x24	14x20	12x18	12x18
29	12x16	12x14	12x18	12x20	12x20	16x16	16x16	12x24	12x24	18x18	18x18
30	14x14	14x14	14x14	12x20	14x20	12x18	12x18	14x20	14x20	12x18	12x18
31	18x18	16x18	18x24	18x18	18x18	18x18	18x18	18x20	18x20	24x20	24x20
32	18x20	18x18	18x22	14x20	14x20	16x18	16x18	14x22	14x22	16x20	16x20
33	12x18	12x14	12x20	12x20	N/A	18x18	N/A	12x20	12x20	18x18	18x18
34	16x16	14x14	18x18	16x20	N/A	16x18	N/A	18x20	N/A	18x20	N/A

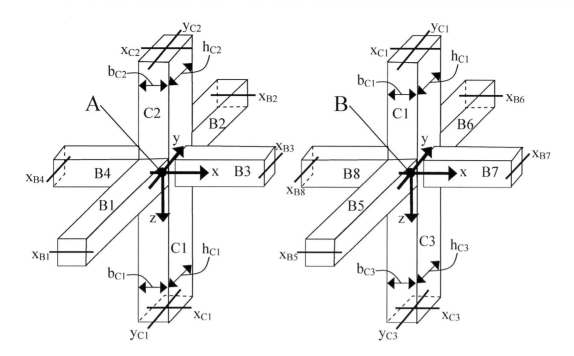

35. Redo Problem 25 assuming that beams B3, B4, B7 and B8 are nonexistent but a 5 in concrete slab is connected to column C1. Use the provisions of T-beams in ACI Code that the effective slab section extends 8 times the slab thickness on each side of the column.

36. Redo Problem 26 assuming that beams B1, B2, B5 and B6 are nonexistent but a 6 in concrete slab is connected to column C1. Use the provisions of T-beams in ACI Code that the effective slab section extends 8 times the slab thickness on each side of the column.

37. Redo Problem 27 assuming that beams B1, B2, B5 and B6 are nonexistent but a 4 in concrete slab is connected to column C1. Use the provisions of T-beams in ACI Code that the effective slab section extends 8 times the slab thickness on each side of the column.

38. Redo Problem 28 assuming that beams B3, B4, B7 and B8 are nonexistent but a 7 in concrete slab is connected to column C1. Use the provisions of T-beams in ACI Code that the effective slab section extends 8 times the slab thickness on each side of the column.

Determine the values of design (factored) moment and axial force for the columns utilizing the moment magnifier method of ACI Code assuming that bucking is one direction only and using the pertinent load combinations by ACI Code. Use $P_u/0.75P_c$ for the column in question in Equation 17.10 in lieu of $\Sigma P_u/0.75P_c$ and use $\beta_{ds} = 0$.

No.	M_D (k.ft)	P_D (k)	M_L (k.ft)	P_L (k)	M_{Lr} (k.ft)	P_{Lr} (k)	M_W (k.ft)	P_W (k)	M_E (k.ft)	P_E (k)	kl_u/r	$b{\times}h$ (in)	f'_c (psi)
39	80	30	65	25	NA	NA	25	8	65	15	48	14x16	3,000
40	60	20	45	15	NA	NA	20	10	55	25	55	12x18	3,500
41	70	190	50	140	NA	NA	85	20	125	55	60	18x18	4,000
42	100	300	80	200	NA	NA	125	30	195	85	40	20x24	4,500
43	55	400	35	250	NA	NA	95	50	120	75	38	18x22	4,000
44	95	280	65	150	NA	NA	45	20	105	35	45	15x28	3,800
45	70	180	45	120	NA	NA	35	7	85	10	70	14x18	4,000
46	80	30	NA	NA	65	25	25	8	65	15	48	16x16	3,000
47	70	190	NA	NA	25	75	45	12	75	18	50	14x18	5,000
48	40	120	NA	NA	20	60	85	9	105	18	65	12x20	5,500

The pertinent load combinations for Problems 39 – 48 to be selected from the following ACI Code equations:

$U1 = 1.4(D + F)$

$U2 = 1.2(D + F + T) + 1.6(L + H) + 0.5 \text{ (Lr or S or R)}$

$U3 = 1.2D + 1.6(\text{Lr or S or R}) + (\text{L or} \pm 0.8W)$

$U4 = 1.2D \pm 1.6W + L + 0.5(\text{Lr or S or R})$

$U5 = 1.2D \pm 1.0E + L + 0.2S$

$U6 = 0.9D \pm 1.6W + 1.6H$

$U7 = 0.9D \pm 1.0E + 1.6H$

Chapter 18

REINFORCED CONCRETE FOUNDATION

18.1 Introduction

Footings or foundation are the elements that transfer loads from buildings or structures to the underlying soil. Compressive strength of concrete typically ranges between 300 and 1,200 ksf (15 to 60 MPa) while bearing (surface compression loading) strength of soil typically ranges between 3 and 12 ksf (0.15 to 0.6 MPa). As such, footings spread building loads over an area of soil so that bearing pressure is safely lower than bearing strength. It is important that the bearing pressure of footings or foundations upon the underlying soil not exceed allowable bearing capacity. Furthermore, soil settlement due to footing load is required to be within acceptable limits. Also, differential settlement among footings supporting the same structure needs to remain within acceptable limits. To limit differential settlement among footings, similar bearing stress is used for design of all the footings. Settlement causes unnecessary additional loads and stresses within the structure.

Reinforced concrete is the optimum material selection for footings. Concrete mixture can be designed to resist any potential harmful chemicals of the bordering soil. Concrete clear cover over reinforcing steel bars in footings is 3 in (80 mm) per ACI Code so that adequate protection is accomplished. Reinforced concrete footings can be designed to resist loading caused stresses as is the case for concrete beams, slabs and columns.

Stresses result in footings as load transfer from the structure to the underlying soil take place. Footings are generally exposed to several types of stress caused ultimate conditions/failure modes:

1. Punching shear caused by column load.

2. One–way shear due to soil pressure.

3. Bending moment.

4. Bearing load.

Footings are designed for the above mentioned failure modes. Such design includes footing dimensions (particularly thickness), concrete strength as well as steel reinforcement. Weight of footing and soil on top of footing are directly transferred to the underlying soil, thus, causing no moment or shear stresses within the footing.

18.2 Types of Footing

Footings are classified into different types depending on their function including wall or continuous footing as shown in Figure 18.1. Another common type is isolated or single footing used for individual columns (Figure 18.1). Combined footing is used for supporting two or more adjacent columns (Figure 18.2). They are utilized (a) near property line, (a) when two adjacent columns are very close, (3) if soil quality underneath a column is questionable.

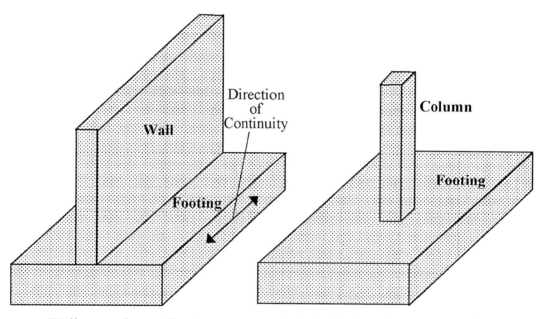

a. Wall or continuous footing.   b. Individual or single column footing.

Figure 18.1: Schematic diagram of wall or continuous footing.

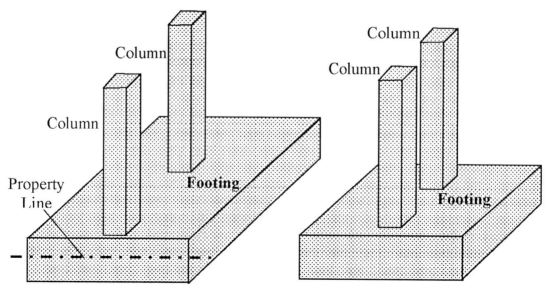

a. Column near property line.   b. Two close adjacent columns.

Figure 18.2: Schematic of combined footing for two columns.

Other types of footing include mat foundation and pile cap. Mat foundation is a combined footing for all the columns and walls supporting the structure. Mat foundation is typically used for buildings with large loads or for sites with low soil bearing capacity. Pile cap is a thick concrete element that transfers loads from the structure to piles extending into the soil.

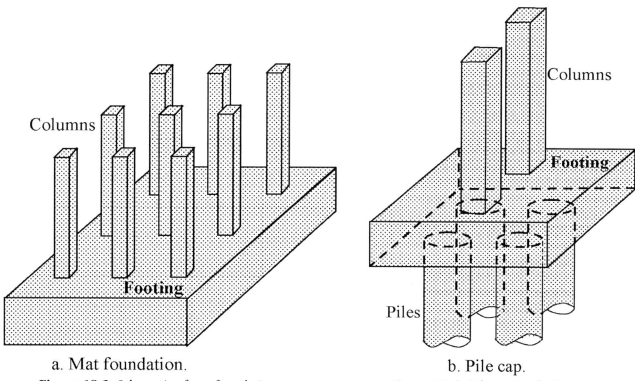

a. Mat foundation.
Figure 18.3: Schematic of mat foundation.

b. Pile cap.
Figure 18.4: Schematic of pile cap.

18.3 Allowable Bearing Capacity of Soil

Prior to structural design, footing area and depth need to be determined. Such determination is based upon a project soil report that contains footing design guidelines. Depth is measured from soil surface to the bottom of footing or bearing plane. Depth in warm climates is typically ≥ 1 ft (300 mm) from finished grade (soil surface). In areas subject to freezing, footing depth is required to be greater than frost depth 3–4 ft (1–1.5 m). The project soil report also includes guidelines for adjusting allowable bearing capacity as footing width or depth increases as well as guidelines for reducing settlement and estimates for projected settlement. Generally, allowable bearing capacity of soil (q_a) is determined as the ultimate bearing capacity divided by a factor of safety of 2.5 to 3.0 for gravity loads (q_{ult}/FS). Allowable bearing capacity for foundation soil or rock typically ranges between 1,000 and 8,000 psf (7 to 60 MPa) depending upon soil type. The International Building Code (IBC) allows design of building or structures on non–expansive soil without a soil report assuming that minimum values of q_a are utilized. These minimum values are included in a table within the IBC and range between 1,000 and 2,000 psf (7 to 15 MPa).

Allowable Stress Design method (ASD) is used in geotechnical engineering. No gravity load factors are used. Thus, for determining the required footing area due to gravity load, the following equation may be used:

$$Areq = \frac{DL+LL}{q_a}$$

(18.1)

Where:

A_{req} = required footing area due to gravity load,

DL,LL = dead and live load, respectively, and

q_a = allowable soil bearing capacity.

Dead and live loads used in computing A_{req} are the superimposed load (including footing weight and soil on footing) upon the footing.

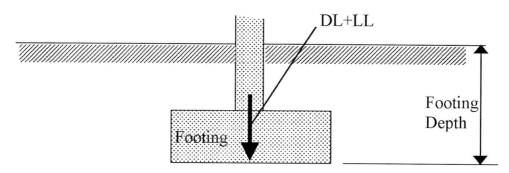

Figure 18.5: Loaded footing.

Under concentric axial load, soil bearing pressure underneath footings is assumed uniform for design purposes. However, studies have shown that soil pressure on footings is not uniform and varies based on soil type as shown in Figure 18.6.

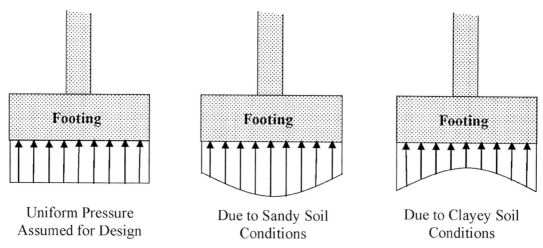

Uniform Pressure
Assumed for Design

Due to Sandy Soil
Conditions

Due to Clayey Soil
Conditions

Figure 18.6: Soil pressure on footings.

Lateral load effect on footing is generally axial load and bending moment that cause a variable soil pressure. It is generally assumed to be linearly variable as in Figure 18.7. The footing is studied under the effect of the following ASD load combinations:

(1)   DL ± (W or 0.7E)

(2)   DL ± (W or 0.7E) + 0.75LL

(3) $0.6DL + W$

(4) $0.6DL + 0.7E$

Where W and E = wind and earthquake loads, respectively, determined based on ASCE 7 as discussed in Chapter 3. Other loads are not shown in the ASD load combinations above. Also, LL is assumed to include roof live load.

The resulting maximum soil bearing pressure cannot exceed q_a for a safe design per IBC Code. Two possible situations could result due to the effect of gravity and lateral load upon footing:

—   Bearing on the entire footing.

—   Bearing on a portion of the footing. This situation is not recommended since it may cause instability and possible excessive deformations.

Both situations are illustrated in Figures 18.7 and 18.8.

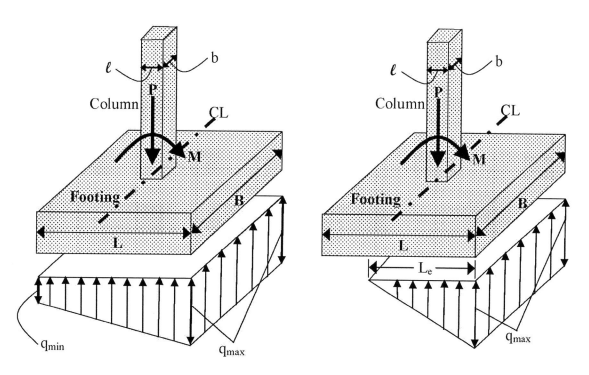

a. Bearing on the entire footing   b. Bearing on a portion of the footing

Figure 18.7: Soil pressure on footings.

The equations for the bearing situations of Figure 18.7 are shown in Figure 18.8 in a two dimensional representation. Case (b) of Figure 18.8 occurs if the maximum tension stress caused by moment exceeds the compression stress caused by axial force. As tension cannot exist between the footing and soil, soil reaction to load and moment is redistributed over a portion of the footing so that tension stress is zero.

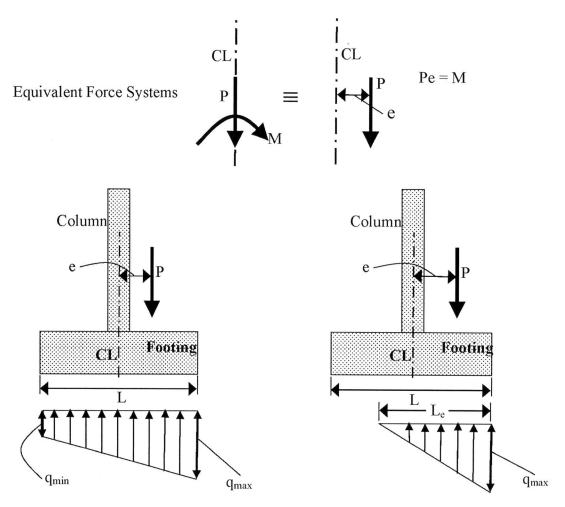

Case (a): e ≤ L/6

$q_{max} = P/A + M/S = P/BL + 6Pe/BL^2$
$q_{min} = P/A + M/S = P/BL + 6Pe/BL^2$

Case (b): e > L/6

$q_{max} = P/A + M/S = P/BL + 6Pe/BL^2$
$q_{max} = 2P/BL_e$ where Le = 3×(L/2 − e)

Figure 18.8: Soil pressure on footings.

In addition to studying concrete footings for soil bearing pressure, overturning and sliding under the effect of lateral load requires analysis to verify that safety factor ≥ 1.5 under the effect of the appropriate load combination per the IBC Code.

18.4 Analysis and Design of Wall/Continuous Footings

For wall or continuous footings, a representative 1 ft linear width of the footing is studied. Footing depth is determined as discussed in Section 18.3. Initial footing width is computed based on the following equation:

$$L = \frac{DL+LL}{q_a}$$

(18.2)

Where:

L = required footing width in ft.

DL,LL = dead and live load (lb/ft or k/ft) including footing and soil on footing; where footing depth and thickness are initially assumed.

q_a = allowable soil bearing capacity based on the soil report in psf or ksf.

Equation 18.2 encompasses footing design from a geotechnical/soil engineering standpoint based on ASD. Consequently, footing structural design can be accomplished using the strength design method of reinforced concrete. Wall footing analysis and design are required for three modes of failure: shear force, bending moment and bearing. Shear force and bending moment analyses of wall footing under concentric load are presented in Figure 18.9 showing footing cross–sections. For shear design, the critical shear planes are at distances equal to d (footing effective depth) from faces of wall. For moment design, the critical bending planes are at wall faces. The weight of footing and soil on footing are typically not included for structural design. They are transferred directly to the underlying soil via small compressive stresses perpendicular to the footing surface without causing bending or shear stresses.

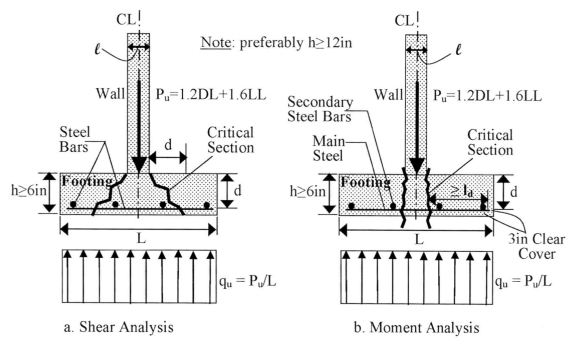

a. Shear Analysis   b. Moment Analysis

Figure 18.9: Soil pressure on footings.

Based on Figure 18.9, the factored (ultimate) shear force affecting the critical section may be computed as follows:

$$Q_u = q_u \times [(L - \ell)/2 - d] \tag{18.3}$$

Where:

Q_u = ultimate or factored shear force in lb/ft or k/ft.

$q_u = P_u/L = (1.2DL + 1.6LL)/L$ = net factored soil bearing pressure (psf or ksf) including wall forces only while disregarding the footing and soil on footing.

L, ℓ = footing width and wall thickness, respectively.

d = footing effective depth = h − 3 in − d_b/2.

h = footing thickness not less than 6 in per ACI Code. Designers recommend using 12 in as minimum footing depth.

3in = concrete clear cover over steel bars in footings per ACI Code.

d_b = diameter of steel bars in the footing.

Footings are typically not reinforced against shear. As such, concrete is required to resist all the shear force of Equation 18.3 utilizing the following equation:

$$\phi V_n = 0.75 \times 2\sqrt{f'_c} \times 12d \qquad (18.4)$$

Where:

ϕ = 0.75 = shear capacity reduction factor of concrete.

V_n = nominal shear resistance of wall footing in lb/ft.

Equation 18.4 assumes units of inch and pound. The critical section dimensions of the wall footing are 12in×d. Where 12in is utilized as a representative linear foot of the footing and d is footing effective depth as explained above. Based on Equations 18.3 and 18.4, minimum footing effective depth can be computed based on shear resistance as shown below:

$$\phi V_n \geq Q_u$$

or $\quad 0.75 \times 2\sqrt{f'_c} \times 12d \geq q_u \times [(L - \ell)/2 - d]$

or $\quad d = \dfrac{q_u\,(L - \ell)/2}{18\sqrt{f'_c} + 1} \qquad (18.5)$

Where Equation 18.5 uses inch and pound as units.

An iterative process is employed for Equations 18.2 and 18.5. Footing thickness is initially assumed in Equation 18.2 and verified via Equation 18.5. This process is repeated until a suitable value of footing thickness (h) is achieved.

Consequent to computing the required footing effective depth for shear resistance, flexural design can be performed to determine the required steel reinforcement as follows:

$$M_u = q_u \times [(L - \ell)/2]^2/2$$

or $\quad M_u = q_u(L - \ell)^2/8 \qquad (18.6)$

Where M_u is the factored (ultimate) bending moment (lb.ft/ft or k.ft/ft) affecting the critical footing section at the faces of wall (Figure 18.9). Reinforcement for the footing can be calculated as explained in flexural design of beams and slabs (Chapters 4 thru 8). Alternatively, the following approximate equation can be used to compute the area of steel reinforcement for bending resistance in wall footings:

$$A_s \approx M_u/(0.9f_y \times 0.9d)$$

or $\quad A_s \approx M_u/0.8df_y \qquad (18.7)$

In Equation 18.7, A_s is the required area of steel per linear ft of the footing (in²/ft). ϕ = 0.9 is used. Furthermore, the moment arm (d − a/2) is assumed equal to 0.9d. Such assumptions are typical for concrete slabs and footings as well as two dimensional elements. ACI Code does not require minimum reinforcement in concrete footings. Typically, prudent designers include minimum reinforcement for concrete footings in the order of 0.0018 of gross footing area similar to the minimum reinforcement in concrete slabs for shrinkage and cracking resistance. Furthermore, steel bars used for footing reinforcement have to be fully developed at the critical bending section at wall face as shown in Figure 18.9. In absence of adequate development length, steel bars may be terminated with a hook. The following equation needs to be satisfied for proper bending design for the case of straight bars:

$$l_d \geq (L - \ell)/2 - 3 \text{ in} \tag{18.8}$$

Where:

l_d = required tension development length of steel bars used for bending moment computed as illustrated in Chapter 13.

3in = concrete clear cover over steel bars in concrete footings as ACI Code requires for elements poured against soil.

Another important item of wall footing design is bearing or transfer of load from the wall to the footing. Load transfer between the wall and footing is a critical issue if the wall is made of improved quality concrete compared to the footing. Bearing strength of transfer area of the footing and wall can be computed as follows:

$$P_n = 0.85 f_c' A_1 \sqrt{\frac{A_2}{A_1}} + A_{sd} f_y \text{ not to exceed } 1.7 f_c' A_1 + A_{sd} f_y \tag{18.9}$$

Where:

P_n = bearing load capacity of the transfer area of the footing.

$A_1 = 12 \times \ell$ = footing bearing area or contact area of the footing and wall.

A_2 = lower cone base area based on 2H:1V side slope (Figure 18.10).

A_{sd} = area of steel dowels or connectors connecting the wall to the footing.

f_c' and f_y = specified compressive strength of the footing and yield strength of the dowels, respectively.

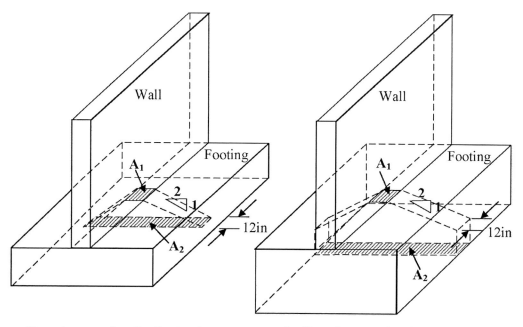

a. Slope intercepting the footing base.   b. Slope intercepting footing sides.

Figure 18.10: Determination of areas A_1 and A_2.

Other ACI Code provisions regarding load transfer between the wall and footing include:

1. Capacity reduction factor for bearing is equal to 0.6 ($\phi = 0.6$).

2. For use in Equation 18.9, the value of $\sqrt{\dfrac{A_2}{A_1}}$ may not exceed 2.0 ($\sqrt{\dfrac{A_2}{A_1}} \leq 2.0$). The second part of Equation 18.9 (not to exceed $1.7f'_c A_1 + A_{sd}f_y$) is used to fulfill this code requirement.

3. Area of steel bar dowels shall be adequate to transfer or support the factored compressive between the wall and footing in excess of concrete bearing strength as indicated in Equation 18.9.

4. Dowel bars shall be designed to resist any tensile force or tension or compression due to bending transferred between the wall and footing.

5. For shear force transfer between the wall and footing, the following shear friction equation at the wall–footing connection needs to be satisfied:

$$\phi V_n = 0.75 A_{sd} f_y \mu \tag{18.10}$$

Where:

V_n = shear resistance of wall–footing connection with $\phi = 0.75$.

A_{sd}, f_y = area and yield strength of dowel bars, respectively.

μ = coefficient of friction at the wall–footing connection:

$\mu = u\lambda$ with $u = 1.4$ for monolithic pour of wall and footing.

$u = 1.0$ for intentionally roughened connection surface.

$u = 0.6$ for other cases.

$\lambda = 1.0, 0.85$ and 0.75 for conventional concrete, sand lightweight concrete and lightweight concrete, respectively.

6. Development length and splicing requirements for dowel bars in compression or tension shall be satisfied per ACI Code requirements as outlined in Chapter 13.

7. The area of dowel bars A_{sd} shall not be less than $0.0015 \times$ the contact area of wall and footing.

Figure 18.11 illustrates a typical layout of dowel bars between walls and footings as well as footing reinforcement. Although short dowels are common as shown in Figure 18.11, dowels extending the entire height of the wall are typical for short and medium height walls. Further illustration in 3–D is presented in Figure 18.12

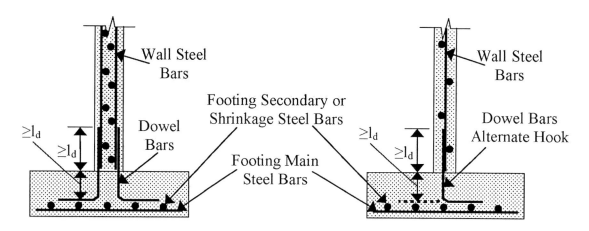

a. Wall with two curtains of steel. b. Wall with one curtain of steel.
Use alternating hooks for dowel bars.

Figure 18.11: Steel bar schematic layout in walls and footings.

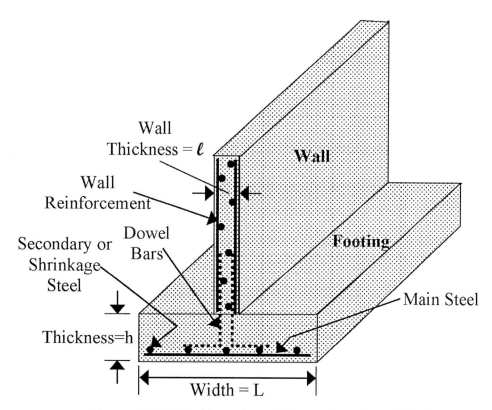

Figure 18.12: Steel bar schematic layout in walls and footings.

Similar to beams and columns, footing design is an iterative process. Soil design precedes structural design. For structural design, shear design is performed to determine footing thickness followed by bending design to compute the required steel bar area. The last step of design is bearing. The completed design is, then, reviewed. The footing is redesigned a number of times until a satisfactory design is reached.

Example 18.1:

Design a continuous wall footing for a 12 in–wide bearing wall with #6@12 in curtain of steel bars. DL = 9 k/ft and LL = 5 k/ft. Allowable soil bearing pressure = q_a = 3 ksf. Footing depth is required to be 4 ft below finished grade and soil unit weigh = 120 pcf.

Solution:

(1)   Geotechnical Design of Footing Width: (Equations 18.2)

Start with an initial assumption for footing width = 6 ft. Assuming a footing depth of 15 in, the required footing width based on service load can be determined as follows:

Footing weight = 6×1.25×0.15 = 1.13 k/ft   where 0.15 k/ft³ is unit weight of concrete

Weight of soil on footing = 6×2.75×0.12 = 1.98 k/ft   where 0.12 k/ft³ is soil unit weight

Use Equation 18.2 ==> $L = \dfrac{DL+LL}{q_a}$

Required footing width = (9+5+1.13+1.98)/3 = 5.70 ft ==> Use Footing Width, L = 6 ft

(2)   Structural Design:

q_u = (1.2×9 + 1.6×5)/6 = 3.13 ksf

(a)   Shear Design:

Use Equation 18.3 ==> $Q_u = q_u×[(L − \ell)/2 − d]$

L = 6 ft ℓ = 1 ft d ≈ 15 − 3 − ½ = 11.5 in assuming #8 bars maximum size

Q_u = 3.13×[(6 − 1)/2 − 11.5/12] = 6.87 k/ft

Use Equation 18.4 ==> $\phi V_n = 0.75×2\sqrt{f'_c}×12d$

Use f'_c = 2,500 psi (typical for footing)

ϕV_n = 0.75×2$\sqrt{2,500}$×12×11.5/1,000 = 10.35 kips/ft > Q_u = 6.87 kips/ft   OK

(b)   Moment Design:

(b.1)   Main Steel Bars: use f_y = 60 ksi

Use Equation 18.6 ==> $M_u = q_u(L − \ell)^2/2 = 3.13×(6 − 1)^2/8 = 9.78$ k.ft/ft

Use Equation 18.7 ==> $A_s ≈ M_u/0.8df_y = 9.78×12/(0.8×11.5×60) = 0.21$ in²/ft

Typically spacing of main steel bars in footing should not exceed footing depth and minimum size of main steel bars should not be less than #5.

Based on Table A.24, use #5@10 in o.c. (A_s = 0.37 in²/ft) for main steel bars.

Main steel should be ≥ 0.0018×gross footing area = 0.0018×12×15 = 0.32 in²/ft

Thus #5@10in o.c. is adequate

Use Equation 18.8 ==> $l_d ≥ (L − \ell)/2 − 3$ in

Available development length = (L − ℓ)/2 − 3 in = (6 ft − 1 ft)/2 − 3 in = 27 in

Table A.31 & Chapter 13, required development length (A_{sreq}/A_{sprov} = 1.0), $l_d ≈ 16$ in < 27 in OK

(b.2)   Secondary Steel Bars:

Secondary Steel $\approx 0.0018 \times (6 \times 12) \times 15 = 1.94$ in$^2 \Longrightarrow$ Use 6#5 bars along the footing length

(b.3)   Bearing or Load Transfer:

Use Equation 18.9 $\Longrightarrow P_n = 0.85 f_c' A_1 \sqrt{\dfrac{A_2}{A_1}} + A_{sd} f_y$ not to exceed $1.7 f_c' A_1 + A_{sd} f_y$

$A_1 = 12 \times 12 = 144$ in$^2 = 1$ ft^2

$A_2 = 12 \times (12 + 15 \times 4) = 864$ in$^2 = 6$ ft^2 follows Figure 18.10.b

$\sqrt{\dfrac{A_2}{A_1}} = 2.44 > 2.0 \Longrightarrow$ use $\sqrt{\dfrac{A_2}{A_1}} = 2.0$ in Equation 18.9 or use the not to exceed part

$A_{st} = $#6@12in o.c. $= 0.44$ in^2/ft

$\phi P_n = \phi(1.7 f_c' A_1 + A_{sd} f_y) = 0.6 \times (1.7 \times 2,500 \times 144 + 0.44 \times 60,000)/1,000 = 383$ kips

$P_u = 1.2DL + 1.6LL = 1.2 \times 9 + 1.6 \times 5 = 18.8$ kips $<< \phi P_n = 383$ kips OK

Compression development length for #6 Dowels ≈ 10.5 in based on Equation 13.3 in Chapter 13 assuming $A_{sprovided}/A_{srequired} = 1.0$ for simplicity

Available dowel compression development length $= 15 - 3 - 2 \times (5/8) = 10.75$ in > 10.5 in OK

Where $2 \times (5/8)$ is the sum of diameters of the top and bottom layer of footing steel reinforcement. It is subtracted from the available development length since dowels typically terminate or bent at the top layer of footing steel reinforcement.

Reinforcement ratio for dowels $= 0.44/(12 \times 12) = 0.003 > 0.0012$ min. for walls OK

(3)   Footing Detail:

Footing design is reasonable and initial assumptions do not require revision.

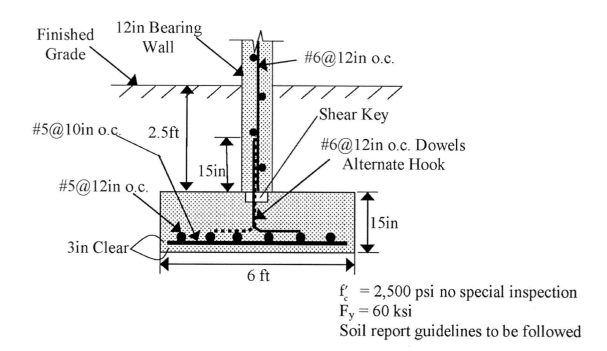

$f_c' = 2,500$ psi no special inspection
$F_y = 60$ ksi
Soil report guidelines to be followed

18.5 Analysis and Design of Single Column Footings

Initial footing area is computed based on the following equation:

$$B \times L = \frac{DL + LL}{q_a}$$

(18.11)

Where:

L,B = footing width and length, respectively, in feet.

DL,LL = dead and live load, respectively, on the column including the weight of footing and soil on footing (initially assumed) in lb or k.

q_a = allowable soil bearing capacity based on the soil report in psf or ksf.

Soil engineering design of footing area utilizes Equation 18.11 based on ASD. Footing reinforced concrete structural design requires analysis and design for several modes of failure: punching or two–way shear, one–way shear, bending moment and bearing or load transfer.

18.5.1 Punching Shear of Single Column Footings

Analysis and design for punching shear is illustrated in Figures 18.13 and 18.4. The critical failure surface for punching shear is located at distance d/2 from face of column in all directions where d is footing effective depth.

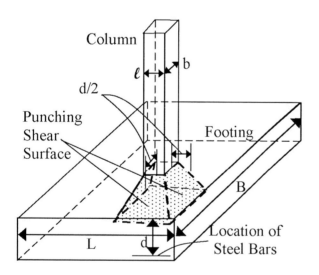

a. 3–D view of punching shear failure surface.

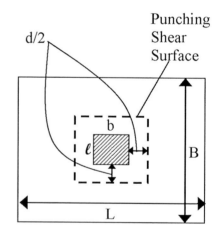

b. Plan view of punching shear failure surface.

Footing dimensions: L = width, B = length, h = thickness.
Column dimensions: ℓ and b = side length opposite to L and B, respectively

Figure 18.13: Punching shear in footings.

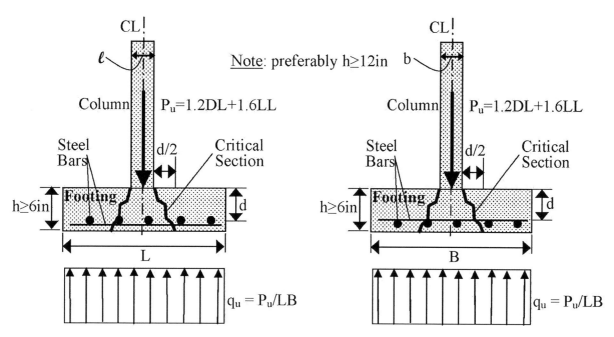

Figure 18.14: Punching shear in footings.

As stated earlier, h_{min} = 6 in per ACI Code. However, h_{min} = 12 in it is recommended. There are two steel bar layers in the footing. The longer steel bars are placed on the bottom to achieve a larger value of d (effective depth) for the longer direction. Then, the shorter bars have a smaller d. Distance d/2 is for punching shear taken as the smaller value so that design is to the safe side. The values of footing effective depth can be determined based on the following equations utilizing 3in clear concrete cover per ACI Code:

$$d_{long} = h - 3in - d_{bl}/2$$

$$d_{short} = h - 3in - d_{bl} - d_{bs}/2 \tag{18.12}$$

Where:

d_{long} and d_{short} = footing effective depth in the long and short directions.

h = footing thickness.

d_{bl} and d_{bs} = steel bars diameter in the long and short directions, respectively.

Based on Figures 18.13 and 18.14, the factored (ultimate) shear force affecting the punching shear critical section may be computed as follows:

$$Q_u = P_u - q_u \times [(\ell + d) \times (b + d)] \tag{18.13}$$

Where:

Q_u = ultimate or factored punching shear force.

q_u = P_u/LB = (1.2DL + 1.6LL)/LB = factored soil pressure (psf or ksf) including column forces only without the footing and soil on footing.

ℓ and b = column cross–sectional dimensions.

d = footing effective depth in the short direction per Equation 18.12.

In Equation 18.13, the value of Q_u is obtained by subtracting from P_u the soil pressure underneath the failure surface which is transferred directly to the column causing no shear stresses. Column footings are not reinforced against shear. Concrete is required to resist the entire shear force of Equation 18.13 utilizing the following equation:

$$\phi V_n = 0.75 \times (2 + \frac{4}{\beta_c})\sqrt{f'_c} \times b_o d \quad \text{not to exceed} \quad 0.75 \times 4\sqrt{f'_c} \times b_o d \qquad (18.14)$$

Where:

$\phi = 0.75$ = shear capacity reduction factor of concrete.

V_n = nominal shear resistance of column footing in lb or kips.

β_c = ratio of long to short side of column section (ℓ/b if $\ell > b$ or b/ℓ if $b > \ell$). ACI Code stipulates zhat $\beta_c \geq 2.0$ for use in this equation by means of the not to exceed equation part. The limit on is not a restriction on column dimensional proportions; it is only a limit on footing punching shear resistance for rectangular columns.

b_o = perimeter of punching shear area = $2 \times (\ell + b + 2d)$.

d = footing minimum effective depth according to Equation 18.12.

Based on Equations 18.13 and 18.14, the minimum footing effective depth can be determined using iteration. A construction suitable value (a multiple of 2 or 3 in) is initially assumed for h, based on which, the value of d is computed using Equation 18.12. h is reassumed so that it matches the required value of computed d using Equation 18.14. This operation is repeated until an acceptable value of h is reached:

$$\phi V_n \text{ (Equation 18.14)} \geq Q_u \text{ (Equation 18.13)}$$

This iterative process of footing design needs to extend to geotechnical engineering design so constant footing thickness is used in all footing design aspects. Thus, for a sound design, footing thickness used in Equations 18.11 – 18.14 are required to be equal.

18.5.2 One–Way Shear of Single Column Footings

Analysis and design for one–way shear is illustrated in Figures 18.15 and 18.16. Critical failure surfaces for one–way shear are located at distances d from face of column in all directions where d is footing effective depth. Thus, there are four potentially critical one–way shear planes, two in each direction on opposite sides of the column. Two of which are shown in Figure 18.15 and 18.16, one in each direction.

For one–way shear analysis of footings, the distance d to critical section is taken as the smaller value (Equation 18.13) so that the design is to the safe side. Based on Figures 18.15 and 18.16, the factored (ultimate) shear force affecting a critical section may be computed as follows:

$$Q_{uB} = q_u \times [(L - \ell)/2 - d] \times B \qquad \text{For critical sections parallel to B}$$

$$(18.15)$$

$$Q_{uL} = q_u \times [(B - b)/2 - d] \times L \qquad \text{For critical sections parallel to L}$$

Where:

Q_{uB}, Q_{uL} = ultimate or factored one–way shear force for the critical sections parallel to B and L, respectively.

In the Equation 18.15, the values of Q_u are obtained by multiplying ultimate soil pressure, q_u, by the area of footing beyond the one–way shear critical section location. Footing concrete is required to resist the entire shear force of Equation 18.15 absent any shear reinforcement determined utilizing the following equation:

$$\phi V_{nB} = 0.75 \times 2\sqrt{f'_c} \times Bd \quad \text{For critical sections parallel to B}$$

(18.16)

$$\phi V_{nL} = 0.75 \times 2\sqrt{f'_c} \times Ld \quad \text{For critical sections parallel to L}$$

Where:

$\phi = 0.75$ = shear capacity reduction factor of concrete.

V_{nB}, V_{nB} = nominal one–way shear resistance of footing for the critical sections parallel to B and L, respectively.

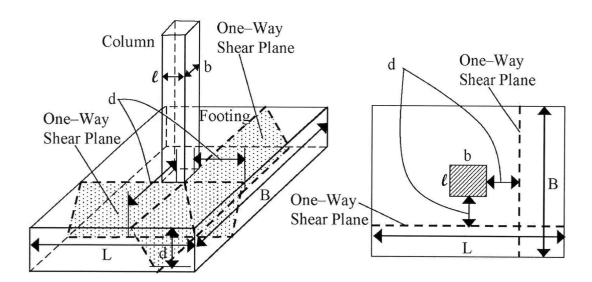

a. 3–D view of one–way shear failure surfaces.

b. Plan view of one–way shear failure surfaces.

Figure 18.15: One–way shear in footings.

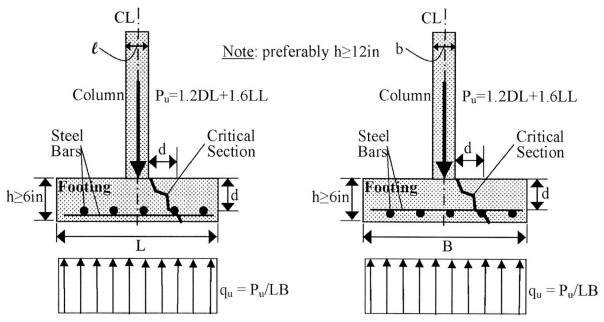

Figure 18.16: One–way shear in footings.

To investigate footing adequacy for one–way shear resistance, the thickness and effective depth selected for the footing based on punching shear are utilized. If these values are adequate, design can proceed to bending moment. Otherwise, footing thickness is reassumed, geotechnical design is repeated, punching shear is reexamined and one–way shear analysis and design is performed. The following equations need to be satisfied so that the footing is presumed adequate for one–way shear resistance:

$$\phi V_{nB} \geq Q_{uB}$$
$$\phi V_{nL} \geq Q_{uL}$$

Upon verification of footing adequacy for punching shear and one–way shear, flexural design can be performed.

18.5.3 Bending Moment Analysis and Design of Single Column Footings

18.5.3.a General Analysis Principles

Figures 18.17 and 18.18 illustrate the analysis and design for bending moment. Critical sections with maximum bending moments are located at the face of column in all directions. There are four critical maximum moment plane, two in each direction on opposite sides of the column. Two sections, one in each direction, are shown in Figure 18.17 and 18.18.

Figures 18.17 and 18.18 assume that L > B. Thus, the bending moment in the long direction (L) is larger than the bending moment in the short direction (B). The larger bending moment requires a larger value of effective depth (d), thus steel bars in the long direction are placed closer to the bottom of the footing to acquire a larger value of d. It is typical to use the smaller value of d (Equation 18.12) for footing moment resistance of both directions as it simplifies analysis and results in a safer design.

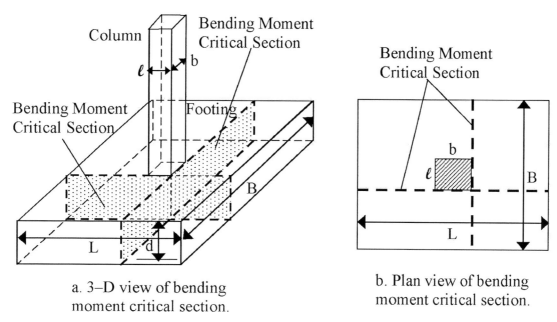

a. 3–D view of bending moment critical section.

b. Plan view of bending moment critical section.

Figure 18.17: Bending moment critical sections in footings.

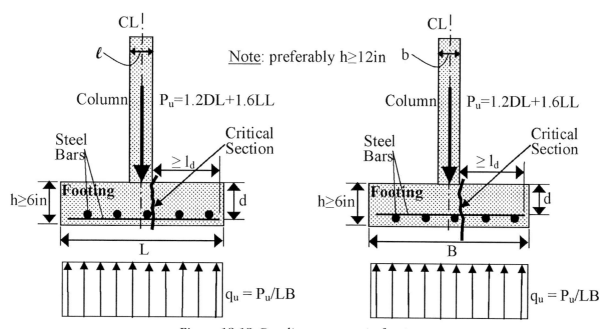

Figure 18.18: Bending moment in footings.

Following the determination of required footing effective depth for shear resistance, flexural design is utilized to compute the required steel reinforcement as follows:

$$M_{uB} = q_u \times B \times [(L - \ell)/2]^2/2 \text{ For critical sections parallel to B}$$

(18.17)

$$M_{uL} = q_u \times L \times [(B - b)/2]^2/2 \text{ For critical sections parallel to L}$$

Where M_{uB} and M_{uL} is the factored (ultimate) bending moment affecting the critical footing section at face of column shown in Figures 18.17 and 18.18. Reinforcement can be calculated utilizing flexural design methods of beams and slabs (Chapters 4 thru 8). Instead, the following equations can be used to compute the required area of steel reinforcement for bending resistance for the footing directions:

$$A_{sB} \approx M_{uB}/0.8df_y \text{ For steel bars distributed within B and parallel to L}$$

$$(18.18)$$

$$A_{sL} \approx M_{uL}/0.8df_y \text{ For steel bars distributed within L and parallel to B}$$

Equation 18.18 is similar Equation 18.7. ACI Code does not require minimum reinforcement in column footings. Generally, some reinforcement is included in column footings of the order of 0.0018 of gross footing area in both directions against shrinkage and cracking.

18.5.3.b Development Lengths

Steel bars used for footing reinforcement are required to be fully developed at the critical bending sections at column faces (Figure 18.18). Steel bars may be terminated with a hook if inadequate development length is available. The following equations are needed to be satisfied for proper bending design:

$$l_d \geq (L - \ell)/2 - 3 \text{ in For bending in L direction}$$

$$(18.19)$$

$$l_d \geq (B - b)/2 - 3 \text{ in For bending in B direction}$$

Where:

l_d = required tension development length (Chapter 13).

3 in = concrete clear cover.

18.5.3.c Distribution of Steel Bars in Column Footings

Steel bars are distributed evenly in square footings. Also, steel bars in the long direction of rectangular footings (parallel to L) are distributed evenly. ACI Code specifies that steel bars in the short direction of rectangular footings (parallel to B) are required to be more concentrated within a band directly at the column location than the rest of footing as explained in Figure 18.19. This ACI Code requirement is due to higher concentration of loads, stresses and moments directly at the column location as compared with other parts of the footing in the short direction. The band width is a portion of footing long side L with length equal to footing short side B centered on the column (Figure 18.19). The band width contains a concentration of steel bars in the short direction as in Equation 18.20:

$$\text{steel area in band width/total steel area in short direction} = 2/(\beta +1)$$

or $$A_{sband} = A_{sL} \times 2/(\beta +1)$$

$$(18.20)$$

Where:

Band width = footing short side = B.

Total steel area in footing short direction = A_{sL}.

A_{sband} = steel area in band width.

$\beta = L/B$.

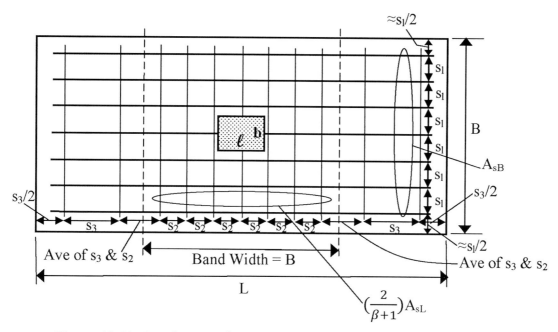

Figure 18.19: Distribution of moment reinforcement in rectangular footings.

ACI Code specifies that A_{sband}, be evenly distributed within the band width. Also, the remainder of A_{sL} is required to be evenly distributed outside the band width as shown in Figure 18.19.

18.5.4 Transfer of Load between the Column and Footings

Bearing or transfer of load from the column to the footing is an important item of footing design. Bearing strength of transfer area of the footing and column can be computed according to Equation 18.9 repeated below:

$$P_n = 0.85f_c'A_1\sqrt{\frac{A_2}{A_1}} + A_{sd}f_y \quad \text{not to exceed} \quad 1.7f_c'A_1 + A_{sd}f_y \quad \text{(18.9)}$$

Where:

P_n = bearing load capacity of the transfer area of the footing.

$A_1 = \ell \times b$ = footing bearing area or column cross–sectional area.

A_2 = lower cone base area based on 2H:1V side slope (Figure 18.20).

A_{sd} = area of steel dowels connecting the column and footing.

f_c' and f_y = compressive strength of footing and yield strength of dowels.

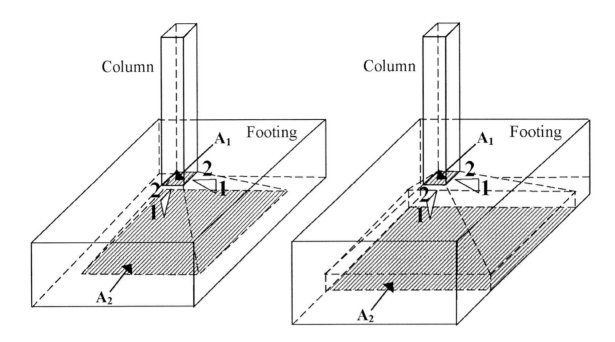

a. Slope intercepting the footing base.   b. Slope intercepting footing sides.

Figure 18.20: Determination of areas A_1 and A_2 for column footing bearing strength.

ACI Code provisions regarding load transfer between columns and footings are very similar to the provisions of load transfer of walls and footings stated in Section 18.4:

1. $\phi = 0.6$,

2. Dowels shall be adequate to transfer factored compression of column and footing in excess of concrete strength per Equation 18.9.

3. Dowel bars shall be designed to resist any tensile force or tension and compression of bending moment of the column and footing.

4. Shear force transfer between the column is via friction based on Equation 18.10.

5. Code development length and splicing requirements for dowels bars in compression or tension shall be satisfied.

6. The area of dowel bars A_{sd} shall not be less than $0.005 \times$ the column cross–sectional area.

Figure 18.21 illustrates a typical layout of dowel bars of columns and footings and footing reinforcement.

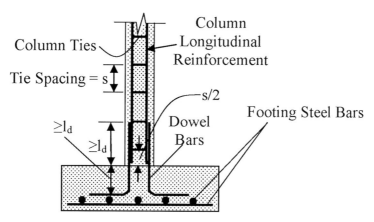

Figure 18.21: Steel bar schematic layout in columns and footings.

Example 18.2

Design a column footing for an 18×18in column containing of 8#7 steel bars. DL = 220 kips and LL = 130 kips. Allowable soil bearing pressure = q_a = 4 ksf. Footing depth is required to be 5 ft below finished grade and soil unit weigh = 120 pcf. Use L = 1.5B.

Solution

(1) Geotechnical Design of Footing: (Equations 18.1 and 18.11)

Neglecting the weight of footing and soil on footing initially:

Use Equation 18.1 ===> $B \times L = \dfrac{DL+LL}{q_a} = \dfrac{220+130}{4} = 87.5$ ft² = $2/3L^2$

Use 14×9 ft footing L = 14 ft and B = 9 ft ===> L/B ≈ 1.5

To include the weight of footing and soil on footing, footing depth of 30 in is assumed:

Footing weight = 9×14×2.5×0.15 = 47.3 k/ft where 0.15 k/ft³ is unit weight of concrete

Weight of soil on footing = 9×14×2.5×0.12 = 37.8 k/ft where 0.12 k/ft³ is soil unit weight

Soil bearing pressure = (220 + 130 + 47.3 + 37.8)/(9×14) = 3.45 ksf < q_a = 4 ksf OK

(2) Structural Design:

P_u = 1.2×220 + 1.6×130 = 472 kips ===> q_u = (1.2×220 + 1.6×130)/(9×14) = 3.75 ksf

(a) Punching Shear Design:

Use Equation 18.13 ===> $Q_u = P_u - q_u \times [(\ell + d) \times (b + d)]$

$\ell = b = 18$ in

Use Equation 18.12   ===> $d_{short} = h - 3\text{in} - d_{bl} - d_{bs}/2$

$d_{short} \approx 30 - 3 - 1 - \frac{1}{2} = 25.5$ in (#8 bars max. assumed)

Q_u = 472 – 3.74×[(18 + 25.5)×(18 + 25.5)]/144 = 422.8 kips

Use Equation 18.14 ===> $\phi V_n = 0.75 \times (2 + \dfrac{4}{\beta c}) \sqrt{f'_c} \times b_o d$ not to exceed $0.75 \times 4\sqrt{f'_c} \times b_o d$

Use f'_c = 2,500 psi (typical for footing)

β_c=1.0(square column)<β_{cmin}=2.0 NG ⟹ Use the not to exceed part of Equation 18.14

$b_o = 2 \times (\ell + b + 2d) = 2 \times (18 + 18 + 2 \times 25.5) = 174$ in

$\phi V_n = 0.75 \times 4 \times \sqrt{2{,}500} \times 174 \times 25.5/1000 = 665.5$ kips

$\phi V_n = 665.5$ kips > $Q_u = 422.8$ kips/ft   OK

(b)   One–Way Shear Design:

Use Equation 18.15 ⟹ $Q_{uB} = q_u \times [(L - \ell)/2 - d] \times B$ For critical sections parallel to B

$Q_{uB} = 3.75 \times [(14 - 18/12)/2 - 25.5/12] \times 9 = 139.2$ kips

Use Equation 18.16 ⟹ $\phi V_n = 0.75 \times 2\sqrt{f_c'} \times Bd$ For critical sections parallel to B

$\phi V_n = 0.75 \times 2\sqrt{2{,}500} \times 9 \times 12 \times 25.5/1{,}000 = 206.6$ kips > 139.2 kips OK

Use Equation 18.15 ⟹ $Q_{uL} = q_u \times [(B - b)/2 - d] \times L$ For critical sections parallel to L

$Q_{uL} = 3.75 \times [(9 - 18/12)/2 - 25.5/12] \times 13 = 79.2$ kips

Use Equation 18.16 ⟹ $\phi V_n = 0.75 \times 2\sqrt{f_c'} \times Ld$ For critical sections parallel to L

$\phi V_n = 0.75 \times 2\sqrt{2{,}500} \times 14 \times 12 \times 25.5/1{,}000 = 321.4$ kips > 79.2 kips OK

(c)   Moment Design:

(c.1) Steel Bars: use $f_y = 60$ ksi

Use Equation 18.17 ⟹ $M_{uB} = q_u \times B \times [(L - \ell)/2]^2/2$ For critical sections parallel to B

$M_{uB} = 3.75 \times 9 \times [(14 - 18/12)/2]^2/2 = 658.8$ k.ft

Equation 18.18 ⟹ $A_{sB} \approx M_{uB}/0.8df_y$ Steel bars distributed within B and parallel to L

$A_{sB} \approx 658.8 \times 12/(0.8 \times 25.5 \times 60) = 6.46$ in^2 parallel to L

Minimum reinforcement distributed within B parallel to L = $0.0018 \times 9 \times 12 \times 30 = 5.83$ in^2

Steel bars in the long direction use 9#9 (#9@12in o.c.) ⟹ As = 9.0 in^2 > 6.46 in^2 OK

Use Equation 18.19 ⟹ $l_d \geq (L - \ell)/2 - 3$ in For bending in L direction

Available development length in the long direction for #9 bars = $(13 \times 12 - 18)/2 - 3 = 66$ in

Required development length for #9, $l_d \approx 56$ in (Table A.31, Chapter 13) $l_d \approx 56$ in < 66 in OK

In the calculation of development length $A_{srequired}/A_{sprovided} = 1.0$ was used for simplicity

Use Equation 18.17 ⟹ $M_{uL} = q_u \times L \times [(B - b)/2]^2/2$ For critical sections parallel to L

$M_{uL} = 3.75 \times 14 \times [(9 - 18/12)/2]^2/2 = 368.1$ k.ft

Equation 18.18 ⟹ $A_{sL} \approx M_{uL}/0.8df_y$ Steel bars distributed within L and parallel to B

$A_{sL} \approx 386.1 \times 12/(0.8 \times 25.5 \times 60) = 3.78$ in^2 parallel to B

Minimum reinforcement distributed within L parallel to B = $0.0018 \times 14 \times 12 \times 30 = 9.07$ in^2

For reinforcement in the short direction use 15#7 ⟹ As = 9.15 in^2 > 9.07 in^2 OK

Use Equation 18.19 ⟹ $l_d \geq (B - b)/2 - 3$ in For bending in B direction

Available development length in the short direction for #7 bars = $(9 \times 12 - 18)/2 - 3 = 42$ in

Required development length for #7, $l_d \approx 26$ in (Table A.31, Chapter 13) $l_d \approx 26$ in < 42 in OK

In the calculation of development length $A_{srequired}/A_{sprovided} = 1.0$ was used for simplicity

Steel bars within the band in the short direction = $A_{sband} = A_{sL} \times 2/(\beta +1)$ Equation 18.20

$\beta = 14/9 = 1.55$ No. of bars within the band = $15 \times 2/(1.55 + 1) = 11.76$ bars ≈ 12 bars

Use #7@9in o.c. within the band (total of 12 #7 bars) and #7@15in o.c. outside the band (total of 4 #7 bars) resulting in a total of 16#7 bars in the short direction.

(b.3) Bearing or Load Transfer:

Use Equation 18.9 $\implies P_n = 0.85 f'_c A_1 \sqrt{\dfrac{A_2}{A_1}} + A_{sd} f_y$ not to exceed $1.7 f'_c A_1 + A_{sd} f_y$

$A_1 = 18 \times 18 = 324$ in^2 = 2.25 ft^2

A_2:   first side = $(18 + 30 \times 4)/12 = 11.5$ ft < 13 ft OK

            other side = $(18 + 30 \times 4)/12 = 11.5$ ft > 9 ft NG $\implies$ Use other side = 9 ft

$A_2 = 11.5 \times 9 = 103.5$ ft^2 See Figure 18.20 for further illustrations

$\sqrt{\dfrac{A_2}{A_1}} = 6.9 > 2.0 \implies$ use $\sqrt{\dfrac{A_2}{A_1}} = 2.0$ in Equation 18.9 Use the not to exceed part

$A_{st} = 8\#7$ bars = 4.8 in^2

$\phi P_n = \phi(1.7 f'_c A_1 + A_{sd} f_y) = 0.6 \times (1.7 \times 2{,}500 \times 324 + 4.8 \times 60{,}000)/1{,}000 = 999$ kips > 472 kips OK

Compression development length for #7 Dowels ≈ 21in (Equation 13.3 Chapter 13)

$A_{srequired}/A_{sprovided} = 1.0$ is used to simplify calculation of compression development length

Available dowel compression development length = $30 - 3 - (7+9)/8 = 27$ in > 21 in OK

(7+9)/8 is the sum of diameters of the top and bottom layer of footing steel reinforcement. It is subtracted from the available development length since dowels typically terminate at the top layer of footing steel reinforcement.

Reinforcement ratio for dowels = $4.88/(18 \times 18) = 0.015 > 0.005$ min. for columns OK

At this stage of footing design, the entire design is evaluated to assess construction suitability. All design aspects are checked including geotechnical design. If it is determined that any part is not suitable, then, the entire design is reviewed until a suitable design is achieved. Such iterative design procedure is typical for reinforced concrete design as well as other structural design processes.

(3)   Footing Detail:

Footing Specifications:

$f'_c = 2{,}500$ psi   no special inspection.

$f_y = 60$ ksi.

Soil report guidelines to be followed.

General project specifications for reinforced concrete shall also apply to footings.

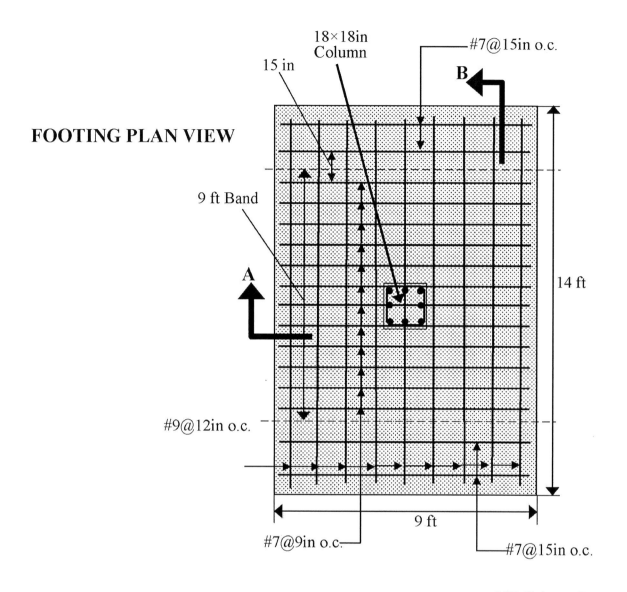

FOOTING PLAN VIEW

18×18in Column

15 in

#7@15in o.c.

B

9 ft Band

14 ft

A

#9@12in o.c.

9 ft

#7@9in o.c.

#7@15in o.c.

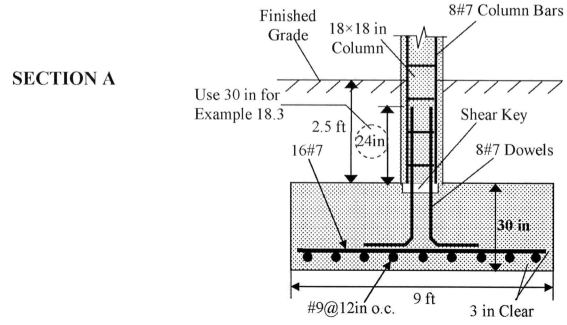

SECTION A

Finished Grade

18×18 in Column

8#7 Column Bars

Use 30 in for Example 18.3

2.5 ft

24in

Shear Key

16#7

8#7 Dowels

30 in

#9@12in o.c.

9 ft

3 in Clear

SECTION B

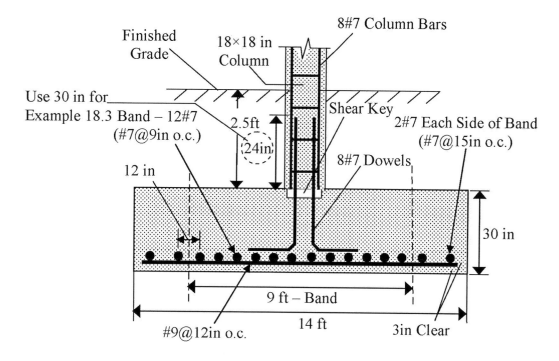

18.6 Column Footings under Eccentric Loading

This section is focused on the differences in analysis and design between eccentrically and concentrically loaded footings. Soil engineering design of footing area utilizes Figures 18.7 and 18.8 and Section 18.3 based on ASD with $q_{max} \le q_a$. Footing dimensions are selected or designed by iteration or trial and error. Design using bearing on the entire footing is generally utilized. With bearing on a portion of the footing, a part of footing is not utilized which may cause instability and, possibly, excessive deformations. Footing reinforced concrete structural design utilizes the strength design method under the effect of factored load to verify footing dimensions, select materials and detail steel bars. Failure modes include: punching shear, one–way shear, bending moment and bearing or load transfer. Consequent to footing structural design, footing design may be adjusted for construction suitability.

In Figures 18.7 and 18.8, two potential loading situations under eccentric loading are exhibited: bearing on the entire footing and bearing on a portion of the footing. Bearing on a portion of the footing is not typically utilized in footing design under service or ultimate loading for the reasons explained above. It is presumed that the bending axis is the centerline axis parallel to footing dimension B resulting in the loading situation shown in Figure 18.22.

The following equations and conditions are applicable to Figure 18.22:

Factored Load Eccentricity = $e_u = M_u/P_u$

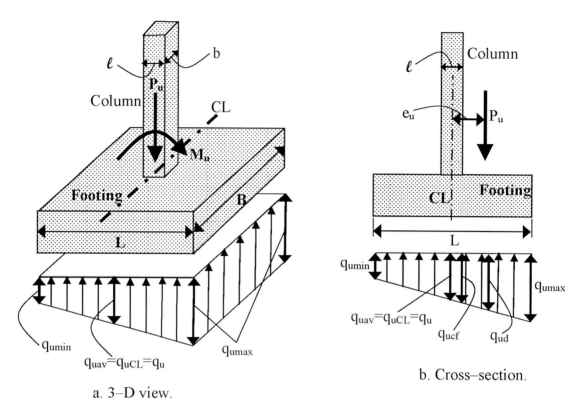

Figure 18.22: Concrete footing under eccentric loading.

Average Factored Soil Pressure = q_{uav} = q_u at CL or q_{uCL} = q_u = P_u/LB

$$\left.\begin{array}{l} \text{Maximum Factored Soil Pressure} = q_{umax} = P_u/A + M_u/S = P_u/BL + 6P_u e_u/BL^2 \\ \text{Minimum Factored Soil Pressure} = q_{umin} = P_u/A + M_u/S = P_u/BL - 6P_u e_u/BL^2 \end{array}\right\} \quad \textbf{(18.21)}$$

Factored Soil Pressure at Column Face = q_{ucf}

$$q_{ucf} = q_{umax} - (L - \ell) \times (q_{umax} - q_{umin})/2L \qquad \textbf{(18.22)}$$

Factored Soil Pressure at d from Column Face = q_{ud}

$$q_{ud} = q_{umax} - (L - \ell - d) \times (q_{umax} - q_{umin})/2L \qquad \textbf{(18.23)}$$

18.6.1 Punching Shear

Analysis and design for punching shear for eccentrically loaded footing using q_{uCL} or q_{uav} (q_{uCL} = q_{uav} = q_u) are identical to the analysis and design for concentrically loaded footings explained in Section 18.5.1. As shown above, average factored soil pressure occurring at column centerline for eccentric loading (q_{uav}) is equal to factored soil pressure for concentric loading (q_u) under the same conditions with no moment.

18.6.2 One–Way Shear

Locations of critical sections for one–way shear are the same as for concentrically loaded footings shown in Figure 18.15. Calculation for factored shear forces Q_{uB} and Q_{uL} is as follows:

$$Q_{uB} = (q_{umax} + q_{ud}) \times [(L - \ell)/2 - d]/2 \text{ For critical sections parallel to B} \qquad \textbf{(18.24)}$$

Where

q_{umax} and q_{ud} are per Equations 18.21 and 18.23. Also see Figure 18.23.

Q_{uL} = same as Equation 18.15 = $q_u \times [(B - b)/2 - d] \times L$ For critical sections parallel to L

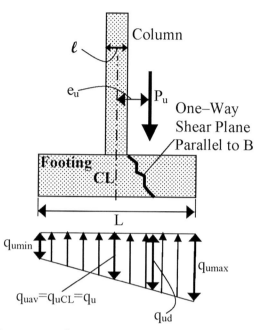

Figure 18.23: One – way shear in concrete footing under eccentric loading.

Q_{uL} is the same as Q_{uL} for concentric loading since the average factored soil pressure does not change for the section parallel to L with the addition of bending moment. For the equation of Q_{uB}, average factored soil pressure for the one–way shear plane parallel to B, $((q_{umax} + q_{ud})/2)$, is used in place of q_u in Equation 18.7 where q_{umax} and q_{ud} are shown above. The remainder of design for one–way shear is the same as Section 18.5.2 for footings under concentric loading.

Typically, bending moments due to lateral forces are alternating. As such, q_{umax} and q_{umin} exchange locations in the footings with moment direction change. Selection of one–way shear plane neighboring to q_{umax} is, thus, appropriate.

Concrete is required to resist the entire shear force of Equations 18.13, 18.15 and 18.24 since shear reinforcement is not typically utilized in footings. Footing reduced nominal shear resistance, ϕV_n may be determined utilizing Equations 18.14 and 18.16 in Section 18.5.2.

18.6.3 Analysis and Design for Bending Moment

Analysis and design of column footings subjected to eccentric loading is the same as the case of concentric loading for all aspects except the value of factored bending moment. Such similar aspects include locations of critical sections, computation of required steel reinforcement and distribution of steel bars. The change in bending moment value is due to the non–uniform distribution of soil pressure caused by bending moment. As L is the long side of footing and load eccentricity is about the centerline axis parallel to B (Figure 18.22), factored bending moments M_{uB} and M_{uL} are as follows:

$$M_{uB} = (2q_{umax} + q_{ucf}) \times B \times [(L - \ell)/2]^2/6 \text{ For critical sections parallel to B} \qquad (18.25)$$

Where

q_{umax} and q_{ud} are per Equations 18.21 and 18.22. Also see Figure 18.24.

M_{uL} = same as Equation 18.17 = $q_u \times L \times [(B - b)/2]^2/2$ For critical sections parallel to L

M_{uL} is the same as M_{uL} for concentric loading since the average factored soil pressure parallel to L since eccentricity about B. For the equation of M_{uB}, trapezoidal soil pressure distribution analysis is utilized as illustrated in Figure 18.24. The remainder of moment design is the same as Section 18.5.3 for footings under concentric loading.

q_{umax} is used in lieu of q_{umin} for calculation of M_{uB} since lateral force caused moment change directions. q_{umax} and q_{umin} exchange locations as moment direction alternate. Thus, utilization of q_{umax} is proper and prudent.

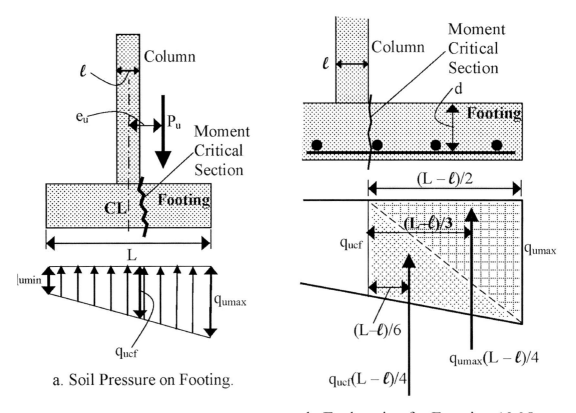

a. Soil Pressure on Footing.

b. Explanation for Equation 18.25.

Figure 18.24: Bending moment analysis in concrete footing under eccentric loading.

18.6.4 Transfer of load between the Column and Footings

Bearing or transfer of eccentric load from the column to the footing follows the same design guidelines of Section 18.5.4 for footings under concentric loading.

Example 18.3

Redo Example 18.2 if the footing is subjected to a seismic moment equal to 350 k.ft in the long direction and a seismic axial force of 40 kips (axial force due to seismic is typically small compared to dead and live loads). Consider only load combinations (2) and (4) in Section 18.3. Assume that the soil engineer has allowed the use of q_a = 5 ksf for short duration loading such earthquake and wind.

Solution

(1)   Geotechnical Design of Footing:

Equations 18.1 and 18.11 for gravity load conditions were verified in Example 18.2. Footing design under the effect of gravity and seismic forces is as follows with positive axial load indicating compression vice versa:

Load Combination (2) DL ± 0.7E + 0.75LL Section 18.3:

Case 1 – seismic downthrust:

Axial load = 220 + 47.3 + 37.8 + 0.7×40 + 0.75×130 = 431 kips

Eccentricity = e = M/P = (0.7×350)/431=0.57 ft<L/6=2.33 ft ===> Case a in Figure 18.8

Maximum soil pressure = $P/BL + 6Pe/BL^2$ = 431/(9×14)+6×(0.7×350)/[9×(14)^2]= 4.3 ksf

q_a = 5 ksf > 4.3 ksf OK

Case 2 – seismic uplift:

Axial load = 220 + 47.3 + 37.8 – 0.7×40 + 0.75×130 = 375 kips

Eccentricity = e = M/P = (0.7×350)/375 =0.65 ft<L/6=2.33 ft ===> Case a in Figure 18.8

Maximum soil pressure = $P/BL + 6Pe/BL^2$ = 375/(9×14)+6×(0.7×350)/[9×(14)^2]= 3.8 ksf

q_a = 5 ksf > 3.8 ksf OK

Load Combination (4) 0.6DL ± 0.7E Section 18.3:

Case 1 – seismic downthrust:

Axial load = 0.6×(220 + 47.3 + 37.8) + 0.7×40 = 211 kips

Eccentricity = e = M/P = (0.7×350)/211 = 1.16 ft<L/6=2.33 ft===> Case a in Figure 18.8

Maximum soil pressure = $P/BL + 6Pe/BL^2$ = 211/(9×14)+6×(0.7×350)/[9×(14)^2]= 2.5 ksf

q_a = 5 ksf > 2.5 ksf OK

Case 2 – seismic uplift:

Axial load = 0.6×(220 + 47.3 + 37.8) – 0.7×40 = 155 kips

Eccentricity = e = M/P = (0.7×350)/155 = 1.58 ft<L/6=2.33 ft===> Case a in Figure 18.8

Maximum soil pressure = $P/BL + 6Pe/BL^2$ = 155/(9×14)+6×(0.7×350)/[9×(14)^2]= 2.1 ksf

q_a = 5 ksf > 2.1 ksf OK

Notes: All four loading combinations above have the same moment but variable axial force with maximum axial force in Combination (2) Case 1, due to which maximum soil bearing pressure occurs. This is confirmed based on the soil bearing pressures computed above for all four load cases. An experienced designer can easily detect that Combination (2) Case 1 – seismic downthrust is the most critical load combination.

All four loading cases above are according to case (a) in Figure 18.8: bearing on the entire footing as required for footing design. As stated earlier, load case (b) of Figure 18.8 is generally not recommended for footing geotechnical engineering design since it may result in instability and possibly excessive deformations.

Based on the most critical loading case that is Load Combination (2) Case 1 – seismic downthrust in Section 18.3, allowable soil bearing pressure is not exceeded. Thus, footing geotechnical engineering design is adequate.

(2)   Structural Design:

In addition to footing analysis under the effect of gravity load (Example 18.2), the following analysis under the effect of gravity and seismic load is required. Two load combinations of interest are (see Chapter 3):

$U5 = 1.2D \pm 1.0E + L + 0.2S$ (no snow load in this example)

$U7 = 0.9D \pm 1.0E + 1.6H$ (no soil load in this example)

Load Combination U5 – seismic downthrust:

$P_u = 1.2 \times 220 + 40 + 130 = 434$ kips $\Longrightarrow$ $q_{uav} = 434/(9 \times 14) = 3.44$ ksf

$e_u = M_u/P_u = 350/434 = 0.81$ ft $< L/6 = 2.33$ ft $\Longrightarrow$ Case a in Figure 18.8 and Figure 18.22

Load Combination U5 – seismic uplift:

$P_u = 1.2 \times 220 - 40 + 130 = 354$ kips $\Longrightarrow$ $q_{uav} = 354/(9 \times 14) = 2.81$ ksf

$e_u = M_u/P_u = 350/354 = 1.00$ ft $< L/6 = 2.33$ ft $\Longrightarrow$ Case a in Figure 18.8 and Figure 18.22

Load Combination U6 – seismic downthrust:

$P_u = 0.9 \times 220 + 40 = 238$ kips $\Longrightarrow$ $q_{uav} = 238/(9 \times 14) = 1.89$ ksf

$e_u = M_u/P_u = 510/238 = 1.47$ ft $< L/6 = 2.33$ ft $\Longrightarrow$ Case a in Figure 18.8 and Figure 18.22

Load Combination U6 – seismic uplift:

$P_u = 0.9 \times 220 - 40 = 158$ kips $\Longrightarrow$ $q_{uav} = 158/(9 \times 14) = 1.25$ ksf

$e_u = M_u/P_u = 350/158 = 2.22$ ft $< L/6 = 2.33$ ft $\Longrightarrow$ Case a in Figure 18.8 and Figure 18.22

The most critical load Combination is U5 – seismic downthrust since it has the highest ultimate average soil bearing stress, q_{uav}, while the moment is the same for all load combinations alternatives. Also, all load combination alternatives follow load case (a) of Figure 18.8. As such, footing dimensions are satisfactory. Similar to geotechnical engineering design, load case (b) of Figure 18.8 is generally not recommended for footing structural engineering design footing for the same reasons.

Note: Typically, the most critical combination for uplift is Combination U6 – seismic uplift that may cause soil pressure distribution according to case (b) of Figure 18.8 which not acceptable for footings. Increasing footing dimensions (particularly L) is utilized so that soil pressure is over the entire footing as in case (a) of Figure 18.8. This situation is not applicable for this example.

(a) Punching Shear Design:

The same procedure of Example 18.2 is applicable using $q_{uav} = 3.44$ ksf (soil pressure directly under the column) of Load Combination U5 – seismic downthrust in lieu of $q_u = 3.75$ ksf in Example 18.2.

$Q_u = 472 - 3.44 \times [(18 + 25.5) \times (18 + 25.5)]/144 = 426.8$ kips

$\phi V_n = 665.5$ kips $> Q_u = 422.8$ kips/ft   OK

(b) One–Way Shear Design:

Maximum Factored Soil Pressure $= q_{umax} = P_u/A + M_u/S = P_u/BL + 6P_u e_u/BL^2$

$q_{umax} = 434/(9 \times 14) + 6 \times 434 \times 0.81/[9 \times (14)^2] = 4.64$ ksf

Minimum Factored Soil Pressure $= q_{umin} = P_u/A + M_u/S = P_u/BL - 6P_u e_u/BL^2$

$q_{umin} = 434/(9 \times 14) - 6 \times 434 \times 0.81/[9 \times (14)^2] = 2.24$ ksf

Equation 18.23 $\Longrightarrow q_{ud} = q_{umax} - (L - \ell - d) \times (q_{umax} - q_{umin})/2L$

$q_{ud} = 3.44 \times [14 - (18 + 25.5)/12] \times (4.64 - 2.24)/(2 \times 14) = 3.06$ ksf

Equation 18.24 ===> $Q_{uB} = (q_{umax} + q_{ud}) \times [(L - \ell)/2 - d] \times B/2$ Critical sections parallel to B

$$Q_{uB} = (4.64 + 3.06) \times [(14 - 18/12)/2 - 25.5/12] \times 9/2 = 142.7 \text{ k}$$

Equation 18.16 ===> $\phi V_n = 0.75 \times 2\sqrt{f'_c} \times Bd$ Critical sections parallel to B

$$\phi V_n = 0.75 \times 2\sqrt{2{,}500} \times 9 \times 12 \times 25.5/1{,}000 = 206.6 \text{ k} > 142.7 \text{ k OK}$$

Equation 18.15 ===> $Q_{uL} = q_u \times [(B - b)/2 - d] \times L$ For critical sections parallel to L

$$Q_{uL} = 3.44 \times [(9 - 18/12)/2 - 25.5/12] \times 13 = 72.7 \text{ k}$$

Use Equation 18.16 ===> $\phi V_n = 0.75 \times 2\sqrt{f'_c} \times Ld$ For critical sections parallel to L

$$\phi V_n = 0.75 \times 2\sqrt{f'_c} \times Ld = 0.75 \times 2\sqrt{2{,}500} \times 14 \times 12 \times 25.5/1{,}000 = 321.4 \text{ k} > 72.7 \text{ k OK}$$

(c) Moment Design:

(c.1) Steel Bars: use $f_y = 60$ ksi

Equation 18.22 ===> $q_{ucf} = q_{umax} - (L - \ell) \times (q_{umax} - q_{umin})/2L$

$$q_{ucf} = 4.64 - [14 - (18/12)] \times (4.64 - 2.24)/(2 \times 14) = 3.57 \text{ ksf}$$

Equation 18.25 ===> $M_{uB} = (2q_{umax} + q_{ucf}) \times B \times [(L - \ell)/2]^2/6$ Critical sections parallel to B

$$M_{uB} = (2 \times 4.64 + 3.57) \times 9 \times [(14 - 18/12)/2]^2/6 = 752.9 \text{ k.ft}$$

Equation 18.18 ===> $A_{sB} \approx M_{uB}/0.8df_y$ Steel bars distributed within B and parallel to L

$$A_{sB} \approx 752.9 \times 12/(0.8 \times 25.5 \times 60) = 7.38 \text{ in}^2 \qquad \text{parallel to L}$$

Use Equation 18.17 ===> $M_{uL} = q_u \times L \times [(B - b)/2]^2/2$ For critical sections parallel to L

$$M_{uL} = 3.44 \times 14 \times [(9 - 18/12)/2]^2/2 = 337.7 \text{ k.ft}$$

Equation 18.18 ===> $A_{sL} \approx M_{uL}/0.8df_y$ Steel bars distributed within L and parallel to B

$$A_{sL} \approx 337.7 \times 12/(0.8 \times 25.5 \times 60) = 3.47 \text{ in}^2 \qquad \text{parallel to B}$$

Steel bar reinforcement selection, distribution and development length verification utilized in Example 18.2 is, thus, satisfactory for this example. The recommended minimum footing reinforcement guidelines are also satisfied.

(b.3) Bearing or Load Transfer:

Bearing capacity of the column footing contact area is determined as in Example 18.2:

$\phi P_n = 999$ kips $> P_{umax} = 434$ kips OK Where P_{umax} is for Combination U5 – seismic downthrust

For this example, due to the presence of bending moment, dowels require tension development. Tension development length for #7 Dowels ≈ 26.3 in using Equation 13.12 and Chapter 13 with $A_{srequired}/A_{sprovided} = 1.0$.

Thus the footing detail of Example 18.2 is suitable for this example with extending dowel development to 30 in. Furthermore, dowel development into the footing is adequate since 90° hook is used. It can be easily verified that adequate hook development length is provided per Chapter 13.

18.7 Design Guidelines for Combined Footings, Mat Foundation and Pile Cap

The general design principles for wall footings and column footings apply to other types of footing as summarized below:

a. Combined footings are designed as beam resting on soil or grade termed grade beam. Thus, concrete clear cover provisions need to be followed. Combined footings dimensions are generally designed so that soil pressure is uniform under the effect of DL and all LL to reduce the possibility of tilting or rotation. Furthermore, the procedures for rectangular beam design (Chapters 4 through 10) are applicable. Thus, combined footings need to be designed for punching shear, one–way shear, bending moment as well as bearing as illustrated in Example 18.4.

b. Mat foundation is designed like an inverted thick concrete slab. It follows the same design principles for footings discussed in this chapter. Shear, moment and bearing are the key failure modes for which design is required. ACI Code states that the direct design method for two–way slabs shall not be used for mat foundation.

c. Pile cap design is similar to mat foundation. However, punching shear and bearing of the piles require special attention. Pile caps are considered not to receive any direct soil pressure. It just receives load from the building walls and columns as well as the piles.

Example 18.4

Design a combined footing for the two adjacent columns based on the preliminary design shown below assuming the both columns are loaded with DL and LL concurrently. $f'_c = 3,000$ psi, $q_a = 7.0$ ksf, $\gamma = 120$ pcf and footing depth = 4 ft.

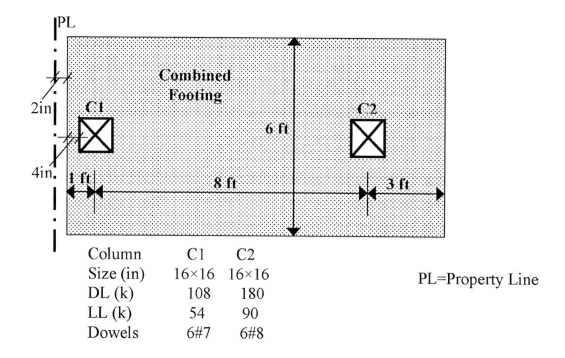

Column	C1	C2
Size (in)	16×16	16×16
DL (k)	108	180
LL (k)	54	90
Dowels	6#7	6#8

PL=Property Line

Solution

(1)   Geotechnical Design of Footing:

Combined footings are designed so that the resultant DL and LL coincide with the footing center line so uniform soil pressure results, thus, reducing the potential for footing tilt. The figure shown below illustrates load and stress distribution of the footing and underlying soil assuming that footing thickness is 32 in.

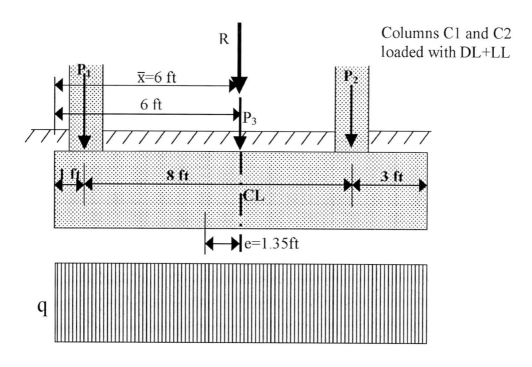

Soil Pressure Diagram for the Combined Footing
Under Service Conditions

P1 = column C1 load = 108 + 54 = 162 k

P2 = column C2 load = 180 + 90 = 270 k

P3 = weight of footing and soil on footing = B×L×[h×0.15 + (footing depth – h)×g]

$$= 6×12×[2.7×0.15 + 1.3×0.12] = 40.4 \text{ k}$$

R = resultant service load on footing = 162 + 270 + 40.4 = 472.4 k

Location of resultant = $\bar{x}$ = (162×1 + 270×9 + 39.9×6)/472.4 = 6 ft OK

q = 472.4/(6×12) = 6.6 ksf < 7.0 ksf OK

It is important that the case of bearing upon the entire footing (Figure 18.8 case a) results for all possible loading cases including:

(a)   Column C1 is subjected to DL and column C2 is subjected to DL and LL, and

(b)   Column C1 is subjected to DL and LL and column C2 is subjected to DL.

(2)   Structural Design:

(a)   Soil Pressure:

The soil pressure diagram for the combined footing under ultimate load conditions 1.2DL + 1.6LL is shown below. It is common that the location of resultant under ultimate condition also coincide with the footing CL as the case for service load. For footing structural design, the weight of footing and soil on top of footing are not included since they do not cause any moment or shear.

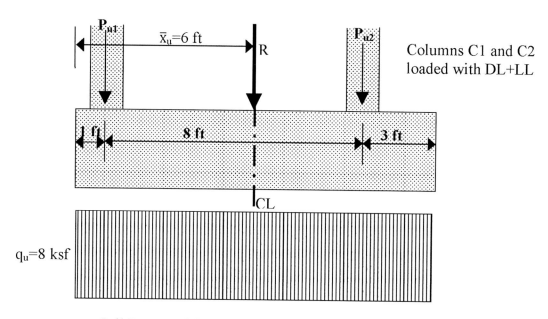

Soil Pressure Diagram for the Combined Footing
Under Ultimate Conditions

P_{u1} = column C1 ultimate load = 1.2×108 + 1.6×54 = 216 k

P_{u2} = column C2 ultimate load = 1.2×180 + 1.6×90 = 360 k

R_{u} = resultant ultimate load on footing = 216 + 360 = 576 k

Location of resultant = $_{u}$ = (216×1 + 360×9)/576 = 6 ft OK q_{u} = 576/(6×12) = 8 ksf

It is important the case of bearing upon the entire footing is (Figure 18.8 case a) results for all ultimate loading cases:

(a)   Column C1 is subjected to DL and column C2 is subjected to DL and LL, and

(b)   Column C1 is subjected to DL and LL and column C2 is subjected to DL.

(b)   Shear Force and Bending Moment Diagrams:

For combined footing structural design, the shear force and bending moment diagrams under the effect of ultimate loads are constructed based on the principles of structural analysis as shown below:

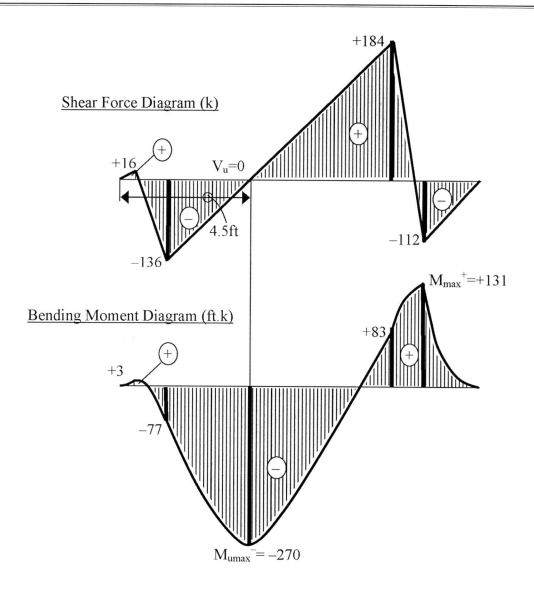

Shear Force Diagram (k)

+184

+16   V$_u$=0

4.5ft

−136   −112

Bending Moment Diagram (ft.k)

M$_{max}^+$=+131

+83

+3

−77

M$_{umax}^-$=−270

(c)   Structural System:

For combined footing structural design, a structural system needs to be selected. It is common that the combined footing is handled as a system of perpendicular beams:

–   A longitudinal wide beam extending between the two columns. Resistance to moment and shear within this beam is achieved by one–way action in the longitudinal direction only where primary reinforcement is used. In the transverse direction, only minimum reinforcement is used.

–   Lateral beams or bands directly at each column location. These beams connect to the longitudinal beam to facilitate the integrated function of the combined footing as a structural element. They resist bending and shear in the transverse direction. The width of each beam is equal to the column width + d/2 on each side of the column where available. As such, the maximum possible width of each beam is column width + d.

The figures that follow illustrate the concept of combined footing structural system for the combined footing. Longitudinal beam main steel is placed at the bottom, thereby, achieving a larger d for resistance of positive bending moments.

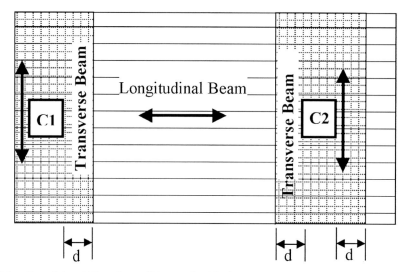

Structural System of the Combined Footing of Example 18.4

(d)   Punching Shear:

Punching shear is required to be checked at each column location. It is handled similar to other examples in this chapter except that column C1 has 3 sides of the concrete footing resisting punching shear rather than 4 sides. Punching shear critical sections are shown in the figure below for columns C1 and C2.

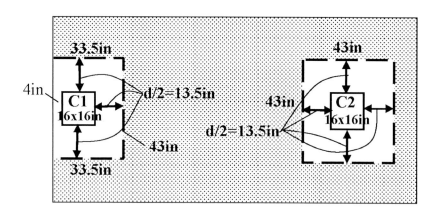

Ultimate Punching Shear Force at Column C1 = 216 − (33.5×43/144)×8 = 136 k

Punching Shear Resistance at Column C1 = 0.75×4×$\sqrt{3,000}$×110×27/1,000 = 488 k

ϕV_n = 488 k >> Q_u = 136 k OK

Ultimate Punching Shear Force at Column C2 = 360 − (43×43/144)×8 = 257 k

Punching Shear Resistance at Column C1 = 0.75×4×$\sqrt{3,000}$×172×27/1,000 = 763 k

ϕV_n = 763 k >> Q_u = 257 k OK

(e)   Longitudinal Beam Design:

One–Way Shear:

Maximum factored one–way shear force at d from column C2 face = Q_u

$Q_u = 184 - 2.25 \times (184 + 136)/6.67 = 76 \text{ k}$

One–way shear resistance $= \phi V_n = 0.75 \times 2 \times \sqrt{3,000} \times 72 \times 27/1,000 = 159 \text{ k}$

$\phi V_n = 159 \text{ kips} \gg Q_u = 76 \text{ kips OK}$

Note: Some designers consider that longitudinal beams in combined footings are required to be designed similar to conventional beams. As such, no lateral steel reinforcement is needed is $Q_u > \phi V_c/2$ which is the reason for increasing the footing depth to 32 in for this example. Footing shear reinforcement is another option instead thickness increase.

Bending Moment:

Maximum factored negative moment $= M_{max}^{-} = -270 \text{ k.ft}$

$A_s^{-} \approx 270 \times 12/(0.8 \times 27 \times 60) = 2.5 \text{ in}^2$

Maximum factored moment $= M_{max}^{+} = +131 \text{ k.ft}$

$A_s^{+} \approx 131 \times 12/(0.8 \times 27 \times 60) = 1.21 \text{ in}^2$

Minimum reinforcement in the longitudinal directions $= 0.0018 \times 72 \times 32 = 4.15 \text{ in}^2$

Use 6#8 for positive and negative reinforcement to extend throughout the whole footing

Use #5@10in o.c. for the top and the bottom secondary reinforcement for the footing.

For computation of secondary reinforcement ratio, the top and bottom secondary reinforcement in the same direction may be added:

$A_{ssecondary} = 0.31 \times 2 \times 12/10 = 0.74 \text{ in}^2/\text{ft} \implies \rho_{secondary} = 0.62/(12 \times 32) = 0.0019 > 0.0018 \text{ OK}$

Note: It is common in concrete footings that shear is more critical than bending moment and minimum steel is typically used as in this example.

(f)   Lateral Beam at C1:

One–Way Shear:

Maximum factored one–way shear force at d from column face $= Q_u$

$Q_u = 36 \times [(6 - 1.33)/2 - 2.25] = 3.1 \text{ k Negligible}$

One–way shear resistance $= \phi V_n = 0.75 \times 2 \times \sqrt{3,000} \times 33.5 \times 27/1,000 = 74 \text{ k}$

$\phi V_n = 75 \text{ k} \gg Q_u = 3.1 \text{ k OK}$

Bending Moment:

Maximum factored moment $= M_{max} = 36 \times [(6 - 1.33)/2]^2/2 = 98 \text{ k.ft}$

$A_s \approx 98 \times 12/(0.8 \times 27 \times 60) = 0.91 \text{ in}^2$

Use 4#5 bottom steel under column C1 in the short direction

(g)   Lateral Beam at C2:

One–Way Shear:

Maximum factored one–way shear force at d from column face $= Q_u$

$Q_u = 60 \times [(6 - 1.33)/2 - 2.25] = 5.2 \text{ k Negligible}$

One–way shear resistance $= \phi V_n = 0.75 \times 2 \times \sqrt{3,000} \times 43 \times 27/1,000 = 95 \text{ k}$

$\phi V_n = 95 \text{ k} \gg Q_u = 5.2 \text{ k OK}$

Bending Moment:

Maximum factored moment $= M_{max} = 60 \times [(6 - 1.33)/2]^2/2 = 160 \text{ k.ft}$

$A_s \approx 160 \times 12/(0.8 \times 27 \times 60) = 1.48$ in^2

Use 6#5 bottom steel under column C2 in the short direction

<u>Note:</u> The steel bars under columns C1 and C2 function as flexural as well as shrinkage steel, thereby, eliminating the need for shrinkage steel at these locations.

Development length of steel bars can be handled similar to Examples 18.2 and 18.3. Termination with hooks for main or primary steel of longitudinal and transverse beams is a common practice.

The footing thickness in Example 18.4 was increased to satisfy beam shear requirements in Chapter 11 requiring shear reinforcement $V_u > \phi V_c/2$ (not included in ACI Code for footings) resulting in a conservative design. An alternative solution for Example 18.4 is presented in the following:

(h)   Bearing:

Bearing for this example can be handled similar to Examples 18.2 and 18.3 where compression development of the dowels is required since the columns are not subjected to bending.

Footing details are shown in the following illustrations:

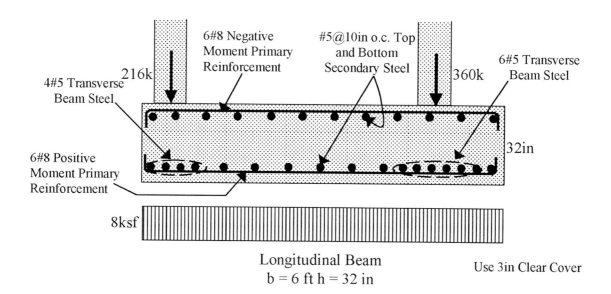

Longitudinal Beam
b = 6 ft h = 32 in

Use 3in Clear Cover

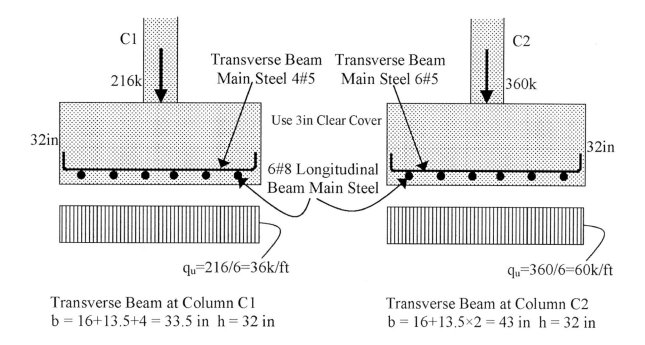

Transverse Beam at Column C1
b = 16+13.5+4 = 33.5 in h = 32 in

Transverse Beam at Column C2
b = 16+13.5×2 = 43 in h = 32 in

Example 18.4 – Alternative Solution to Example 18.3

Assume footing thickness = 24 in while all other parameter remain the same.

Structural Design:

(a) Punching Shear:

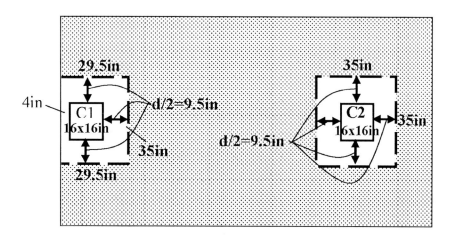

Ultimate Punching Shear Force at Column C1 = 216 − (29.5×35/144)×8 = 158 k

Punching Shear Resistance at Column C1 = 0.75×4×$\sqrt{3,000}$×94×19/1,000 = 293 k

ϕV_n = 293 k >> Q_u = 158 k OK

Ultimate Punching Shear Force at Column C2 = 360 − (35×35/144)×8 = 292 k

Punching Shear Resistance at Column C1 = 0.75×4×$\sqrt{3,000}$×140×19/1,000 = 437 k

ϕV_n = 437 k >> Q_u = 292 k OK

(b)   Longitudinal Beam Design:

One–Way Shear:

Maximum factored one–way shear force at d from column C2 face = Q_u

$Q_u = 184 - 1.58 \times (184 + 136)/6.67 = 108$ k

One–way shear resistance = $\phi V_n = 0.75 \times 2 \times \sqrt{3,000} \times 72 \times 19/1,000 = 112$ k

$\phi V_n = 112$ kips $\gg Q_u = 108$ kips OK

Bending Moment:

Maximum factored negative moment = $M_{max}^- = -270$ k.ft

$A_s^- \approx 270 \times 12/(0.8 \times 19 \times 60) = 3.55$ in^2

Maximum factored moment = $M_{max}^+ = +131$ k.ft

$A_s^+ \approx 131 \times 12/(0.8 \times 19 \times 60) = 1.72$ in^2

Minimum reinforcement in the longitudinal directions = $0.0018 \times 72 \times 24 = 3.11$ in^2

Use 6#7 for positive and negative reinforcement to extend throughout the whole footing

Use #5@12in o.c. for the top and the bottom secondary reinforcement for the footing.

$A_{ssecondary} = 0.31 \times 2 = 0.62$ in2/ft $\Longrightarrow \rho_{secondary} = 0.62/(12 \times 24) = 0.0022 > 0.0018$ OK

(c)   Lateral Beam at C1:

One–Way Shear:

Maximum factored one–way shear force at d from column face = Q_u

$Q_u = 36 \times [(6 - 1.33)/2 - 1.58] = 27$ k

One–way shear resistance = $\phi V_n = 0.75 \times 2 \times \sqrt{3,000} \times 29.5 \times 19/1,000 = 46$ k

$\phi V_n = 46$ kips $> Q_u = 27$ k OK

Bending Moment:

Maximum factored moment = $M_{max} = 36 \times [(6 - 1.33)/2]^2/2 = 98$ k.ft

$A_s \approx 98 \times 12/(0.8 \times 19 \times 60) = 1.28$ in^2

Use 5#5 bottom steel under column C1 in the short direction

(d)   Lateral Beam at C2:

One–Way Shear:

Maximum factored one–way shear force at d from column face = Q_u

$Q_u = 60 \times [(6 - 1.33)/2 - 1.58] = 45$ k

One–way shear resistance = $\phi V_n = 0.75 \times 2 \times \sqrt{3,000} \times 35 \times 19/1,000 = 54$ k

$\phi V_n = 54$ k $> Q_u = 45$ k OK

Bending Moment:

Maximum factored moment = $M_{max} = 60 \times [(6 - 1.33)/2]^2/2 = 160$ k.ft

$A_s \approx 160 \times 12/(0.8 \times 19 \times 60) = 2.10$ in^2

Use 7#5 bottom steel under column C2 in the short direction

The initial design of this footing (Example 18.3) presents, in some parts, trivial calculations as it is very conservative. The alternative solution (Example 18.4) illustrates non trivial analysis and design computation of combined footings per ACI Code. Based on the designer's experience in combined footing, selection of footing thickness may be performed. It is in the author's opinion that increasing footing thickness by 6 in to enhance safety against one–way shear is a prudent approach and should be followed.

Example 18.4 Alternative Solution Summary Sketch

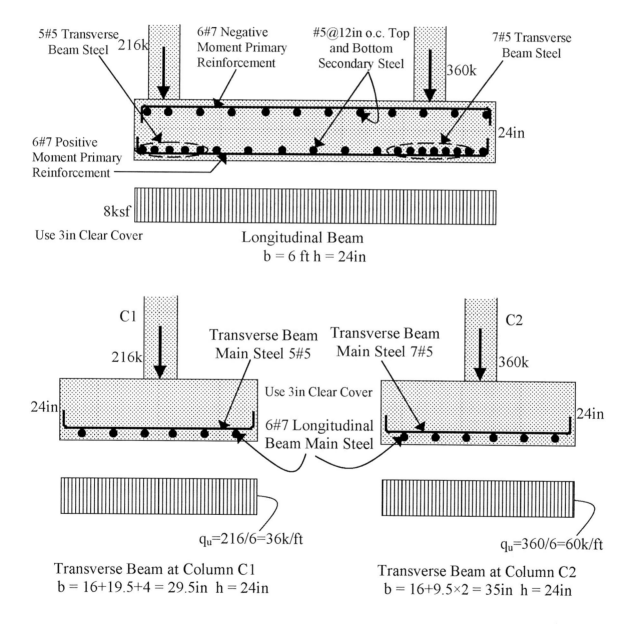

5#5 Transverse Beam Steel   216k   6#7 Negative Moment Primary Reinforcement   #5@12in o.c. Top and Bottom Secondary Steel   7#5 Transverse Beam Steel   360k

6#7 Positive Moment Primary Reinforcement

24in

8ksf

Use 3in Clear Cover

Longitudinal Beam
b = 6 ft h = 24in

C1   216k   Transverse Beam Main Steel 5#5   Transverse Beam Main Steel 7#5   C2   360k

24in   Use 3in Clear Cover   6#7 Longitudinal Beam Main Steel   24in

$q_u=216/6=36k/ft$

$q_u=360/6=60k/ft$

Transverse Beam at Column C1
b = 16+19.5+4 = 29.5in h = 24in

Transverse Beam at Column C2
b = 16+9.5×2 = 35in h = 24in

18.8 Plain Concrete Footings

ACI Code permits the use of plain concrete footings provided that continuous support is provided by the underlying soil. Thus, plain concrete may not be used for pile caps. Plain concrete footings are occasionally utilized for light loads. Omitting steel bars necessitates larger plain concrete footing size. Cost saving and construction simplification could be accomplished with plain footings with light load. Furthermore, the following provisions apply to the design of plain concrete footings:

a.   Plain concrete footings may be designed for support of walls, columns or pedestals. Further explanation of pedestals is presented in Section 18.8(i).

b.   Capacity reduction factor for plain concrete footing shall equal 0.55 ($\phi = 0.55$)

c.   Plain concrete footings shall have a specified compressive strength equal or more than 2,500 psi ($f'_c \geq 2,500$ psi).

d.   Thickness of plain concrete footings shall not be less than 8 in. As footings are placed against soil, ACI Code specifies that, for flexural and shear design, effective thickness (h) utilized shall be 2 in less than actual thickness.

e.   Critical section for one–way shear is at h from the face of column, pedestal or wall. Plain concrete footing strength in one–way shear is:

$$V_n = \frac{4}{3}\sqrt{f'_c}\ b_v h \qquad\qquad (18.26)$$

Where:

   V_n = nominal one–way shear resistance of plain concrete.

   $b_v h$ = footing one–way shear critical section's dimensions. Effective thickness (h) shall be taken as actual thickness minus 2 in. b_v = L or B for single footings. b_v for wall footings is 1 ft or 12 in to simplify analysis as explained earlier in this chapter.

f.   Critical section for punching shear is at h/2 from the face of column or pedestal. Plain footing strength in punching shear is:

$$V_n = \left[\frac{4}{3} + \frac{8}{3\beta_c}\right]\sqrt{f'_c}\,b_0 h \ \ \text{not to exceed} \ \ 2.66\sqrt{f'_c}\,b_0 h \qquad\qquad (18.27)$$

Where:

   V_n = nominal punching shear resistance of plain concrete.

   b_0 = perimeter of punching shear area at distance d/2 from column face,

   β_c = ratio of column or pedestal section long side to short side = ℓ/b as explained earlier in this chapter.

   h = footing effective thickness as explained above.

g.   Critical section for bending moment is at the face of column, pedestal or wall. Plain footing strength in bending is:

$$M_n = 5\ \sqrt{f'_c}\ b_v h^2/6 \qquad\qquad (18.28)$$

Where:

   M_n = nominal moment resistance of plain concrete.

   $b_v h$ = footing moment critical section's dimensions. Effective thickness (h) shall be taken as actual thickness minus 2 in. b_v = L or B for single footings. b_v for wall footings is 1 ft or 12 in to simplify analysis as explained earlier in this chapter.

h.   Although steel is not required for plain footings, many designers utilize longitudinal steel bars in wall footings for shrinkage resistance, and

i.   Plain concrete footings are used as pedestal support as shown in Figure 18.25. The exposed height of pedestal shall not exceed three times its average cross–sectional dimension ($Y \leq 3X$ in Figure 18.25).

Analysis and design of plain concrete footings is a special case of reinforced concrete footings. This section outlines the particulars of plain footings from the standpoint of dissimilarity with reinforced footings. Review should be combined with other sections in this chapter.

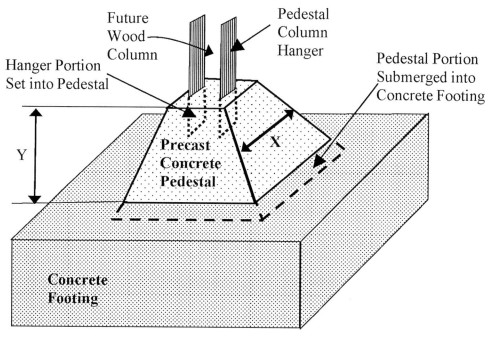

Figure 18.25: Precast concrete pedestal with hanger for wood column set into and supported by a cast–in–place plain concrete footing.

18.9 Concluding Remarks

Analysis Remarks

1.   The complexity of design of footings subjected to bending and axial force is revealed in Example 18.3. Most designers tend to design buildings or structures so that only axial forces are applied to footings due lateral loading (wind or earthquake). With this, efficient footing design can be accomplished.

2.   The presentation of combined footings in this chapter is somewhat brief as it assumes certain knowledge of the principles of structural analysis of the reader.

3.   It is important to design combined footings to achieve a uniform soil pressure. This can be done by adjusting the footing dimensions so that the resultant force location coincides with the footing center of area. Such design precaution reduces tilting of footings and possible uneven deformations.

4. Mat foundation and pile caps are generally designed as series of beams receiving building loads and transferring these loads to the underlying soil or piles. These beam series form a perpendicular grid to achieve combined force resisting action of the mat foundation or pile cap.

Design Remarks

1. In footing design for punching shear and bearing, commonly designers use the "not to exceed part" of the design equations by ACI Code for simplification.

2. For deep footings (about 36in and greater), designers commonly utilize two mats of steel to reduce shrinkage cracking. The added top mat typically consists of two perpendicular layers of steel bars with each complying with the recommended minimum reinforcement ratio.

3. Based on the discussion of single concrete footings in this chapter, one may conclude that stresses due to moment and shear are maximized near the column location. For materials saving, some designers utilize pear–shaped or stepped footings. Such footings have a larger depth near column location compared to the remainder of footing (Figures 18.26). However, these types of footings require extra care during construction.

4. For footing that are subjected to eccentric axial loading due to dead load, designers commonly design a footing with eccentrically loaded column to achieve a uniform soil pressure. Thus, tilting and uneven footing deformations can be avoided. An example of this situation is shown in Figure 18.27.

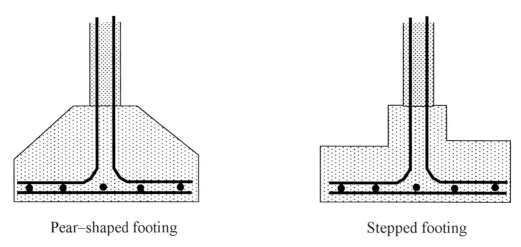

Pear–shaped footing   Stepped footing

Figure 18.26: Pear–shaped and stepped reinforced concrete footing.

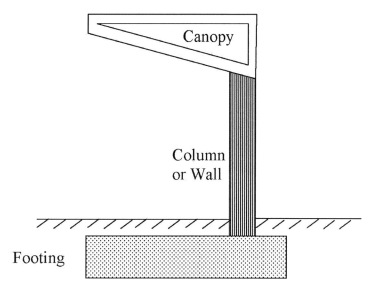

Figure 18.27: Reinforced concrete footing under eccentric dead load.

5. In addition to rectangular combined footings, designers also use trapezoidal combined footings and combined footings connected with a grade beam (beam at soil or grade level). See Figure 18.28 for illustration. Trapezoidal shaped footings are typically reinforced with fan shaped steel. Grade beams in combined footings are designed similar to the longitudinal beam in Example 18.4

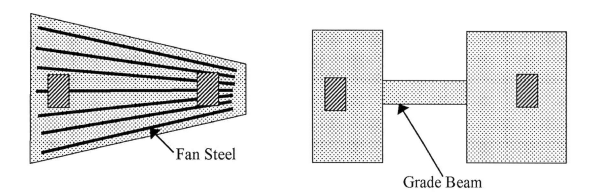

Trapezoidal footing with fan shaped steel   Combined footing with grade beam

Figure 18.28: Examples of combined footings.

6. In longitudinal beams of combined footings, designers tend to utilize the design principles of conventional beams where shear reinforcement is needed if $V_u > \phi V_c/2$. Thus footing thickness requires increase or transverse reinforcement (typically closed stirrups) is needed in longitudinal beams if $V_u > \phi V_c/2$. See Example 18.4 for further details.

7. Combined footings are typically designed to be at 2 in from Property Line (PL) and not directly at PL to avoid encroachment into the neighboring property since PL survey includes error or variation. Furthermore, with such minimal setback, footing construction does not need to disturb the adjacent property.

Construction Remarks

1.   In footing construction, it is typical to add a shear key between the footing and column or wall. A piece of wood is used to form the shear key which is removed before setting of concrete. This wood piece is painted with a debonding agent to facilitate removal from fresh concrete. Furthermore, this shear key is used as a leveling and positioning guide for the footing.

Test your knowledge

1.   What are the common types of reinforced concrete footings?

2.   In addition to soil characteristics, what are the other factors affecting allowable soil bearing pressure?

3.   Why are the weights of footing and soil on footing not included in structural design?

4.   What are the failure modes for design of wall or continuous footings?

5.   What are the failure modes for design of column or isolated footings?

6.   How many ASD load combinations are required for the geotechnical design of footing under axial load and moment?

7.   How many LRFD load combinations are required for the structural design of footing under axial load and moment?

8.   Why do designers adjust the dimensions of combined footings to achieve a uniform soil pressure?

9.   Explain the structural system of combined footings?

10.   Explain the discrepancy of shear longitudinal beam in combined footings compared to conventional beams?

11.   What is the required minimum reinforcement ratio for concrete footings by ACI Code?

12   What is the recommended minimum reinforcement ratio in concrete footings?

13.   Explain the reason for using pear–shaped or stepped reinforced concrete footings.

Design wall footings for the following conditions:

No.	DL (lb/ft)	LL (lb/ft)	Footing Depth (ft)	t (in)	q_a (ksf)	γ (pcf)	f'_c (ksi)	f_y (ksi)
14	6	4	3	6	2.5	120	2.5	40
15	8	5	2	8	3	110	2.5	60
16	10	6	3	10	4	120	3	60
17	12	7	4	12	3.5	115	3.5	60
18	14	6	3	12	2.5	120	2.5	60
19	15	8	3	12	3	125	3	60
20	7	5	3	8	2	120	2.5	40
21	20	12	4	14	4	115	3	60
22	7	5	3	8	1.5	120	2.5	40
23	3	2	2	6	1	120	2.5	40

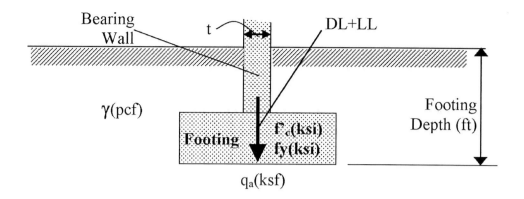

Design column footings for the following conditions:

No.	DL (kips)	LL (kips)	Footing Depth (ft)	$\ell \times b$ (in$\times$in)	q_a (ksf)	γ (pcf)	f'_c (ksi)	f_y (ksi)
24	50	30	2	12×12	2	120	2.5	40
25	80	40	2.5	12×12	1.5	115	2.5	60
26	40	20	2	10×10	1	110	2.5	60
27	150	120	3	12×12	4	120	2.5	60
28	160	90	3	14×14	4.5	125	3	60
29	180	110	3	14×14	5	120	4	60
30	200	150	4	15×15	4.5	120	3	60
31	130	70	3	12×12	3	120	2.5	60
32	140	110	2.5	12×12	3.5	110	3	60
33	30	20	1.5	8×8	1	120	2.5	40

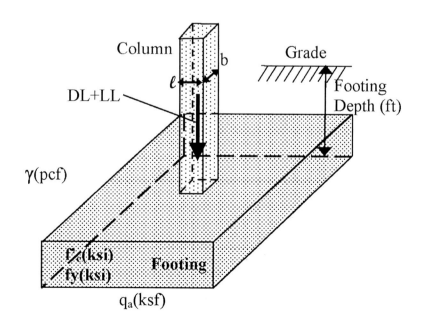

Design column footings for the following conditions with $f_c' = 3$ ksi and $f_y = 60$ ksi:

No.	DL (kips)	LL (kips)	PE (kips)	ME (ft.k)	PW (kips)	MW (ft.k)	Footing Depth(ft)	$\ell \times b$ (in)	q_a (ksf)	γ (pcf)
34	50	30	25	0	0	0	2	10×10	4	120
35	80	40	20	80	0	0	2.5	12×12	5.5	110
36	110	75	35	140	0	0	3	12×12	6	115
37	150	120	40	250	0	0	2.5	12×12	6	120
38	220	160	60	300	0	0	4	16×16	7	125
39	190	120	120	0	0	0	3	14×14	5	110
40	70	40	0	0	40	0	2	10×10	4	120
41	90	40	0	0	50	20	3	12×12	5	130
42	190	100	0	0	50	120	4	14×14	6	115
43	130	90	0	0	30	90	3	14×14	6	125

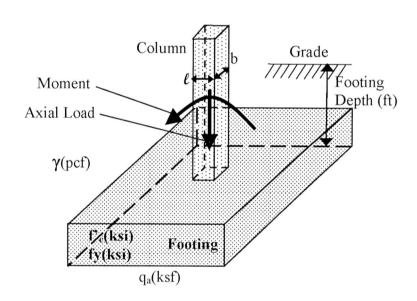

Design combined footings for the following conditions with $f_c' = 4$ ksi $f_y = 60$ ksi and $\gamma = 120$ pcf:

No.	C_1			C_2			Distance C_1 to C_2 (ft)	C_1 Edge to PL (in)	Footing Depth(ft)	q_a (ksf)
	DL (kips)	LL (kips)	l×b (in)	DL (kips)	LL (kips)	l×b (in)				
44	150	110	12×12	70	50	12×12	6	6	3	5
45	220	140	14×14	90	60	12×12	7	12	4	6
46	180	100	14×14	80	40	10×10	8	9	3.5	5.5
47	350	140	18×18	140	90	12×12	7.5	10	4	6
48	400	170	20×20	150	100	12×12	5	7	4	7
49	450	230	22×22	180	110	14×14	8	8	5	6.5
50	250	130	16×16	100	50	12×12	6	6	4	6

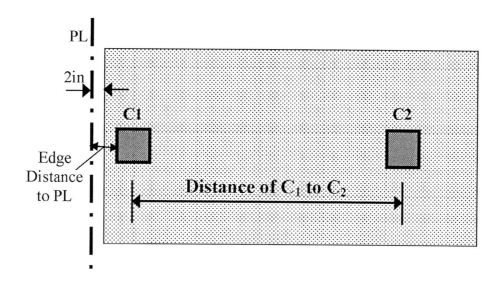

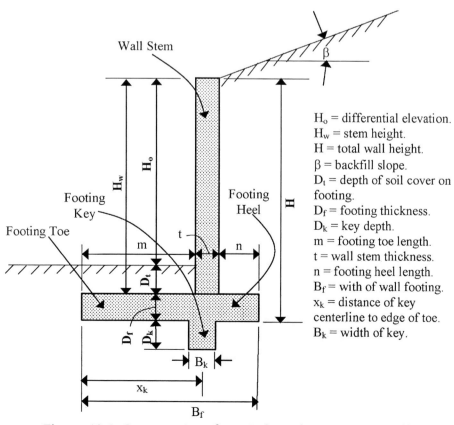

Chapter 19

REINFORCED CONCRETE CANTILEVER RETAINING WALLS

19.1 Introduction

Retaining walls are used to expand flat or usable areas of land. They are used around highways, streets and railroads to facilitate construction and utilization of such transportation systems. Soil is filled behind the retaining wall to achieve the required objective of expanding usable areas. Figure 19.1 illustrates the different parts and nomenclature used for cantilever retaining walls.

Wall Stem

β

Footing
Key

Footing
Heel

Footing Toe

H_w

H_o

H

t

m

n

D_t

D_f

D_k

B_k

x_k

B_f

H_o = differential elevation.
H_w = stem height.
H = total wall height.
β = backfill slope.
D_t = depth of soil cover on footing.
D_f = footing thickness.
D_k = key depth.
m = footing toe length.
t = wall stem thickness.
n = footing heel length.
B_f = with of wall footing.
x_k = distance of key centerline to edge of toe.
B_k = width of key.

Figure 19.1: Cross–section of a typical cantilever retaining wall.

Cantilever retaining walls can be constructed on a cut or fill in the soil. Walls on a cut (termed restraining walls) are designated by long toe and short heel while walls on a fill (termed retaining walls) are designated by short toe and long heel.

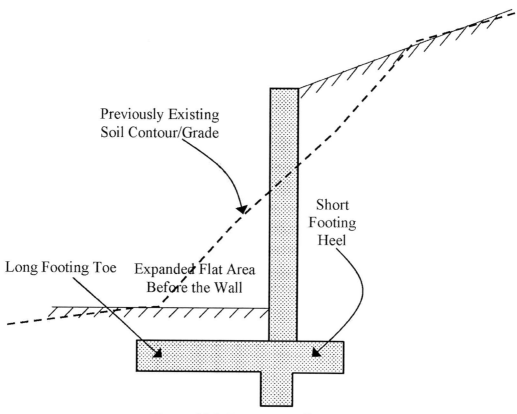

Figure 19.2: Retaining wall at a cut.

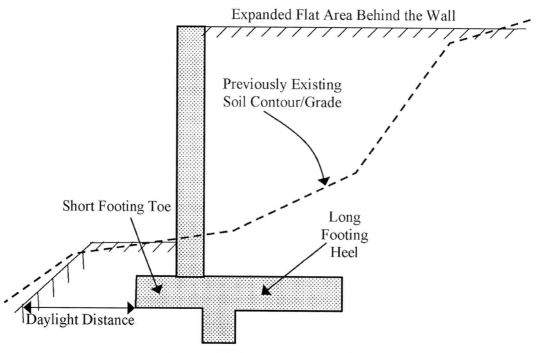

Figure 19.3: Retaining wall at a fill.

19.2 Geotechnical Engineering Design Aspects of Cantilever Retaining Walls

Retaining walls are designed to withstand soil pressure caused by the retained earth. The soil or geotechnical project engineer prepares a report describing the soils at the site from various aspects. The following information within the soils report is important for retaining wall design:

1.   Soil unit weight = γ

2.   Soil angle of internal friction = Φs

3.   Allowable soil bearing pressure = q_a

4 .   Coefficient of friction between the soil and concrete = μ

5.   Reduction factor for use of friction and passive resistance = R

6.   Required minimum footing depth = $D_t + D_f$

7.   Minimum distance to daylight

8.   Required drainage behind the retaining wall

The soil forces acting on cantilever retaining walls are illustrated in Figure 19.4 where the following notations apply:

1. P_a, P_{as} = resultant active soil loads due to soil weight and surcharge applied at H/3 and H/2, respectively.

2. W_s = surcharge load behind the retaining wall normally applied at flat backfill.

3. K_a, K_p = coefficients of active and passive soil pressure, respectively.

4. δ = angle of friction of the soil and wall.

5. P_p = resultant passive soil pressure.

6. $D = D_t + D_f + D_k$ = depth of footing and key used for passive resistance.

7.   1 ft in dashed line is neglected passive resistance of top soil due to potential disturbance.

8.   F = friction resistance between the footing and soil.

9. q_t, q_h = soil pressure on the footing toe and footing heel.

The active soil pressure behind the wall is resisted by the passive resistance before the wall, soil pressure and friction on the footing in addition to wall and soil weight. Active soil pressure behind the retaining wall (P_a and P_{as}) is applied at angle δ with the surface of the wall due to friction between the wall and soil. Angle δ ranges between $\Phi_s/2$ and $2\Phi_s/3$. The friction component of active soil pressure is advantageous to wall stability as it thrusts the wall towards the underlying soil. Designers commonly assume that $\delta = 0$ as such assumption simplifies design and results in enhanced safety.

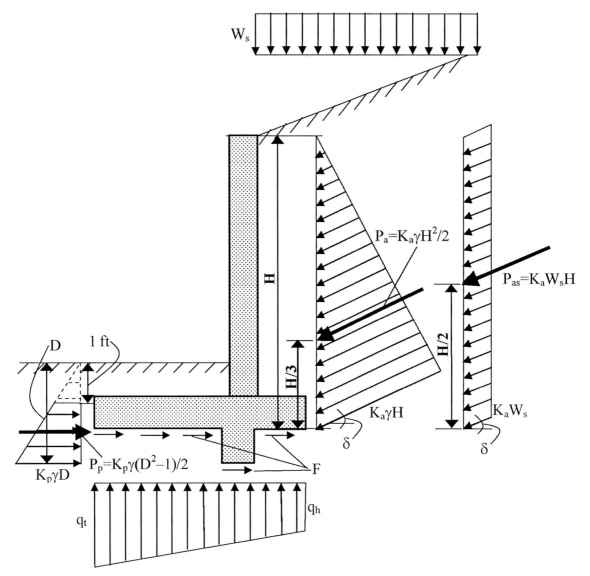

Figure 19.4: Soil forces on cantilever retaining wall.

There are several methods for determining the value of K_a and K_p using Rankine's theory, Coulomb's theory or others. Rankine's theory assumes that soil pressure (active or passive) is inclined at angle β (backfill slope) with the horizontal direction. Typically the soil before the retaining wall is flat resulting in a horizontal passive resistance (Figure 19.5). K_a and K_p can be determined according to the following equations:

$$K_a = \cos \beta \, \frac{\cos \beta - \sqrt{\cos^2\beta \, - \cos^2\phi s}}{\cos \beta + \sqrt{\cos^2\beta \, - \cos^2\phi s}}$$

$$K_p = \cos \beta_2 \, \frac{\cos \beta_2 - \sqrt{\cos^2\beta_2 \, - \cos^2\phi s}}{\cos \beta_2 + \sqrt{\cos^2\beta_2 \, - \cos^2\phi s}}$$

Rankine's equations

Where β and β_2 are slope angles for backfill and soil before the wall, respectively.

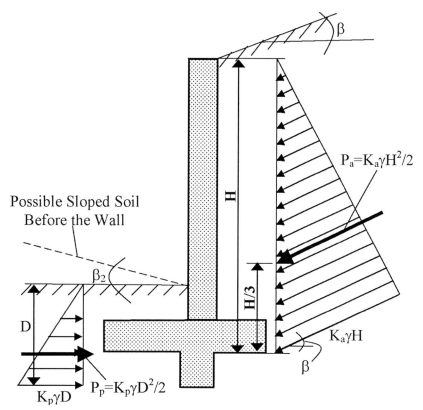

Figure 19.5: Active and passive soil pressure based on Rankine's theory.

According to Rankine's theory, the soil pressure caused by surcharge is also inclined at angle β with respect to the horizontal direction as shown in Figure 19.5. For the special case of horizontal backfill, the equations for K_a and K_p using Rankine's theory can be rewritten as follows:

$$\left. \begin{array}{l} K_a = \dfrac{1 - \sin\phi_s}{1 + \sin\phi_s} \\[2em] K_p = \dfrac{1 + \sin\phi_s}{1 - \sin\phi_s} \end{array} \right\} \quad \text{Simplified Rankine's equations for flat backfill}$$

The following equations for K_a and K_p and Figure 19.6 are based on Coulomb's theory for active and passive soil pressure:

$$\left. \begin{array}{l} K_a = \dfrac{\cos^2(\phi_s - \theta)}{\cos^2\theta \, \cos(\delta + \theta) \left[1 + \sqrt{\dfrac{\sin(\phi_s + \delta)\sin(\phi_s - \beta)}{\cos(\delta + \theta)\cos(\beta - \theta)}} \right]^2} \\[3em] K_p = \dfrac{\cos^2(\phi_s - \theta_2)}{\cos^2\theta_2 \, \cos(\delta - \theta_2) \left[1 - \sqrt{\dfrac{\sin(\phi_s + \delta)\sin(\phi_s + \beta_2)}{\cos(\delta - \theta_2)\cos(\beta - \theta_2)}} \right]^2} \end{array} \right\} \quad \text{Coulomb's equations}$$

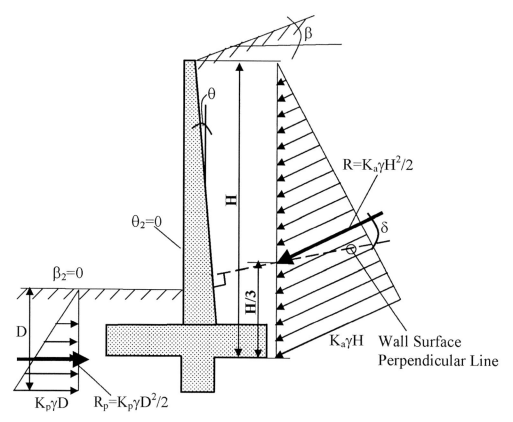

Figure 19.6: Active and passive soil pressure based on Coulomb's theory.

θ and θ_2, in Coulomb's equations for K_a and K_p, are the inclination angles of the wall face at the backfill side and front, respectively. Coulomb's theory assumes that soil pressure is at angle δ from the wall face due to friction. In Figure 19.6, it is assumed that $\theta_2 = \beta_2 = 0$ and that passive resistance is horizontal for simplification.

Coulomb's theory for active and passive pressure can be simplified if $\beta = \theta = 0$ assuming a flat backfill and vertical stem. Friction is a stabilizing force in retaining walls. Neglecting friction in retaining wall design results in safer designs ($\delta = 0$). With these simplifying assumptions, Coulomb's equations for K_a and K_p can be reduced as follows:

$$\left. \begin{array}{l} K_a = \dfrac{\cos^2 \phi_s}{(1 + \sin \phi_s)^2} \\[2em] K_p = \dfrac{\cos^2 \phi_s}{(1 - \sin \phi_s)^2} \end{array} \right\}$$ Simplified Coulomb's equations with flat backfill vertical wall surface and zero friction

Rankine's and Coulomb's simplified equations for K_a and K_p are the same but written in alternative trigonometric forms. Thus, Rankine's and Coulomb's theories match for the conditions of zero friction, vertical wall surface and flat backfill.

To simplify wall design, geotechnical engineers commonly include in the project soils report the equivalent fluid pressure (EFP) and passive resistance (p) representing active and passive earth pressure on retaining walls, respectively. The following equations are considered the theoretical basis for these design parameters:

$$EFP = \gamma \cdot Ka \tag{19.1}$$

$$p = \gamma \cdot K_p \qquad \qquad (19.2)$$

Where K_a and K_p are the coefficients of active and passive soil pressure, respectively.

Important Comments

1. The previous discussion of retaining walls ignores the effect of cohesion on soil pressure. Such practice is common among soil engineers to safeguard against surface tension cracks in cohesive soil due to drying (Figure 19.7).

2. Passive resistance against the retaining wall key is considered by certain theories to extend along an influence line of Φ_s. With this, key location near the footing heel is more advantageous for passive resistance. This effect is neglected by most designers as a safety precaution.

3. For surcharge load at a distance from the wall stem, a 45° influence line is used to define the range of application of soil pressure due to such loading (Figure 19.7).

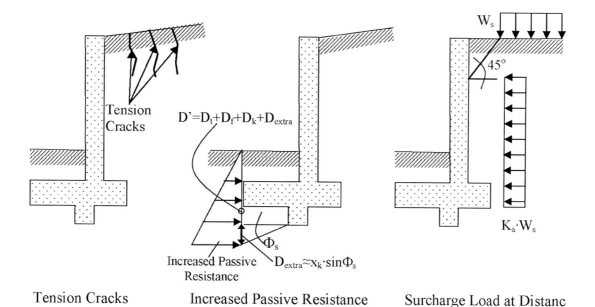

Tension Cracks   Increased Passive Resistance   Surcharge Load at Distanc

Figure 19.7: Special considerations for retaining walls.

Geotechnical engineering considerations for retaining walls design include computing the safety factor against the potential failure modes listed below and shown in Figure 19.8:

1. Overturning about point O at the edge of the footing toe.

2. Sliding at the bottom of the footing.

3. Bearing failure due to soil pressure at the edge of the footing toe.

4. Global soil failure within the retained earth.

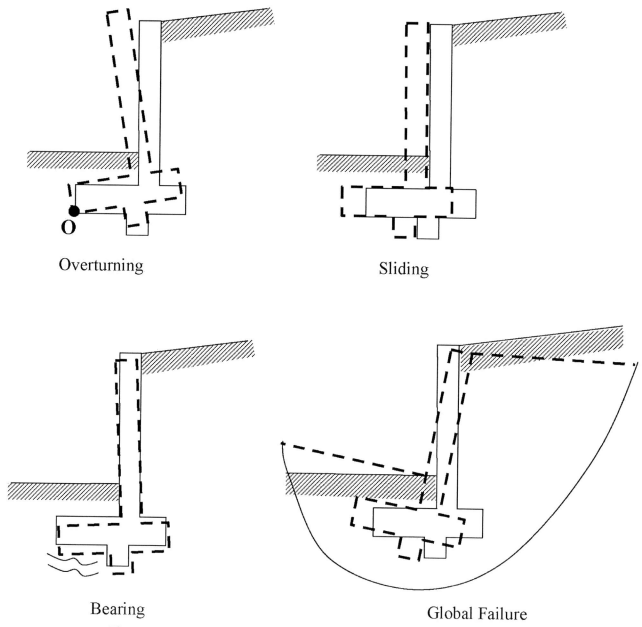

Overturning

Sliding

Bearing

Global Failure

Figure 19.8: Soil engineering failure modes for cantilever retaining walls.

Most designers use the simplified force diagram for geotechnical analysis cantilever retaining walls shown in Figure 19.9 where soil pressure is horizontal (no friction). Figure 19.9 will be used in this textbook.

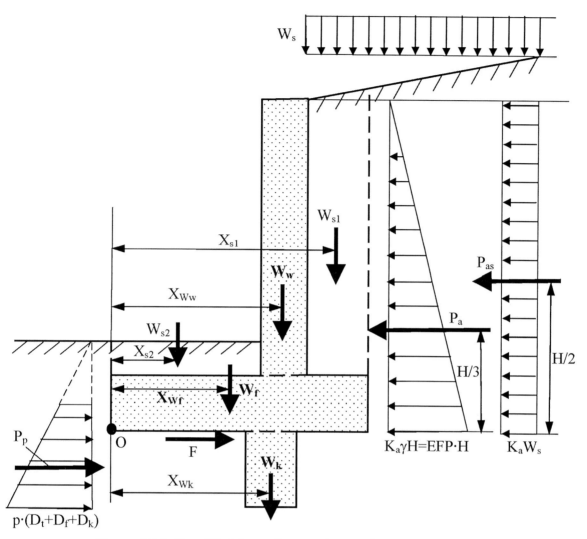

Figure 19.9: Simplified force diagram for cantilever retaining wall.

Figure 19.9 requires review in conjunction with Figure 19.1 which shows the general dimensions of a cantilever retaining wall. Based on these figures, the following force equations for a representative 1 linear foot section of the wall can be obtained:

W_w = weight of wall stem = $t \cdot H_w \cdot \gamma_c$

W_f = weight of footing = $D_f \cdot B_f \cdot \gamma_c$

W_k = weight of key = $B_k \cdot D_k \cdot \gamma_c$

W_{s1} = weight of soil behind the wall $\approx n \cdot H_w \cdot \gamma + W_s \cdot n$

W_{s2} = weight of soil in front of the wall = $D_t \cdot a \cdot \gamma$

P_a = active soil pressure resultant due to soil weight = $EFP \cdot H^2/2$

P_{as} = active soil pressure resultant due to surcharge = $K_a \cdot W_s \cdot H$

P_p = passive soil pressure resultant = $p \cdot [(D_t + D_f + D_k)^2 - 1]/2$ where the top 1 ft is ignored

F = friction force between the wall footing and soil = $(Ww + Wf + Wk + Ws1 + Ws2) \cdot \mu$

Where γ and γ_c are unit weights of soil and concrete. While γ is included in the soils report, γ_c = 145 to 150 pcf, typically

Furthermore, force moment arms about point O are listed below:

X_{Ww} = moment arm for weight of wall stem = a + t/2

X_{Wf} = moment arm for weight of footing = $B_f/2$

X_{Wk} = moment arm for weight of key = x_k = $2B_f/3$ typically

X_{s1} = moment arm for weight of soil behind the wall ≈ $B_f - n/2$

X_{s2} = moment arm for weight of soil in front of the wall = a/2

Y_{Pa} = moment arm for active soil pressure due to soil weight = H/3

Y_{Pas} = moment arm for active soil pressure due to surcharge = H/2

The following equations are used for cantilever retaining wall geotechnical engineering design:

$$\text{Overturning Moment} = \text{OTM} = P_a \cdot Y_{Pa} + P_{as} \cdot Y_{Pas}$$

$$\text{OTM} = \text{EFP} \cdot H^3/6 + K_a \cdot W_s \cdot H^2/2 \tag{19.3}$$

$$\text{Resisting Moment} = \text{RM} = W_w \cdot X_{Ww} + W_f \cdot X_{Wf} + W_k \cdot X_{Wk} + W_{s1} \cdot X_{s1} + W_{s2} \cdot X_{s2}$$

$$\text{RM} = [t \cdot H_w(a+t/2) + D_f \cdot B_f^2/2 + 2B_k \cdot D_k \cdot B_f/3]\gamma_c + (n \cdot H_w \cdot \gamma + W_s \cdot n)(B_f - n/2) + D_t \cdot \gamma \cdot a^2/2 \tag{19.4}$$

<u>Safety Factor against Overturning</u> = RM/OTM ≥ 1.5 per the IBC Code

$$\text{Total Weight} = W_w + W_f + W_k + W_{s1} + W_{s2}$$

$$\text{Total Weight} = (t \cdot H_w + D_f \cdot B_f + B_k \cdot D_k)\gamma_c + n \cdot H_w \cdot \gamma + W_s \cdot n + D_t \cdot a \cdot \gamma = \Sigma W \tag{19.5}$$

$$\text{Sliding Force} = P_a + P_{as} = \text{EFP} \cdot H^2/2 + K_a \cdot W_s \cdot H \tag{19.6}$$

$$\text{Resisting Force} = R \cdot (P_p + F) = R \cdot \{p \cdot [(D_t + D_f + D_k)^2 - 1]/2 + \Sigma W \cdot \mu\} \tag{19.7}$$

<u>Safety Factor against Sliding</u> = $R \cdot (P_p + \Sigma W \cdot \mu)/(P_a + P_{as})$ ≥ 1.5 per the IBC Code

For determining the soil bearing pressure on cantilever retaining wall footing, total weight eccentricity requires computation as illustrated in Equation 19.8 and Figure 19.10 based on the concept of equivalent force systems in Chapter 12 Figure 18.8:

$$e = B_f/2 - X_{\Sigma W} = B_f/2 - (\text{RM} - \text{OTM})/\Sigma W \tag{19.8}$$

Soil bearing pressure at the footing toe and heel, q_t and q_h, respectively, can be determined as follows:

$$\left.\begin{array}{l} q_t = \Sigma W/B_f + 6\Sigma W \cdot e/B_f^2 \\ q_h = \Sigma W/B_f - 6\Sigma W \cdot e/B_f^2 \end{array}\right\} \tag{19.9}$$

It is required that e < $B_f/6$ so that the case of bearing is on the entire footing results rather than on a portion of the footing as discussed in Chapter 18 and illustrated in Figures 18.7 and 18.8.

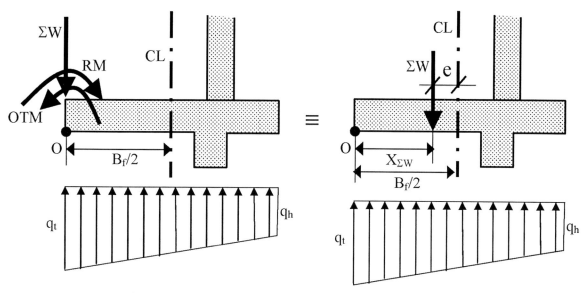

Figure 19.10: Eccentric loading on cantilever wall footing.

For a safe design, the following equations require satisfaction:

$$\text{and} \quad \left. \begin{array}{l} q_t \le q_{all} \\[2mm] q_h > 0 \quad \text{or} \quad e < B_f/6 \end{array} \right\} \qquad (19.10)$$

From a geotechnical engineering standpoint, the factors of safety against overturning and sliding are required to be greater than 1.5 (IBC Code), soil pressure at the toe is required to be less than allowable bearing pressure, and soil pressure at the heel is required to be greater than 0. Furthermore, the factor of safety against global failure should also be greater than 1.5. However, global failure is not discussed in this textbook as it pertains more to soil engineering textbooks.

Important Comment

A value of weight eccentricity less than zero (e < 0) from Equation 19.18 indicates that wall failure is towards the retained soil which is unrealistic. With this, the retaining wall resisting moment is significantly greater than the overturning moment and, consequently, the wall is safe. For calculation purposes it is assumed that e = 0.

19.3 Structural Design of Cantilever Retaining Walls

19.3.1 Stem Analysis and Design

Figure 19.11 illustrates soil forces on the stem of a cantilever retaining wall under service conditions. Section 1–1 is the most critical section for bending moment and shear. Axial forces from wall weight and soil–wall friction are generally neglected by designers. Axial forces in the stem are typically small and increase moment and shear resistance. Thus, neglecting axial forces enhances design safety.

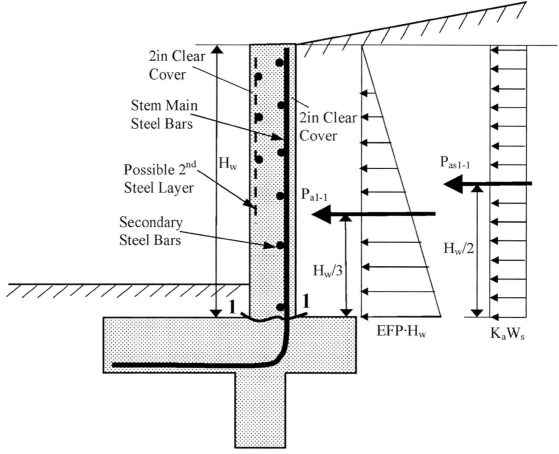

Figure 19.11: Forces acting on a stem of a cantilever retaining wall.

Based on Figure 19.11, magnified shear and moment at section 1-1 can be expressed as follows:

$$V_{u1\text{-}1} = 1.6(EFPH_w^2/2 + K_aW_sH_w)$$

$$M_{u1\text{-}1} = 1.6(EFPH_w^3/6 + K_aW_sH_w^2/2) \qquad (19.11)$$

Retaining wall stems are generally not reinforced against shear. Thus, concrete is required to resist all shear forces as shown below:

$$\phi V_n = 0.75 \times 2 \times \sqrt{f_c'} \times 12d > V_{u1\text{-}1} = 1.6(EFPH_w^2/2 + K_aW_sH_w) \qquad (19.12)$$

Equation 19.12 uses units of in and lb. Stem effective depth d is calculated based on following equation:

$$d = t - 2\ in - d_b/2$$

Where:

t = stem thickness.

2 in = ACI Code required minimum clear cover in concrete in contact with soil or under exterior exposure.

d_b = diameter of main steel bars in the stem.

To determine the required main steel reinforcement in the stem based on bending moment, the following equation discussed in Chapter 18 may be used:

$$A_s = M_u/0.8df_y \qquad (19.13)$$

Where A_s is stem reinforcement per linear foot. ACI Code specifies that $A_s/12t > 0.0012$ for steel bars < #5 with $f_y = 60$ ksi or welded wire fabric and $A_s/12t > 0.0015$ for other cases. 12 in is used since A_s is reinforcement per linear foot. Furthermore, ACI Code states that the spacing of main steel in walls cannot be grater 3t or 18 in whichever is smaller. Main reinforcement need not be enclosed with stirrups or ties if $A_s/12t < 0.01$. ACI Code also states that walls with stems > 10in thick require two layers of steel bars near the back of the wall and near the front of the wall. The reinforcement ratio of the sum both layers to gross wall area shall be greater than minimum reinforcement ratio of 0.0012 or 0.0015.

ACI Code requires secondary reinforcement in walls with a reinforcement ratio based on gross stem area not less than 0.002 for steel bars < #5 with $f_y = 60$ ksi or welded wire fabric and 0.0025 for other cases. Spacing of secondary steel bars cannot be grater 3t or 18 in whichever is smaller. Walls with stems > 10 in thick require two layers of secondary steel bar reinforcement near the back of the wall and near the front of the wall. The reinforcement ratio of the sum both layers to gross wall area shall be greater than minimum reinforcement ratio of 0.0020 or 0.0025.

Development length requirements by ACI Code must be satisfied for stem main steel bars within the stem and footing.

19.3.2 Footing Analysis and Design

Figure 19.12 illustrates soil bearing pressure on the footing of a cantilever retaining wall under service conditions. Sections 2–2 and 3-3 are the most critical sections for bending moment and shear for the toe and heel, respectively.

The values of q_{t1} and q_{h1} (soil bearing pressure at the wall face and back, respectively, as shown in Figure 19.12) are required for calculating the shear and moment at the toe and heel critical section and may be determined as follows:

$$q_{t1} = q_t - (q_t - q_h)a/B_f$$

$$q_{h1} = q_h + (q_t - q_h)n/B_f$$

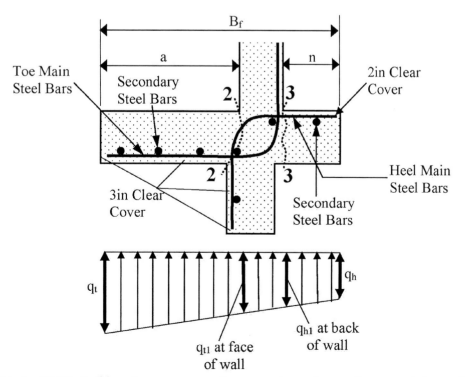

Figure 19.12: Soil bearing pressure acting on a stem of a cantilever retaining wall.

It is common that designers multiply the soil pressure on footing under service conditions (obtained in geotechnical design) with 1.6 to obtain ultimate soil pressure for structural design of footings. This simplification results in a safer design. This procedure will be followed. Using the method of determining shear and moment for trapezoidal pressure distribution discussed in Chapter 12 Sections 18.6.2 and 18.6.3, the following equations can be obtained:

$$\left.\begin{aligned} V_{u2\text{-}2} &= 1.6a(q_t + q_{t1})/2 \\ M_{u2\text{-}2} &= 1.6(2q_t + q_{t1}){\cdot}a^2/6 \end{aligned}\right\} \tag{19.14}$$

$$\left.\begin{aligned} V_{u3\text{-}3} &= 1.6n(q_h + q_{h1})/2 \\ M_{u3\text{-}3} &= 1.6(2q_h + q_{h1}){\cdot}n^2/6 \end{aligned}\right\} \tag{19.15}$$

As footings are typically not reinforced against shear, concrete is required to resist $V_{u2\text{-}2}$ and $V_{u3\text{-}3}$ as expressed below:

$$\left.\begin{aligned} \phi V_n &= 0.75{\times}2{\times}\sqrt{f'_c}{\times}12d_{toe} > V_{u2\text{-}2} = 1.6a(q_t + q_{t1})/2 \\ \phi V_n &= 0.75{\times}2{\times}\sqrt{f'_c}{\times}12d_{heel} > V_{u3\text{-}3} = 1.6n(q_h + q_{h1})/2 \end{aligned}\right\} \tag{19.16}$$

Equations 19.16 use units of in and lb. Footing effective depth d is calculated based on the following equations:

$$d_{toe} = D_f - 3 \text{ in} - d_{btoe}/2$$

$$d_{heel} = D_f - 2 \text{ in} - d_{bheel}/2$$

Where:

D_f = footing thickness.

3 in = ACI Code required minimum clear cover in concrete poured against soil.

2 in = ACI Code required minimum clear cover for the heel similar to the stem.

d_{btoe}, d_{bheel} = diameter of main steel bars in the toe and the heel, respectively.

To determine the required main steel reinforcement in the footing based on bending moment, the following equation discussed in Chapter 18 may be used:

$$\left.\begin{aligned} A_{stoe} &= M_{u2\text{-}2}/0.8d_{toe}f_y \\ A_{sheel} &= M_{u3\text{-}3}/0.8d_{heel}f_y \end{aligned}\right\} \tag{19.17}$$

Where A_{stoe} and A_{sheel} are the toe and heel reinforcement per linear foot. Minimum recommended steel reinforcement in footings is 0.0018 based on the gross area of the footing ($12D_f$). This minimum recommended reinforcement applies to A_{stoe}, A_{sheel} and longitudinal secondary reinforcement. The longitudinal steel used near the top of footing and near the bottom of footing can be added to satisfy minimum secondary reinforcement ratio. Further discussion of concrete footings is presented in Chapter 18. Development length requirements by ACI Code must be satisfied for footing main steel bars within the stem key and footing.

19.3.3 Key Analysis and Design

Figure 19.13 illustrates soil pressure on the key of a cantilever retaining wall under service conditions. Similar to the procedure followed for the footing, service soil pressure is multiplied by 1.6 to achieve ultimate soil pressure required for structural design. Section 4–4 is the most critical section for moment and shear:

Using the method of determining shear and moment for trapezoidal pressure distribution discussed in Chapter 12 Sections 18.6.2 and 18.6.3, the following equations can be obtained:

$$V_{u4\text{-}4} = 1.6D_k(p_1 + p_2)/2$$

Where p_1 and p_2 are shown in Figure 19.13 above. Concrete in the key is required to resist $V_{u4\text{-}4}$:

$$\phi V_n = 0.75 \times 2 \times \sqrt{f'_c} \times 12 d_{key} > V_{u4\text{-}4} = 1.6D_k(p_1 + p_2)/2 \qquad (19.18)$$

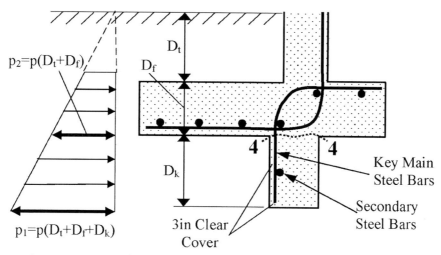

Figure 19.13: Soil pressure on the key of a cantilever retaining wall.

Equations 19.18 use units of in and lb. Key effective depth d is calculated as follows:

$$d_{key} = B_k - 3\text{ in} - d_{bkey}/2$$

Where:

  B_k = key width.

  3 in = ACI Code required minimum clear cover in concrete poured against soil.

  d_{key} = diameter of main steel bars in the key.

To determine the required main steel reinforcement in the footing based on bending moment, the following equation discussed in Chapter 18 may be used:

$$A_{skey} = M_{u4\text{-}4}/0.8d_{key}f_y = 0.33(2p_1 + p_2)\cdot D_k^2/d_{key}f_y \qquad (19.19)$$

Minimum recommended steel reinforcement in footings is 0.0018 based on the gross area of the key ($12B_k$). This minimum recommended reinforcement applies to the secondary steel in the heel. Further discussion of concrete footings is presented in Chapter 18.

Development length requirements by ACI Code must be satisfied for key main steel bars within the key and footing.

From Figure 19.7, it is important to note that Equations 19.18 and 19.19 can be considered an underestimate of the shear and moment in the key. As such, it is important that key structural design be significantly to the safe side.

19.4 Seismic Soil Pressure on Cantilever Retaining Walls

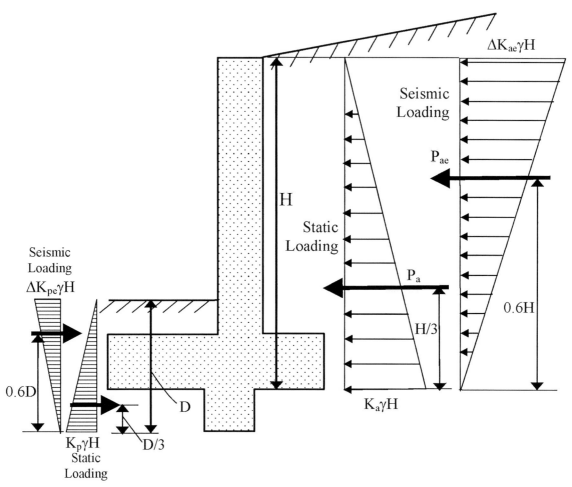

Figure 19.14: Static and seismic loading on cantilever retaining wall.

19.4.1 Extra Soil Pressure due to Earthquake Motion

Typical static and seismic load due to soil weight on a cantilever retaining wall are shown in Figure 19.14. Earthquake loading is inverted triangular (base is on top) with the resultant located at 0.6H from the wall base (bottom of footing). There are two common methods for evaluating earthquake loading on cantilever retaining:

The Mononobe–Okabe Method.

The Seed–Whitman Method.

19.4.2 The Mononobe–Okabe Method for Retaining Wall Seismic Load

The equations for seismic active and passive earth pressure coefficients under dynamic earthquake motion according to the Mononobe–Okabe method are shown below:

$$K_{ae} = \cfrac{\cos^2(\phi_s - \theta - \psi)}{\cos\psi \cos^2\theta \cos(\delta + \theta + \psi) \left[1 + \sqrt{\cfrac{\sin(\phi_s + \delta)\sin(\phi_s - \beta - \psi)}{\cos(\delta + \theta + \psi)\cos(\beta - \theta)}}\right]^2}$$

$$K_{pe} = \cfrac{\cos^2(\phi_s + \theta - \psi)}{\cos\psi \cos^2\theta_2 \cos(\delta - \theta_2 + \psi) \left[1 - \sqrt{\cfrac{\sin(\phi_s + \delta)\sin(\phi_s - \beta_2 - \psi)}{\cos(\delta + \theta_2 + \psi)\cos(\beta - \theta_2)}}\right]^2}$$

$$(19.20)$$

Where:

$$\psi = \tan^{-1}\left(\frac{k_h}{1-k_v}\right)$$

$k_h = a_h/g$

$k_v = a_v/g$

$a_h = 1/3$ to $\frac{1}{2} S_s =$ peak horizontal ground acceleration

$S_s =$ short period maximum probable horizontal spectral earthquake acceleration.

$a_v =$ peak horizontal ground acceleration.

$g =$ gravity acceleration.

$\Delta K_{ae} = K_{ae} - K_a$

$\Delta K_{pe} = K_{pe} - K_p$

K_a and K_p are based on Coulomb's theory as explained earlier.

Seismic passive soil resistance due to soil weight follows the same distribution as the active seismic pressure as shown in Figure 19.14. Equations 19.20 represent total earth pressure coefficients under seismic conditions. Total pressure coefficient is comprised of two parts:

(1)   Static part with K_a and K_p.

(2)   Dynamic part with $\Delta K_{ae} = K_{ae} - K_a$ and $\Delta K_{pe} = K_{pe} - K_p$.

Equations 19.20 can be simplified for the case of $\beta = \theta = \delta = 0$ as follows:

$$K_{ae} = \cfrac{\cos^2(\phi_s - \psi)}{\cos^2\psi \left[1 + \sqrt{\cfrac{\sin\phi_s \sin(\phi_s + \psi)}{\cos\psi}}\right]^2}$$

$$K_{pe} = \cfrac{\cos^2(\phi_s - \psi)}{\cos^2\psi \left[1 - \sqrt{\cfrac{\sin\phi_s \sin(\phi_s + \psi)}{\cos\psi}}\right]^2}$$

$$(19.21)$$

19.4.3 The Seed–Whitman Approach for Retaining Wall Seismic Load

This is a simplified approach for seismic earth pressure on retaining walls. Based on the Seed–Whitman approach, active seismic earth pressure is horizontal. ASCE-SEA 7 has adopted this approach with minor modifications using the following equation:

$$\Delta K_{ae} = 0.75(a_h/g) \qquad\qquad (19.22)$$

Where a_h peak horizontal ground acceleration, g is gravity acceleration.

Passive soil resistance and surcharge earth pressure due to seismic activity are not addressed in this simplified approach and typically remain unchanged.

Since seismic load is of short duration, the recommended safety factor for retaining walls under the effect of proper static and seismic loading combination is ≥1.2 rather than ≥1.5 used for static loading only. Structural design of retaining walls requires inclusion of seismic load with the proper LRFD load combination shown in the following examples.

For seismic analysis of retaining walls, it is also common that designers ignore friction and assume that all soil loading on the retaining wall are horizontal.

19.5 Load Combinations

The Working Stress Design (WSD) method is typically used for retaining wall geotechnical design or stability. IBC implements special load combinations for retaining walls: static load combination and special seismic load combination as shown below. The required safety factor is 1.5 for the static load combination. The required safety factor is 1.1 (1.2 is recommended) for the retaining wall special load combination. For both load combinations, allowable soil bearing stress q_a remains unchanged as specified in the soils report:

Static Load Combinations = DL + LL + H

Special Retaining Wall Load Combination = DL + LL + H + 0.7E

For the wall structural design of the stem and footing, the ultimate strength design method or LRFD is used. The following load combinations are critical for retaining walls (see Chapter 2):

(1)   U2 = 1.2DL + 1.6(LL + H)

(2)   U5 = 1.2DL + 1.0E + LL

(3)   U6 = 0.9DL + 1.6W + 1.6H

(4)   U7 = 0.9DL + 1.0E + 1.6H

Where:

DL = Dead Load.

H = Lateral Earth Pressure.

LL = Live Load Surcharge.

W = Wind Load applied to wall extension above ground.

E = Seismic Dead Load.

Example 19.1

Determine the factors of safety for overturning, sliding and soil bearing under static and static and seismic loading for a retaining wall under the following conditions:

H_o = 7 ft	H_w = 8 ft	H = 9.25 ft	β = 2h:1v = 26.6° (W_s=0)
t = 10 in	a = 5.5 ft	n = 1.67 ft	B_f = 8.00 ft
D_k = 2 ft	B_k = 1 ft	x_k = 2/3Bf = 5.33 ft	

Soil design parameters in the soils report:

$\Phi_s = 32°$ $\gamma = 120$ pcf $\mu = 0.35$ $R = 1.0$

$q_a = 2,000$ psf $EFL =$ 60 pcf $p = 350$ pcf $S_s = 0.20g$ $a_v = 0$

Solution:

Under Static Conditions:

Study 1 ft representative length of the wall (LL = 0 & H = 0).

Soil Pressure:

(1)   using Rankine's method with $\beta_2 = 0$:

$$K_a = \cos\beta \, \frac{\cos\beta - \sqrt{\cos^2\beta - \cos^2\phi_s}}{\cos\beta + \sqrt{\cos^2\beta - \cos^2\phi_s}} = \cos 26.6 \, \frac{\cos 26.6 - \sqrt{\cos^2 26.6 - \cos^2 32}}{\cos 26.6 + \sqrt{\cos^2 26.6 - \cos^2 32}} = 0.46$$

$$K_p = \frac{1 + \sin\phi_s}{1 - \sin\phi_s} = \frac{1 + \sin 32}{1 + \sin 32} = 3.25$$

(2)   using Coulomb's method with $\theta = \delta = \beta_2 = 0$:

$$K_a = \frac{\cos^2(\phi_s - \theta)}{\cos^2\theta \, \cos(\delta + \theta) \left[1 + \sqrt{\dfrac{\sin(\phi_s + \delta)\sin(\phi_s - \beta)}{\cos(\delta + \theta)\cos(\beta - \theta)}}\right]^2} = \frac{\cos^2 32}{\left[1 + \sqrt{\dfrac{\sin 32 \sin(32 - 26.6)}{\cos 26.6}}\right]^2} = 0.47$$

$$K_p = \frac{\cos^2\phi_s}{(1 - \sin\phi_s)^2} = \frac{\cos^2 32}{(1 - \sin 32)^2} = 3.25$$

Assuming no friction and horizontal soil pressure, the three methods of evaluating soil pressure are comparable: Soils Report, Rankine's and Coulomb's:

Using the soils report:

$P_a = 60 \times (9.25)^2/2 = 2,570$ lb/ft

$P_p = 350 \times [(4.75)^2 - 1]/2 = 3,773$ lb/ft   Ignoring the top 1 ft of the soil

Using Rankine's method:

$P_a = 0.46 \times 120 \times (9.25)^2/2 = 2,360$ lb/ft

$P_p = 3.25 \times 120 \times [(4.25)^2 - 1]/2 = 4,204$ lb/ft   Also ignoring the top 1 ft of the soil

Using Rankine's method:

$P_a = 0.47 \times 120 \times (9.25)^2/2 = 2,410$ lb/ft

$P_p = 3.25 \times 120 \times [(4.75)^2 - 1]/2 = 4,204$ lb/ft   Also ignoring the top 1 ft of the soil

For the static analysis of this wall, the parameters in the soil report will be used. The forces affecting this wall are as shown in Section 19.2 and illustrated as follows:

$W_w = t \cdot H_w \cdot \gamma_c = (10/12) \times 8 \times 0.15 = 1.00$ k/ft

$W_f = D_f \cdot B_f \cdot \gamma_c = 1.25 \times 8.00 \times 0.15 = 1.50$ k/ft

$W_k = B_k \cdot D_k \cdot \gamma_c = 1 \times 2 \times 0.15 = 0.30$ k/ft

$W_{s1} \approx n \cdot H_w \cdot \gamma + W_s \cdot n = 1.67 \times 8 \times 0.12 + 0 = 1.60$ k/ft

$W_{s2} = D_t \cdot a \cdot \gamma = 1 \times 5.5 \times 0.12 = 0.66$ k/ft

$P_a = EFP \cdot H^2/2 = 2.57$ k/ft

$P_{as} = K_a \cdot H = 0$

$P_p = p \cdot [(D_t + D_f + D_k)^2 - 1]/2 = 2.99$ k/ft

$F = (Ww + Wf + Wk + Ws_1 + Ws_2) \cdot \mu = \Sigma W \cdot \mu$

$F = (1.00 + 1.50 + 0.30 + 1.60 + 0.66) \times 0.35 = 5.06 \times 0.35 = 1.77$ k/ft

Also, moment arms for the above forces are (Section 19.2):

$X_{Ww} = a + t/2 = 5.5 + 10/12/2 = 5.92$ ft

$X_{Wf} = B_f/2 = 8.0/2 = 4.00$ ft

$X_{Wk} = x_k = 2B_f/3 = 5.33$ ft

$X_{s1} \approx B_f - n/2 = 8.0 - 1.67/2 = 7.17$ ft

$X_{s2} = a/2 = 5.5/2 = 2.75$ ft

$Y_{Pa} = H/3 = 9.25/3 = 3.08$ ft

$Y_{Pas} = H/2 = 9.25/2 = 4.63$ ft

Overturning and resisting moments (Equations 19.3 – 19.4):

$OTM = EFP \cdot H^3/6 + K_a \cdot H^2/2 = 2.57 \times 3.08 + 0 = 7.92$ k.ft/ft

$RM = W_w \cdot X_{Ww} + W_f \cdot X_{Wf} + W_k \cdot X_{Wk} + W_{s1} \cdot X_{s1} + W_{s2} \cdot X_{s2}$
$= 1.00 \times 5.92 + 1.50 \times 4.00 + 0.30 \times 5.33 + 1.60 \times 7.17 + 0.66 \times 2.75 = 26.8$ k.ft/ft

Safety Factor against Overturning = RM/OTM = 26.8/7.92 = 3.38 ≥ 1.5 OK

Sliding and resisting forces (Equations 19.6 – 19.7):

Sliding Force = $P_a + P_{as} = 2.57 + 0 = 2.57$ k/ft

Resisting Force = $R \cdot (P_p + F) = 1 \times [2.99 + 1.77] = 4.76$ k/ft

Safety Factor against Sliding = 4.76/2.57 = 1.85 ≥ 1.5 OK

Weight eccentricity and soil bearing pressure:

$e = B_f/2 - x_{\Sigma W} = B_f/2 - (RM-OTM)/\Sigma W = 8.00/2 - (26.8-7.92)/5.06 = 0.27 < B_f/6 = 1.33$ OK

$q_t = \Sigma W/B_f + 6\Sigma W \cdot e/B_f^2 = 5.06/8.00 + 6 \times 5.06 \times 0.27/8.00^2 = 761$ psf < 2,000 psf OK

$q_h = \Sigma W/B_f - 6\Sigma W \cdot e/B_f^2 = 5.06/8.00 - 6 \times 5.06 \times 0.27/8.00^2 = 504$ psf > 0 OK

With $q_h > 0$, the footing loading condition remained according to case (a) of Figure 18.8 Chapter 8 as recommended to reduce tilting.

Under Seismic Conditions:

Soil Pressure:

(1) using the Mononobe–Okabe method with $\theta = \delta = \beta_2 = 0$ (Equations 19.20):

Use $a_h = 0.4S_s = 0.4 \times 0.2 = 0.08$

$\psi = \tan^{-1}\left(\dfrac{k_h}{1-k_v}\right) = \tan^{-1}\left(\dfrac{0.08}{1+0}\right) = 4.6°$

$$K_{ae} = \frac{\cos^2(\phi_s + \theta - \psi)}{\cos\psi \cos^2\theta_2 \cos(\delta - \theta + \psi)\left[1 - \sqrt{\dfrac{\sin(\phi_s + \delta)\sin(\phi_s - \beta - \psi)}{\cos(\delta + \theta + \psi)\cos(\beta - \theta)}}\right]^2}$$

$$K_{ae} = \frac{\cos^2 27.4}{\cos^2 4.6\left[1 + \sqrt{\dfrac{\sin 32 \sin(32 - 26.6 - 4.6)}{\cos 4.6 \cos 26.6}}\right]^2} = 0.71$$

Active seismic soil pressure coefficient = $\Delta K_{ae} = K_{ae} - K_a = 0.71 - 0.47 = 0.24$

Active seismic soil pressure = $\Delta K_{ae} \cdot \gamma \cdot H^2/2 = 0.24 \times 120 \times (9.25)^2/2 = 1{,}230$ lb/ft $= 1.23$ k/ft

Active seismic force moment = $0.6 \cdot \Delta K_{ae} \cdot \gamma \cdot H^2/2 = 1.23 \times 0.6 \times 9.25 = 6.83$ k.ft/ft

Typically, designers ignore the effect of seismic motion in increasing passive soil resistance as a measure of added safety. The IBC special seismic load combination for retaining walls applies requiring multiplication of seismic force and moment with 0.7 (Section 19.4). The factors of safety under the effect of earthquake load are:

Safety Factor against Overturning = RM/OTM = $26.8/(7.92 + 0.7 \times 6.83) = 2.11 \geq 1.2$ OK

Safety Factor against Sliding = $4.76/(2.57 + 0.7 \times 1.23) = 1.38 \geq 1.2$ OK

Weight eccentricity and soil bearing pressure due to seismic loading:

$e_s = B_f/2 - x_{\Sigma W} = B_f/2 - (RM - OTM)/\Sigma W$

$e_s = 8.00/2 - (26.8 - 7.92 - 0.7 \times 6.83)/5.06 = 1.21 < B_f/6 = 1.33$ OK

$q_t = \Sigma W/B_f + 6\Sigma W \cdot e/B_f^2 = 5.06/8.00 + 6 \times 5.06 \times 1.21/8.00^2 = 1{,}206$ psf $< 2{,}000$ psf OK

$q_h = \Sigma W/B_f - 6\Sigma W \cdot e/B_f^2 = 5.06/8.00 - 6 \times 5.06 \times 1.21/8.00^2 = 59$ psf > 0 OK

With $q_h > 0$, the footing loading condition remains according to case (a) of Figure 18.8 Chapter 8 as recommended to reduce tilting.

(2)   using the Seed–Whitman method with (Equations 19.22):

Active seismic soil pressure coefficient = $\Delta K_{ae} = 0.75(a_h/g) = 0.75 \times 0.08 = 0.06$

Active seismic soil pressure = $\Delta K_{ae} \cdot \gamma \cdot H^2/2 = 0.06 \times 120 \times (9.25)^2/2 = 2{,}260$ lb/ft $= 0.31$ k/ft

Active seismic force moment = $0.6 \cdot \Delta K_{ae} \cdot \gamma \cdot H^2/2 = 0.31 \times 0.6 \times 9.25 = 1.72$ k.ft/ft

The seismic force and moment computed above requires multiplication with 0.7 as soil analysis uses ASD method. The factors of safety under the effect of earthquake load are:

Safety Factor against Overturning = RM/OTM = $26.8/(7.92 + 0.7 \times 1.72) = 2.94 \geq 1.2$ OK

Safety Factor against Sliding = $4.76/(2.57 + 0.7 \times 0.31) = 1.71 \geq 1.2$ OK

Weight eccentricity and soil bearing pressure:

$e_s = B_f/2 - X_{\Sigma W} = B_f/2 - (RM - OTM)/\Sigma W$

$e_s = 8.00/2 - (26.8 - 7.92 - 0.7 \times 1.72)/5.06 = 0.51 < B_f/6 = 1.33$ OK

$q_t = \Sigma W/B_f + 6\Sigma W \cdot e/B_f^2 = 5.06/8.00 + 6 \times 5.06 \times 0.51/8.00^2 = 874$ psf $< 2{,}000$ psf OK

$q_h = \Sigma W/B_f - 6\Sigma W \cdot e/B_f^2 = 5.06/8.00 - 6 \times 5.06 \times 0.51/8.00^2 = 391$ psf > 0 OK

With $q_h > 0$, the footing loading condition remained according to case (a) of Figure 18.8 Chapter 8 as recommended to reduce tilting.

Example 19.2:

Determine the factors of safety for overturning, sliding and soil bearing under static and static and seismic loading for a retaining wall under the following conditions:

$H_o = 6$ ft	$H_w = 6.5$ ft	$H = 7.5$ ft	$W_s = 250$ psf ($\beta = 0$)
$t = 8$ in	$a = 4.5$ ft	$n = 1.33$ ft	$B_f = 6.50$ ft
$D_k = 2$ ft	$B_k = 1$ ft	$x_k = 2/3Bf = 4.33$ ft	

Soil design parameters in the soils report:

$\Phi s = 35°$	$\gamma = 120$ pcf	$\mu = 0.40$	$R = 1.0$	
$q_a = 2{,}500$ psf	EFL =	40 pcf	$p = 350$ pcf	$S_s = 0.8g$ $a_v = 0$

Solution:

<u>Under Static Conditions:</u> Study 1 ft representative length of the wall (H =0).

<u>Soil Pressure:</u>

(1)   using Rankine's method with $\beta_2 = 0$:

$$K_a = \frac{1 - \sin\phi_s}{1 + \sin\phi_s} = \frac{1 - \sin 35}{1 + \sin 35} = 0.31$$

$$K_p = \frac{1 + \sin\phi_s}{1 - \sin\phi_s} = \frac{1 + \sin 35}{1 - \sin 35} = 3.20$$

(2)   using Coulomb's method with $\theta = \delta = \beta_2 = 0$:

$$K_a = \frac{\cos^2\phi_s}{(1 + \sin \phi_s)^2} = \frac{\cos^2 35}{(1 + \sin 35)^2} = 0.31$$

$$K_p = \frac{\cos^2\phi_s}{(1 - \sin \phi_s)^2} = \frac{\cos^2 35}{(1 - \sin 35)^2} = 3.20$$

<u>Using the soils report:</u>

$P_a = 40 \times (7.5)^2/2 = 1{,}125$ lb/ft

$P_{as} = 0.31 \times 250 \times 7.5 = 581$ lb/ft

$P_p = 350 \times [(3.5)^2 - 1]/2 = 1{,}969$ lb/ft   Ignoring the top 1 ft of the soil

<u>Using Rankine's or Coulomb's method:</u>

$P_a = 0.31 \times 120 \times (7.5)^2/2 = 1{,}046$ lb/ft

$P_{as} = 0.31 \times 250 \times 7.5 = 581$ lb/ft

$P_p = 3.20 \times 120 \times [(3.5)^2 - 1]/2 = 2{,}160$ lb/ft   Also ignoring the top 1 ft of the soil

For the static analysis of this wall, the parameters in the soil report will be used. The forces affecting this wall are as follows (Section 19.2):

$W_w = t \cdot H_w \cdot \gamma_c = (8/12) \times 6.5 \times 0.15 = 0.65$ k/ft

$W_f = D_f \cdot B_f \cdot \gamma_c = 1.00 \times 6.50 \times 0.15 = 0.98$ k/ft

$W_k = B_k \cdot D_k \cdot \gamma_c = 1 \times 2 \times 0.15 = 0.30$ k/ft

$W_{s1} \approx n \cdot H_w \cdot \gamma + W_s \cdot n = 1.33 \times 6.5 \times 0.12 + 0.25 \times 1.33 = 1.37$ k/ft

$W_{s2} = D_t \cdot a \cdot \gamma = 1 \times 4.5 \times 0.12 = 0.54$ k/ft

$P_a = EFP \cdot H^2/2 = 1.13$ k/ft

$P_{as} = K_a \cdot H = 0.58$ k/ft

$P_p = p \cdot [(D_t + D_f + D_k)^2 - 1]/2 = 1.97$ k/ft

$F = (Ww + Wf + Wk + Ws_1 + Ws_2) \cdot \mu = \Sigma W \cdot \mu$

$F = (0.65 + 0.98 + 0.30 + 1.37 + 0.54) \times 0.40 = 3.85 \times 0.40 = 1.54$ k/ft

Also, moment arms for the above forces are (Section 19.2):

$X_{Ww} = a + t/2 = 4.5 + 8/12/2 = 4.83$ ft

$X_{Wf} = B_f/2 = 6.5/2 = 3.25$ ft

$X_{Wk} = x_k = 2B_f/3 = 4.33$ ft

$X_{s1} \approx B_f - n/2 = 6.5 - 1.33/2 = 5.84$

$X_{s2} = a/2 = 4.5/2 = 2.25$ ft

$Y_{Pa} = H/3 = 7.5/3 = 2.50$ ft

$Y_{Pas} = H/2 = 7.5/2 = 3.75$ ft

Overturning and resisting moments (Equations 19.3 – 19.4):

$OTM = EFP \cdot H^3/6 + K_a \cdot H^2/2 = 1.13 \times 2.50 + 0.58 \times 3.75 = 2.83 + 2.18 = 5.00$ k.ft/ft

$RM = W_w \cdot X_{Ww} + W_f \cdot X_{Wf} + W_k \cdot X_{Wk} + W_{s1} \cdot X_{s1} + W_{s2} \cdot X_{s2}$

$= 0.65 \times 4.83 + 0.98 \times 3.25 + 0.30 \times 4.33 + 1.37 \times 5.84 + 0.54 \times 2.25 = 16.8$ k.ft/ft

Safety Factor against Overturning = RM/OTM = 16.8/5.00 = 3.37 ≥ 1.5 OK

Sliding and resisting forces (Equations 19.6 – 19.7):

Sliding Force = $P_a + P_{as}$ = 1.13 + 0.58 = 1.71 k/ft

Resisting Force = $R \cdot (P_p + F)$ = 1 × [1.97 + 1.54] = 3.51 k/ft

Safety Factor against Sliding = 3.51/1.71 = 2.05 ≥ 1.5 OK

Weight eccentricity and soil bearing pressure:

$e = B_f/2 - X_{\Sigma W} = B_f/2 - (RM - OTM)/\Sigma W = 6.50/2 - (16.8 - 5.00)/3.85 = 0.19 < B_f/6 = 1.08$ OK

$q_t = \Sigma W/B_f + 6\Sigma W \cdot e/B_f^2 = 3.85/6.5 + 6 \times 3.85 \times 0.19/6.50^2 = 696$ psf < 2,500 psf OK

$q_t = \Sigma W/B_f + 6\Sigma W \cdot e/B_f^2 = 3.57/5.5 - 6 \times 3.57 \times 0.45/5.50^2 = 488$ psf > 0 OK

With $q_h > 0$, the footing loading condition remained according to case (a) of Figure 18.8 Chapter 8 as recommended to reduce tilting.

Under Seismic Conditions:

Soil Pressure:

(1) using the Mononobe–Okabe method with $\theta = \delta = \beta_2 = 0$ (Equations 19.20):

Use $a_h = 0.4S_s = 0.4 \times 0.8 = 0.32$

$$\psi = \tan^{-1}\left(\frac{k_h}{1-k_v}\right) = \tan^{-1}\left(\frac{0.32}{1+0}\right) = 19.7°$$

$$K_{ae} = \frac{\cos^2(\phi_s - \psi)}{\cos^2\psi\left[1 + \sqrt{\dfrac{\sin\phi_s \sin(\phi_s + \psi)}{\cos\psi}}\right]^2} = \frac{\cos^2 15.3}{\cos^2 19.7\left[1 + \sqrt{\dfrac{\sin 35 \sin 15.3}{\cos 19.7}}\right]^2} = 0.56$$

Active seismic soil pressure coefficient = $\Delta K_{ae} = K_{ae} - K_a = 0.56 - 0.31 = 0.25$

Active seismic soil pressure = $\Delta K_{ae} \cdot \gamma \cdot H^2/2 = 0.25 \times 120 \times (7.5)^2/2 = 844$ lb/ft = 0.84 k/ft

Active seismic force moment = $0.6 \cdot \Delta K_{ae} \cdot \gamma \cdot H^2/2 = 0.84 \times 0.6 \times 7.5 = 3.78$ ft.k/ft

Ignore passive resistance increase under seismic conditions. The IBC special seismic load combination for retaining walls applies requiring multiplication of seismic force and moment with 0.7 (Section 19.4). The factors of safety under the effect of earthquake load are:

Safety Factor against Overturning = RM/OTM = $16.8/(2.83+0.7\times3.78)=2.11 \geq 1.2$ OK

Safety Factor against Sliding $\approx 3.51/(1.13 + 0.7\times0.84) = 2.04 \geq 1.2$ OK

Weight eccentricity and soil bearing pressure due to seismic loading:

$e_s = B_f/2 - X_{\Sigma W} = B_f/2 - (RM - OTM)/\Sigma W$

$e_s = 6.5/2 - (16.8 - 2.83 - 0.7\times3.78)/3.85 = 0.30 < B_f/6 = 1.08$ OK

$q_t = \Sigma W/B_f + 6\Sigma W \cdot e/B_f^2 = 3.85/6.50 + 6\times3.85\times0.30/6.50^2 = 756$ psf < 2,000 psf OK

$q_t = \Sigma W/B_f + 6\Sigma W \cdot e/B_f^2 = 3.85/6.50 - 6\times3.85\times0.30/6.50^2 = 428$ psf > 0 OK

With $q_h > 0$, the footing loading condition remains according to case (a) of Figure 18.8 Chapter 8 as recommended to reduce tilting.

(2)   using the Seed–Whitman method with (Equations 19.22):

Active seismic soil pressure coefficient = $\Delta K_{ae} = 0.75(a_h/g) = 0.75\times0.32 = 0.24$

Active seismic soil pressure = $\Delta K_{ae} \cdot \gamma \cdot H^2/2 = 0.24\times120\times(7.5)^2/2 = 810$ lb/ft = 0.81 k/ft

Active seismic force moment = $0.6 \cdot \Delta K_{ae} \cdot \gamma \cdot H^2/2 = 0.81\times0.6\times7.5 = 3.65$ ft.k/ft

The values of active seismic soil pressure force and moment computed using the Mononobe–Okabe method and the Seed–Whitman method are very comparable. Analysis results of both methods are also comparable.

The geotechnical engineering design for Example 19.2 is conservative as safety factors under the effect of seismic load as considerably greater than 1.2.

Example 19.3

Perform the structural design of the retaining wall in Example 19.1 using $f'_c = 3,000$ psi and $f_y = 60,000$ psi.

Solution

(1) Stem Analysis:

The following load combinations are critical for stem design in this example:

U2 = 1.2DL + 1.6(LL + H)

U5 = 1.2DL + 1.0E + LL

U7 = 0.9DL + 1.0E + 1.6H

(no live load in this example)

As mentioned earlier, the weight of stem is generally neglected as a measure of safety.

Retaining wall stems are designed for shear and moment. Based on the analysis presented in Example 19.1 utilizing Coulomb and Mononobe–Okabe methods for static and seismic loading, respectively, the following table can be assembled:

For the following calculations, EFP = $Ka \cdot \gamma$ = 0.47×120 using Coulomb's

$$\Delta K_{ae} = 0.24 \text{ using Mononobe–Okabe}$$

<u>Shear Forces:</u>

Due to DL = 0

Due to H = $EFPH_w^2/2$=0.47×120×(8)^2/2=1.80 k/ft using Figure 19.11

Due to E = $\Delta K_{ae}H_w^2/2$ = 0.24×120×(8)^2/2=0.92 k/ft Figure 19.14 with H_w in lieu of H

<u>Bending Moments:</u>

Due to DL = 0

Due to H = $EFPH_w^3/6$=0.47×120×(8)^3/6=4.81 k.ft/ft using Figure 19.11

Due to E=0.6H_w[$\Delta K_{ae}Hw^2/2$]=0.6×8×[0.24×120×(8)^2/2]=4.42 k.ft/ft Figure 19.14 with H_w

Load	DL	H	E	U2	U5	U7
V (k/ft)	0	1.80	0.92	2.88	0.92	4.77
M (k.ft/ft)	0	4.81	4.42	9.70	4.42	12.12
V_{umax} (k/ft)						4.77
M_{umax} (k.ft/ft)						12.12

d = 10 in – 2 in – ½ in = 7.5 in using Section 19.3 and assuming #8 bars maximum

ϕV_n = 0.75×2×√3,000×12×7.5/1,000 = 7.39 k/ft > V_{umax} = 4.77k/ft using Equation 19.12

A_s = 12.12×12/0.8×7.5×60 = 0.4 in^2/ft   using Equation 19.13

Minimum stem reinforcement = 0.0015×10×12 = 0.18 in^2/ft

Use #6 @12in o.c.(As=0.44 in^2/ft) main/vertical steel, steel ratio=0.0037 of gross section.

Use #5@12in o.c. (As = 0.31 in^2/ft) for secondary/horizontal steel, ratio=0.0026 of gross section.

Development Length for #6 bars based on Chapter 13l_d = 22.2 in l_{dh} = 10.5 in

Available straight bar development length in stem = 8 ft >> 22.2 in OK

Available hooked bar development length in footing ≈ 11 in > 10.5 in OK

(2)   Footing Design:

q_t and q_h for structural analysis are adopted as q_{tmax} and q_{hmax} from geotechnical design under service load combinations:

$q_t = q_{tmax}$ from geotechnical design = 1,206 psf

$q_h = q_{hmax}$ from geotechnical design = 504 psf

Section 19.3.2: $q_{t1} = q_t - (q_t - q_h)a/B_f = 1,206 - (1,206 - 504)\times5.5/8.0 = 723$ psf

Section 19.3.2: $q_{h1} = q_h + (q_t - q_h)n/B_f = 504 + (1,206 - 504)\times1.67/8.0 = 651$ psf

Toe Design:

Section 19.3.2: $d_{toe} = D_f - 3$ in $- d_{btoe}/2$

$d_{toe} = 15$ in$- 3$ in-1.06 in ≈ 11in   using #6 bars (stem steel bars) for main and #5 for secondary

Shear:

Equation 19.16: $\phi V_n = 0.75\times2\times\sqrt{f'_c}\times12d_{toe} > V_{u2\text{-}2} = 1.6a(q_t + q_{t1})/2$

$0.75\times2\times\sqrt{3,000}\times12\times11/1,000=10.8$ k/ft $> 1.6\times5.5\times(1,206 + 723)/2/1,000 = 8.5$ k/ft OK

Moment:

Equation 19.14: $M_{u2\text{-}2}=1.6(2q_t + q_{t1})\cdot a^2/6=1.6\times(2\times1,206+723)\times(5.5)^2/6/1,000=25.3$ k.ft/ft

Equation 19.17: $A_{stoe} = M_{u2\text{-}2}/0.8d_{toe}f_y = 25.3\times12/(0.8\times11\times60) = 0.575$ in^2/ft

A_{stoe} is greater than the required steel bars in the stem (0.40 in^2/ft). However, both steel bars should be the same for construction suitability (the same steel bars are bend with a 90° hook to act as stem bars and toe bars). Thus, use #6@8in o.c. ($A_s = 0.66$ in^2/ft) for both stem bars and toe bars. Another solution is to use #6@8in o.c. for the toe and #5@8in o.c. for the stem. #5 and #6 should then be spliced at the base of the stem.

It is easily shown that development length for toe bars is satisfied as they extend in the toe (5.5 ft straight) and terminate with a hook at the stem.

Heel Design:

Section 19.3.2: $d_{heel}=D_f-2$ in$-d_{bheel}/2 = 15$ in-2 in-1 in ≈ 12 in use #5 main & secondary

Shear:

Equation 19.16: $\phi V_n = 0.75\times2\times\sqrt{f'_c}\times12d_{heel} > V_{u3\text{-}3} = 1.6n(q_h + q_{h1})/2$

$0.75\times2\times\sqrt{3,000}\times12\times12/1,000=11.8$ k/ft $> 1.6\times1.67\times(504 + 651)/2/1,000=1.5$ k/ft OK

Moment:

Equation 19.15: $M_{u3\text{-}3} = 1.6(2q_h+q_{h1})\cdot n^2/6=1.6\times(2\times504+651)\times(1.67)^2/6/1,000=1.2$ k.ft/ft

Equation 19.17: $A_{sheel} = M_{u3\text{-}3}/0.8d_{heel}f_y = 1.2\times12/(0.8\times12\times60) = 0.025$ in^2/ft

Use #5@8in o.c. ($A_s = 0.46$ in^2/ft) for main steel in the heel and #5@12 in o.c. ($A_s = 0.31$ in^2/ft) for secondary steel in the entire footing. Bas sizes are comparable to stem and toe.

Steel ratio in the heel = 0.0026 for main steel bars and 0.0017 for secondary steel bars.

Spacing of main bars in the toe and heel should be the same to facilitate construction.

Key Design:

Section 19.3.2: $d_{key} = B_k - 3$ in $- d_{bkey}/2 = 12$ in $- 3$ in $- 1$ in ≈ 8 in use #5 main & secondary

$p_1 = p \cdot (D_t + D_f + D_k) = 350 \times (1 + 1.25 + 2) = 1,488$ psf

$p_2 = p \cdot (D_t + D_f) = 350 \times (1 + 1.25) = 788$ psf

Shear:

Equation 19.18: $\phi V_n = 0.75 \times 2 \times \sqrt{f'_c} \times 12 d_{key} > V_{u4-4} = 1.6 D_k (p_1 + p_2)/2$

$0.75 \times 2 \times \sqrt{3,000} \times 12 \times 8/1,000 = 7.9$ k/ft $> 1.6 \times 2 \times (1,488 + 788)/2/1,000 = 3.6$ k/ft OK

Moment:

Equation 19.19: $A_{skey} = M_{u4-4}/0.8 d_{key} f_y = 0.33(2p_1 + p_2) \cdot D_k^2/d_{key} f_y$

$A_{skey} = 0.33 \times (2 \times 1,488 + 788) \times (2)^2/(8 \times 60) \times (12/1000) = 0.12$ in²/ft

Use #5@8in o.c. ($A_s = 0.46$ in²/ft) for main steel in the key (continuation of heel steel) and #5@12in o.c. ($A_s = 0.31$ in²/ft) for secondary steel in the key.

Steel ratio in the key = 0.0032 for main steel bars and 0.0022 for secondary steel bars.

Development length for straight #5 bars is 16.4 in and for hooked #5 bars is about 10in. Clearly, they are both satisfied for the heel and key.

Wall detail is shown in the following page:

The detail presented includes all pertinent information to the construction of the retaining wall including:

-    Free board

-    Filter fabric

-    Perforated drainage pipe

-    Crushed rock backfill

-    Water proofing

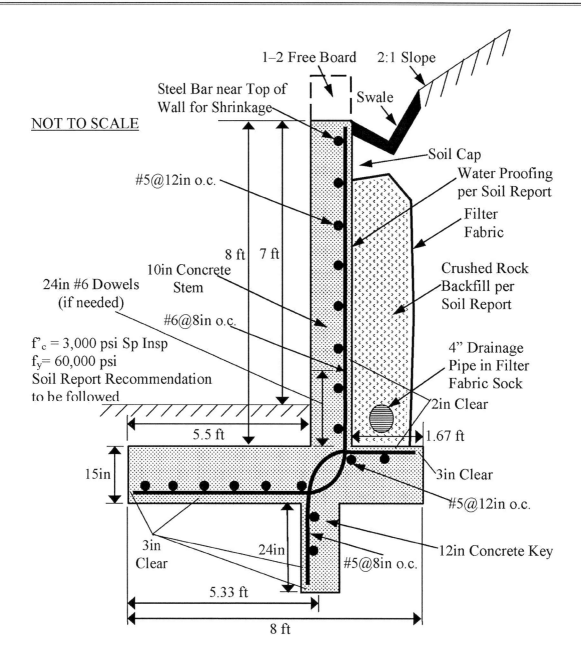

NOT TO SCALE

1–2 Free Board

2:1 Slope

Steel Bar near Top of
Wall for Shrinkage

Swale

Soil Cap

Water Proofing
per Soil Report

#5@12in o.c.

Filter
Fabric

8 ft   7 ft

10in Concrete
Stem

24in #6 Dowels
(if needed)

Crushed Rock
Backfill per
Soil Report

#6@8in o.c.

f'_c = 3,000 psi Sp Insp
f_y= 60,000 psi
Soil Report Recommendation
to be followed

4" Drainage
Pipe in Filter
Fabric Sock

2in Clear

5.5 ft

1.67 ft

15in

3in Clear

#5@12in o.c.

3in
Clear

24in

12in Concrete Key

#5@8in o.c.

5.33 ft

8 ft

Example 19.4

Perform the structural design of the retaining wall in Example 19.2 using f'_c = 4,000 psi and f_y = 60,000 psi.

Solution

(3)   Stem Analysis:

Use the following load combinations:

$$U2 = 1.2DL + 1.6(LL + H)$$

$$U5 = 1.2DL + 1.0E + LL$$

$$U7 = 0.9DL + 1.0E + 1.6H$$

Based on the analysis presented in Example 19.2 utilizing Coulomb and Mononobe–Okabe methods for static and seismic loading, respectively, the following table can be assembled:

For the following calculations, $EFP = Ka \cdot \gamma = 0.31 \times 120$ using Coulomb's

$$\Delta K_{ae} = 0.25 \text{ using Mononobe–Okabe}$$

Shear Forces:

Due to $DL = 0$

Due to $H = EFPH_w^2/2 = 0.31 \times 120 \times (6.5)^2/2 = 0.79$ k/ft using Figure 19.11

Due to $LL = K_a \cdot W_s \cdot H_w = 0.31 \times 250 \times (6.5) = 0.51$ k/ft   using Figure 19.11

Due to $E = \Delta K_{ae} Hw^2/2 = 0.25 \times 120 \times (6.5)^2/2 = 0.63$ k/ft Figure 19.14 with H_w in lieu of H

Bending Moments:

Due to $DL = 0$

Due to $H = EFPH_w^3/6 = 0.31 \times 120 \times (6.5)^3/6 = 1.70$ k.ft/ft using Figure 19.11

Due to $LL = K_a \cdot W_s \cdot (H_w)^2/2 = 0.31 \times 250 \times (6.5)^2/2 = 1.64$ k/ft using Equation 19.24

Due to $E = 0.6 H_w[\Delta K_{ae} Hw^2/2] = 0.6 \times 6.5 \times [0.25 \times 120 \times (6.5)^2/2] = 2.47$ k.ft/ft Figure 19.14 with H_w

Load	DL	H	LL	E	U2	U5	U7
V (k/ft)	0	0.79	0.51	0.63	2.08	1.14	1.89
M (k.ft/ft)	0	1.70	1.64	2.47	5.34	4.11	5.19
V_{umax} (k/ft)					2.08		
M_{umax} (k.ft/ft)					5.34		

$d = 8in - 2in - 3/8n = 5.6in$ using Section 19.3 and assuming #6 bars maximum

$\phi V_n = 0.75 \times 2 \times \sqrt{4,000} \times 12 \times 5.6/1,000 = 6.38$ k/ft $> V_{umax} = 4.77$k/ft using Equation 19.12

$A_s = 5.34 \times 12/0.8 \times 5.6 \times 60 = 0.23$ in^2/ft   using Equation 19.13

Minimum stem reinforcement $= 0.0015 \times 8 \times 12 = 0.14$ in^2/ft

Use #5 @12in o.c.(As=0.31 in^2/ft) main/vertical steel, steel ratio=0.0032 of gross section.

Use #4@12in o.c. (As = 0.20 in^2/ft) for secondary/horizontal steel, ratio=0.0021 of gross section.

Development Length for #5 bars based on Chapter 13$l_d = 16.4$ in $l_{dh} = 8$ in

Available straight bar development length in stem = 6.5 ft >> 16.4 in OK

Available hooked bar development length in footing ≈ 8 in ≥ 8 in OK

(4)   Footing Design:

q_t and q_h for structural analysis are adopted as q_{tmax} and q_{hmax} from geotechnical design under service load combinations:

$q_t = q_{tmax}$ from geotechnical design = 898 psf

$q_h = q_{hmax}$ from geotechnical design = 488 psf

Section 19.3.2: $q_{t1} = q_t - (q_t - q_h)a/B_f = 898 - (898 - 488) \times 4.5/6.5 = 614$ psf

Section 19.3.2: $q_{h1} = q_h + (q_t - q_h)n/B_f = 488 + (898 - 488) \times 1.33/6.5 = 572$ psf

Toe Design:

Section 19.3.2: $d_{toe} = D_f - 3\text{ in} - d_{btoe}/2$

$d_{toe} = 12\text{ in} - 3\text{ in} - 0.88\text{ in} \approx 8.1\text{ in}$ using #5 bars (stem steel bars) for

main and #4 for secondary

Shear:

Equation 19.16: $\phi V_n = 0.75 \times 2 \times \sqrt{f'_c} \times 12 d_{toe} > V_{u2\text{-}2} = 1.6a(q_t + q_{t1})/2$

$0.75 \times 2 \times \sqrt{4,000} \times 12 \times 8.1/1,000 = 9.2\text{ k/ft} > 1.6 \times 4.5 \times (898 + 614)/2/1,000 = 5.4\text{ k/ft OK}$

Moment:

Equation 19.14: $M_{u2\text{-}2} = 1.6(2q_t + q_{t1}) \cdot a^2/6 = 1.6 \times (2 \times 898 + 614) \times (4.5)^2/6/1,000 = 13.0\text{ k.ft/ft}$

Equation 19.17: $A_{stoe} = M_{u2\text{-}2}/0.8 d_{toe} f_y = 13.0 \times 12/(0.8 \times 8.1 \times 60) = 0.40\text{in}^2/\text{ft}$

A_{stoe} is greater than the required steel bars in the stem (0.23 in²/ft). However, both steel bars should be the same for construction suitability (the same steel bars are bend with a 90° hook to act as stem bars and toe bars). Thus, use #5@8in o.c. ($A_s = 0.47$ in²/ft) for both stem bars and toe bars.

It is easily shown that development length for toe bars is satisfied as they extend in the toe (4.5 ft straight) and terminate with a hook at the stem.

Heel Design:

Section 19.3.2: $d_{heel} = D_f - 2\text{ in} - d_{bheel}/2 = 12\text{ in} - 2\text{ in} - 0.9\text{ in} \approx 9\text{ in}$ #5 main & secondary

Shear:

Equation 19.16: $\phi V_n = 0.75 \times 2 \times \sqrt{f'_c} \times 12 d_{heel} > V_{u3\text{-}3} = 1.6n(q_h + q_{h1})/2$

$0.75 \times 2 \times \sqrt{4,000} \times 12 \times 9/1,000 = 10.2\text{ k/ft} > 1.6 \times 1.33 \times (488 + 572)/2/1,000 = 1.1\text{ k/ft OK}$

Moment:

Equation 19.15: $M_{u3\text{-}3} = 1.6(2q_h + q_{h1}) \cdot n^2/6 = 1.6 \times (2 \times 488 + 572) \times (1.33)^2/6/1,000 = 0.73\text{ k.ft/ft}$

Equation 19.17: $A_{sheel} = M_{u3\text{-}3}/0.8 d_{heel} f_y = 0.73 \times 12/(0.8 \times 9 \times 60) = 0.02\text{ in}^2/\text{ft}$

Use #4@8in o.c. ($A_s = 0.30$ in²/ft) for main steel in the heel and #4@12in o.c.($A_s = 0.20$ in²/ft) for secondary steel in the entire footing. Bas sizes are comparable to stem and toe.

Steel ratio in the heel = 0.0021 for main steel bars and 0.0014 for secondary steel bars.

Spacing of main bars in the toe and heel should be the same to facilitate construction.

Key Design:

Section 19.3.2: $d_{key} = B_k - 3\text{ in } d_{bkey}/2 = 12\text{ in} - 3\text{ in} - 1\text{ in} \approx 8\text{ in}$ use #5 main and secondary

$p_1 = p \cdot (D_t + D_f + D_k) = 350 \times (1 + 1 + 2) = 1,400\text{ psf}$

$p_2 = p \cdot (D_t + D_f) = 350 \times (1 + 1) = 700\text{ psf}$

Shear: Equation 19.18: $\phi V_n = 0.75 \times 2 \times \sqrt{f'_c} \times 12 d_{key} > V_{u4\text{-}4} = 1.6 D_k(p_1 + p_2)/2$

$0.75 \times 2 \times \sqrt{4,000} \times 12 \times 8/1,000 = 9.1\text{ k/ft} > 1.6 \times 2 \times (1,400 + 700)/2/1,000 = 3.4\text{ k/ft OK}$

Moment:

Equation 19.19: $A_{skey} = M_{u4-4}/0.8d_{key}f_y = 0.33(2p_1 + p_2) \cdot D_k^2/d_{key}f_y$

$A_{skey} = 0.33 \times (2 \times 1,400 + 700) \times (2)^2/(8 \times 60) \times (12/1000) = 0.12$ in²/ft

Use #4@8in o.c. ($A_s = 0.30$ in²/ft) for main steel in the key (continuation of heel steel) and #4@12in o.c.($A_s = 0.20$ in²/ft) for secondary steel in the key.

Steel ratio in the key = 0.0021 for main steel bars and 0.0014 for secondary steel bars.

Development length for straight #4 bars is 13.1 in and for hooked #4 bars is about 8in. Clearly, they are both satisfied for the heel and key.

Wall detail is shown below:

The following detail includes all pertinent information as explained in Example 19.3

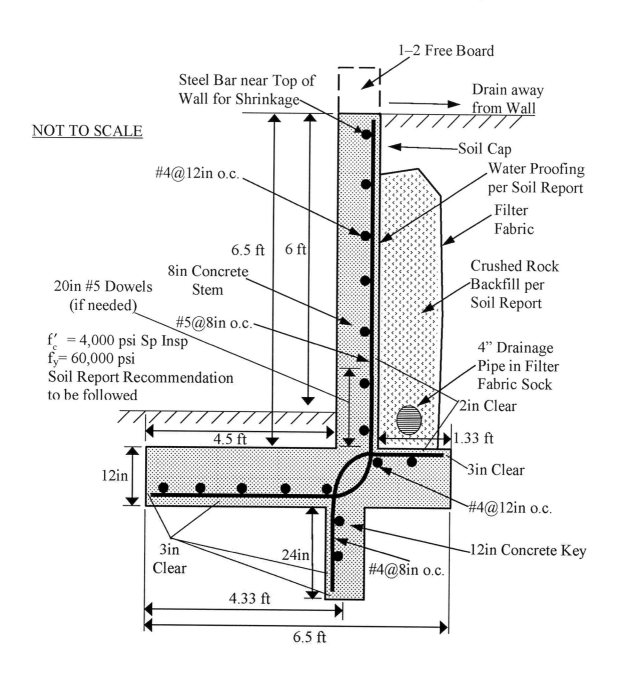

19.6 Concluding Remarks

Static Design Remarks

The design procedures included in this chapter are primarily intended for cohesionless well drained soil. Cohesive or expansive soils or saturated soils require special treatment in design.

Traffic surcharge on retaining walls is typically kept at a distance of 3 ft minimum from the top of wall since to reduce possible local stresses in the soil and wall.

Typically for static design, footing width (B_f) is about 2/3 the height (H_w). In Examples 19.1 and 19.2, due to seismic loading B_f is increased.

In design, increasing a helps in increasing safety factor against overturning. Increasing D_k and n help in increasing sliding safety factor.

Development length of steel bars in walls is a significant design item and requires design. In case the development length is not satisfied, bars can be terminated with a hook.

Seismic Design Remarks

Seismic design of retaining walls is continuously being updated in codes and standards. There remain inconsistencies. Designers are required to pay special attention to seismic design of retaining walls.

Horizontal earthquake acceleration for design of retaining wall, a_h, is site specific and typically provided by the soil engineer.

Construction and Detailing Remarks

For short and medium height retaining walls, instead of using dowel bars connecting the stem to the footing, it is typical to insert the entire main steel bars of the stem in the footing to avoid bar splicing. When the footing is poured, the stem steel bars extend to the top of the proposed retaining wall.

Contractors use 2×4in piece of wood tied to the dowels (or stem main steel bars) as a guide for the top surface of the footing during construction. Upon removal of this wood piece (about 2 hours after pouring the footing) a shear key is created that increases shear resistance of the stem footing connection as shown below.

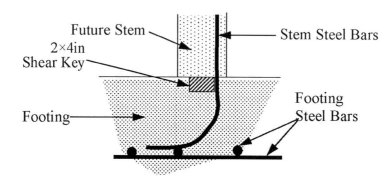

Designers used different thickness of soil on the footing toe (D_t). Typical numbers are between 4in and 12in.

1 to 2 ft of free board is commonly used for retaining walls (see the details in Examples 19.3 and 19.4).

The soil behind the wall requires drainage. Saturation of backfill soil increases the pressure on the retaining wall and can cause wall failure. Several methods of drainage of the backfill soil are used by designers including the method presents in Examples 19.3 and 19.4.

Backfill behind retaining walls requires compaction so that future settlement does not affect the retaining wall performance. Crush rock backfill does not typically require compaction.

There are two methods of water proofing retaining walls:

Asphalt cement coating of the backside of the wall (two coats are required). The asphalt requires a styrofoam board for protection against damage due to soil pressure. This type of water proofing is excellent but requires protection.

Low permeability cement based coating of the backside of the wall (two coats are required). This type of water proofing is adequate and does not require protection since cement based water proofing materials can handle soil pressure.

The perforated drainage pipe must be placed with the perforations (holes) pointing downward to avoid their plugging with settling backfill particles.

Test your knowledge

Design the following retaining walls for the conditions shown:

1. $H_o = 4$ ft $\beta = 3h{:}1v$ ($W_s = 0$) $\Phi s = 30°$ $\gamma = 120$ pcf
    $\mu = 0.35$ $R = 1.0$ $q_a = 2,000$ psf $p = 300$ pcf
    $a_v = 0.4_g$ $a_v = 0$

2. $H_o = 6$ ft $\beta = 4h{:}1v$ ($W_s = 0$) $\Phi s = 28°$ $\gamma = 115$ pcf
    $\mu = 0.35$ $R = 1.0$ $q_a = 1,500$ psf $p = 300$ pcf
    $a_v = 0.25g$ $a_v = 0.02g$

3. $H_o = 5$ ft $\beta = 1.5h{:}1v$ ($W_s = 0$) $\Phi s = 33°$ $\gamma = 125$ pcf
    $\mu = 0.40$ $R = 0.8$ $q_a = 3,000$ psf $p = 400$ pcf
    $a_v = 0.5g$ $a_v = 0$

4. $H_o = 8$ ft $\beta = 4h{:}1v$ ($W_s = 0$) $\Phi s = 25°$ $\gamma = 110$ pcf
    $\mu = 0.30$ $R = 1.0$ $q_a = 2,000$ psf $p = 300$ pcf
    $a_v = 0.35$ g $a_v = 0.015g$

5. $H_o = 10$ ft $\beta = 3h{:}1v$ ($W_s = 0$) $\Phi s = 38°$ $\gamma = 130$ pcf
    $\mu = 0.45$ $R = 1.0$ $q_a = 5,000$ psf $p = 450$ pcf
    $a_v = 0.6$ g $a_v = 0$

6. $H_o = 12$ ft $\beta = 5h{:}1v$ ($W_s = 0$) $\Phi s = 30°$ $\gamma = 120$ pcf
     $\mu = 0.35$ $R = 1.0$ $q_a = 3,000$ psf $p = 350$ pcf
     $a_v = 0.8$ g $a_v = 0$

7. $H_o = 18$ ft $\beta = 2h{:}1v$ ($W_s = 0$) $\Phi s = 34°$ $\gamma = 120$ pcf
     $\mu = 0.40$ $R = 1.0$ $q_a = 2,000$ psf $p = 400$ pcf
     $a_v = 0.7$ g $a_v = 0.025g$

8. $H_o = 4$ ft $W_s = 250$ psf ($\beta = 0$) $\Phi s = 30°$ $\gamma = 120$ pcf
     $\mu = 0.35$ $R = 1.0$ $q_a = 2,000$ psf $p = 300$ pcf
     $a_v = 0.4_g$ $a_v = 0$

9. $H_o = 6$ ft $W_s = 250$ psf ($\beta = 0$) $\Phi s = 28°$ $\gamma = 115$ pcf
     $\mu = 0.35$ $R = 1.0$ $q_a = 1,500$ psf $p = 300$ pcf
     $a_v = 0.25g$ $a_v = 0.02g$

10. $H_o = 5$ ft $W_s = 250$ psf ($\beta = 0$) $\Phi s = 33°$ $\gamma = 125$ pcf
     $\mu = 0.40$ $R = 0.8$ $q_a = 3,000$ psf $p = 400$ pcf
     $a_v = 0.5g$ $a_v = 0$

11. $H_o = 8$ ft $W_s = 250$ psf ($\beta = 0$) $\Phi s = 25°$ $\gamma = 110$ pcf
     $\mu = 0.30$ $R = 1.0$ $q_a = 2,000$ psf $p = 300$ pcf
     $a_v = 0.35$ g $a_v = 0.015g$

12. $H_o = 10$ ft $W_s = 250$ psf ($\beta = 0$) $\Phi s = 38°$ $\gamma = 130$ pcf
     $\mu = 0.45$ $R = 1.0$ $q_a = 5,000$ psf $p = 450$ pcf
     $a_v = 0.6$ g $a_v = 0$

13. $H_o = 12$ ft $W_s = 250$ psf ($\beta = 0$) $\Phi s = 30°$ $\gamma = 120$ pcf
     $\mu = 0.35$ $R = 1.0$ $q_a = 3,000$ psf $p = 350$ pcf
     $a_v = 0.8$ g $a_v = 0$

14. $H_o = 18$ ft $W_s = 250$ psf ($\beta = 0$) $\Phi s = 34°$ $\gamma = 120$ pcf
     $\mu = 0.40$ $R = 1.0$ $q_a = 2,000$ psf $p = 400$ pcf
     $a_v = 0.7$ g $a_v = 0.025g$

15. $H_o = 4$ ft $W_s = 50$ psf ($\beta = 0$) $\Phi s = 30°$ $\gamma = 120$ pcf
     $\mu = 0.35$ $R = 1.0$ $q_a = 2,000$ psf $p = 300$ pcf
     $a_v = 0.4_g$ $a_v = 0$

16. $H_o = 6$ ft $W_s = 50$ psf ($\beta = 0$) $\Phi s = 28°$ $\gamma = 115$ pcf
     $\mu = 0.35$ $R = 1.0$ $q_a = 1,500$ psf $p = 300$ pcf
     $a_v = 0.25g$ $a_v = 0.02g$

17. $H_o = 5$ ft $W_s = 50$ psf $(\beta = 0)$ $\Phi s = 33°$ $\gamma = 125$ pcf

      $\mu = 0.40$ $R = 0.8$ $q_a = 3,000$ psf $p = 400$ pcf

      $a_v = 0.5g$ $a_v = 0$

18. $H_o = 8$ ft $W_s = 50$ psf $(\beta = 0)$ $\Phi s = 25°$ $\gamma = 110$ pcf

      $\mu = 0.30$ $R = 1.0$ $q_a = 2,000$ psf $p = 300$ pcf

      $a_v = 0.35$ g $a_v = 0.015g$

19. $H_o = 10$ ft $W_s = 50$ psf $(\beta = 0)$ $\Phi s = 38°$ $\gamma = 130$ pcf

      $\mu = 0.45$ $R = 1.0$ $q_a = 5,000$ psf $p = 450$ pcf

      $a_v = 0.6$ g $a_v = 0$

20. $H_o = 12$ ft $W_s = 50$ psf $(\beta = 0)$ $\Phi s = 30°$ $\gamma = 120$ pcf

      $\mu = 0.35$ $R = 1.0$ $q_a = 3,000$ psf $p = 350$ pcf

      $a_v = 0.8$ g $a_v = 0$

21. $H_o = 18$ ft $W_s = 50$ psf $(\beta = 0)$ $\Phi s = 34°$ $\gamma = 120$ pcf

      $\mu = 0.40$ $R = 1.0$ $q_a = 2,000$ psf $p = 400$ pcf

      $a_v = 0.7$ g $a_v = 0.025g$

Appendix A

USEFUL TABLES AND FIGURES

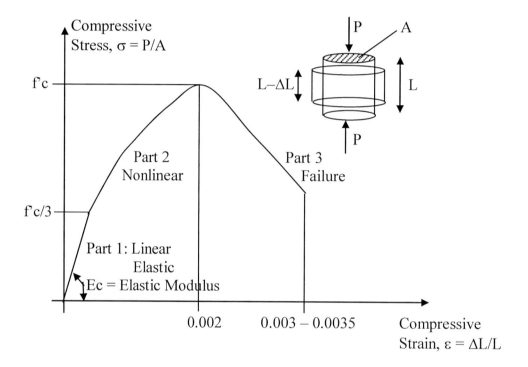

Simplified compressive Stress – Strain Behavior of Concrete
Figure A.1 (Duplicate of Figure 2.9)

Table A.2
Modulus of Elasticity and Tensile Strength of Plain Concrete

f'c (psi)	Modulus of Elasticity[1] (ksi)	Direct Tensile Strength[2] (psi)	Splitting Tensile Strength[3] (psi)	Flexural Strength[4] MOR[5] (psi)
2,500	2,850	290	340	380
3,000	3,120	310	370	410
3,500	3,370	340	400	440
4,000	3,600	360	420	470
4,500	3,820	380	450	500
5,000	4,030	400	470	530
5,500	4,230	420	490	550
6,000	4,420	440	520	580
6,500	4,600	460	540	600
7,000	4,770	480	560	620
8,000	5,100	510	600	670

1 $E_c = 57{,}000\sqrt{f'_c}$

2 Direct Tensile Strength, $f_c = 5.7\sqrt{f'_c}$

3 Splitting Tensile Strength, $f_{ct} = 6.7\sqrt{f'_c}$

4 Flexural Strength, $f_{cf} = 7.5\sqrt{f'_c}$

5 Modulus of Rupture, MOR = Flexural Strength, f_{cf}

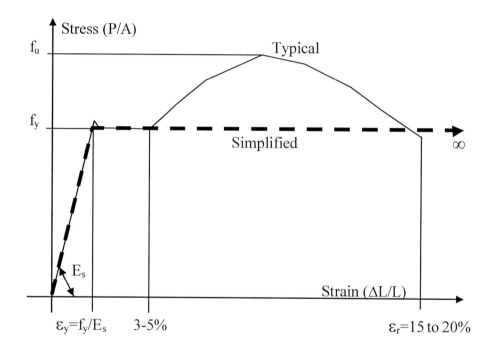

Simplified Tensile Stress – Strain Diagram of Steel
Figure A.3 (Duplicate of Figure 2.16)

Table A.4 (Duplicate of Tables 2.1 and 2.2)

Summary of ASTM A615 Requirements for Steel Reinforcing Bars

Bar No.	Diameter (in)	Area (in^2)	Deformation Requirements		
			Max Ave Spacing (in)	Max Ave Height (in)	Max Gap (in)
3	$^3/_8$	0.11	0.26	0.015	0.14
4	$^1/_2$	0.20	0.35	0.020	0.19
5	$^5/_8$	0.31	0.44	0.028	0.24
6	$^3/_4$	0.44	0.52	0.038	0.29
7	$^7/_8$	0.61	0.61	0.044	0.33
8	1	0.79	0.70	0.050	0.38
9	$1\,^1/_8$	1.00	0.79	0.056	0.43
10	$1\,^1/_4$	1.27	0.89	0.064	0.49
11	$1\,^3/_8$	1.56	0.99	0.071	0.54
14	$1\,^3/_4$	2.25	1.19	0.085	0.65
18	$2\,^1/_4$	4.00	1.60	0.102	0.86

Table 2.2: Common Sizes/Types of Welded Wire Fabric.

WWF Designation	Steel Area (in^2/ft)		Weight (lb/100 ft^2)
	Longitudinal	Transverse	
Rolls			
$6 \times 6 - W1.4 \times W1.4$	0.028	0.028	21
$6 \times 6 - W2 \times W2$	0.040	0.040	29
$6 \times 6 - W2.9 \times W2.9$	0.058	0.058	42
$6 \times 6 - W4 \times W4$	0.080	0.080	58
$4 \times 4 - W1.4 \times W1.4$	0.042	0.042	31
$4 \times 4 - W2 \times W2$	0.060	0.060	43
$4 \times 4 - W2.9 \times W2.9$	0.087	0.087	62
$4 \times 4 - W4 \times W4$	0.120	0.120	86
Sheets			
$6 \times 6 - W2.9 \times W2.9$	0.058	0.058	42
$6 \times 6 - W4 \times W4$	0.080	0.080	58
$6 \times 6 - W5.5 \times W5.5$	0.110	0.110	80
$4 \times 4 - W4 \times W4$	0.120	0.120	86

Table A.5
Area of Steel Bars

No of Bars Bar Size	2	3	4	5	6	7	8	9
3	0.22	0.33	0.44	0.55	0.66	0.77	0.88	0.99
4	0.40	0.60	0.80	1.00	1.20	1.40	1.60	1.80
5	0.62	0.92	1.22	1.53	1.83	2.14	2.44	2.75
6	0.88	1.32	1.76	2.20	2.64	3.08	3.52	3.96
7	1.22	1.83	2.44	3.05	3.66	4.27	4.88	5.49
8	1.58	2.37	3.16	3.95	4.74	5.53	6.32	7.11
9	2.00	3.00	4.00	5.00	6.00	7.00	8.00	9.00
10	2.54	3.81	5.08	6.35	7.62	8.89	10.16	11.43
11	3.12	4.68	6.24	7.80	9.36	10.92	12.48	14.04
14	4.50	6.75	9.00	11.25	13.50	15.75	18.00	20.25
18	8.00	12.00	16.00	20.00	24.00	28.00	32.00	36.00

Table A.6
Strength design Method (LRFD) Equations

General Equation:

$$\phi R_n \geq U$$

Where:

R_n = nominal member resistance. It is termed nominal since it is based on assumptions and structural analysis method selected by the designer and as such, there is a degree of uncertainty in its value. For reinforced concrete, these methods are included in ACI 318,

ϕ = capacity or resistance reduction factor utilized to account for unforeseen circumstances, and

U = Ultimate or maximum applied load or load effect as specified by ACI 318 or ASCE 7 according to the following equations:

$$U1 = 1.4(D + F)$$

$$U2 = 1.2(D + F + T) + 1.6(L + H) + 0.5 \ (Lr \ or \ S \ or \ R)$$

$$U3 = 1.2D + 1.6(Lr \ or \ S \ or \ R) + (L \ or \pm 0.8W)$$

$$U4 = 1.2D \pm 1.6W + L + 0.5(Lr \ or \ S \ or \ R)$$

$$U5 = 1.2D \pm 1.0E + L + 0.2S$$

$$U6 = 0.9D \pm 1.6W + 1.6H$$

$$U7 = 0.9D \pm 1.0E + 1.6H$$

Diagram A.7 (Duplicate of Figure 4.7 and Text)
Simplified Failure Analysis in Bending

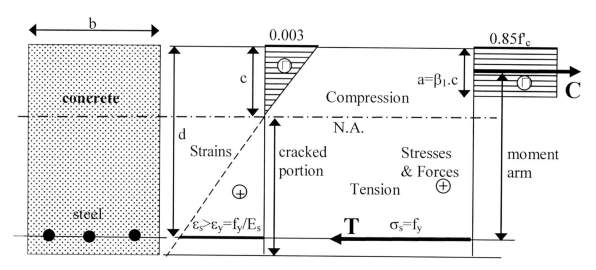

Where:

 c = depth of compression zone,

 a = depth of Whitney's compression block (a = β_1 x c),

 0.003 = crushing strain of concrete,

 0.85f'c = average compressive strength in concrete,

 ε_s = tensile strain of steel,

 ε_y = f_y/E_s = yield strain of steel,

 E_s = modulus of elasticity of steel = 29 x 10^6 psi,

 f_y = yield stress of steel,

 C = resultant compressive force in concrete,

 T = resultant tensile force in steel, and

 moment arm = d – a/2.

 β_1 =   0.85   for f'$_c$ $\le$ 4,000 psi,
          0.80   for f'$_c$ = 5,000 psi,
          0.75   for f'$_c$ = 6,000 psi,
          0.70   for f'$_c$ = 7,000 psi, and
          0.65   for f'$_c$ = 8,000 psi.

 C = T   ====> 0.85f'$_c$.a.b = A$_s$.f$_y$
          ====> a = A$_s$.f$_y$ /0.85 f'$_c$.b
 M$_n$ = A$_s$.f$_y$.(d – a/2) = 0.85f'$_c$.a.b.(d – a/2)

Table A.8 (Duplicate of Table 5.1)
The Capacity or Resistance Reduction Factor in Bending

ε_t	ϕ
≥ 0.0050	0.900
0.0049	0.887
0.0048	0.878
0.0047	0.870
0.0046	0.862
0.0045	0.854
0.0044	0.845
0.0043	0.837
0.0042	0.829
0.0041	0.820
0.0040	0.812
< 0.0040	Not Allowed by ACI 318

The equation for ϕ is:

$$\phi = 0.48 + 83\varepsilon_t$$

Where:

ϕ = capacity or resistance reduction factor in bending, and

ε_t = tensile strain in steel at failure.

Table A.9 (Duplicate of Table 6.1)
Minimum and Maximum Reinforcement Ratios (%)
For
Concrete Beams According to ACI 318 Code

f'$_c$ (psi) f$_y$(psi)		3,000	4,000	5,000	6,000	7,000	8,000
40,000	ρ_{min}	0.50	0.50	0.53	0.58	0.63	0.67
	ρ_{maxp}	2.04	2.72	3.20	3.60	3.92	4.16
	ρ_{max}	2.30	3.06	3.60	4.05	4.41	4.68
	ρ_b	3.71	4.95	5.82	6.55	7.13	7.57
60,000	ρ_{min}	0.33	0.33	0.35	0.39	0.42	0.45
	ρ_{maxp}	1.36	1.81	2.13	2.40	2.61	2.77
	ρ_{max}	1.53	2.04	2.40	2.70	2.94	3.12
	ρ_b	2.14	2.85	3.35	3.77	4.11	4.36
75,000	ρ_{min}	0.27	0.27	0.28	0.31	0.33	0.36
	ρ_{maxp}	1.09	1.45	1.71	1.92	2.09	2.22
	ρ_{max}	1.22	1.63	1.92	2.16	2.35	2.50
	ρ_b	1.55	2.07	2.43	2.74	2.98	3.16
90,000	ρ_{min}	0.22	0.22	0.24	0.26	0.28	0.30
	ρ_{maxp}	0.91	1.21	1.42	1.60	1.74	1.85
	ρ_{max}	1.02	1.36	1.60	1.80	1.96	2.08
	ρ_b	1.18	1.58	1.86	2.09	2.27	2.41

ρ_{maxp} is for tensile strain in steel at failure of 0.005
ρ_{max} is for tensile strain in steel at failure of 0.004

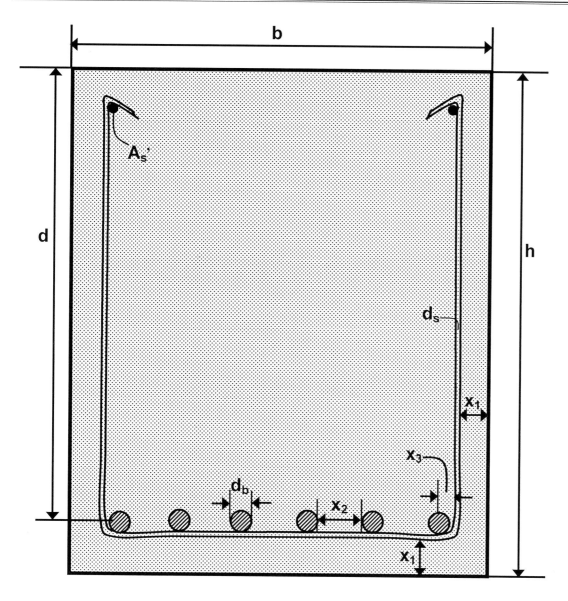

x_1 = concrete clear cover thickness = 1.5 in min for interior exposure
x_2 = clear distance between bars = max of 1 inch, d_b and 4/3 times max aggregate size
x_3 = distance from the inner side of stirrup to centroid of first bar = max of $2d_s$ and $d_b/2$
d_b = diameter of reinforcing bar
d_s = diameter of stirrup = $^3/_8$ or ½ in
n = number of steel bars
h = total beam depth or thickness
d = effective beam depth ===> $d = h - x_1 - d_s - d_b/2$
b = beam width $\geq nd_b + (n - 1)x_2 + 2(x_3 + d_s + x_1)$

Details of a Reinforced Concrete Beam Section with One Layer of Steel Bars
Figure A.10 (Duplicate of Figure 6.2)

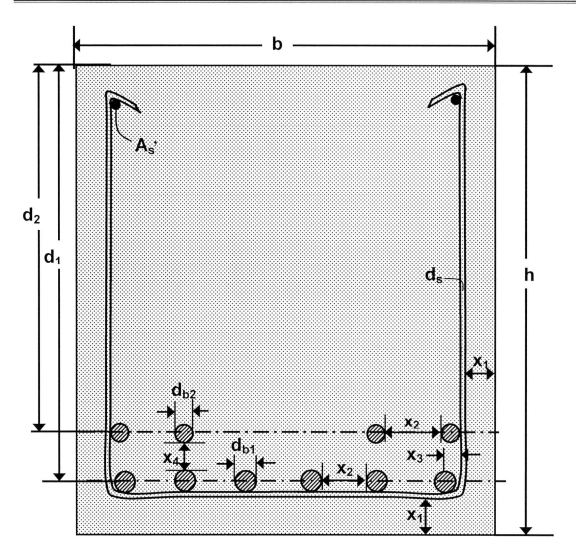

x_1 = concrete clear cover thickness = 1.5 in min for interior exposure

x_2 = clear distance between bars = max of 1 inch, d_b and 4/3 times max aggregate size

x_3 = distance from the inner side of stirrup to centroid of first bar = max of $2d_s$ and $d_b/2$

x_4 = clear distance between lower and upper bars = max of 1 inch d_{b1}, and d_{b2}

d_{b1} = diameter of reinforcing bar in the lower layer

d_{b2} = diameter of reinforcing bar in the upper layer

d_s = diameter of stirrup = $^3/_8$ or ½ in

h = total beam height

d_1 = effective beam depth for the lower layer = $h - x_1 - d_s - d_{b1}/2$

d_2 = effective beam depth for the upper layer = $h - x_1 - d_s - d_{b1} - x_4 - d_{b2}/2$

A_{s1} = the total area of steel bars in the lower layer

A_{s2} = the total area of steel bars in the upper layer

d = effective depth for the beam as a whole = $(A_{s1}.d_1 + A_{s2}.d_2) / (A_{s1} + A_{s2})$

Details of a Reinforced Concrete Beam Section with Two Layers of Steel Bars
Figure A.11 (Duplicate of Figure 6.3)

Table A.12 (Duplicate of Table 6.2)
Minimum Required Beam Width per ACI 318 to Fit Steel Bars[1]

No of Bars Bar Size	2	3	4	5	6	7	8	Added for One Extra Bar
4	6.8	8.3	9.8	11.3	12.8	14.3	15.6	1.5
5	6.9	8.5	10.2	11.8	13.4	15.0	16.7	1.7
6	7.0	8.8	10.5	12.3	14.0	15.8	17.5	1.8
7	7.2	9.0	10.9	12.8	14.7	16.5	18.4	1.9
8	7.3	9.3	11.3	13.3	15.3	17.3	19.6	2.0
9	7.5	9.8	12.0	14.3	16.5	18.9	21.0	2.3
10	7.8	10.3	12.8	15.3	17.8	20.3	22.8	2.5
11	8.0	10.8	13.5	16.3	19.0	21.8	24.5	2.8
14	9.0	12.5	16.0	19.5	23.0	26.5	30	3.5
18	10.5	15.0	19.5	24.0	28.5	33.0	37.5	4.5

1   Interior exposure with 1.5 in min clear cover.
    Use 3/8 in stirrups.

Table A.12 (Duplicate of Tables 6.3 and 8.1)
Minimum thickness (h) for reinforced concrete beams and one – way slabs
unless deflections are computed[1]

Members	Simply Supported	One-End Continuous	Both-Ends Continuous	Cantilever
Beam and Ribbed One – Way Slabs	Span/16	Span/18.5	Span/21	Span/8
Solid One – Way Slabs	Span/20	Span/24	Span/28	Span/10

1 This table is for normal weight concrete (unit weight, γ = 145 lb/ft^3 or pcf) and 60,000 psi yield strength of steel, f_y. For light weight concrete and steel with yield strength (f_y) different that 60,000 psi, the numbers in the table need to be multiplied by (1.65 – 0.005γ) and (0.4 + fy/100,000), respectively. Also, this table does not apply to members supporting construction that is likely to be damaged by large deflections.

Table A.14

Resistance Factor R_n (psi) as a Function of Reinforcement Ratio ρ

f'c =	2,500	psi		fy =	40,000	psi	
ρ	$R_n=M_n/bd^2$	ρ	$R_n=M_n/bd^2$	ρ	$R_n=M_n/bd^2$	ρ	$R_n=M_n/bd^2$
0.0020	78.5	0.0056	212.2	0.0092	336.1	0.0156	532.4
0.0021	82.3	0.0057	215.8	0.0093	339.4	0.0158	538.0
0.0022	86.2	0.0058	219.3	0.0094	342.7	0.0160	543.6
0.0023	90.0	0.0059	222.9	0.0095	346.0	0.0162	549.2
0.0024	93.8	0.0060	226.4	0.0096	349.3	0.0164	554.7
0.0025	97.6	0.0061	230.0	0.0097	352.6	0.0166	560.3
0.0026	101.5	0.0062	233.5	0.0098	355.8	0.0168	565.7
0.0027	105.3	0.0063	237.1	0.0099	359.1	0.0170	571.2
0.0028	109.0	0.0064	240.6	0.0100	362.4		
0.0029	112.8	0.0065	244.1	0.0102	368.8		
0.0030	116.6	0.0066	247.6	0.0104	375.3		
0.0031	120.4	0.0067	251.1	0.0106	381.7		
0.0032	124.1	0.0068	254.6	0.0108	388.1		
0.0033	127.9	0.0069	258.1	0.0110	394.4		
0.0034	131.6	0.0070	261.6	0.0112	400.8		
0.0035	135.4	0.0071	265.0	0.0114	407.1		
0.0036	139.1	0.0072	268.5	0.0116	413.3		
0.0037	142.8	0.0073	271.9	0.0118	419.6		
0.0038	146.6	0.0074	275.4	0.0120	425.8		
0.0039	150.3	0.0075	278.8	0.0122	432.0		
0.0040	154.0	0.0076	282.3	0.0124	438.1		
0.0041	157.7	0.0077	285.7	0.0126	444.2		
0.0042	161.4	0.0078	289.1	0.0128	450.3		
0.0043	165.0	0.0079	292.5	0.0130	456.4		
0.0044	168.7	0.0080	295.9	0.0132	462.4		
0.0045	172.4	0.0081	299.3	0.0134	468.4		
0.0046	176.0	0.0082	302.7	0.0136	474.4		
0.0047	179.7	0.0083	306.1	0.0138	480.3		
0.0048	183.3	0.0084	309.4	0.0140	486.2		
0.0049	187.0	0.0085	312.8	0.0142	492.1		
0.0050	190.6	0.0086	316.2	0.0144	497.9		
0.0051	194.2	0.0087	319.5	0.0146	503.8		
0.0052	197.8	0.0088	322.8	0.0148	509.5		
0.0053	201.4	0.0089	326.2	0.0150	515.3		
0.0054	205.0	0.0090	329.5	0.0152	521.0		
0.0055	208.6	0.0091	332.8	0.0154	526.7		

$\rho_{min\ temp\&shrink}$ = 0.0020

$\rho_{min\ bending}$ = 0.0050

ρ_{maxp} = 0.0170 at ε_t = 0.005

Table A.15

Resistance Factor R_n (psi) as a Function of Reinforcement Ratio ρ

		f'c =	3,000	psi		fy =	40,000	psi	
ρ	$R_n=M_n/bd^2$	ρ	$R_n=M_n/bd^2$	ρ	$R_n=M_n/bd^2$	ρ	$R_n=M_n/bd^2$
0.0020	78.7	0.0056	214.2	0.0092	341.4	0.0156	547.7
0.0021	82.6	0.0057	217.8	0.0093	344.9	0.0158	553.7
0.0022	86.5	0.0058	221.4	0.0094	348.3	0.0160	559.7
0.0023	90.3	0.0059	225.1	0.0095	351.7	0.0162	565.7
0.0024	94.2	0.0060	228.7	0.0096	355.1	0.0164	571.6
0.0025	98.0	0.0061	232.3	0.0097	358.5	0.0166	577.5
0.0026	101.9	0.0062	235.9	0.0098	361.9	0.0168	583.5
0.0027	105.7	0.0063	239.5	0.0099	365.3	0.0170	589.3
0.0028	109.5	0.0064	243.1	0.0100	368.6	0.0172	595.2
0.0029	113.4	0.0065	246.7	0.0102	375.4	0.0174	601.0
0.0030	117.2	0.0066	250.3	0.0104	382.1	0.0176	606.8
0.0031	121.0	0.0067	253.9	0.0106	388.7	0.0178	612.6
0.0032	124.8	0.0068	257.5	0.0108	395.4	0.0180	618.4
0.0033	128.6	0.0069	261.1	0.0110	402.0	0.0182	624.1
0.0034	132.4	0.0070	264.6	0.0112	408.6	0.0184	629.8
0.0035	136.2	0.0071	268.2	0.0114	415.2	0.0186	635.5
0.0036	139.9	0.0072	271.7	0.0116	421.8	0.0188	641.1
0.0037	143.7	0.0073	275.3	0.0118	428.3	0.0190	646.7
0.0038	147.5	0.0074	278.8	0.0120	434.8	0.0192	652.3
0.0039	151.2	0.0075	282.4	0.0122	441.3	0.0194	657.9
0.0040	155.0	0.0076	285.9	0.0124	447.8	0.0196	663.5
0.0041	158.7	0.0077	289.4	0.0126	454.2	0.0198	669.0
0.0042	162.5	0.0078	292.9	0.0128	460.6	0.0200	674.5
0.0043	166.2	0.0079	296.4	0.0130	467.0	0.0202	680.0
0.0044	169.9	0.0080	299.9	0.0132	473.3	0.0204	685.4
0.0045	173.6	0.0081	303.4	0.0134	479.7	$\rho_{min\ temp\&shrink}$ = 0.0020	
0.0046	177.4	0.0082	306.9	0.0136	486.0	$\rho_{min\ bending}$ = 0.0050	
0.0047	181.1	0.0083	310.4	0.0138	492.3		
0.0048	184.8	0.0084	313.9	0.0140	498.5	ρ_{maxp} = 0.0204	
0.0049	188.5	0.0085	317.3	0.0142	504.7	at ε_t = 0.005	
0.0050	192.2	0.0086	320.8	0.0144	510.9		
0.0051	195.8	0.0087	324.3	0.0146	517.1		
0.0052	199.5	0.0088	327.7	0.0148	523.3		
0.0053	203.2	0.0089	331.1	0.0150	529.4		
0.0054	206.9	0.0090	334.6	0.0152	535.5		
0.0055	210.5	0.0091	338.0	0.0154	541.6		

Table A.16

Resistance Factor R_n (psi) as a Function of Reinforcement Ratio ρ

	f'c =	4,000	psi		fy =	40,000	psi	
ρ	$R_n=M_n/bd^2$	ρ	$R_n=M_n/bd^2$	ρ	$R_n=M_n/bd^2$	ρ	$R_n=M_n/bd^2$
0.0020	79.1	0.0056	216.6	0.0104	390.6	0.0215	751.2
0.0021	83.0	0.0057	220.4	0.0106	397.6	0.0220	766.1
0.0022	86.9	0.0058	224.1	0.0108	404.6	0.0225	780.9
0.0023	90.8	0.0059	227.8	0.0110	411.5	0.0230	795.5
0.0024	94.6	0.0060	231.5	0.0112	418.5	0.0235	810.1
0.0025	98.5	0.0061	235.2	0.0114	425.4	0.0240	824.5
0.0026	102.4	0.0062	239.0	0.0116	432.3	0.0245	838.8
0.0027	106.3	0.0063	242.7	0.0118	439.2	0.0250	852.9
0.0028	110.2	0.0064	246.4	0.0120	446.1	0.0255	867.0
0.0029	114.0	0.0065	250.1	0.0122	453.0	0.0260	880.9
0.0030	117.9	0.0066	253.8	0.0124	459.8	0.0265	894.8
0.0031	121.7	0.0067	257.4	0.0126	466.6	0.0270	908.5
0.0032	125.6	0.0068	261.1	0.0128	473.4	0.0272	913.9
0.0033	129.4	0.0069	264.8	0.0130	480.2	$\rho_{min\ temp\&shrink}$ = 0.0020	
0.0034	133.3	0.0070	268.5	0.0132	487.0	$\rho_{min\ bending}$ = 0.0050	
0.0035	137.1	0.0071	272.1	0.0134	493.8		
0.0036	141.0	0.0072	275.8	0.0136	500.5	ρ_{maxp} = 0.0272	
0.0037	144.8	0.0073	279.5	0.0138	507.2	at ε_t = 0.005	
0.0038	148.6	0.0074	283.1	0.0140	513.9		
0.0039	152.4	0.0075	286.8	0.0142	520.6		
0.0040	156.2	0.0076	290.4	0.0144	527.2		
0.0041	160.0	0.0077	294.0	0.0146	533.8		
0.0042	163.8	0.0078	297.7	0.0148	540.5		
0.0043	167.6	0.0079	301.3	0.0150	547.1		
0.0044	171.4	0.0080	304.9	0.0155	563.5		
0.0045	175.2	0.0082	312.2	0.0160	579.8		
0.0046	179.0	0.0084	319.4	0.0165	595.9		
0.0047	182.8	0.0086	326.6	0.0170	612.0		
0.0048	186.6	0.0088	333.8	0.0175	627.9		
0.0049	190.4	0.0090	340.9	0.0180	643.8		
0.0050	194.1	0.0092	348.1	0.0185	659.5		
0.0051	197.9	0.0094	355.2	0.0190	675.1		
0.0052	201.6	0.0096	362.3	0.0195	690.5		
0.0053	205.4	0.0098	369.4	0.0200	705.9		
0.0054	209.1	0.0100	376.5	0.0205	721.1		
0.0055	212.9	0.0102	383.5	0.0210	736.2		

Table A.17

Resistance Factor R_n (psi) as a Function of Reinforcement Ratio ρ

| | | | fc = | 2,500 | psi | | | | | fy = | 60,000 | psi |

ρ	$R_n = M_n/bd^2$	ρ	$R_n = M_n/bd^2$	ρ	$R_n = M_n/bd^2$
0.0018	105.3	0.0054	299.3	0.0090	471.4
0.0019	110.9	0.0055	304.4	0.0091	475.9
0.002	116.6	0.0056	309.4	0.0092	480.3
0.0021	122.3	0.0057	314.5	0.0093	484.7
0.0022	127.9	0.0058	319.5	0.0094	489.2
0.0023	133.5	0.0059	324.5	0.0095	493.6
0.0024	139.1	0.0060	329.5	0.0096	497.9
0.0025	144.7	0.0061	334.5	0.0097	502.3
0.0026	150.3	0.0062	339.4	0.0098	506.6
0.0027	155.8	0.0063	344.4	0.0099	511.0
0.0028	161.4	0.0064	349.3	0.0100	515.3
0.0029	166.9	0.0065	354.2	0.0101	519.6
0.003	172.4	0.0066	359.1	0.0102	523.9
0.0031	177.9	0.0067	364.0	0.0103	528.1
0.0032	183.3	0.0068	368.8	0.0104	532.4
0.0033	188.8	0.0069	373.7	0.0105	536.6
0.0034	194.2	0.0070	378.5	0.0106	540.8
0.0035	199.6	0.0071	383.3	0.0107	545.0
0.0036	205.0	0.0072	388.1	0.0108	549.2
0.0037	210.4	0.0073	392.9	0.0109	553.4
0.0038	215.8	0.0074	397.6	0.0110	557.5
0.0039	221.1	0.0075	402.4	0.0111	561.6
0.004	226.4	0.0076	407.1	0.0112	565.7
0.0041	231.8	0.0077	411.8	0.0113	569.8
0.0042	237.1	0.0078	416.5	$\rho_{\text{min temp\&shrink}}$ = 0.0018	
0.0043	242.3	0.0079	421.1	$\rho_{\text{min bending}}$ = 0.0033	
0.0044	247.6	0.0080	425.8		
0.0045	252.8	0.0081	430.4	ρ_{maxp} = 0.0113	
0.0046	258.1	0.0082	435.0	at ε_t = 0.005	
0.0047	263.3	0.0083	439.6		
0.0048	268.5	0.0084	444.2		
0.0049	273.7	0.0085	448.8		
0.005	278.8	0.0086	453.4		
0.0051	284.0	0.0087	457.9		
0.0052	289.1	0.0088	462.4		
0.0053	294.2	0.0089	466.9		

Table A.18

Resistance Factor R_n (psi) as a Function of Reinforcement Ratio ρ

	f'c =	3,000	psi		fy =	60,000	psi

ρ	$R_n=M_n/bd^2$	ρ	$R_n=M_n/bd^2$	ρ	$R_n=M_n/bd^2$	ρ	$R_n=M_n/bd^2$
0.0018	105.7	0.0054	303.4	0.0090	482.8	0.0126	643.9
0.0019	111.5	0.0055	308.6	0.0091	487.5	0.0127	648.1
0.002	117.2	0.0056	313.9	0.0092	492.3	0.0128	652.3
0.0021	122.9	0.0057	319.1	0.0093	496.9	0.0129	656.5
0.0022	128.6	0.0058	324.3	0.0094	501.6	0.0130	660.7
0.0023	134.3	0.0059	329.4	0.0095	506.3	0.0131	664.9
0.0024	139.9	0.0060	334.6	0.0096	510.9	0.0132	669.0
0.0025	145.6	0.0061	339.7	0.0097	515.6	0.0133	673.1
0.0026	151.2	0.0062	344.9	0.0098	520.2	0.0134	677.3
0.0027	156.9	0.0063	350.0	0.0099	524.8	0.0135	681.4
0.0028	162.5	0.0064	355.1	0.0100	529.4	0.0136	685.4
0.0029	168.1	0.0065	360.2	0.0101	534.0	$\rho_{\text{min temp\&shrink}}$ = 0.0018	
0.003	173.6	0.0066	365.3	0.0102	538.6	$\rho_{\text{min bending}}$ = 0.0033	
0.0031	179.2	0.0067	370.3	0.0103	543.1		
0.0032	184.8	0.0068	375.4	0.0104	547.7	ρ_{maxp} = 0.0136	
0.0033	190.3	0.0069	380.4	0.0105	552.2	at ε_t = 0.005	
0.0034	195.8	0.0070	385.4	0.0106	556.7		
0.0035	201.4	0.0071	390.4	0.0107	561.2		
0.0036	206.9	0.0072	395.4	0.0108	565.7		
0.0037	212.3	0.0073	400.4	0.0109	570.1		
0.0038	217.8	0.0074	405.3	0.0110	574.6		
0.0039	223.3	0.0075	410.3	0.0111	579.0		
0.004	228.7	0.0076	415.2	0.0112	583.5		
0.0041	234.1	0.0077	420.1	0.0113	587.9		
0.0042	239.5	0.0078	425.1	0.0114	592.3		
0.0043	244.9	0.0079	429.9	0.0115	596.6		
0.0044	250.3	0.0080	434.8	0.0116	601.0		
0.0045	255.7	0.0081	439.7	0.0117	605.4		
0.0046	261.1	0.0082	444.5	0.0118	609.7		
0.0047	266.4	0.0083	449.4	0.0119	614.0		
0.0048	271.7	0.0084	454.2	0.0120	618.4		
0.0049	277.1	0.0085	459.0	0.0121	622.7		
0.005	282.4	0.0086	463.8	0.0122	626.9		
0.0051	287.6	0.0087	468.6	0.0123	631.2		
0.0052	292.9	0.0088	473.3	0.0124	635.5		
0.0053	298.2	0.0089	478.1	0.0125	639.7		

Table A.19

Resistance Factor R_n (psi) as a Function of Reinforcement Ratio ρ

	f'c =	4,000	psi		fy =	60,000	psi	
ρ	$R_n=M_n/bd^2$	ρ	$R_n=M_n/bd^2$	ρ	$R_n=M_n/bd^2$	ρ	$R_n=M_n/bd^2$
0.0018	106.3	0.0054	308.6	0.0090	497.1	0.0152	789.7
0.0019	112.1	0.0055	314.0	0.0091	502.2	0.0154	798.4
0.002	117.9	0.0056	319.4	0.0092	507.2	0.0156	807.2
0.0021	123.7	0.0057	324.8	0.0093	512.2	0.0158	815.8
0.0022	129.4	0.0058	330.2	0.0094	517.2	0.0160	824.5
0.0023	135.2	0.0059	335.6	0.0095	522.2	0.0162	833.1
0.0024	141.0	0.0060	340.9	0.0096	527.2	0.0164	841.6
0.0025	146.7	0.0061	346.3	0.0097	532.2	0.0166	850.1
0.0026	152.4	0.0062	351.6	0.0098	537.2	0.0168	858.6
0.0027	158.1	0.0063	357.0	0.0099	542.1	0.0170	867.0
0.0028	163.8	0.0064	362.3	0.0100	547.1	0.0172	875.4
0.0029	169.5	0.0065	367.6	0.0102	556.9	0.0174	883.7
0.003	175.2	0.0066	372.9	0.0104	566.7	0.0176	892.0
0.0031	180.9	0.0067	378.2	0.0106	576.5	0.0178	900.3
0.0032	186.6	0.0068	383.5	0.0108	586.2	0.0180	908.5
0.0033	192.2	0.0069	388.8	0.0110	595.9	0.0181	912.6
0.0034	197.9	0.0070	394.1	0.0112	605.6		
0.0035	203.5	0.0071	399.3	0.0114	615.2	$\rho_{min\,temp\&shrink} = 0.0018$	
0.0036	209.1	0.0072	404.6	0.0116	624.8	$\rho_{min\,bending} = 0.0033$	
0.0037	214.8	0.0073	409.8	0.0118	634.3		
0.0038	220.4	0.0074	415.0	0.0120	643.8	$\rho_{maxp} = 0.0181$	
0.0039	225.9	0.0075	420.2	0.0122	653.2	at $\varepsilon_t = 0.005$	
0.004	231.5	0.0076	425.4	0.0124	662.6		
0.0041	237.1	0.0077	430.6	0.0126	672.0		
0.0042	242.7	0.0078	435.8	0.0128	681.3		
0.0043	248.2	0.0079	441.0	0.0130	690.5		
0.0044	253.8	0.0080	446.1	0.0132	699.8		
0.0045	259.3	0.0081	451.3	0.0134	708.9		
0.0046	264.8	0.0082	456.4	0.0136	718.1		
0.0047	270.3	0.0083	461.5	0.0138	727.2		
0.0048	275.8	0.0084	466.6	0.0140	736.2		
0.0049	281.3	0.0085	471.8	0.0142	745.2		
0.005	286.8	0.0086	476.8	0.0144	754.2		
0.0051	292.2	0.0087	481.9	0.0146	763.2		
0.0052	297.7	0.0088	487.0	0.0148	772.0		
0.0053	303.1	0.0089	492.1	0.0150	780.9		

Table A.20

Resistance Factor R_n (psi) as a Function of Reinforcement Ratio ρ

	f'c =	5,000	psi		fy =	60,000	psi

ρ	$R_n=M_n/bd^2$	ρ	$R_n=M_n/bd^2$	ρ	$R_n=M_n/bd^2$	ρ	$R_n=M_n/bd^2$
0.0018	106.6	0.0054	311.6	0.0090	505.7	0.0155	828.2
0.0019	112.5	0.0055	317.2	0.0091	510.9	0.0160	851.6
0.002	118.3	0.0056	322.7	0.0092	516.2	0.0165	874.7
0.0021	124.1	0.0057	328.2	0.0093	521.4	0.0170	897.6
0.0022	130.0	0.0058	333.8	0.0094	526.6	0.0175	920.3
0.0023	135.8	0.0059	339.3	0.0095	531.8	0.0180	942.8
0.0024	141.6	0.0060	344.8	0.0096	537.0	0.0185	965.0
0.0025	147.4	0.0061	350.2	0.0097	542.2	0.0190	987.1
0.0026	153.1	0.0062	355.7	0.0098	547.3	0.0195	1009.0
0.0027	158.9	0.0063	361.2	0.0099	552.5	0.0200	1030.6
0.0028	164.7	0.0064	366.7	0.0100	557.6	0.0205	1052.0
0.0029	170.4	0.0065	372.1	0.0102	567.9	0.0210	1073.2
0.003	176.2	0.0066	377.6	0.0104	578.2	0.0213	1085.8
0.0031	181.9	0.0067	383.0	0.0106	588.4	$\rho_{min\ temp\&shrink} = 0.0018$	
0.0032	187.7	0.0068	388.4	0.0108	598.6	$\rho_{min\ bending} = 0.0035$	
0.0033	193.4	0.0069	393.8	0.0110	608.8		
0.0034	199.1	0.0070	399.2	0.0112	618.9		
0.0035	204.8	0.0071	404.6	0.0114	629.0	$\rho_{maxp} = 0.0213$	
0.0036	210.5	0.0072	410.0	0.0116	639.0	at $\varepsilon_t = 0.005$	
0.0037	216.2	0.0073	415.4	0.0118	649.0		
0.0038	221.9	0.0074	420.8	0.0120	659.0		
0.0039	227.6	0.0075	426.2	0.0122	669.0		
0.004	233.2	0.0076	431.5	0.0124	678.9		
0.0041	238.9	0.0077	436.9	0.0126	688.8		
0.0042	244.5	0.0078	442.2	0.0128	698.6		
0.0043	250.2	0.0079	447.6	0.0130	708.4		
0.0044	255.8	0.0080	452.9	0.0132	718.2		
0.0045	261.4	0.0081	458.2	0.0134	728.0		
0.0046	267.0	0.0082	463.5	0.0136	737.7		
0.0047	272.6	0.0083	468.8	0.0138	747.3		
0.0048	278.2	0.0084	474.1	0.0140	757.0		
0.0049	283.8	0.0085	479.4	0.0142	766.6		
0.005	289.4	0.0086	484.7	0.0144	776.2		
0.0051	295.0	0.0087	489.9	0.0146	785.7		
0.0052	300.5	0.0088	495.2	0.0148	795.2		
0.0053	306.1	0.0089	500.5	0.0150	804.7		

Table A.21

Resistance Factor R_n (psi) as a Function of Reinforcement Ratio ρ

	f'c =	6,000	psi		fy =	60,000	psi	
ρ	$R_n=M_n/bd^2$	ρ	$R_n=M_n/bd^2$	ρ	$R_n=M_n/bd^2$	ρ	$R_n=M_n/bd^2$
0.0018	106.9	0.0054	313.7	0.0090	511.4	0.0155	845.2
0.0019	112.7	0.0055	319.3	0.0091	516.8	0.0160	869.6
0.002	118.6	0.0056	324.9	0.0092	522.1	0.0165	893.9
0.0021	124.4	0.0057	330.5	0.0093	527.5	0.0170	918.0
0.0022	130.3	0.0058	336.1	0.0094	532.8	0.0175	941.9
0.0023	136.1	0.0059	341.7	0.0095	538.1	0.0180	965.6
0.0024	142.0	0.0060	347.3	0.0096	543.5	0.0185	989.2
0.0025	147.8	0.0061	352.9	0.0097	548.8	0.0190	1012.6
0.0026	153.6	0.0062	358.4	0.0098	554.1	0.0195	1035.8
0.0027	159.4	0.0063	364.0	0.0099	559.4	0.0200	1058.8
0.0028	165.2	0.0064	369.5	0.0100	564.7	0.0205	1081.7
0.0029	171.0	0.0065	375.1	0.0102	575.3	0.0210	1104.4
0.003	176.8	0.0066	380.6	0.0104	585.8	0.0215	1126.9
0.0031	182.6	0.0067	386.2	0.0106	596.3	0.0220	1149.2
0.0032	188.4	0.0068	391.7	0.0108	606.8	0.0225	1171.3
0.0033	194.2	0.0069	397.2	0.0110	617.3	0.0230	1193.3
0.0034	199.9	0.0070	402.7	0.0112	627.7	0.0235	1215.1
0.0035	205.7	0.0071	408.2	0.0114	638.1	0.0240	1236.7
0.0036	211.4	0.0072	413.7	0.0116	648.5		
0.0037	217.2	0.0073	419.2	0.0118	658.9		
0.0038	222.9	0.0074	424.7	0.0120	669.2		
0.0039	228.6	0.0075	430.1	0.0122	679.5		
0.004	234.4	0.0076	435.6	0.0124	689.7		
0.0041	240.1	0.0077	441.1	0.0126	700.0		
0.0042	245.8	0.0078	446.5	0.0128	710.2		
0.0043	251.5	0.0079	452.0	0.0130	720.4		
0.0044	257.2	0.0080	457.4	0.0132	730.5		
0.0045	262.9	0.0081	462.8	0.0134	740.6		
0.0046	268.5	0.0082	468.3	0.0136	750.7		
0.0047	274.2	0.0083	473.7	0.0138	760.8		
0.0048	279.9	0.0084	479.1	0.0140	770.8		
0.0049	285.5	0.0085	484.5	0.0142	780.8		
0.005	291.2	0.0086	489.9	0.0144	790.8		
0.0051	296.8	0.0087	495.3	0.0146	800.8		
0.0052	302.5	0.0088	500.7	0.0148	810.7		
0.0053	308.1	0.0089	506.0	0.0150	820.6		

$\rho_{min\ temp\&shrink} = 0.0018$

$\rho_{min\ bending} = 0.0039$

$\rho_{maxp} = 0.0240$

at $\varepsilon_t = 0.005$

Table A.22

Resistance Factor R_n (psi) as a Function of Reinforcement Ratio ρ

	f'c =	5,000	psi		fy =	80,000	psi

ρ	$R_n=M_n/bd^2$	ρ	$R_n=M_n/bd^2$	ρ	$R_n=M_n/bd^2$	ρ	$R_n=M_n/bd^2$
0.0014	110.5	0.0050	381.2	0.0086	632.3	0.0148	1019.1
0.0015	118.3	0.0051	388.4	0.0087	639.0	0.0150	1030.6
0.0016	126.1	0.0052	395.6	0.0088	645.7	0.0152	1042.0
0.0017	133.8	0.0053	402.8	0.0089	652.4	0.0154	1053.4
0.0018	141.6	0.0054	410.0	0.0090	659.0	0.0156	1064.8
0.0019	149.3	0.0055	417.2	0.0091	665.6	0.0158	1076.0
0.002	157.0	0.0056	424.4	0.0092	672.3	0.0160	1087.2
0.0021	164.7	0.0057	431.5	0.0093	678.9		
0.0022	172.4	0.0058	438.7	0.0094	685.5		
0.0023	180.0	0.0059	445.8	0.0095	692.0		
0.0024	187.7	0.0060	452.9	0.0096	698.6		
0.0025	195.3	0.0061	460.0	0.0098	711.7		
0.0026	202.9	0.0062	467.1	0.0100	724.7		
0.0027	210.5	0.0063	474.1	0.0102	737.7		
0.0028	218.1	0.0064	481.2	0.0104	750.6		
0.0029	225.7	0.0065	488.2	0.0106	763.4		
0.003	233.2	0.0066	495.2	0.0108	776.2		
0.0031	240.8	0.0067	502.2	0.0110	788.9		
0.0032	248.3	0.0068	509.2	0.0112	801.6		
0.0033	255.8	0.0069	516.2	0.0114	814.1		
0.0034	263.3	0.0070	523.1	0.0116	826.7		
0.0035	270.8	0.0071	530.0	0.0118	839.2		
0.0036	278.2	0.0072	537.0	0.0120	851.6		
0.0037	285.7	0.0073	543.9	0.0122	863.9		
0.0038	293.1	0.0074	550.8	0.0124	876.2		
0.0039	300.5	0.0075	557.6	0.0126	888.5		
0.004	308.0	0.0076	564.5	0.0128	900.6		
0.0041	315.3	0.0077	571.4	0.0130	912.8		
0.0042	322.7	0.0078	578.2	0.0132	924.8		
0.0043	330.1	0.0079	585.0	0.0134	936.8		
0.0044	337.4	0.0080	591.8	0.0136	948.7		
0.0045	344.8	0.0081	598.6	0.0138	960.6		
0.0046	352.1	0.0082	605.4	0.0140	972.4		
0.0047	359.4	0.0083	612.1	0.0142	984.2		
0.0048	366.7	0.0084	618.9	0.0144	995.9		
0.0049	373.9	0.0085	625.6	0.0146	1007.5		

$\rho_{min\ temp\&shrink} = 0.0014$

$\rho_{min\ bending} = 0.0027$

$\rho_{maxp} = 0.0160$

at $\varepsilon_t = 0.005$

Table A.23

Resistance Factor R_n (psi) as a Function of Reinforcement Ratio ρ

	f'c =	6,000	psi		fy =	80,000	psi	
ρ	$R_n=M_n/bd^2$	ρ	$R_n=M_n/bd^2$	ρ	$R_n=M_n/bd^2$	ρ	$R_n=M_n/bd^2$
0.0014	110.8	0.0050	384.3	0.0086	641.6	0.0148	1046.6
0.0015	118.6	0.0051	391.7	0.0087	648.5	0.0150	1058.8
0.0016	126.4	0.0052	399.0	0.0088	655.4	0.0152	1071.0
0.0017	134.2	0.0053	406.4	0.0089	662.3	0.0154	1083.2
0.0018	142.0	0.0054	413.7	0.0090	669.2	0.0156	1095.3
0.0019	149.7	0.0055	421.0	0.0091	676.0	0.0158	1107.4
0.002	157.5	0.0056	428.3	0.0092	682.9	0.0160	1119.4
0.0021	165.2	0.0057	435.6	0.0093	689.7	0.0162	1131.3
0.0022	173.0	0.0058	442.9	0.0094	696.6	0.0164	1143.2
0.0023	180.7	0.0059	450.2	0.0095	703.4	0.0166	1155.1
0.0024	188.4	0.0060	457.4	0.0096	710.2	0.0168	1166.9
0.0025	196.1	0.0061	464.7	0.0098	723.7	0.0170	1178.7
0.0026	203.8	0.0062	471.9	0.0100	737.3	0.0172	1190.4
0.0027	211.4	0.0063	479.1	0.0102	750.7	0.0174	1202.0
0.0028	219.1	0.0064	486.3	0.0104	764.1	0.0176	1213.6
0.0029	226.7	0.0065	493.5	0.0106	777.5	0.0178	1225.2
0.003	234.4	0.0066	500.7	0.0108	790.8	0.0180	1236.7
0.0031	242.0	0.0067	507.8	0.0110	804.1		
0.0032	249.6	0.0068	515.0	0.0112	817.3		
0.0033	257.2	0.0069	522.1	0.0114	830.5		
0.0034	264.7	0.0070	529.3	0.0116	843.6		
0.0035	272.3	0.0071	536.4	0.0118	856.6		
0.0036	279.9	0.0072	543.5	0.0120	869.6		
0.0037	287.4	0.0073	550.6	0.0122	882.6		
0.0038	294.9	0.0074	557.6	0.0124	895.5		
0.0039	302.5	0.0075	564.7	0.0126	908.4		
0.004	310.0	0.0076	571.8	0.0128	921.2		
0.0041	317.5	0.0077	578.8	0.0130	934.0		
0.0042	324.9	0.0078	585.8	0.0132	946.7		
0.0043	332.4	0.0079	592.8	0.0134	959.3		
0.0044	339.9	0.0080	599.8	0.0136	971.9		
0.0045	347.3	0.0081	606.8	0.0138	984.5		
0.0046	354.7	0.0082	613.8	0.0140	997.0		
0.0047	362.1	0.0083	620.8	0.0142	1009.5		
0.0048	369.5	0.0084	627.7	0.0144	1021.9		
0.0049	376.9	0.0085	634.7	0.0146	1034.3		

$\rho_{min\ temp\&shrink} = 0.0014$

$\rho_{min\ bending} = 0.0029$

$\rho_{maxp} = 0.0180$

at $\varepsilon_t = 0.005$

Table A.24 (Duplicate of Table 8.2)
Equivalent Area of Steel Bars in Concrete Slabs (in^2/ft)

Bar # Spacing	3	4	5	6	7	8	9	10	11
3	0.44	0.80	1.24	1.76	2.40	3.20	4.00	5.08	5.64
4	0.33	0.60	0.93	1.32	1.80	2.37	3.00	3.81	4.23
5	0.26	0.48	0.74	1.06	1.44	1.90	2.40	3.05	3.38
6	0.22	0.40	0.62	0.88	1.20	1.58	2.00	2.54	2.82
7	0.19	0.34	0.53	0.75	1.03	1.35	1.71	2.18	2.42
8	0.17	0.30	0.47	0.66	0.90	1.19	1.50	1.91	2.12
9	0.15	0.27	0.41	0.59	0.80	1.05	1.33	1.69	1.88
10	0.13	0.24	0.37	0.53	0.72	0.95	1.20	1.52	1.69
12	0.11	0.20	0.31	0.44	0.60	0.79	1.00	1.27	1.41
14	0.094	0.17	0.27	0.38	0.51	0.68	0.86	1.09	1.21
16	0.083	0.15	0.23	0.33	0.45	0.59	0.75	0.95	1.06
18	0.073	0.13	0.21	0.29	0.40	0.53	0.67	0.85	0.94

Table A.25 (Duplicate of Table 8.3)
Summary of Live Load on Slabs

Use of Structure	Uniform Live Load (psf)	Concentrated Live Load (lb) on 2.5 ft Square Area
Balconies	100	N/A
Decks	100	N/A
Residential Dwellings	40	N/A
Offices	50	2,000
Office Corridors & Computers	100	2,000
Schools	40	1,000
Corridors	100	1,000
Light Storage	125	N/A
Heavy Storage	250	N/A
Retail Store	100	1,000
Whole Sale Store	125	1,000
Light Manufacturing	125	2,000
Heavy Manufacturing	250	3,000
Restaurant and Ballroom	100	N/A
Hospitals	80	1,000
Hotel – Rooms	40	N/A
Corridors and Public Rooms	100	N/A
Passenger Garage	40	N/A
Sidewalks and Driveways	250	8,000
Roofs	20	N/A

Table A.26: T–beam Design Aid (compression in flange)

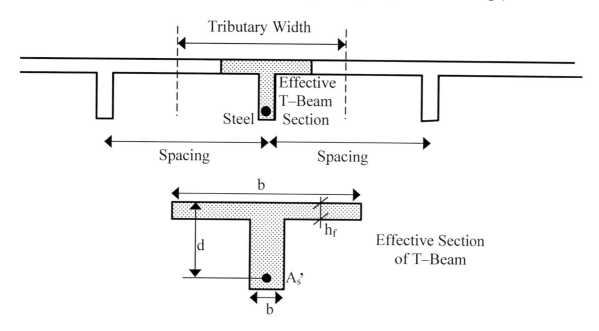

b = the smallest value among b_1, b_2 and b_3,
   b_1 = T–beam span/4
   b_2 = tributary width
   b_3 = $b_w + 16h_f$

For isolated T–beams (not a part of floor system) code: requires:
$h_f \geq b_w/2$ and $b \leq 4b_w$

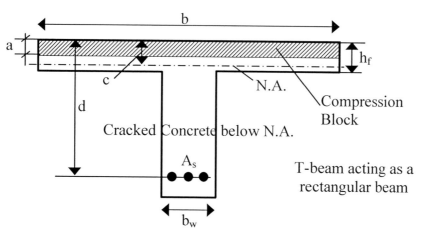

$a = A_s \cdot f_y/(0.85f'_c \cdot b) < h_f$

$\varepsilon_t = 0.003(d - c)/c \geq 0.004$ (Code)   Where $c = a/\beta_1$ ===> $\phi = 0.48 + 83\varepsilon_t \leq 0.90$

$A_s \geq \rho_{min} \cdot b_w \cdot d$ (Code)

$\phi M_n = \phi A_s \cdot f_y \cdot (d - a/2)$

Notation is illustrated in the figures above.

Table A.27: T–beam Design Aid (compression in web)

Also see Table A.26 for effective section.

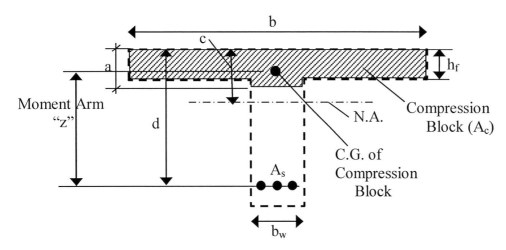

$a = A_s \cdot f_y/(0.85f'_c \cdot b) > h_f$ ===>   This equation is not applicable use the following procedure

$A_c = A_s \cdot f_y/0.85f'_c$

$a = [A_c - h_f \cdot (b - b_w)]/b_w$

$z = d - [(b - b_w) \cdot h_f^2/2 + b_w \cdot a^2/2]/A_c$

$\varepsilon_t = 0.003(d - c)/c \geq 0.004$ (Code)   Where $c = a/\beta_1$ ===> $\phi = 0.48 + 83\varepsilon_t \leq 0.90$

$A_s \geq \rho_{min} \cdot b_w \cdot d$ (Code)

$\phi M_n = \phi A_s \cdot f_y \cdot z$

Notation is illustrated in the figures above.

Table A.28: Doubly Reinforced Concrete Beam Design Aid
(compression steel has yielded)

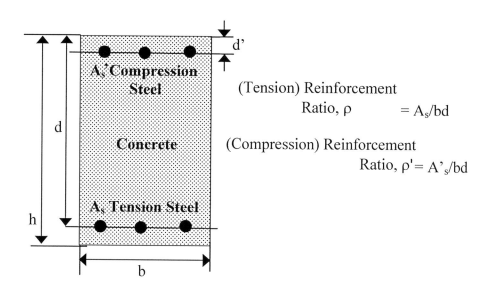

(Tension) Reinforcement
Ratio, ρ $= A_s/bd$

(Compression) Reinforcement
Ratio, $\rho' = A'_s/bd$

If (a) or (b) is satisfied, then beam can be analyzed as singly reinforced.
   a. $A'_s < 0.20 A_s$
   b. $\rho = A_s/bd < \rho_{maxp,singly}$

Correction

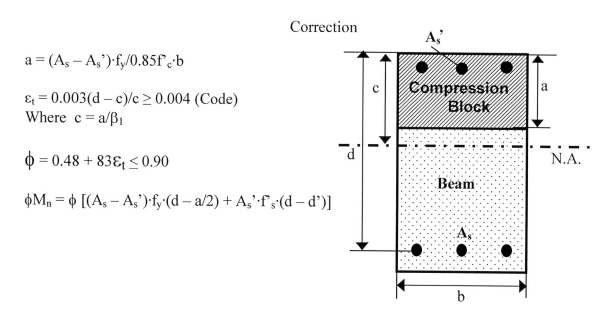

$a = (A_s - A_s') \cdot f_y / 0.85 f'_c \cdot b$

$\varepsilon_t = 0.003(d - c)/c \geq 0.004$ (Code)
Where $c = a/\beta_1$

$\phi = 0.48 + 83\varepsilon_t \leq 0.90$

$\phi M_n = \phi [(A_s - A_s') \cdot f_y \cdot (d - a/2) + A_s' \cdot f'_s \cdot (d - d')]$

Notation is illustrated in the figures above.

Table A.29: Shear Design Aid

$$\phi V_n \geq V_u$$

$$V_n = V_c + V_s$$

Where:

$\phi = 0.75$,

V_n = nominal beam shear resistance,

V_u = factored applied (design) shear force,

V_c = concrete shear resistance = $2\sqrt{f'_c} \cdot b_w \cdot d$ with $\sqrt{f'_c} \leq 100$ psi,

V_s = steel shear resistance = $V_s = a_v \cdot f_y \cdot d/s$ with $a_v = n.a_s$

n = number of stirrup vertical branches and a_s = stirrup bar area, and

s = stirrup spacing.

Shear reinforcement is required if:

$$V_u > \phi V_c/2$$

Maximum stirrup spacing:

$$V_u \leq 6\phi\sqrt{f'_c} \cdot b_w d \quad ===> \quad s \leq d/2 \text{ or } 24" \text{ whichever is smaller}$$

$$V_s > 6\phi\sqrt{f'_c} \cdot b_w d \quad ===> \quad s \leq d/4 \text{ or } 12" \text{ whichever is smaller}$$

Beam is adequacy for shear resistance:

$$V_u \leq 10\phi\sqrt{f'_c} \cdot b_w d \quad ===> \quad \text{Beam may be designed resist applied shear force}$$

$$V_u > 10\phi\sqrt{f'_c} \cdot b_w d \quad ===> \quad \text{Beam is inadequate} \quad ===> \quad \text{Enlarge beam section}$$

Minimum shear reinforcement area:

$$A_{vmin} = \quad 0.75\sqrt{f'_c} \cdot b_w \cdot s/f_y \quad \text{or} \quad 50b_w \cdot s/f_y \quad \text{whichever is larger}$$

Design of stirrup spacing:

$$s = a_v \cdot f_y \cdot d/ V_s = a_v \cdot f_y \cdot d/ (V_u - V_c) \qquad A_v = 0.22 \text{ in}^2 \text{ \#3–U stirrups or } 0.40 \text{ in}^2 \text{ for \#4–U stirrups}$$

Note for shear reinforcement design:

If $V_u \leq \phi(V_c + A_v \cdot f_y \cdot d/s_{max}) = \phi(V_c + 2A_v \cdot f_y) ===>$ Minimum shear reinforcement is required

Table A.30: Tension development length for deformed steel bars (f_y = 40,000 psi)

Assumptions		Bar Number								
		5	6	7	8	9	10	11	14	18
c		1.31	1.38	1.44	1.50	1.56	1.63	1.69	1.88	2.25
K_{tr}		0.196	0.196	0.196	0.196	0.196	0.196	0.356	0.356	0.356
$(c+K_{tr})/d_b$		2.41	2.09	1.87	1.70	1.56	1.46	1.49	1.28	1.16
f'_c (ksi)		Bottom Bars								
3	l_d/d_b	18.2	20.9	29.3	32.3	35.0	37.6	36.9	43.0	47.3
	l_d(in)	11.3	15.7	25.7	32.3	39.4	47.0	50.7	75.2	106.4
4	l_d/d_b	15.7	18.1	25.4	28.0	30.4	32.6	31.9	37.2	41.0
	l_d(in)	9.8	13.6	22.2	28.0	34.1	40.7	43.9	65.1	92.2
5	l_d/d_b	14.1	16.2	22.7	25.0	27.1	29.1	28.6	33.3	36.6
	l_d(in)	8.8	12.2	19.9	25.0	30.5	36.4	39.3	58.3	82.4
6	l_d/d_b	12.8	14.8	20.8	22.8	24.8	26.6	26.1	30.1	33.4
	l_d(in)	8.0	11.1	18.2	22.8	27.9	33.2	35.8	53.2	75.3
f'_c (ksi)		Top Bars								
3	l_d/d_b	23.6	27.2	38.2	42.0	45.6	48.9	47.9	55.9	61.5
	l_d(in)	14.8	20.8	33.4	42.0	51.3	61.1	65.9	97.8	138.3
4	l_d/d_b	20.4	23.6	33.0	36.4	39.5	42.3	41.5	48.4	53.2
	l_d(in)	12.8	17.7	28.9	36.4	44.4	52.9	57.1	84.7	119.8
5	l_d/d_b	18.3	21.1	29.6	32.5	35.3	37.9	37.1	43.3	47.6
	l_d(in)	11.4	15.8	25.9	32.5	39.7	47.3	51.0	75.7	107.2
6	l_d/d_b	16.7	19.2	27.0	29.7	32.2	34.6	33.9	39.5	43.5
	l_d(in)	10.4	14.4	23.6	29.7	36.2	43.2	46.6	69.1	97.8

Assumptions: 1. Clear spacing of steel bars = 2 inch (2.25 inch for #18 bars).
2. Steel bar cover = 2.5 inch from center of steel bar.
3. Stirrup or tie diameter = #3 for bars smaller than #11 and #4 for #11 bars and larger.
4. Stirrup or tie spacing = 6 inch
5. Stirrup or tie yield strength = 60 ksi
6. No. of bars being developed = 5

Table A.31: Tension development length for deformed steel bars (f_y = 60,000 psi)

Assumptions		Bar Number								
		5	6	7	8	9	10	11	14	18
c		1.31	1.38	1.44	1.50	1.56	1.63	1.69	1.88	2.25
K_{tr}		0.293	0.293	0.293	0.293	0.293	0.293	0.533	0.533	0.533
$(c+K_{tr})/d_b$		2.50	2.22	1.98	1.79	1.65	1.54	1.62	1.38	1.24
f'_c (ksi)		Bottom Bars								
3	l_d/d_b	25.6	29.5	41.5	45.8	49.8	53.5	50.9	59.7	66.4
	l_d(in)	16.0	22.2	26.3	45.8	56.0	66.9	69.9	104.5	149.4
4	l_d/d_b	22.2	25.6	36.0	39.7	43.1	46.4	44.1	51.7	57.5
	l_d(in)	13.8	19.2	31.5	39.7	48.5	58.0	60.6	90.5	129.4
5	l_d/d_b	19.8	22.9	32.2	35.5	38.6	41.5	39.4	46.2	51.4
	l_d(in)	12.4	17.2	28.2	35.3	43.4	51.8	54.2	80.9	115.8
6	l_d/d_b	18.1	20.9	29.4	32.4	35.2	37.9	36.0	42.2	47.0
	l_d(in)	11.3	15.7	25.7	32.4	39.6	47.3	49.5	73.9	105.7
f'_c (ksi)		Top Bars								
3	l_d/d_b	33.3	38.4	54.0	59.6	64.7	69.6	66.1	77.6	86.3
	l_d(in)	20.8	28.8	47.2	59.6	72.8	87.0	90.9	135.8	194.3
4	l_d/d_b	28.8	33.3	46.8	51.6	56.1	60.3	57.3	67.2	74.8
	l_d(in)	18.0	24.9	40.9	51.6	63.1	75.3	78.7	117.6	168.2
5	l_d/d_b	25.8	29.8	41.8	46.1	50.2	53.9	51.2	60.1	66.9
	l_d(in)	16.1	22.3	36.6	46.1	56.4	67.4	70.4	105.2	150.5
6	l_d/d_b	23.5	27.2	38.2	42.1	45.8	49.2	46.8	54.9	61.1
	l_d(in)	14.7	20.4	33.4	42.1	51.5	61.5	64.3	96.0	137.4

Assumptions: 1. Clear spacing of steel bars = 2 inch (2.25 inch for #18 bars).
2. Steel bar cover = 2.5 inch from center of steel bar.
3. Stirrup or tie diameter = #3 for bars smaller than #11 and #4 for #11 bars and larger.
4. Stirrup or tie spacing = 6 inch
5. Stirrup or tie yield strength = 60 ksi
6. No. of bars being developed = 5

Table A.32: Tension development length for deformed steel bars (f_y = 80,000 psi)

Assumptions		Bar Number								
		5	6	7	8	9	10	11	14	18
c		1.31	1.38	1.44	1.50	1.56	1.63	1.69	1.88	2.25
K_{tr}		0.293	0.293	0.293	0.293	0.293	0.293	0.533	0.533	0.533
$(c+K_{tr})/d_b$		2.50	2.22	1.98	1.79	1.65	1.54	1.62	1.38	1.24
f'_c (ksi)		Bottom Bars								
3	l_d/d_b	34.1	39.4	55.4	61.1	66.4	71.4	67.8	79.6	88.6
	l_d(in)	21.3	29.5	48.5	61.1	74.7	89.2	93.3	139.3	199.2
4	l_d/d_b	29.5	34.1	48.0	52.9	57.5	61.8	58.7	68.9	76.7
	l_d(in)	18.5	25.6	42.0	52.9	64.7	77.3	80.8	120.6	172.6
5	l_d/d_b	26.4	30.5	42.9	47.3	51.4	55.3	52.5	61.7	68.6
	l_d(in)	16.5	22.9	37.5	47.3	57.9	69.1	72.2	107.9	154.3
6	l_d/d_b	24.1	27.9	39.2	43.2	47.0	50.5	48.0	56.3	62.6
	l_d(in)	15.1	20.9	34.3	43.2	52.8	63.1	65.9	98.5	140.9
f'_c (ksi)		Top Bars								
3	l_d/d_b	44.3	51.2	72.0	79.4	86.3	92.8	88.2	103.5	115.1
	l_d(in)	27.7	38.4	63.0	79.4	97.1	116.0	121.2	181.1	259.0
4	l_d/d_b	38.4	44.4	62.3	68.8	74.8	80.4	76.4	89.6	99.7
	l_d(in)	24.0	33.3	54.6	68.8	84.1	100.5	105.0	156.8	224.3
5	l_d/d_b	34.4	39.7	55.8	61.5	66.9	71.9	68.3	80.2	89.2
	l_d(in)	21.5	29.8	48.8	61.5	75.2	89.8	93.9	140.3	200.6
6	l_d/d_b	31.4	36.2	50.9	56.2	61.0	65.6	62.3	73.2	81.4
	l_d(in)	19.6	27.2	44.5	56.2	68.7	82.0	85.7	128.0	183.2

Assumptions: 1. Clear spacing of steel bars = 2 inch (2.25 inch for #18 bars).
2. Steel bar cover = 2.5 inch from center of steel bar.
3. Stirrup or tie diameter = #3 for bars smaller than #11 and #4 for #11 bars and larger.
4. Stirrup or tie spacing = 6 inch
5. Stirrup or tie yield strength = 60 ksi
6. No. of bars being developed = 5

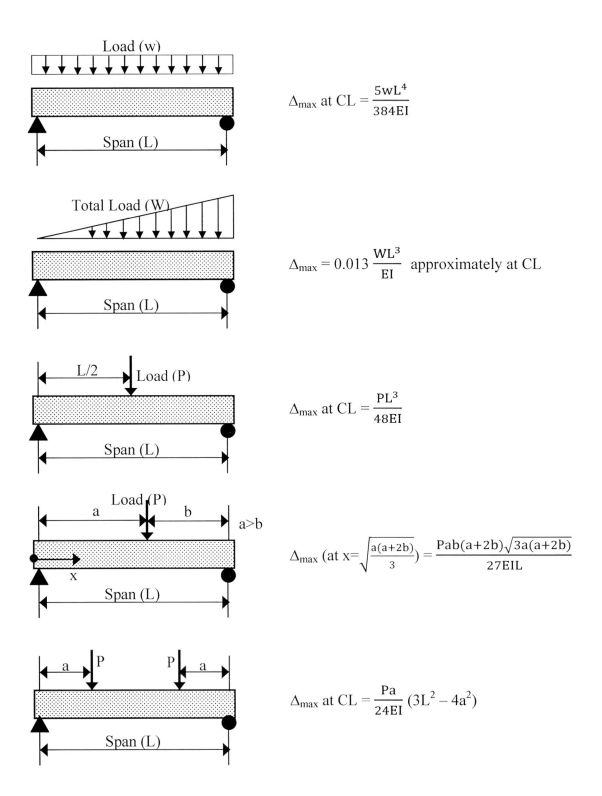

Figure A.33: Elastic deflection equations for simply supported beams.

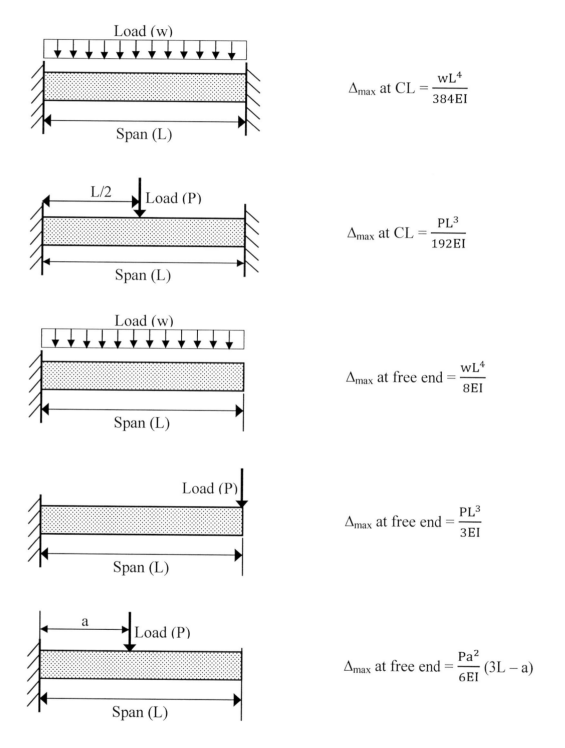

Figure A.34: Elastic deflection equations for fixed–end and cantilever beams.

$$\Delta_{max} \text{ at CL} = \frac{wL^4}{384EI}$$

$$\Delta_{max} \text{ at CL} = \frac{PL^3}{192EI}$$

$$\Delta_{max} \text{ at free end} = \frac{wL^4}{8EI}$$

$$\Delta_{max} \text{ at free end} = \frac{PL^3}{3EI}$$

$$\Delta_{max} \text{ at free end} = \frac{Pa^2}{6EI}(3L - a)$$

Table A.35: Deflection Equations

$$I_e = \left(\frac{M_{cr}}{M_a}\right)^3 I_g + \left[1 - \left(\frac{M_{cr}}{M_a}\right)^3\right] I_{cr}$$

Where:

I_e = effective moment of inertia for computation of deflection,

M_{cr} = the bending moment that causes tension cracking in concrete determined as shown:

$$M_{cr} = \frac{f_r I_g}{y_t}$$

f_r = modulus of rupture of concrete as computed as shown multiplied by 1.0 for normal–weight concrete, 0.75 for all light–,weight concrete and 0.85 for sand light – weight concrete, as specified by ACI:

$$f_r = 7.5\sqrt{f_c'}$$

I_g = gross moment of inertia of the uncracked beam neglecting steel reinforcement,

y_t = distance between the neutral axis and the extreme tension fiber in the uncracked section, and

M_a = maximum bending moment within the member at deflection calculation.

Total service deflection of reinforced concrete element can be expressed as follows:

$$\Delta_{TL} = \Delta_{iDL}(1 + \lambda_\infty) + \Delta_{iLL} + \lambda_{SLL}\Delta_{iSLL}$$

Where:

Δ_{TL} = total element service deflection,

Δ_{iDL} = immediate dead load deflection,

λ_∞ = long–term deflection factor for dead load (Table A.36)

Δ_{iLL} = immediate live load deflection,

Δ_{iSLL} = immediate deflection due to sustained live load. Part of live load is typically sustained for a period of time, and

λ_{SLL} = long–term deflection factor for sustained live load. λ_{SLL} is determined utilizing Table A.36.

Table A.36: Long – Term Deflection Coefficient.

Pursuant to the ACI Code, following equations can be used for determining the value of long term deflection of reinforced concrete:

$$\lambda = 1/(1 + 50\rho')$$

Where:
λ = long–term deflection factor,
ξ = sustained load duration factor, and
$\rho' = A'_s/bd$ = compression steel reinforcement ratio at midspan for simply supported and continuous beams and at support for cantilever beams.
ξ can be determined based on the duration of load utilizing the following table.

Load Duration Factor (ξ)

Duration of sustained load	Value of load duration factor (ξ)
$\geq$ 5 years	2.0
12 months	1.4
6 months	1.2
3 months	1.0

The equations for long–term and total deflection due to a defined load:

$$\Delta_{LT} = \lambda\Delta_i$$

$$\Delta_T = \Delta_i(1 + \lambda)$$

Where:
Δ_{LT} = long–term deflection due to the load under consideration,
λ = long–term deflection factor based on the duration of load under consideration,
Δ_i = immediate deflection due to the same load under consideration, and
Δ_T = total deflection (immediate + long–term) due to the load under consideration,

Table A.37: Maximum permissible computed deflections

Type of member	Deflection to be considered	Deflection limitation
Flat roofs not supporting or attached to nonstructural elements likely to be damaged by large deflections	Immediate deflection due to live load, Δ_{iLL}	$\ell/180$*
Floors not supporting or attached to nonstructural elements likely to be damaged by large deflections	Immediate deflection due to live load, Δ_{iLL}	$\ell/360$
Roof or floor construction supporting or attached to nonstructural elements likely to be damaged by large deflections	That part of total deflection occurring after attachment of nonstructural elements,	$\ell/480$‡
Roof or floor construction not supporting or attached to nonstructural elements likely to be damaged by large deflection	Δ_{ATT} (sum of the long–term deflection due to all sustained loads and the immediate deflection due to any additional live load) †	$\ell/240$§

Notes:

*Limit not intended to safeguard against ponding. Ponding should be checked by suitable calculations of deflection, including added deflections due to ponded water, and considering long–term effects of all sustained loads, camber, construction tolerances, and reliability of provisions for drainage.

†Long – term deflection shall be determined per ACI Code provisions, but may be reduced by amount of deflection calculated to occur before attachment of nonstructural elements.

‡Limit may be exceeded if adequate measures are taken to prevent damage to supported or attached elements.

§Limit shall not be greater than tolerance provided for nonstructural elements. Limit may be exceeded if camber is provided so that total deflection minus camber does not exceed limit.

ℓ = member span defined as clear span + member depth not to exceed center–to–center span of simply supported and continuous beams and clear span of cantilevers,

Immediate deflection due to live load is as discussed earlier referred to as Δ_{iLL},

Deflection that occurs after attachment of nonstructural members, Δ_{ATT}, is the immediate deflection of live load + long term deflection of dead load + long term deflection of sustained live load and may be described in the following equation:

$$\Delta_{ATT} = \Delta_{iLL}, + \lambda_{\infty}\Delta_{iDL} + \lambda_{SLL}\Delta_{iSLL}$$

Δ_{ATT} = deflection after attachment of nonstructural elements, and

Δ_{iDL} = immediate dead load deflection,

λ_{∞} = long–term deflection factor for dead load (infinite duration in Table A.36),

Δ_{iLL} = immediate live load deflection,

Δ_{iSLL} = immediate deflection due to sustained live load.

λ_{SLL} = long–term deflection factor for sustained live load (Table A.36).

Ponding is accumulation of rain water on roofs if roof slope is not adequate to allow water to flow towards drains, and

Camber is negative deflection built in the concrete formwork so that it compensates for anticipated future deflection.

Table A.38: Crack Width Equation.

$$w_{max} = 0.076\beta f_s \sqrt[3]{d_c A}$$

Where:

w_{max} = maximum flexural crack width in 0.001 in,

β = effective depth factor $= \dfrac{h - \bar{y}}{d - \bar{y}}$

Where $\bar{y}$ is the distance between extreme compression fiber and the neutral axis for the elastic cracked section under service condition. Or the following approximate values may be used: $\beta = 1.2$ for beams $\beta = 1.35$ for slabs

f_s = stress in steel bars under service conditions or service moment causing flexural cracks in ksi. For computation of f_s, the reinforced concrete section is considered cracked and elastic. The transformed section is utilized for section stress determination. To assemble the transformed section, steel bar area is magnified by factor $n = E_s/E_c$. As such, computation of f_s also requires multiplication by n since the stress is applied to a hypothesized magnified steel area:

$$f_s = n \times (M_{service}/I_{tr}) \times y_{steelbars} = n \times (M_{service}/I_{tr}) \times (d - \bar{y})$$

Or the following simplified equation may be used: $f_s = 0.6f_y$ (ksi)

d_c = concrete cover over steel bars measured from the extreme tension fiber to the centroid of steel bar closest to concrete outer surface (inch), and

A = area of concrete beam centered around (concentric with) the steel bars divided by the number of tension steel bars (in^2). A is the largest concrete area that have the same centroid as steel bar reinforcement divided by the number of bars. The following equation may be used for A:

$A = 2b(h - d)/\text{number of bars}$ (in^2)

Where b is the beam's width at the tension side, h is total beam depth/height and d is beam's effective depth.

Recommended limits of crack width for various exposure conditions.

Exposure Condition	Recommended Max. Crack Width
Dry air	0.016 inch
Moist air, soil	0.012 inch
Deicing chemicals	0.007 inch
Seawater and seawater spray	0.006 inch
Water retaining structures	0.004 inch

Table A.39: ACI Code Provisions for Crack Width

ACI 318 Code includes general guidelines for beams and one–way slabs not subjected to aggressive exposure (maximum crack width = w_{max} = 0.016 inch). Special investigations and precautions for other cases are required. ACI Code specifies the maximum spacing of steel reinforcing bars so that the above crack width is not exceeded. Such specifications are best suited to one–way slabs and flanges of T–beams subjected to negative bending moments:

$$s = 540/f_s - 2.5C_c \qquad \text{or} \qquad s = 12(36/f_s) \qquad \text{whichever is smaller}$$

Where:

s = center–to–center spacing of flexural tension steel bars closes to the extreme tension face or the entire width of tension face for the case of single steel bar,

f_s = stress in steel bars under service conditions or service moment causing flexural cracks in ksi ($f_s \approx 0.6f_y$), and

C_c = clear cover of reinforcing steel bars nearest to the extreme tension face.

Table A.40: ACI Code Short Column Design Equations.

$$\phi P_n = \phi\alpha[0.85f'_c(A_g - A_{st}) + A_{st}f_y]$$

Where:

ϕ = capacity reduction factor 0.65 for tied columns and 0.70 for spiral columns,

P_n = column nominal axial compressive force capacity,

α = minimum eccentricity factor 0.80 for tied columns and 0.85 for spiral columns,

A_g = column gross cross–sectional area, and

A_{st} = area of longitudinal steel bars of the column.

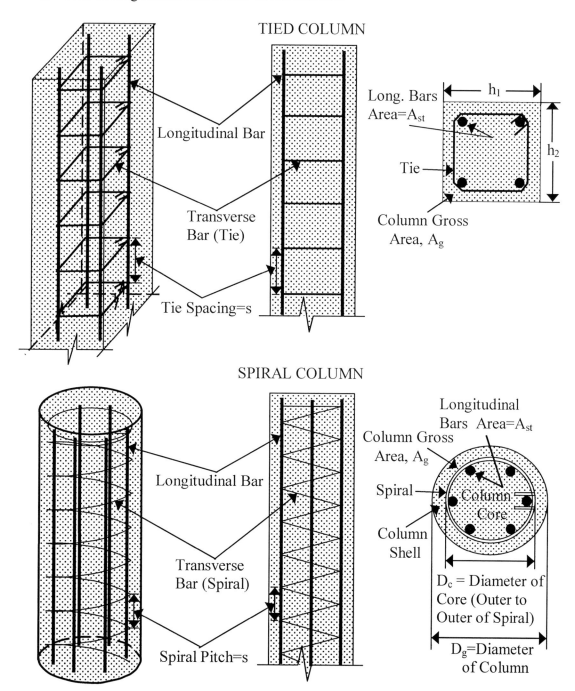

Table A.41: ACI Code Provisions for Columns.

Column Reinforcement Ratio = $\rho_g = A_{st}/A_g$ $1\% \leq \rho_g \leq 8\%$
Where: A_{st} = steel area, and A_g = column gross area.

Minimum No. of Longitudinal Steel Bars	Column Cross – Section Shape
3	Triangular
4	Rectangular
6	Circular

Tie Diameter and Spacing

Longitudinal Bar Size	Tie Bar Size
$\leq$ #10	#3
$\geq$ #11	#4

Tie vertical spacing, s

$$s \leq \begin{cases} s_1 = 16 \times \text{longitudinal bar diameter} \\ s_2 = 48 \times \text{tie bar diameter} \\ s_3 = \text{least dimension of column cross–section} \end{cases}$$

Tie Geometric Configuration

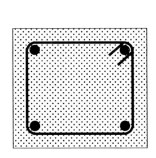
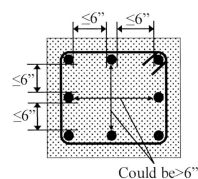

Could be>6”

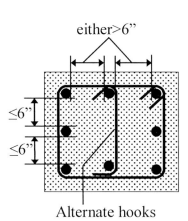

Alternate hooks

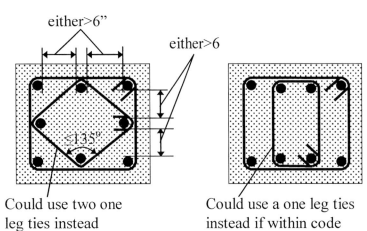

Could use two one leg ties instead

Could use a one leg ties instead if within code

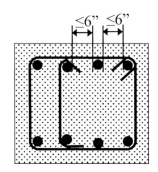

Table A.42: ACI Code Provisions for Spirally Reinforced Columns.

Spiral Diameter and Pitch

1. Spiral bar diameter may not be less than #3,
2. Clear spacing of spiral bar loops may not be less than 1 or greater than 3 inch,
3. Anchorage of the spiral bar shall be accomplished by 1.5 extra revolution (turn) at each end of the column, and
4. Lap splice length of uncoated deformed spiral bar may not be less than 12 inch or $48d_b$ (d_b = spiral bar diameter) whichever is greater.

Spiral Reinforcement Ratio, ρ_s

$$\rho_s = 4a_s/sD_c$$

Where: D_c = diameter of column core (outer–to–outer of spiral),
d_b = diameter of spiral bar,
a_s = cross – sectional area of spiral bar, and
s = spiral pitch.

Minimum Spiral Reinforcement Ratio in Columns, ρ_{smin}

$$\rho_{smin} = 0.45(A_g/A_c - 1)\cdot f'_c/f_y$$

Where:
A_g = gross cross – sectional area of the spiral column,
A_c = core area of the column,
f'_c = specified/design compressive strength of concrete, and
f_y = yield strength of steel.

Table A.43: Columns under Eccentric Loading

The Plastic Centroid

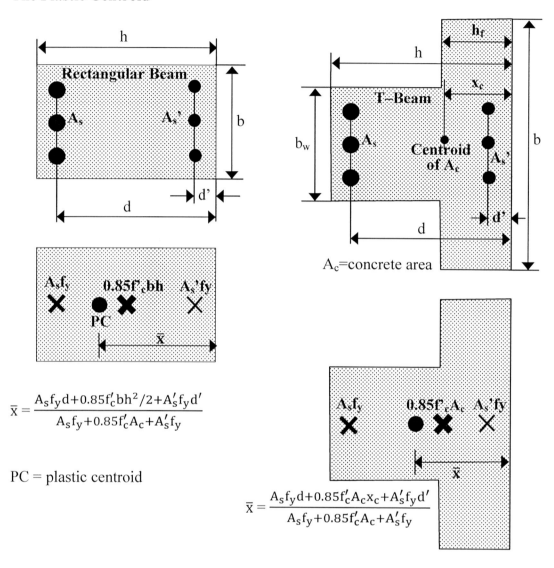

$$\bar{x} = \frac{A_s f_y d + 0.85 f'_c b h^2 / 2 + A'_s f_y d'}{A_s f_y + 0.85 f'_c A_c + A'_s f_y}$$

PC = plastic centroid

$$\bar{x} = \frac{A_s f_y d + 0.85 f'_c A_c x_c + A'_s f_y d'}{A_s f_y + 0.85 f'_c A_c + A'_s f_y}$$

Table A.44: Columns under Eccentric Loading

Nominal force P_n and bending moment M_n equations:

$$P_n = C_c + C_s - T = 0.85f'_cab + A_s'f_s' - A_sf_s$$

$$M_n = 0.85f'_cab(h - a/2) + A_s'f_s'(h/2 - d') + A_sf_s(d - h/2)$$

$$e = [0.85f'_cab(h - a/2) + A_s'f_s'(h/2 - d') + A_sf_s(d - h/2)] / [0.85f'_cab + A_s'f_s' - A_sf_s]$$

Where the stresses in tension and compression steel may be expressed as follows:

$$f_s = (d - c) \times (0.003/c) \times E_s \leq f_y$$

$$f_s' = (c - d') \times (0.003/c) \times E_s \leq f_y$$

Where E_s is the modulus of elasticity of steel.

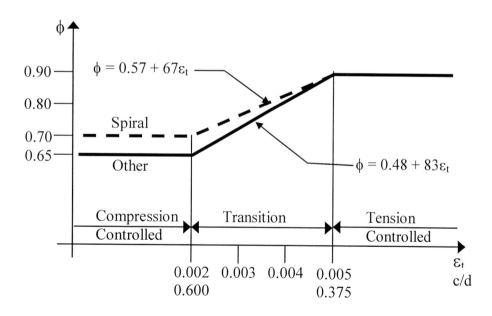

The relationship of strength reduction factor (ϕ) and strain in extreme tension steel (ε_t) and ratio of compression zone depth to effective depth in columns (c/d).

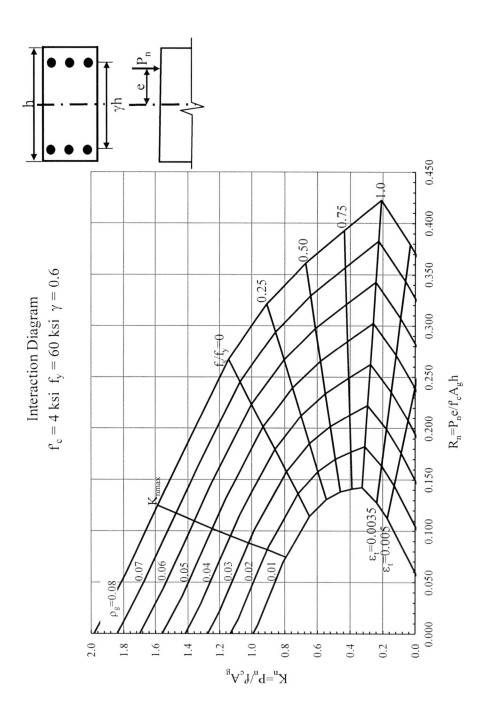

Figure A.45: Column interaction diagram.

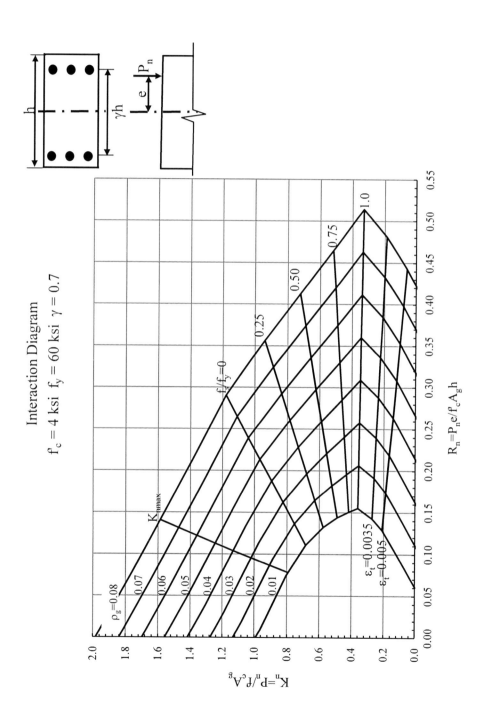

Interaction Diagram

$f_c = 4$ ksi $f_y = 60$ ksi $\gamma = 0.7$

$K_n = P_n / f_c A_g$

$R_n = P_n e / f_c A_g h$

Figure A.46: Column interaction diagram.

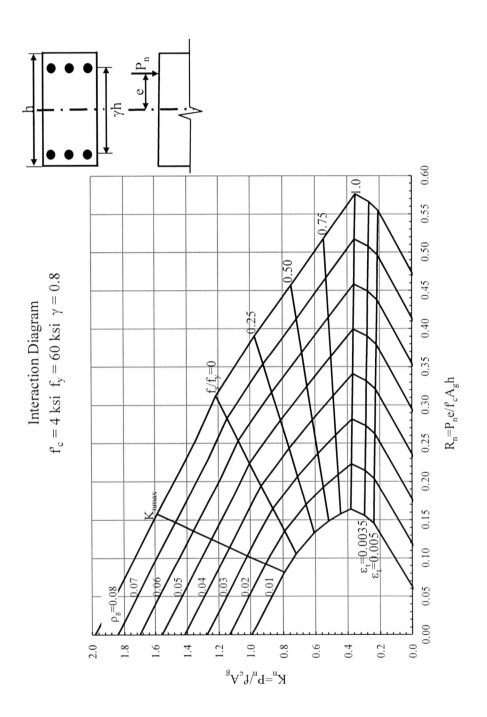

Figure A.47: Column interaction diagram.

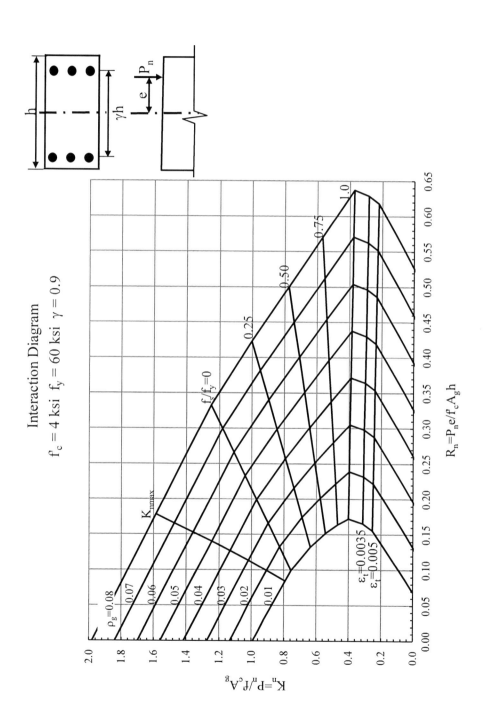

Interaction Diagram

$f_c = 4$ ksi $f_y = 60$ ksi $\gamma = 0.9$

Figure A.48: Column interaction diagram.

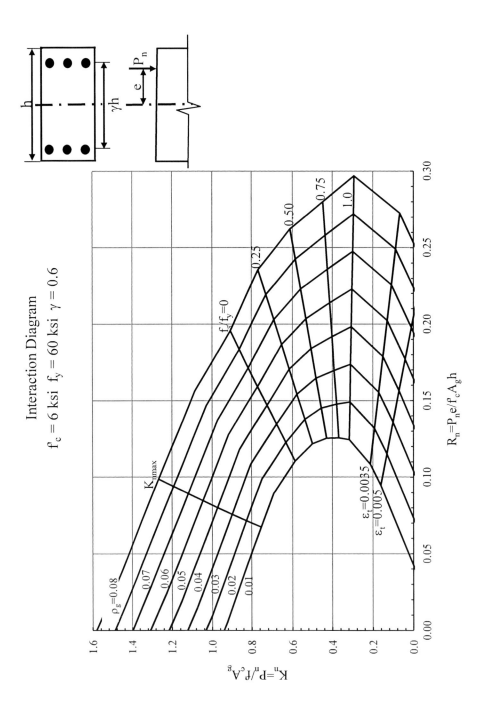

Figure A.49: Column interaction diagram.

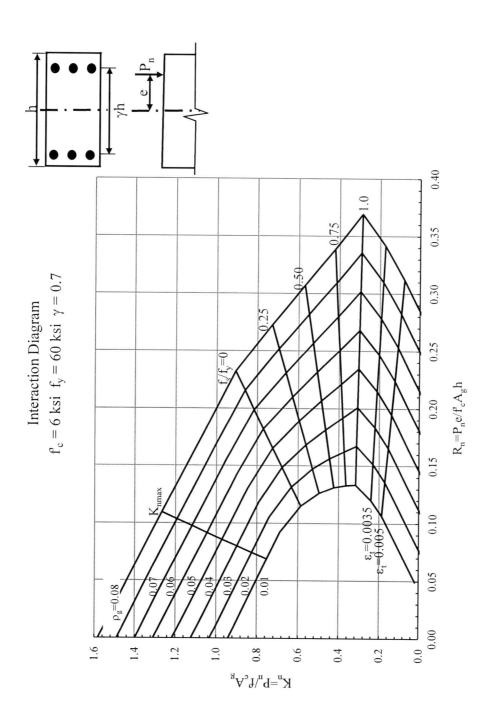

Figure A.50: Column interaction diagram.

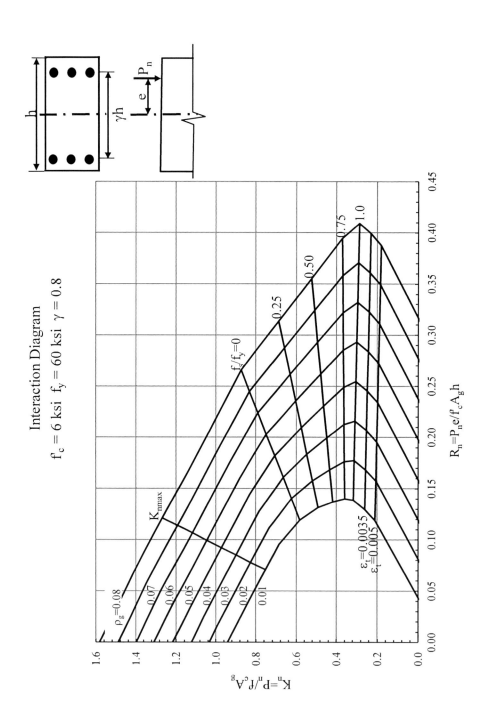

Figure A.51: Column interaction diagram.

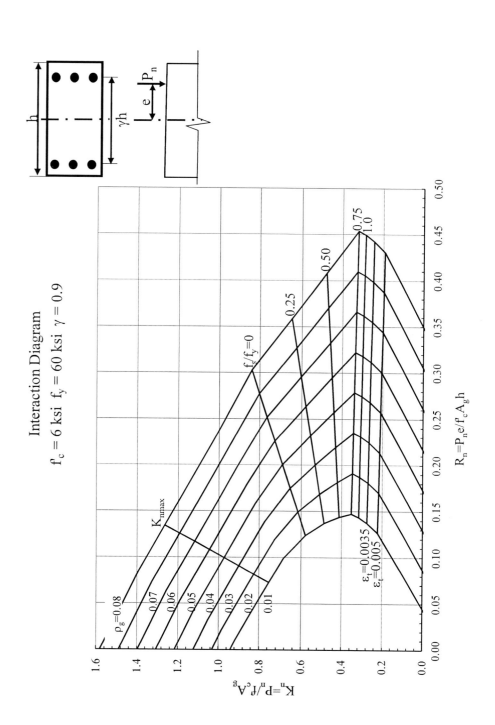

Figure A.52: Column interaction diagram.

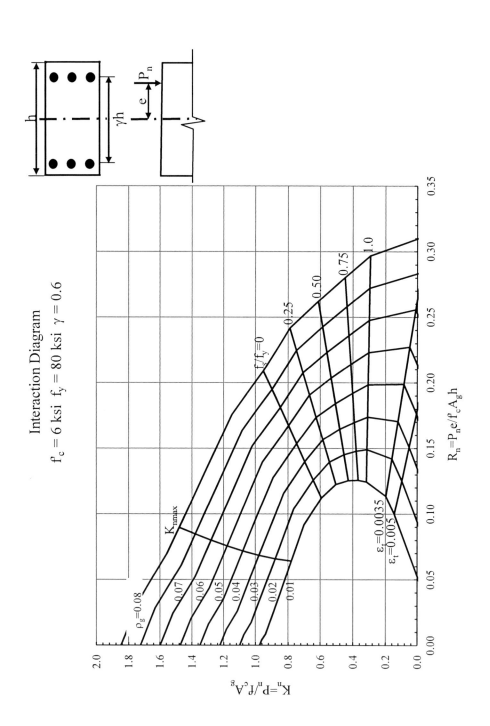

Figure A.53: Column interaction diagram.

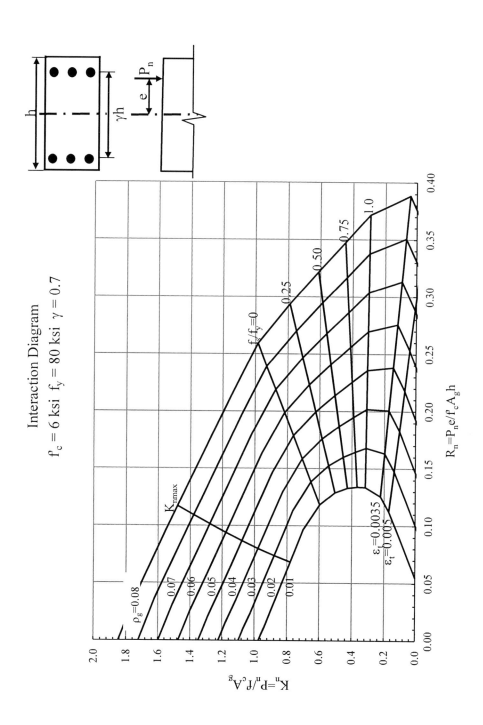

Interaction Diagram

$f_c = 6$ ksi $\quad f_y = 80$ ksi $\quad \gamma = 0.7$

Figure A.54: Column interaction diagram.

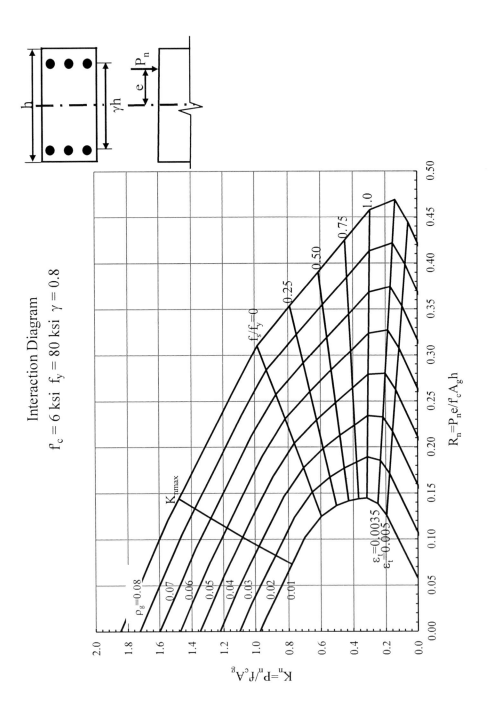

Interaction Diagram

$f_c = 6$ ksi $f_y = 80$ ksi $\gamma = 0.8$

Figure A.55: Column interaction diagram.

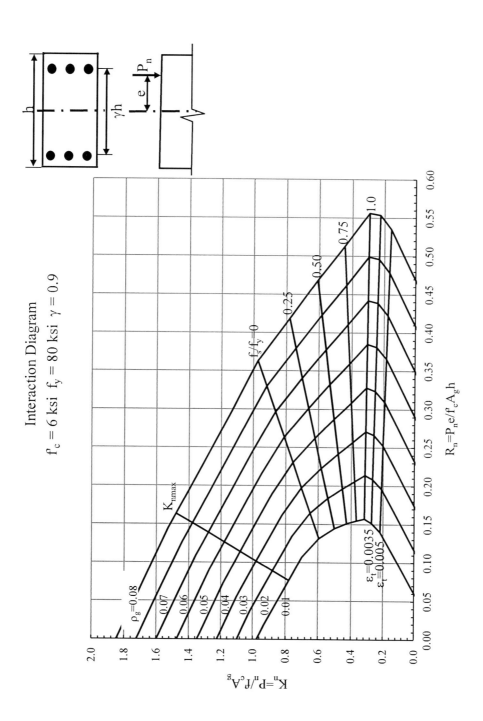

Interaction Diagram

$f'_c = 6$ ksi $f_y = 80$ ksi $\gamma = 0.9$

Figure A.56: Column interaction diagram.

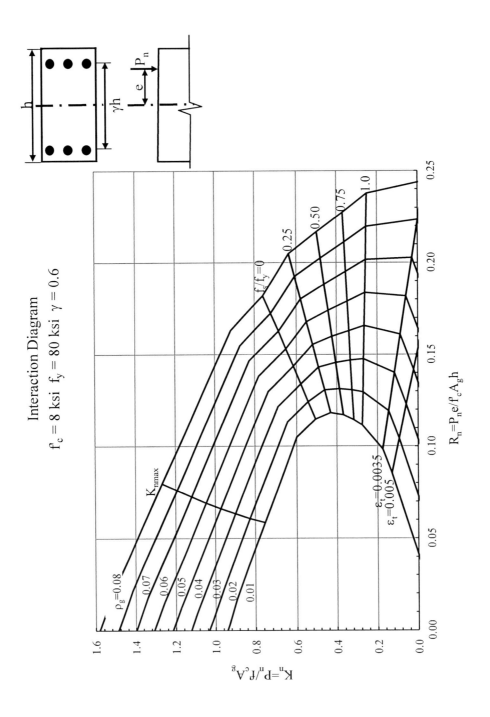

Interaction Diagram

$f_c = 8$ ksi $f_y = 80$ ksi $\gamma = 0.6$

$R_n = P_n e / f_c' A_g h$

$K_n = P_n / f_c' A_g$

Figure A.57: Column interaction diagram.

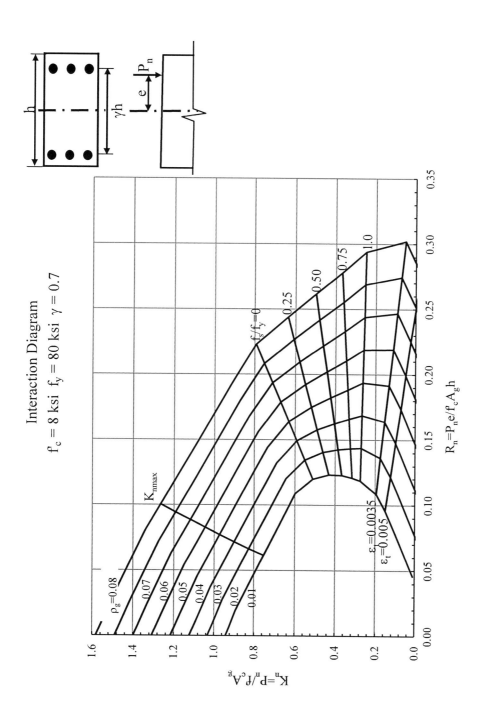

Interaction Diagram
$f_c = 8$ ksi $f_y = 80$ ksi $\gamma = 0.7$

Figure A.58: Column interaction diagram.

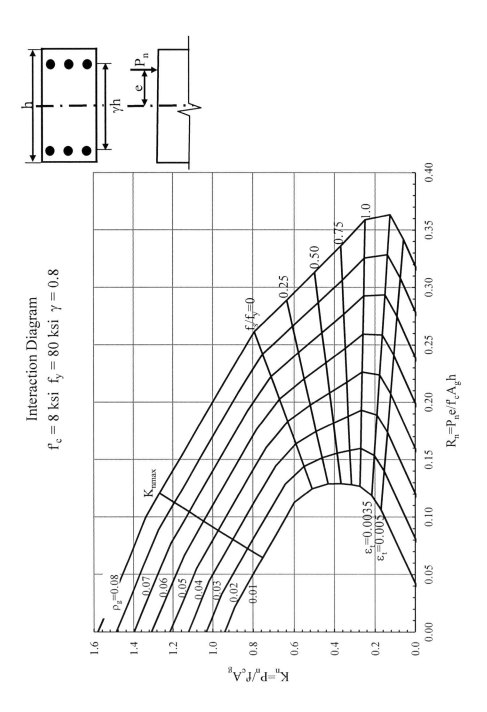

Figure A.59: Column interaction diagram.

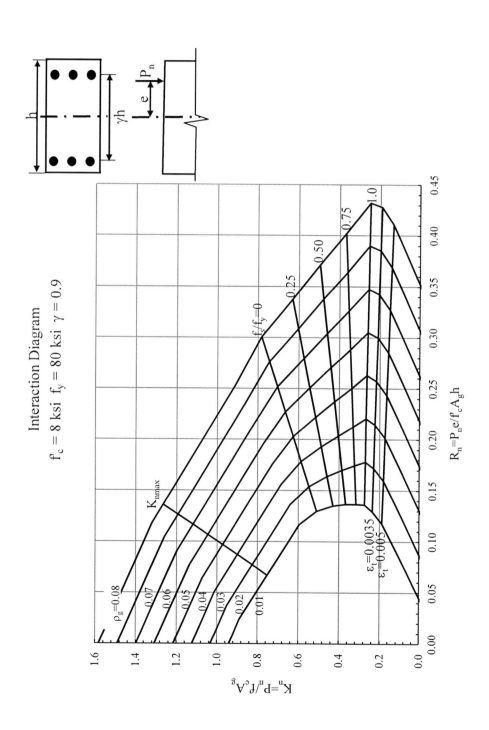

Interaction Diagram
$f_c = 8$ ksi $f_y = 80$ ksi $\gamma = 0.9$

Figure A.60: Column interaction diagram.

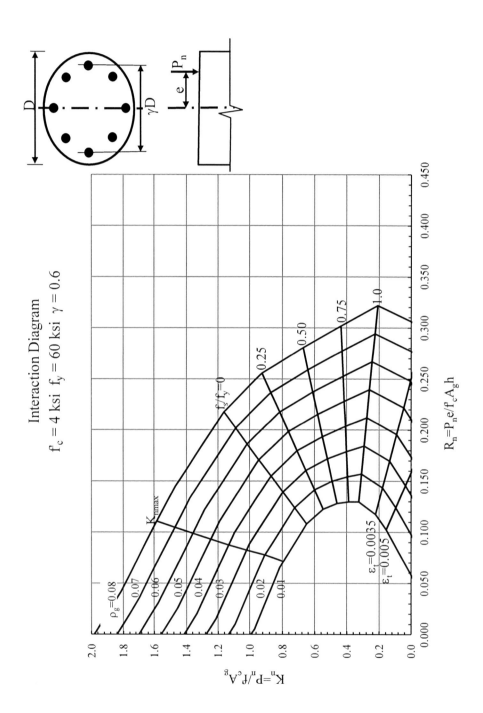

Figure A.61: Column interaction diagram.

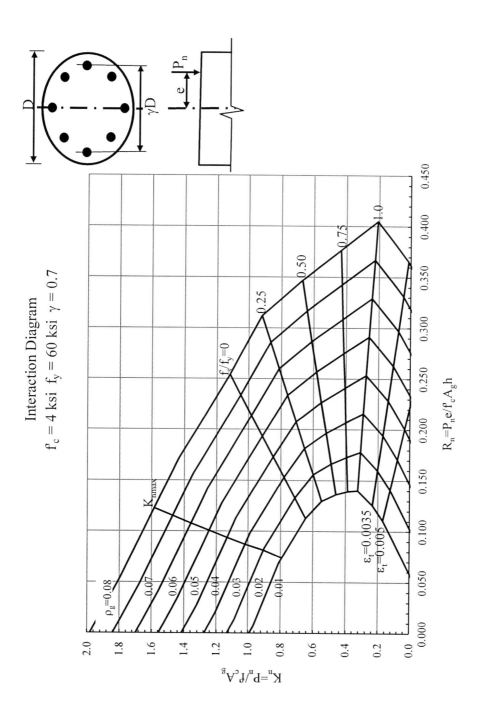

Figure A.62: Column interaction diagram.

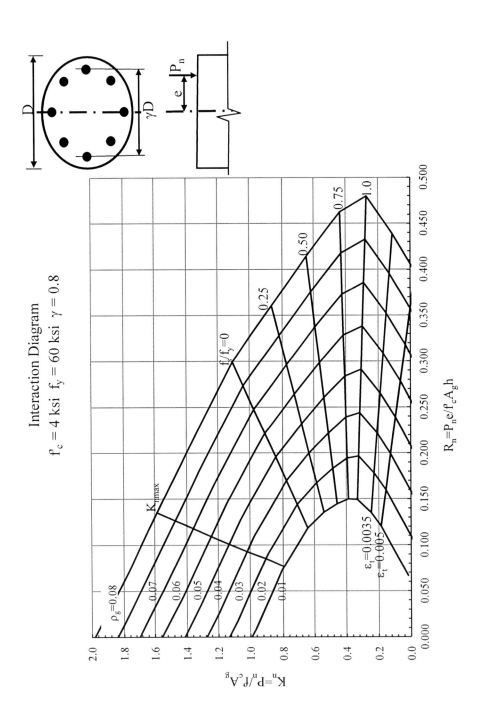

Figure A.63: Column interaction diagram.

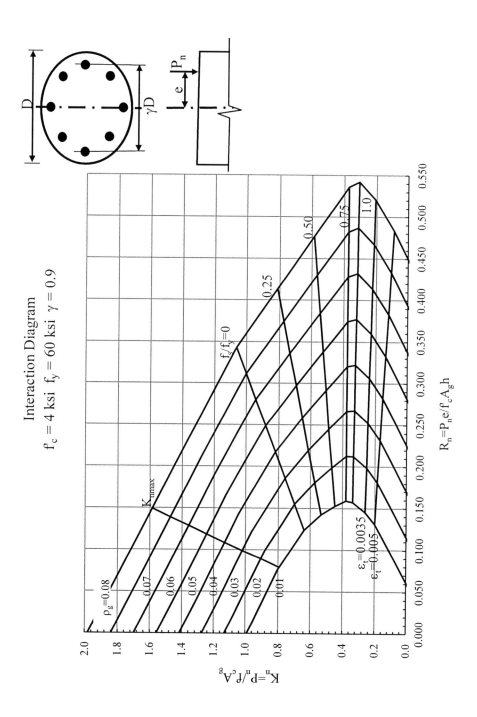

Figure A.64: Column interaction diagram.

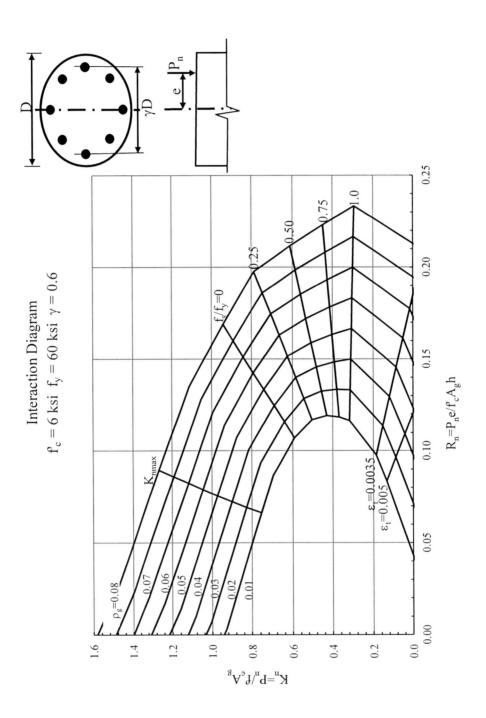

Figure A.65: Column interaction diagram.

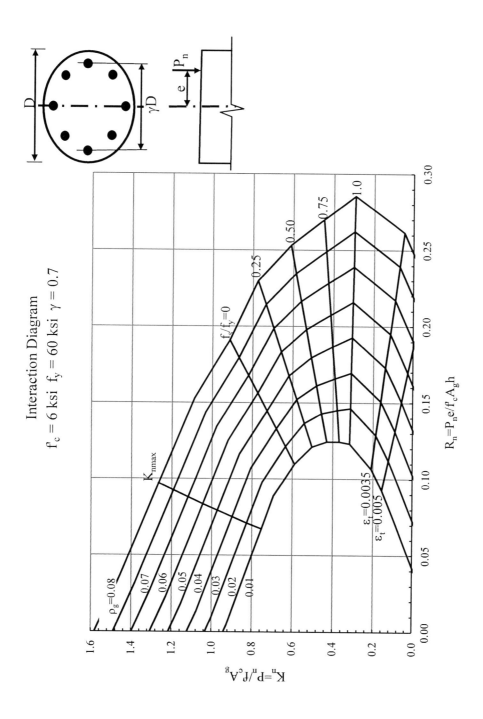

Figure A.66: Column interaction diagram.

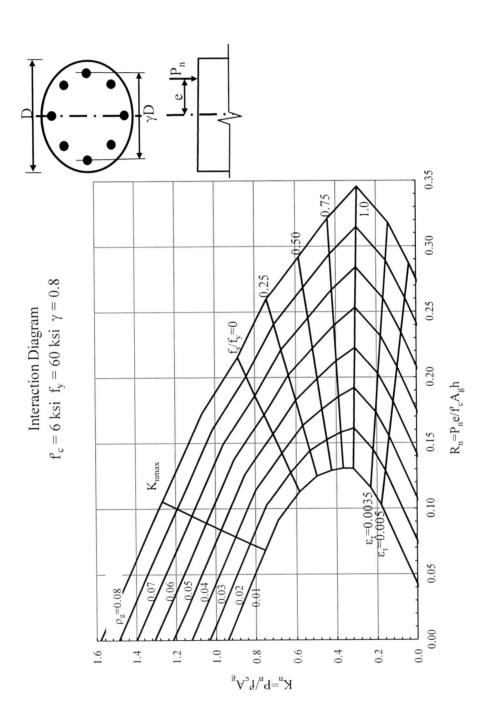

Figure A.67: Column interaction diagram.

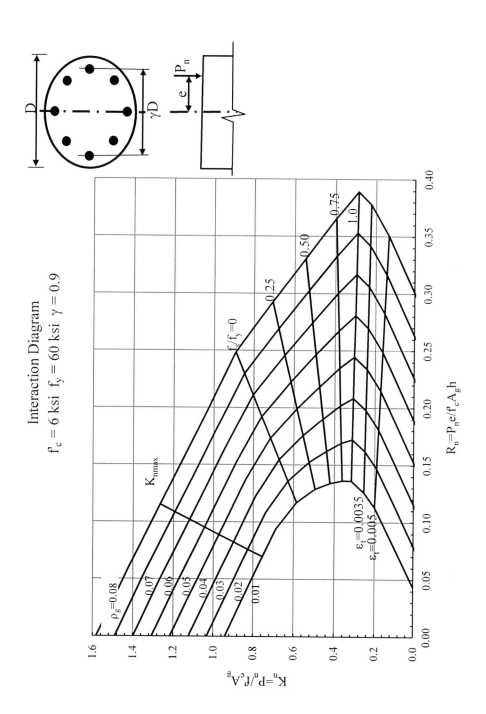

Figure A.68: Column interaction diagram.

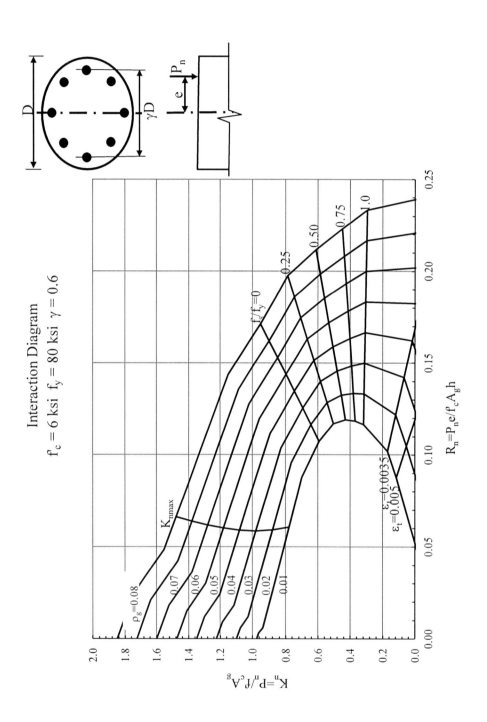

Figure A.69: Column interaction diagram.

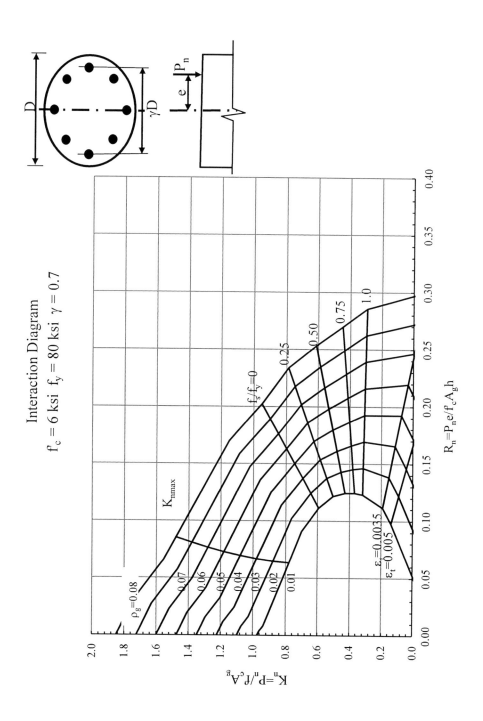

Figure A.70: Column interaction diagram.

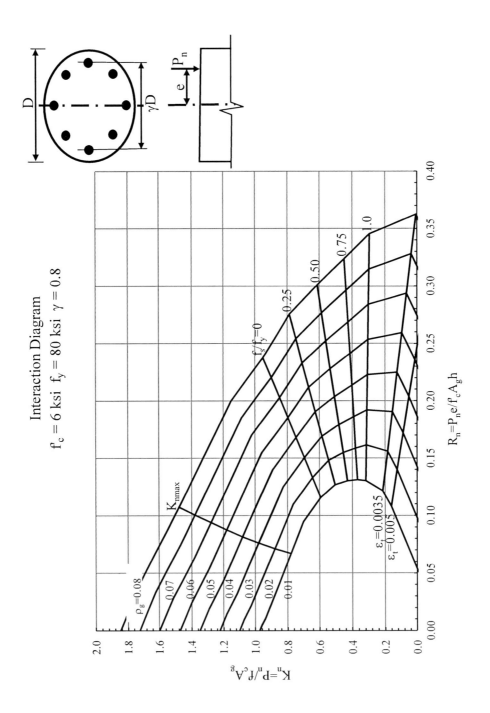

Figure A.71: Column interaction diagram.

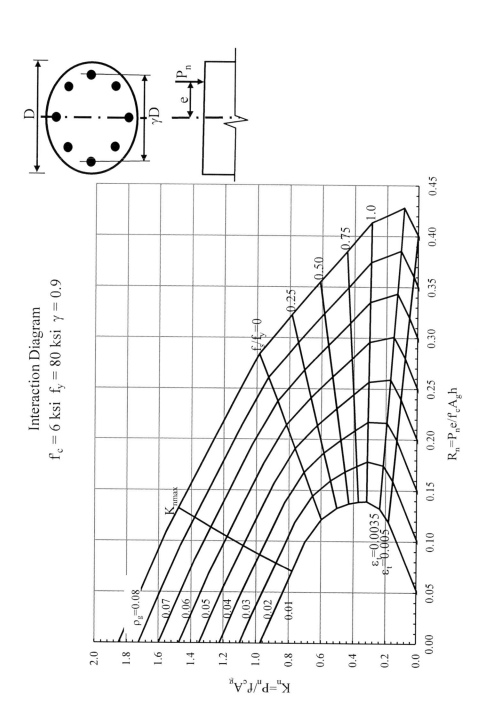

Figure A.72: Column interaction diagram.

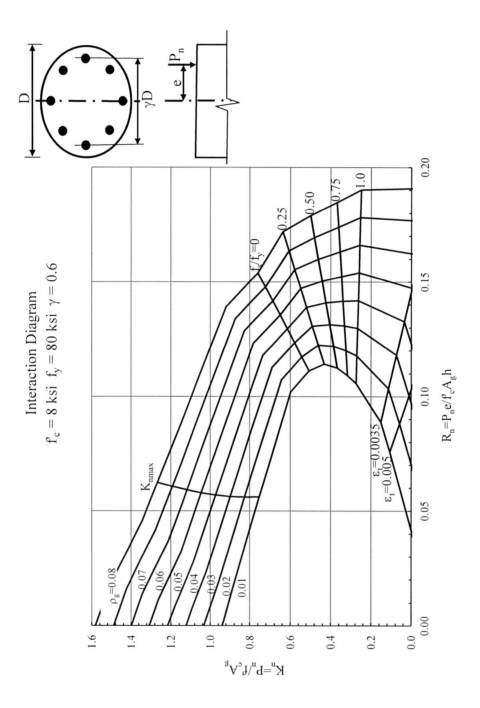

Figure A.73: Column interaction diagram.

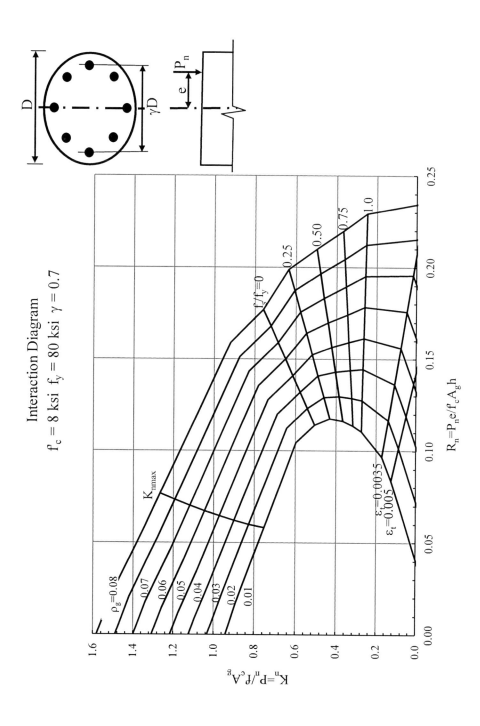

Interaction Diagram

$f'_c = 8$ ksi $f_y = 80$ ksi $\gamma = 0.7$

Figure A.74: Column interaction diagram.

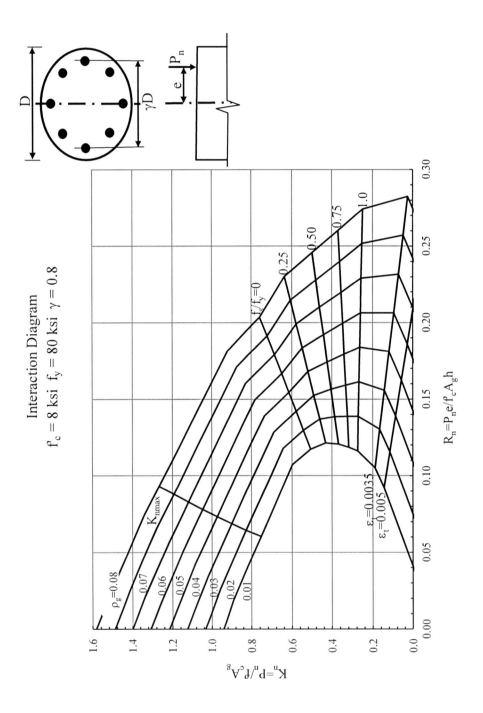

Figure A.75: Column interaction diagram.

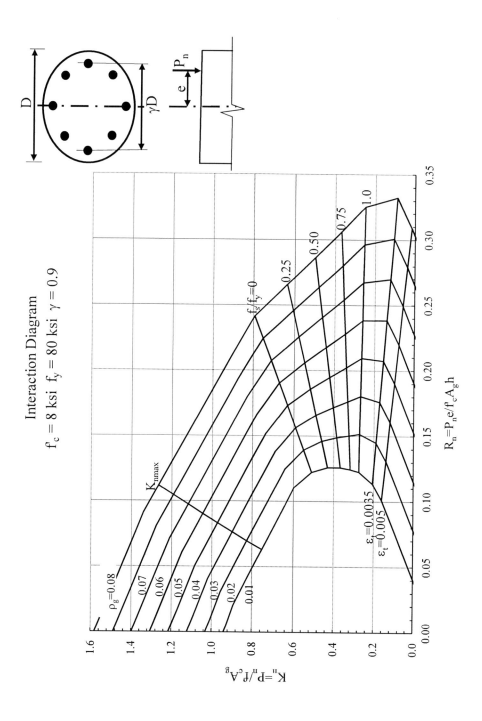

Interaction Diagram

$f_c = 8$ ksi $f_y = 80$ ksi $\gamma = 0.9$

Figure A.76: Column interaction diagram.

Figure A.77: Common column sections and reinforcement lay-outs.

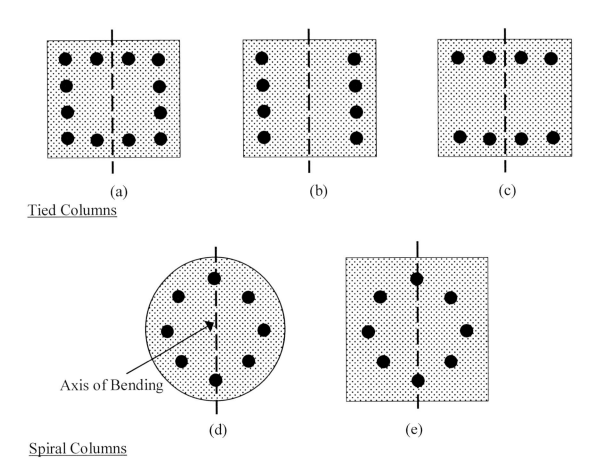

(a)   (b)   (c)

Tied Columns

Axis of Bending

(d)   (e)

Spiral Columns

Conversion of circular column section into equivalent rectangular section.

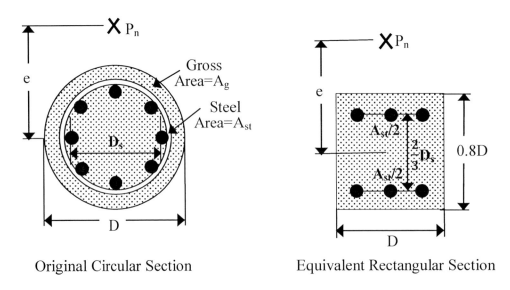

Original Circular Section   Equivalent Rectangular Section

Figure A.78: Biaxial bending in columns.

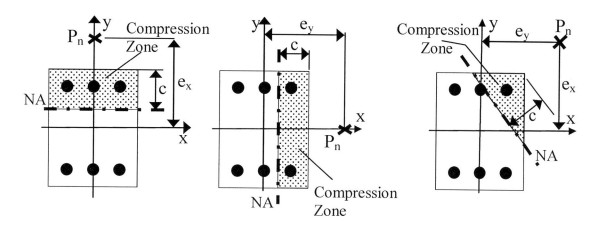

Bending about x   Bending about y   Biaxial bending about x and y

Uniaxial and biaxial bending of columns.

ACI Code introduces a conservative approximate solution for doubly symmetric sections based on the following equation:

$$\frac{1}{P_{ni}} = \frac{1}{P_{nx}} + \frac{1}{P_{ny}} - \frac{1}{P_o}$$

Where:

P_{ni} = section nominal axial load capacity with eccentricities e_x and e_y about x and y axes, respectively.

P_{nx} = section nominal axial load capacity with eccentricity e_x about x axis and zero eccentricity about y axis.

P_{ny} = section nominal axial load capacity with eccentricity e_y about x axis and zero eccentricity about x axis.

P_o = section nominal axial load capacity with zero eccentricity about both axes.

The value of ϕP_{ni} may not exceed $\phi \alpha P_n$ of Appendix A40.

Figure A.79: Effective length factor (k) in columns.

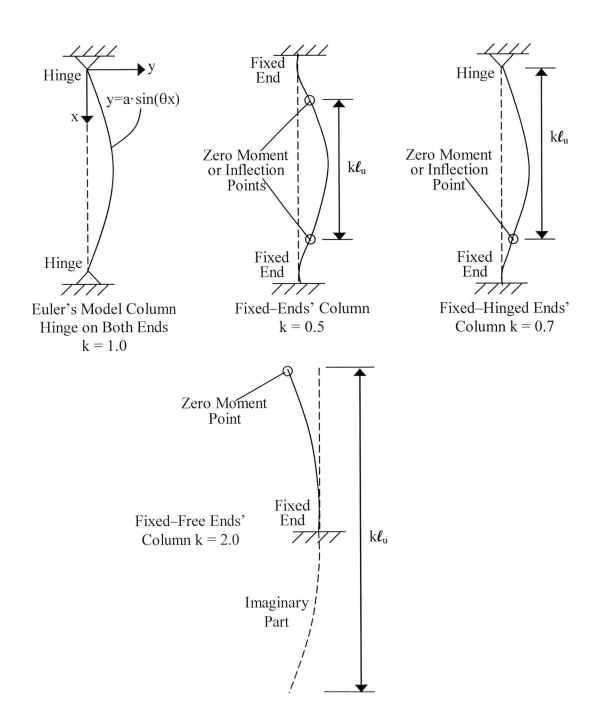

Figure A.80: Slenderness ratio in columns.

Slenderness Ratio = $k\ell_{ui}/r_i$

Radius of Gyration = $\sqrt{I_{ui}/A}$

Where:

ℓ_{ui} = column unsupported length with respect to axis i.

k = effective length factor for buckling about axis i.

$k\ell_{ui}$ = column effective length for buckling about axis i.

I_i = moment of inertia of column cross–section about axis i.

A = column cross–sectional area.

Every column is analyzed for buckling about its cross–section two principal axes (x and y) as shown below:

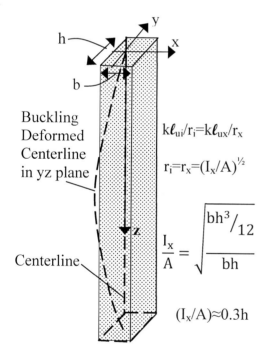

$k\ell_{ui}/r_i = k\ell_{ux}/r_x$

$r_i = r_x = (I_x/A)^{\frac{1}{2}}$

$\dfrac{I_x}{A} = \sqrt{\dfrac{bh^3/12}{bh}}$

$(I_x/A) \approx 0.3h$

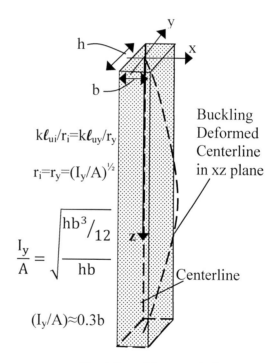

$k\ell_{ui}/r_i = k\ell_{uy}/r_y$

$r_i = r_y = (I_y/A)^{\frac{1}{2}}$

$\dfrac{I_y}{A} = \sqrt{\dfrac{hb^3/12}{hb}}$

$(I_y/A) \approx 0.3b$

Buckling About x Axis
Deformed Centerline in yz Plane
Deformations Parallel to y Axis
and Perpendicular to x Axis
Bending Moment about x Axis

Buckling About y Axis
Deformed Centerline in xz Plane
Deformations Parallel to x Axis
and Perpendicular to y Axis
Bending Moment about y Axis

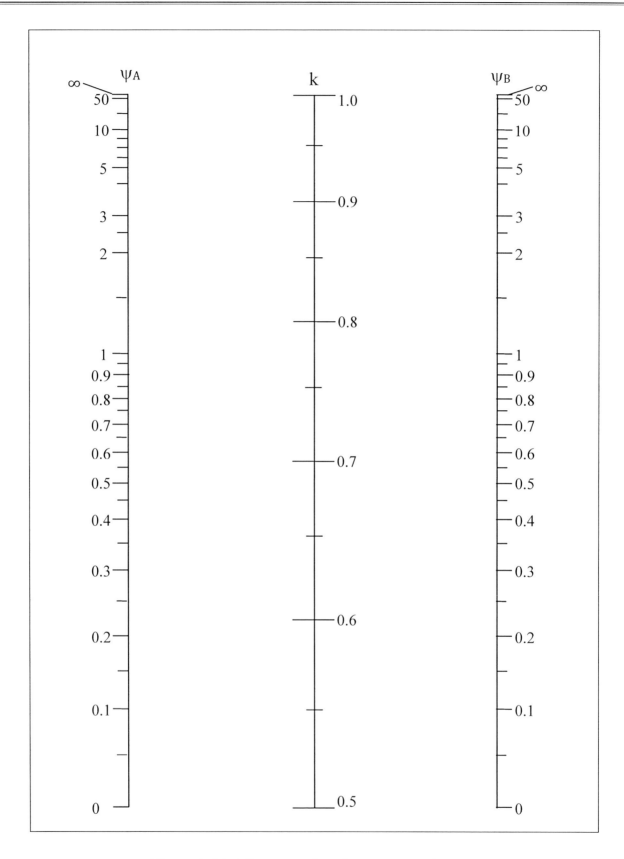

Figure A.81: Alignment chart for non-sway frames.

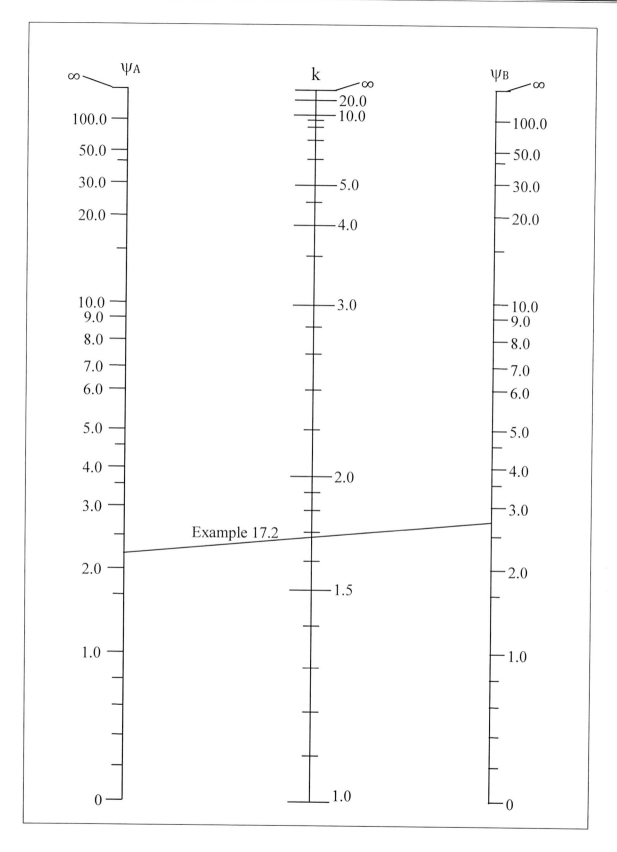

Figure A.82: Alignment chart for sway frames.

A.83: Alignment chart for sway frames.

$$\psi_{Ax} = 2 \times (E_{cc}I_{gC1x}/\ell_{uC1x} + E_{cc}I_{gC2x}/\ell_{uC2x})/ (E_{cb}I_{gB1}/\ell_{B1} + E_{cb}I_{gB2}/\ell_{B2})$$

$$\psi_{Ay} = 2 \times (E_{cc}I_{gC1y}/\ell_{uC1y} + E_{cc}I_{gC2y}/\ell_{uC2y})/ (E_{cb}I_{gB3}/\ell_{B3} + E_{cb}I_{gB4}/\ell_{B4})$$

Where:

E_{cc} and E_{cb} = as explained for Equation 17.3.

I_{gC1x} and I_{gC1y} = gross moments of inertia of column C1 about axes x and y, respectively.

I_{gC2x} and I_{gC2y} = gross moments of inertia of column C2 about axes x and y, respectively.

ℓ_{uC1x} and ℓ_{uC1y} = unsupported length of column C1 for buckling about axes x (in yz plane) and y (in xz plane), respectively.

ℓ_{uC2x} and ℓ_{uC2y} = unsupported length of column C2 for buckling about axes x (in yz plane) and y (in xz plane), respectively.

I_{gB1}, I_{gB2}, I_{gB3} and I_{gB4} = gross moment of inertia of beams B1, B2, B3 and B4 about axes x_{B1}, x_{B2}, x_{B3} and x_{B4}, respectively.

ℓ_{B1}, ℓ_{B2}, ℓ_{B3} and ℓ_{B4} = length of beams B1, B2, B3 and B4, respectively.

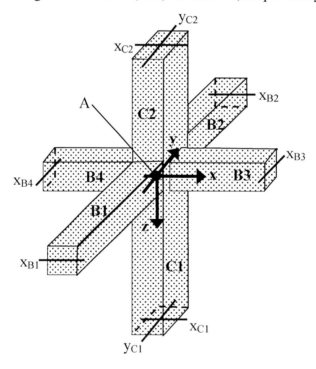

Joint A of beams and columns.

Moments of Inertia:

Beams	$I = 0.35I_g$
Columns	$I = 0.70I_g$

Determine ψ_{Bx} and ψ_{By} for the bottom joint (joint B) of the column similarly. Then, the alignment charts can be used.

Appendix B

METRIC UNITS

Unit Conversion

Length:
1 in = 25.4 mm = 0.0254 m
1 ft = 304.8 mm = 0.3048 m
1 yard = 914.4 mm = 0.9144 m

Area:
$1 \text{ in}^2 = 645.16 \text{ mm}^2$
$1 \text{ ft}^2 = 0.0929 \text{ m}^2$

Volume/Section Modulus:
$1 \text{ in}^3 = 16387 \text{ mm}^3$
$1 \text{ ft}^3 = 0.0283 \text{ m}^3$

Moment of Inertia:
$1 \text{ in}^4 = 416231 \text{ mm}^4$
$1 \text{ ft}^4 = 0.00863 \text{ m}^3$

Force or Load:
1 lb = 4.448 N = 0.4536 kgf
1 kip = 4.448 kN = 0.4536 metric ton
1 lb/ft = 14.59 N/m = 1.488 kgf/m
1 k/ft = 14.59 kN/m = 1.488 metric ton/m

Stress and Modulus of Elasticity:
$1 \text{ psi} = 6.895 \text{ kPa} = 0.703 \text{ kgf/m}^2 = 0.0703 \text{ gf/cm}^2$
$1 \text{ ksi} = 6.895 \text{ MPa} = 0.703 \text{ metric ton/m}^2$
$1 \text{ psf} = 48.77 \text{ Pa} = 4.97 \text{ kgf/m}^2$

Unit Weight/Density:
$1 \text{ lb/ft}^3 = 0.157 \text{ kN/m}^3 = 0.016 \text{ kgf/m}^3$

Common Metric Bar Sizes (Europe)	Common Metric Bar Sizes (Canada)
Bar No.	Bar No.
6	10
8	15
10	20
12	25
14	30
16	
18	
20	
22	
25	
28	
32	
40	

Bar Diameter (mm) = Bar No.